PLGA Based Drug Carrier and Pharmaceutical Applications

PLGA Based Drug Carrier and Pharmaceutical Applications

Editors

Maria Carmo Pereira
Joana A. Loureiro

MDPI • Basel • Beijing • Wuhan • Barcelona • Belgrade • Manchester • Tokyo • Cluj • Tianjin

Editors

Maria Carmo Pereira
Biotechnology and Energy
Portugal

Joana A. Loureiro
Biotechnology and Energy
Portugal

Editorial Office
MDPI
St. Alban-Anlage 66
4052 Basel, Switzerland

This is a reprint of articles from the Special Issue published online in the open access journal *Pharmaceutics* (ISSN 1999-4923) (available at: https://www.mdpi.com/journal/pharmaceutics/special_issues/PLGA_Pharm).

For citation purposes, cite each article independently as indicated on the article page online and as indicated below:

LastName, A.A.; LastName, B.B.; LastName, C.C. Article Title. *Journal Name* **Year**, *Volume Number*, Page Range.

ISBN 978-3-0365-0220-5 (Hbk)
ISBN 978-3-0365-0221-2 (PDF)

Contents

About the Editors . . . . **ix**

Joana Angélica Loureiro and Maria Carmo Pereira
PLGA Based Drug Carrier and Pharmaceutical Applications: The Most Recent Advances
Reprinted from: *Pharmaceutics* **2020**, *12*, 903, doi:10.3390/pharmaceutics12090903 . . . . **1**

Myoung Jin Ho, Hoe Taek Jeong, Sung Hyun Im, Hyung Tae Kim, Jeong Eun Lee, Jun Soo Park, Ha Ra Cho, Dong Yoon Kim, Young Wook Choi, Joon Woo Lee, Yong Seok Choi and Myung Joo Kang
Design and In Vivo Pharmacokinetic Evaluation of Triamcinolone Acetonide Microcrystals-Loaded PLGA Microsphere for Increased Drug Retention in Knees after Intra-Articular Injection
Reprinted from: *Pharmaceutics* **2019**, *11*, 419, doi:10.3390/pharmaceutics11080419 . . . . **7**

Teresa del Castillo-Santaella, Inmaculada Ortega-Oller, Miguel Padial-Molina, Francisco O'Valle, Pablo Galindo-Moreno, Ana Belén Jódar-Reyes and José Manuel Peula-García
Formulation, Colloidal Characterization, and In Vitro Biological Effect of BMP-2 Loaded PLGA Nanoparticles for Bone Regeneration
Reprinted from: *Pharmaceutics* **2019**, *11*, 388, doi:10.3390/pharmaceutics11080388 . . . . **23**

Silvia Minardi, Joseph S. Fernandez-Moure, Dongmei Fan, Matthew B. Murphy, Iman K. Yazdi, Xuewu Liu, Bradley K. Weiner and Ennio Tasciotti
Biocompatible PLGA-Mesoporous Silicon Microspheres for the Controlled Release of BMP-2 for Bone Augmentation
Reprinted from: *Pharmaceutics* **2020**, *12*, 118, doi:10.3390/pharmaceutics12020118 . . . . **41**

Patricia García-García, Ricardo Reyes, Elisabet Segredo-Morales, Edgar Pérez-Herrero, Araceli Delgado and Carmen Évora
PLGA-BMP-2 and PLA-17β-Estradiol Microspheres Reinforcing a Composite Hydrogel for Bone Regeneration in Osteoporosis
Reprinted from: *Pharmaceutics* **2019**, *11*, 648, doi:10.3390/pharmaceutics11120648 . . . . **53**

Sharif Md Abuzar, Jun-Hyun Ahn, Kyung Su Park, Eun Jung Park, Seung Hyuk Baik and Sung-Joo Hwang
Pharmacokinetic Profile and Anti-Adhesive Effect of Oxaliplatin-PLGA Microparticle-Loaded Hydrogels in Rats for Colorectal Cancer Treatment
Reprinted from: *Pharmaceutics* **2019**, *11*, 392, doi:10.3390/pharmaceutics11080392 . . . . **73**

Yujin Kim, Moritz Beck-Broichsitter and Ajay K. Banga
Design and Evaluation of a Poly(Lactide-*co*-Glycolide)-Based In Situ Film-Forming System for Topical Delivery of Trolamine Salicylate
Reprinted from: *Pharmaceutics* **2019**, *11*, 409, doi:10.3390/pharmaceutics11080409 . . . . **87**

Lili Duse, Michael Rene Agel, Shashank Reddy Pinnapireddy, Jens Schäfer, Mohammed A. Selo, Carsten Ehrhardt and Udo Bakowsky
Photodynamic Therapy of Ovarian Carcinoma Cells with Curcumin-Loaded Biodegradable Polymeric Nanoparticles
Reprinted from: *Pharmaceutics* **2019**, *11*, 282, doi:10.3390/pharmaceutics11060282 . . . . **101**

Maria João Ramalho, Joana A. Loureiro, Manuel A. N. Coelho and Maria Carmo Pereira
Factorial Design as a Tool for the Optimization of PLGA Nanoparticles for the Co-Delivery of Temozolomide and O6-Benzylguanine
Reprinted from: *Pharmaceutics* **2019**, *11*, 401, doi:10.3390/pharmaceutics11080401 . . . . **119**

Urszula Bazylińska, Julita Kulbacka and Grzegorz Chodaczek
Nanoemulsion Structural Design in Co-Encapsulation of Hybrid Multifunctional Agents: Influence of the Smart PLGA Polymers on the Nanosystem-Enhanced Delivery and Electro-Photodynamic Treatment
Reprinted from: *Pharmaceutics* **2019**, *11*, 405, doi:10.3390/pharmaceutics11080405 . . . . **137**

Norbert Varga, Árpád Turcsányi, Viktória Hornok and Edit Csapó
Vitamin E-Loaded PLA- and PLGA-Based Core-Shell Nanoparticles: Synthesis, Structure Optimization and Controlled Drug Release
Reprinted from: *Pharmaceutics* **2019**, *11*, 357, doi:10.3390/pharmaceutics11070357 . . . . **153**

Lucia Morelli, Sara Gimondi, Marta Sevieri, Lucia Salvioni, Maria Guizzetti, Barbara Colzani, Luca Palugan, Anastasia Foppoli, Laura Talamini, Lavinia Morosi, Massimo Zucchetti, Martina Bruna Violatto, Luca Russo, Mario Salmona, Davide Prosperi, Miriam Colombo and Paolo Bigini
Monitoring the Fate of Orally Administered PLGA Nanoformulation for Local Delivery of Therapeutic Drugs
Reprinted from: *Pharmaceutics* **2019**, *11*, 658, doi:10.3390/pharmaceutics11120658 . . . . **167**

Christopher Janich, Andrea Friedmann, Juliana Martins de Souza e Silva, Cristine Santos de Oliveira, Ligia E. de Souza, Dan Rujescu, Christian Hildebrandt, Moritz Beck-Broichsitter, Christian E. H. Schmelzer and Karsten Mäder
Risperidone-Loaded PLGA–Lipid Particles with Improved Release Kinetics: Manufacturing and Detailed Characterization by Electron Microscopy and Nano-CT
Reprinted from: *Pharmaceutics* **2019**, *11*, 665, doi:10.3390/pharmaceutics11120665 . . . . **185**

Woo Mi Ryu, Se-Na Kim, Chang Hee Min and Young Bin Choy
Dry Tablet Formulation of PLGA Nanoparticles with a Preocular Applicator for Topical Drug Delivery to the Eye
Reprinted from: *Pharmaceutics* **2019**, *11*, 651, doi:10.3390/pharmaceutics11120651 . . . . **201**

Emilie M. André, Gaëtan J. Delcroix, Saikrishna Kandalam, Laurence Sindji and Claudia N. Montero-Menei
A Combinatorial Cell and Drug Delivery Strategy for Huntington's Disease Using Pharmacologically Active Microcarriers and RNAi Neuronally-Committed Mesenchymal Stromal Cells
Reprinted from: *Pharmaceutics* **2019**, *11*, 526, doi:10.3390/pharmaceutics11100526 . . . . **215**

Yajie Zhang, Miguel García-Gabilondo, Anna Rosell and Anna Roig
MRI/Photoluminescence Dual-Modal Imaging Magnetic PLGA Nanocapsules for Theranostics
Reprinted from: *Pharmaceutics* **2020**, *12*, 16, doi:10.3390/pharmaceutics12010016 . . . . **235**

Tivadar Feczkó, Albrecht Piiper, Thomas Pleli, Christian Schmithals, Dominic Denk, Stephanie Hehlgans, Franz Rödel, Thomas J. Vogl and Matthias G. Wacker
Theranostic Sorafenib-Loaded Polymeric Nanocarriers Manufactured by Enhanced Gadolinium Conjugation Techniques
Reprinted from: *Pharmaceutics* **2019**, *11*, 489, doi:10.3390/pharmaceutics11100489 . . . . **251**

Rana Bakhaidar, Joshua Green, Khaled Alfahad, Shazia Samanani, Nabeehah Moollan, Sarah O'Neill and Zebunnissa Ramtoola
Effect of Size and Concentration of PLGA-PEG Nanoparticles on Activation and Aggregation of Washed Human Platelets
Reprinted from: *Pharmaceutics* **2019**, *11*, 514, doi:10.3390/pharmaceutics11100514 . . . . **267**

Maria Camilla Operti, Yusuf Dölen, Jibbe Keulen, Eric A. W. van Dinther, Carl G. Figdor and Oya Tagit
Microfluidics-Assisted Size Tuning and Biological Evaluation of PLGA Particles
Reprinted from: *Pharmaceutics* **2019**, *11*, 590, doi:10.3390/pharmaceutics11110590 . . . . **281**

Talia A. Shmool, Philippa J. Hooper, Gabriele S. Kaminski Schierle, Christopher F. van der Walle and J. Axel Zeitler
Terahertz Spectroscopy: An Investigation of the Structural Dynamics of Freeze-Dried Poly Lactic-co-glycolic Acid Microspheres
Reprinted from: *Pharmaceutics* **2019**, *11*, 291, doi:10.3390/pharmaceutics11060291 . . . . **299**

About the Editors

Maria Carmo Pereira received her PhD in Chemical Engineering from the University of Porto, Portugal, in 1998. She is a Professor at the Department of Chemical Engineering, Faculty of Engineering, University of Porto, Portugal. Currently, she coordinates the "Supramolecular Assemblies" research group at LEPABE. Her research fields span biophysics, supramolecular interactions, including novel nano-engineered biomaterials for therapeutic applications, and environmental sciences. Her scientific activity is focused on the design and preparation of drug delivery systems—polymeric nanoparticles, liposomes, and conjugated systems for biomedical applications—and the development of immunosensors as biomarkers for neurodegenerative diseases. M.C. Pereira has published more than 160 papers in international peer-reviewed journals and 50 conference proceedings.

Joana A. Loureiro is an MSc in Chemical Engineering and in Pharmaceutical Sciences and received her PhD in Chemical Engineering from the University of Porto (Portugal) in 2013. She is a researcher at LEPABE, Department of Chemical Engineering, Faculty of Engineering, University of Porto, Portugal. In recent years, her scientific research has been directed at the design and production of drug delivery systems (DDS) functionalized with targeting molecules for drug targeting and pharmaceutical applications, mainly related to neurological diseases. Her scientific activity also focuses on the study of drug–membrane interactions using biomimetic in vitro models to predict drug behavior. J.A. Loureiro is the author or co-author of 40 publications in high impact international peer-reviewed journals and 32 conference proceedings.

pharmaceutics

MDPI

Editorial

PLGA Based Drug Carrier and Pharmaceutical Applications: The Most Recent Advances

Joana Angélica Loureiro * and Maria Carmo Pereira *

LEPABE, Department of Chemical Engineering, Faculty of Engineering of the University of Porto, s/n, R. Dr. Roberto Frias, 4200-465 Porto, Portugal

* Correspondence: joana.loureiro@fe.up.pt (J.A.L.); mcsp@fe.up.pt (M.C.P.)

Received: 17 September 2020; Accepted: 21 September 2020; Published: 22 September 2020

Poly(lactic-*co*-glycolic acid) (PLGA) is one of the most successful polymers that has been used to produce medicines, such as drug carriers (DC). This is one of the few polymers that the Food and Drug Administration (FDA) has approved for human administration due to its biocompatibility and biodegradability [1]. DCs produced with PLGA have gained enormous attention over recent years for their ability to be versatile vehicles to transport different type of drugs, e.g., hydrophilic or hydrophobic small molecules or macromolecules, and protect them from degradation and uncontrolled release [2–6]. These drug delivery systems (DDS), including micro and nanoparticles, have the potential to modify their surface properties and improve interactions with biological materials. Furthermore, they can also be conjugated with specific target molecules to reach specific tissues or cells [7,8]. They are being used for different therapeutic applications, such as vaccinations or as treatments for cancer, neurological disorders, inflammation, and other diseases [9–12].

This Special Issue aims to focus on the recent progress of PLGA as a drug carrier and its new pharmaceutical applications. It comprises an exciting series of 19 research articles on the recent advances in the field.

In the first research study presented in this Special Issue, Ho et al. developed polymeric microspheres which contain micronized triamcinolone acetonide (TA) in order to increase the drug retention time in joints after intra-articular administration [13]. Poly(lactic-*co*-glycolic acid)/poly(lactic acid) (PLGA/PLA) carriers were prepared through spray-drying to incorporate the microcrystals that were previously prepared by ultra-sonication. In vivo testing in rat models was demonstrated to prolong drug retention in joints. The TA remained there for over 28 days, which was more 21 days compared with the TA-free group. Furthermore, these nanocarriers were demonstrated to be stable for one year.

The group of Peula-García used PLGA nanoparticles to carry bone morphogenetic protein (BMP-2) [14]. The nanocarriers were synthetized by a double-emulsion (water/oil/water, W/O/W) solvent evaporation technique, using the surfactant Pluronic F68 as a stabilizer. The BMP2-loaded nanocarriers presented positive results when evaluated using mesenchymal stromal cells from human alveolar bone regarding their proliferation, migration, and osteogenic differentiation. Another strategy to encapsulate BMP-2 was conducted by Minardi et al. [15]. PLGA multistage vector composite microspheres were used as carriers that demonstrated a good capacity for BMP-2 encapsulation and did not present toxicity for the rat mesenchymal stem cells. García-García et al. applied a combined strategy to regenerate tissue defects [16]. They used BMP-2- and 17β-estradiol-loaded microspheres, PLGA-based, in a sandwich-like system produced by a hydrogel core.

In another study, Hwang et al. fabricated PLGA carriers combined also with a hydrogel matrix. They produced oxaliplatin-loaded PLGA microparticles using a double emulsion technique and then loaded them into hyaluronic acid and carboxymethyl cellulose sodium-based cross-linked hydrogels [17]. This drug delivery system was analyzed in rat models and a substantial improvement was observed in terms of bioavailability and the mean residence time of the microparticle-loaded hydrogels.

Kim et al. developed an original system to be used in the topical delivery of trolamine salicylate (TS), a topical anti-inflammatory analgesic used for the treatment of small joint pain [18]. Here, the optimization process was done using different amounts of PLGA, ethyl 2-cyanoacrylate, poly (ethylene glycol) (PEG) 400, and TS. The researchers proved that the produced formulations enhanced the delivery of TS into and across the skin.

Duse et al. used PLGA nanoparticles to encapsulate curcumin, a well know natural compound that present anticancer benefits [19]. It was shown that the use of PLGA nanoparticles improves the bioavailability and site-specific drug uptake. The nanoparticles' efficacy was tested using SK-OV-3 human ovarian adenocarcinoma cells and demonstrated to be very efficient in transporting curcumin. Furthermore, with the same objective to treat the cancer, our research group used factorial design as a tool to optimize the co-encapsulation of temozolomide and o6-benzylguanine in PLGA nanoparticles [20]. The produced nanoparticles, rather than demonstrating stability for several days, presented optimal physicochemical properties for brain delivery, including a size lower than 200 nm and a negative surface charge. In the same research line, demonstrating the potential of the co-encapsulation, Bazylińska et al. encapsulated a hydrophobic porphyrin photosensitizing dye—verteporfin—in combination with low-dose cisplatin, a hydrophilic cytostatic drug [21]. Different coatings were applied to the PLGA nanoparticles, PEG, or folic acid functionalized. Those nanoparticles proved to have an increased internalization and efficiency regarding anticancer activity.

Another interesting study proposed by Varga and colleagues, who contributed with an interesting study of nanoparticle design and optimization where the (±)-α-Tocopherol (TP) with vitamin E activity was encapsulated in PLA and PLGA nanoparticles [22]. To stabilize the nanoparticles, the non-ionic stabilizing surfactant Pluronic F127 was used. Several techniques were used to characterize these novel nanoparticles, such as transmission electron microscopy (TEM), dynamic light scattering (DLS), and infrared spectroscopy (FT-IR).

Morelli et al. improved paclitaxel delivery in the gastro-intestinal tract by encapsulating the drug in PLGA nanoparticles coated with PEG [23]. The nanoparticles demonstrated stability in the gastric tract and simply penetrated inside carcinoma colon 2 ($CaCo_2$) cells.

With the objective to overcome the undesired lag time of the commercially available risperidone, Janich et al. encapsulated this drug in PLGA–lipid microcapsules and PLGA–lipid microgels [24]. The carriers were evaluated regarding their physio-chemical properties and the first formulation was demonstrated to achieve a nearly zero order release without a lag time over 2 months.

A research work using PLGA nanoparticles for ocular application was also collected. Ryu et al. produced rapidly dissolving dry tablets containing alginate and dexamethasone-loaded PLGA nanoparticles [25]. These nanoparticles presented sustained drug release for 10 h. In vivo experiments showed their efficiency and make this DDS a promising strategy for aseptic and accurate dose delivery of ophthalmic drugs.

An interesting approach based on a combination of cell and drug delivery for the treatment of Huntington's disease (HD) was proposed by André et al. [26]. The authors used laminin-coated PLGA nanoparticles to transport brain-derived neurotrophic factor (BDNF). The nanoparticles/cell complexes were evaluated in an ex vivo model of HD. Promising results were obtained by the researchers, encouraging them to go further in their research with this system.

Two works lead by Roing and Wacker present new theranostic PLGA-based nanoparticles. In the first one, biodegradable and photoluminescent polyester (BPLP) with PLGA polymer was used to fabricate biocompatible photoluminescent nanocapsules [27]. Additionally, superparamagnetic iron oxide nanoparticles (SPIONs) were incorporated into the polymeric shell to transform the particles into a magnetic resonance/photoluminescence dual-model imaging theranostic platform. The particles demonstrated good uptake and biocompatibility with hCMEC/D3 endothelial cells. In the second study, three different technologies for the encapsulation of sorafenib into PLGA and PLGA–PEG copolymers were adopted [28]. Those nanoparticles presented size ranges between 220 and 240 nm. In order to transform those nanoparticles in a theranostic medicine, gadolinium complexes were covalently attached to the nanoparticles' surface. That way, the nanoparticles could be located using magnetic resonance imaging.

PLGA toxicity was investigated by Bakhaidar et al. [29]. Here, the researchers studied the impact of size-selected PLGA–PEG nanoparticles on platelet activation and aggregation. The results demonstrated that nanoparticles of all sizes are associated with the surface of platelets leading to possible internalization. Furthermore, the NP–platelet interaction proved to not conduct platelet aggregation, making these PLGA nanoparticles promising delivery systems for targeted drug delivery to platelets.

Another relevant study was performed by Operti et al., who used microfluidics technology as a tool to manufacture particles in a highly controllable way [30]. In their study, they produced PLGA particles at diameters ranging from sub-micron to micron using a single microfluidics device. Through modification of flow and formulation parameters, the nanoparticle size changed substantially. Furthermore, in this study, the researchers proved how the particle size influences the release characteristics, cellular uptake, and in vivo clearance of these particles.

Finally, a research study regarding the importance of new techniques to characterize PLGA nanoparticles was included in this special edition. Shmool et al. investigated the dynamics of PLGA microspheres prepared by freeze-drying [31]. The water-oil-water (w/o/w) double-emulsion technique was selected for the production of the microspheres. Their molecular mobility at lower temperatures, leading to the glass transition temperature, using temperature-variable terahertz time-domain spectroscopy (THz-TDS), was evaluated. THz-TDS records show distinct transition processes, one in the range of 167–219 K, associated with local motions, and the other in the range of 313–330 K, associated with large-scale motions.

The papers presented in this Special Issue represent a small part of the research that is ongoing in the field of PLGA nanocarriers all over the world. The huge potential of PLGA nanoparticles make them a promising drug delivery system with outstanding properties and with much more potential for exploring in the coming years. With this Special Issue, the editors expect that the readers from the field find it stimulating and contributing more ideas or methodologies for their future work.

Conflicts of Interest: The authors declare no conflict of interest.

References

1. Makadia, H.K.; Siegel, S.J. Poly Lactic-*co*-Glycolic Acid (PLGA) as Biodegradable Controlled Drug Delivery Carrier. *Polymers* **2011**, *3*, 1377. [CrossRef] [PubMed]
2. Ramalho, M.J.; Sevin, E.; Gosselet, F.; Lima, J.; Coelho, M.; Loureiro, J.A.; Pereira, M. Receptor-mediated PLGA nanoparticles for glioblastoma multiforme treatment. *Int. J. Pharm.* **2018**, *545*, 84–92. [CrossRef] [PubMed]
3. Olivier, J.-C. Drug Transport to Brain with Targeted Nanoparticles. *NeuroRX* **2005**, *2*, 108–119. [CrossRef]
4. Rezvantalab, S.; Drude, N.I.; Moraveji, M.K.; Güvener, N.; Koons, E.K.; Shi, Y.; Lammers, T.; Kiessling, F. PLGA-Based Nanoparticles in Cancer Treatment. *Front. Pharmacol.* **2018**, *9*, 9. [CrossRef] [PubMed]
5. Ramalho, M.J.; Loureiro, J.A.; Gomes, B.; Frasco, M.F.; Coelho, M.A.N.; Pereira, M.D.C. PLGA nanoparticles as a platform for vitamin D-based cancer therapy. *Beilstein J. Nanotechnol.* **2015**, *6*, 1306–1318. [CrossRef]
6. Lu, B.; Lv, X.; Le, Y. Chitosan-Modified PLGA Nanoparticles for Control-Released Drug Delivery. *Polymers* **2019**, *11*, 304. [CrossRef]
7. Jose, S.; Cinu, T.A.; Sebastian, R.; Shoja, M.H.; Aleykutty, N.A.; Durazzo, A.; Lucarini, M.; Santini, A.; Souto, E.B. Transferrin-Conjugated Docetaxel–PLGA Nanoparticles for Tumor Targeting: Influence on MCF-7 Cell Cycle. *Polymers* **2019**, *11*, 1905. [CrossRef] [PubMed]
8. Loureiro, J.A.; Gomes, B.; Fricker, G.; Coelho, M.A.N.; Rocha, S.; Pereira, M.D.C. Cellular uptake of PLGA nanoparticles targeted with anti-amyloid and anti-transferrin receptor antibodies for Alzheimer's disease treatment. *Colloids Surf. B Biointerfaces* **2016**, *145*, 8–13. [CrossRef] [PubMed]

9. Guarecuco, R.; Lu, J.; McHugh, K.J.; Norman, J.J.; Thapa, L.S.; Lydon, E.; Langer, R.; Jaklenec, A. Immunogenicity of pulsatile-release PLGA microspheres for single-injection vaccination. *Vaccine* **2018**, *36*, 3161–3168. [CrossRef] [PubMed]
10. Shen, X.; Li, T.; Xie, X.; Feng, Y.; Chen, Z.; Yang, H.; Wu, C.; Deng, S.; Liu, Y. PLGA-Based Drug Delivery Systems for Remotely Triggered Cancer Therapeutic and Diagnostic Applications. *Front. Bioeng. Biotechnol.* **2020**, *8*. [CrossRef]
11. Rigon, L.; Salvalaio, M.; Pederzoli, F.; Legnini, E.; Duskey, J.T.; D'Avanzo, F.; De Filippis, C.; Ruozi, B.; Marin, O.; Vandelli, M.A.; et al. Targeting Brain Disease in MPSII: Preclinical Evaluation of IDS-Loaded PLGA Nanoparticles. *Int. J. Mol. Sci.* **2019**, *20*, 2014. [CrossRef] [PubMed]
12. Deng, M.; Tan, J.; Hu, C.; Hou, T.; Peng, W.; Liu, J.; Yu, B.; Dai, Q.; Zhou, J.; Yang, Y.; et al. Modification of PLGA Scaffold by MSC-Derived Extracellular Matrix Combats Macrophage Inflammation to Initiate Bone Regeneration via TGF- β -Induced Protein. *Adv. Heal. Mater.* **2020**, e2000353. [CrossRef] [PubMed]
13. Ho, M.J.; Jeong, H.T.; Im, S.H.; Kim, H.T.; Lee, J.E.; Park, J.S.; Cho, H.R.; Kim, D.Y.; Choi, Y.W.; Lee, J.; et al. Design and In Vivo Pharmacokinetic Evaluation of Triamcinolone Acetonide Microcrystals-Loaded PLGA Microsphere for Increased Drug Retention in Knees after Intra-Articular Injection. *Pharmaceutics* **2019**, *11*, 419. [CrossRef] [PubMed]
14. Del Castillo, T.; Ortega-Oller, I.; Padial-Molina, M.; O'Valle, F.; Galindo-Moreno, P.; Jodar-Reyes, A.B.; Peula, J. Formulation, Colloidal Characterization, and In Vitro Biological Effect of BMP-2 Loaded PLGA Nanoparticles for Bone Regeneration. *Pharmaceutics* **2019**, *11*, 388. [CrossRef] [PubMed]
15. Minardi, S.; Fernandez-Moure, J.S.; Fan, D.; Murphy, M.B.; Yazdi, I.K.; Liu, X.; Weiner, B.K.; Tasciotti, E. Biocompatible PLGA-Mesoporous Silicon Microspheres for the Controlled Release of BMP-2 for Bone Augmentation. *Pharmaceutics* **2020**, *12*, 118. [CrossRef]
16. García-García, P.; Reyes, R.; Segredo-Morales, E.; Herrero, E.; Delgado, A.; Évora, C. PLGA-BMP-2 and PLA-17β-Estradiol Microspheres Reinforcing a Composite Hydrogel for Bone Regeneration in Osteoporosis. *Pharmaceutics* **2019**, *11*, 648. [CrossRef]
17. Abuzar, S.M.; Ahn, J.-H.; Park, K.S.; Park, E.; Baik, S.; Hwang, S.-J. Pharmacokinetic Profile and Anti-Adhesive Effect of Oxaliplatin-PLGA Microparticle-Loaded Hydrogels in Rats for Colorectal Cancer Treatment. *Pharmaceutics* **2019**, *11*, 392. [CrossRef]
18. Kim, Y.; Beck-Broichsitter, M.; Banga, A.K. Design and Evaluation of a Poly(Lactide-*co*-Glycolide)-Based In Situ Film-Forming System for Topical Delivery of Trolamine Salicylate. *Pharmaceutics* **2019**, *11*, 409. [CrossRef]
19. Duse, L.; Agel, M.R.; Pinnapireddy, S.R.; Schäfer, J.; Selo, M.A.; Ehrhardt, C.; Bakowsky, U. Photodynamic Therapy of Ovarian Carcinoma Cells with Curcumin-Loaded Biodegradable Polymeric Nanoparticles. *Pharmaceutics* **2019**, *11*, 282. [CrossRef]
20. Ramalho, M.J.; Loureiro, J.A.; Coelho, M.A.N.; Pereira, M.D.C. Factorial Design as a Tool for the Optimization of PLGA Nanoparticles for the Co-Delivery of Temozolomide and O6-Benzylguanine. *Pharmaceutics* **2019**, *11*, 401. [CrossRef]
21. Bazylinska, U.; Kulbacka, J.; Chodaczek, G. Nanoemulsion Structural Design in Co-Encapsulation of Hybrid Multifunctional Agents: Influence of the Smart PLGA Polymers on the Nanosystem-Enhanced Delivery and Electro-Photodynamic Treatment. *Pharmaceutics* **2019**, *11*, 405. [CrossRef] [PubMed]
22. Varga, N.; Turcsányi, Á.; Hornok, V.; Csapó, E. Vitamin E-Loaded PLA- and PLGA-Based Core-Shell Nanoparticles: Synthesis, Structure Optimization and Controlled Drug Release. *Pharmaceutics* **2019**, *11*, 357. [CrossRef] [PubMed]
23. Morelli, L.; Gimondi, S.R.; Sevieri, M.; Salvioni, L.; Guizzetti, M.; Colzani, B.; Palugan, L.; Foppoli, A.; Talamini, L.; Morosi, L.; et al. Monitoring the Fate of Orally Administered PLGA Nanoformulation for Local Delivery of Therapeutic Drugs. *Pharmaceutics* **2019**, *11*, 658. [CrossRef]
24. Janich, C.; Friedmann, A.; Silva, J.M.D.S.E.; De Oliveira, C.S.; De Souza, L.E.; Rujescu, D.; Hildebrandt, C.; Beck-Broichsitter, M.; Schmelzer, C.E.; Mäder, K. Risperidone-Loaded PLGA–Lipid Particles with Improved Release Kinetics: Manufacturing and Detailed Characterization by Electron Microscopy and Nano-CT. *Pharmaceutics* **2019**, *11*, 665. [CrossRef] [PubMed]
25. Ryu, W.M.; Kim, S.-N.; Min, C.H.; Bin Choy, Y. Dry Tablet Formulation of PLGA Nanoparticles with a Preocular Applicator for Topical Drug Delivery to the Eye. *Pharmaceutics* **2019**, *11*, 651. [CrossRef]

26. André, E.M.; Delcroix, G.J.; Kandalam, S.; Sindji, L.; Montero-Menei, C.N. A Combinatorial Cell and Drug Delivery Strategy for Huntington's Disease Using Pharmacologically Active Microcarriers and RNAi Neuronally-Committed Mesenchymal Stromal Cells. *Pharmaceutics* **2019**, *11*, 526. [CrossRef]
27. Zhang, Y.; García-Gabilondo, M.; Rosell, A.; Roig, A. MRI/Photoluminescence Dual-Modal Imaging Magnetic PLGA Nanocapsules for Theranostics. *Pharmaceutics* **2019**, *12*, 16. [CrossRef]
28. Feczkó, T.; Piiper, A.; Pleli, T.; Schmithals, C.; Denk, D.; Hehlgans, S.; Rödel, F.; Vogl, T.J.; Wacker, M.G.; Denk, D.; et al. Theranostic Sorafenib-Loaded Polymeric Nanocarriers Manufactured by Enhanced Gadolinium Conjugation Techniques. *Pharmaceutics* **2019**, *11*, 489. [CrossRef]
29. Bakhaidar, R.; Green, J.; Alfahad, K.; Samanani, S.; Moollan, N.; O'Neill, S.; Ramtoola, Z.; Neill, O. Effect of Size and Concentration of PLGA-PEG Nanoparticles on Activation and Aggregation of Washed Human Platelets. *Pharmaceutics* **2019**, *11*, 514. [CrossRef]
30. Operti, M.C.; Dölen, Y.; Keulen, J.; Van Dinther, E.A.W.; Figdor, C.G.; Tagit, O. Microfluidics-Assisted Size Tuning and Biological Evaluation of PLGA Particles. *Pharmaceutics* **2019**, *11*, 590. [CrossRef]
31. Shmool, T.; Hooper, P.J.; Schierle, G.S.K.; Van Der Walle, C.F.; Zeitler, J.A. Terahertz Spectroscopy: An Investigation of the Structural Dynamics of Freeze-Dried Poly Lactic-*co*-glycolic Acid Microspheres. *Pharmaceutics* **2019**, *11*, 291. [CrossRef] [PubMed]

pharmaceutics

MDPI

Article

Design and In Vivo Pharmacokinetic Evaluation of Triamcinolone Acetonide Microcrystals-Loaded PLGA Microsphere for Increased Drug Retention in Knees after Intra-Articular Injection

Myoung Jin Ho [1], Hoe Taek Jeong [1], Sung Hyun Im [1], Hyung Tae Kim [1], Jeong Eun Lee [1], Jun Soo Park [1], Ha Ra Cho [1], Dong Yoon Kim [1], Young Wook Choi [2], Joon Woo Lee [3], Yong Seok Choi [1,*] and Myung Joo Kang [1,*]

1 College of Pharmacy, Dankook University, 119, Dandae-ro, Dongnam-gu, Cheonan-si, Chungcheongnam-do 31116, Korea
2 College of Pharmacy, Chung-Ang University, 84, Heukseok-ro, Dongjak-gu, Seoul 06974, Korea
3 Department of Radiology, Seoul National University Bundang Hospital, 82, Gumi-ro 173beon-gil, Bundang-gu, Seongnam-si, Gyeonggi-do 13620, Korea
* Correspondence: analysc@dankook.ac.kr (Y.S.C.); kangmj@dankook.ac.kr (M.J.K.); Tel.: +82-41-550-1439 (Y.S.C.); +82-41-550-1446 (M.J.K.)

Received: 22 July 2019; Accepted: 12 August 2019; Published: 19 August 2019

Abstract: A novel polymeric microsphere (MS) containing micronized triamcinolone acetonide (TA) in a crystalline state was structured to provide extended drug retention in joints after intra-articular (IA) injection. Microcrystals with a median diameter of 1.7 μm were prepared by ultra-sonication method, and incorporated into poly(lactic-*co*-glycolic acid)/poly(lactic acid) (PLGA/PLA) MSs using spray-drying technique. Cross-sectional observation and X-ray diffraction analysis showed that drug microcrystals were evenly embedded in the MSs, with a distinctive crystalline nature of TA. In vitro drug release from the novel MSs was markedly decelerated compared to those from the marketed crystalline suspension (Triam inj.®), or even 7.2 μm-sized TA crystals-loaded MSs. The novel system offered prolonged drug retention in rat joints, providing quantifiable TA remains over 28 days. Whereas, over 95% of IA TA was removed from joints within seven days, after injection of the marketed product. Systemic exposure of the steroidal compound was drastically decreased with the MSs, with <50% systemic exposure compared to that with the marketed product. The novel MS was physicochemically stable, with no changes in drug crystallinity and release profile over 12 months. Therefore, the TA microcrystals-loaded MS is expected to be beneficial in patients especially with osteoarthritis, with reduced IA dosing frequency.

Keywords: triamcinolone acetonide; microcrystal; PLGA microsphere; local delivery; spray-drying technique; intra-articular injection; joint retention; systemic exposure

1. Introduction

Intra-articular (IA) injection of corticosteroids, such as triamcinolone acetonide (TA) crystalline suspension, is commonly recommended to alleviate pain and inflammation in knee joints [1,2]. Marketed injectable TA suspensions are intended to be slowly dissolved in the synovial fluid, and the glucocorticoid molecules steadily bind to and activate the glucocorticoid receptors, obstructing the production of inflammation mediators, including prostaglandins, leukotrienes, and pro-inflammatory cytokines [3]. Nevertheless, the analgesic and/or anti-inflammatory effect of the TA crystalline suspension are reported to be weakened within two weeks following IA injection because of a rapid efflux of the drug from the arthritic joint [4–6]. As the synovial lining is ultra-structured with permeable intercellular gaps measuring 0.1–5.5 μm, therapeutic agents injected via the IA route tend to easily escape from the joint [7,8]. Moreover, the steroidal compound is quite soluble in aqueous media (21 μg/mL in phosphate buffered saline at 25 °C) [9], and TA crystals that have decreased to a few microns in size might be readily translocated into systemic circulation. In vitro release experiments showed that the marketed TA crystalline suspension was completely dissolved within 2 h [10].

Several pharmaceutical approaches, including hydrogels, liposomes, nanoparticles, and microparticles (MPs), have been explored to prolong the retention time of steroidal compounds in the synovial tissues, minimizing systemic exposure following IA injection [2,10–15]. When considering the leaky structure of the synovium, one of the sound strategies for localized delivery to the synovial tissues is administering the therapeutic agents to IA after entrapping in a micro-sized carrier system. Biocompatible and biodegradable polymeric MPs generally larger than 10 μm have been reported to be effective for remaining in the synovial cavity and providing a sustained-release profile in the joint [2,9,16]. Actually, a single IA injection of the PLGA MPs was clinically demonstrated to be effective in providing an extended retention time of the corticosteroid in joints while reducing the drug distribution in the bloodstream [17].

We previously formulated a PLGA MS system containing TA as in a stable crystalline form, to achieve the sustained-release profile in joints after IA injection [18]. Compared to the polymeric MS containing TA in an amorphous state, the TA crystals-loaded MSs prepared by layering the suspended drug crystals with PLGA polymer exhibited excellent physicochemical stability under storage condition (25 °C/60% R.H.) in terms of drug crystallinity and drug content in MSs. However, the MS system containing TA crystals with a median diameter of 7 μm could not effectively retard drug release, as the coating layer on the irregular TA crystals was unfair, with insufficient coating thickness; the extent of drug released from the MSs reached 75% for 12 h under sink condition. Thus, we assumed that the additional micronization process of the TA crystals might be beneficial, prior to encapsulation process, to obtain satisfactory coating thickness on the TA crystals, providing sustained-release profile in joints.

Herein, the goals of this study were to construct uniform TA microcrystals and embed the microcrystals in the polymeric MS, providing a prolonged retention profile in the joint following IA injection. The uniform drug microcrystals were prepared by an ultra-sonication method and then encapsulated into the polymeric MS by a spray-drying technique. The physicochemical characteristics of the microcrystals-loaded MSs were evaluated in terms of outer and inner structures, particle size, the drug loading amount, and loading efficiency. In vitro drug release patterns from the MSs were regulated by adjusting the ratio of PLGA to PLA polymers and the ratio of drug to the polymer. The in vivo concentration profile of TA in the plasma and joint tissue following IA injection of the microcrystals-loaded MSs were comparatively evaluated with those of the marketed TA crystalline suspension in rats.

2. Materials and Methods

2.1. Materials

TA powder and Triam inj.® were kindly provided by Shinpoong Pharmaceutical Co. (Seoul, Korea). PLGA polymer with a lactide/glycolide ratio of 50:50 (5050DLG 4A, molecular weight 38,000–54,000 kDa) and PLA polymer (R203H, 18,000–24,000 kDa) were purchased from Lakeshore Biomaterials (Birmingham, AL, USA). Polysorbate 20, polysorbate 80, Sorbian monolaurate (Span 20), poly(ethyleneglycol) 4000 (PEG 4000), polyethylene-polypropylene glycol 188 (Poloxamer 188), cholesterol, benzalkonium chloride (BKC), sodium lauryl sulfate (SLS), and phosphate buffered saline tablets were purchased from Sigma Chemical Co. (St. Louis, MO, USA). Lecithin (L-α-phosphatidylcholine) was obtained from Avanti Polar Lipids, Inc. (Alabaster, AL, USA). Gelatin and glycerol were provided from TCI Chemicals, Co. (Tokyo, Japan). Acetonitrile (ACN), ethyl alcohol, and methanol of HPLC grade were obtained from J.T. Baker (Phillipsburg, NJ, USA). All other reagents were analytical grade.

2.2. Preparation of TA Microcrystals Using an Ultra-Sonication Method

Drug microcrystal suspension was fabricated using an ultra-sonication method previously reported with slight modifications [19]. Different kinds of stabilizers (polysorbate 20, polysorbate 80, span 20, PEG 4000, poloxamer 188, cholesterol, and BKC) were dissolved in ACN in the concentration ranges from 0.05 to 0.5 *w/v* % as shown is Table 1. TA powder (100 mg) was then added to the solution and vigorously vortexed for 5 min to disperse the drug powder homogeneously. The ultra-sonicator (Model Vibracell VC-505, Sonics and Materials Inc., Newtown, NC, USA) equipped with a 1/2-inch (13 mm) probe was placed into the TA suspension and it was sonicated for 3 min at 40% amplitude with 3 s pulses (on/off alteration). To prevent temperature elevation, the samples were located inside an ice bath during the sonication procedure. Prepared TA microcrystal suspensions were then stored at room temperature for further experiments.

2.3. Preparation of TA Microcrystals-Loaded MSs Using a Spray-Drying Technique

TA microcrystals-loaded MSs were fabricated by a spray-drying technique with a Buchi mini-spray dryer (Model B-290, Buchi Labortechnik AG, Flawil, Switzerland). The feeding solution was prepared by subsequently dissolving PLGA/PLA polymers and lecithin into the BKC-stabilized TA microcrystal suspension. The composition of each microcrystals-embedded MSs is represented in Table 2. The feeding solution was then pumped into the spray dryer nozzle at a feeding rate of 3 mL/min and a stirring rate of 250 rpm. The inlet and outlet temperatures were set to 70 °C and 45 °C, respectively, to evaporate the organic solvent. The atomizing air flow was 246 L/h and the aspirator capacity was 100%. Prepared microcrystals-loaded MS powders were collected and stored in a desiccator (Model OH-3S, As-one, Seoul, Korea) at 25 °C for 24 h to remove the residual organic solvent.

2.4. Morphological Features of TA Microcrystals and Microcrystals-Loaded MSs

2.4.1. Appearance f TA Microcrystals and Microcrystals-Loaded MSs

Morphological features of raw material, TA microcrystals, and polymeric MSs were observed by SEM (Model Sigma 500, Carl Zeiss, Oberkochen, Germany). Drug powder and MSs samples were placed on a carbon tape and fixed onto an aluminum stub. The TA microcrystal suspension was dropwise loaded on the carbon tape and then dried for 6 h at room temperature to remove the aqueous vehicle. The platinum coating procedure was then conducted using an automatic sputter coater (Model 108Auto, Cressington, UK) at 15 mA. Appearance of samples was scrutinized by an electron microscope at an accelerated voltage of 15 kV.

2.4.2. Cross-Sectional Image of TA Microcrystals-Loaded MSs

The internal structure of the microcrystals-loaded MSs was scrutinized by SEM after fixing the MSs in the gelatin blocks. At first, the gelatin medium was prepared by dissolving gelatin (20 *w/v* %) and glycerin (5 *w/v* %) in distilled water [20]. Approximately 40 mg of the MSs powder was dispersed in the 3 mL of gelatin medium inside the polystyrene disposable base mold (Tissue-Tek®, 15 × 15 × 5 mm) at 37 °C. The mold was then placed in a deep freezer maintained at −70 °C for 12 h. The frozen gelatin block was mounted to the cryostat stub (Model CM3050S, Leica Microsystems, Wetzlar, Germany) using an optimum cutting temperature compound (Sakura Finetechnical Co., Ltd., Tokyo, Japan). The MS-loaded block was then sectioned at a thickness of 20 μm at −20 °C and was immediately placed on the double-sided carbon tape. Samples were defrosted for 1 h at room temperature and coating and observation procedures were conducted using the same method as described above.

2.4.3. Hyperspectral Mapping Images of TA Microcrystals-Loaded MSs

A hyperspectral microscopy imaging system (Model CytoViva®, Cytoviva Inc., Auburn, AL, USA) was employed to visualize TA microcrystals inside the MSs. The Cytoviva® system included a BX-41 microscope (Olympus Corporation, Tokyo, Japan), a visible-near infrared hyperspectral imaging system, dual fluorescence module, and high-resolution adaptor. Approximately 10 μL of TA microcrystal suspension, blank MSs, and TA microcrystals-loaded MSs suspended in 1 mL of 0.5 *w/v* % polysorbate 80 solution was dropped onto a cover glass and the hyperspectral spectra were analyzed. The mapping process was performed on the TA microcrystals-loaded MS image, with acquired spectra of TA microcrystals and blank MSs (HyperVisual Software ENVI 4.8, ITT Visual Information Solutions, Boulder, CO, USA). The spectra corresponding to TA microcrystals and the blank MSs were expressed as red and yellow, respectively, in the hyperspectral image of TA microcrystals-loaded MSs.

2.5. Physicochemical Characterization TA Microcrystals and Microcrystals-Loaded MSs

2.5.1. Crystallinity Analysis

The crystalline state of TA powder, TA microcrystals, blank MSs, and microcrystals-loaded MSs was analyzed using an X-ray diffractometer (XRD, Model Ultima IV, Rigaku, Tokyo, Japan) at 25 °C. For the TA microcrystal suspension, the aqueous vehicle was removed by centrifuging the suspension at 3500 g for 10 min, and subsequently, oven drying at 40 °C for 12 h. Each sample was put on the glass sample plate and the diffraction pattern over a 2θ range of 5–35° was determined using a step size of 0.02°. Voltage, current, and scan speed were set to 40 kV, 30 mA, and 1 s/step, respectively.

2.5.2. Particle Size Analysis

Size distributions of TA microcrystals and MSs were determined by Mastersizer MS 2000 (Malvern Instruments Ltd., Worcestershire, UK) equipped with a Hydro 2000 S automatic dispersion unit. Prior to analysis, MSs powder was suspended in the aqueous medium consisting of 1 *w/v* % PEG 4000 and 0.5 *w/v* % polysorbate 20. The suspended samples were then dropwise added to an automatic dispersion unit to obtain a 10–15% range of obscuration. Sample and background measurement times were set to 5 s and 10 s, respectively, and 5 runs were conducted for each measurement. Mie theory was applied to calculate the size distribution by volume with the refractive index value of 1.52. The resultant particle sizes of the three batches were averaged and presented as mean ± standard deviation (SD) (n = 3). The $d_{0.5}$, $d_{0.9}$, and $d_{0.1}$ indicated the median value defined as the diameter where 50%, 90%, and 10% of the population were below this value, respectively. SPAN value was an indicator representing the homogeneity of the particle size and was calculated by dividing the difference of $d_{0.9}$ and $d_{0.1}$ by $d_{0.5}$ [21].

2.5.3. Determination of Loading Amount and Efficiency of TA Microcrystals in MSs

To dissolve TA microcrystals-loaded MSs, 10 mg of MSs were added to 1 mL of dimethyl sulfoxide and then sonicated with a bath-type sonicator (Model 5510E-DTH, Bransonic, USA) for 10 min. The opaque solution was diluted with ACN and distilled water mixture (3:2 *v/v*) and was subsequently centrifuged at 16,000 *g* for 10 min to remove the precipitates. The concentration of TA in the supernatant was determined by a Waters HPLC system (Waters Corporation, Milford MA, USA) comprised of a pump (Model 515), auto sampler (Model 717 plus), UV detector (Model 486), and equipped with a Capcell Pak C18 column (150 mm × 2.0 mm, 3 μm, Shiseido, Tokyo, Japan). The mobile phase consisted of ACN and distilled water at a volume ratio of 3:2 and was eluted with a flow rate of 1.0 mL/min. The detection wavelength was set to 254 nm. The calibration curve of TA was linear in the concentration range of 1–100 μg/mL, with r^2 values of 0.999. The drug loading amount and loading efficiency were calculated as follows [22]:

$$\text{Drug loading amount} = \text{WL}/\text{WT}, \quad (1)$$

$$\text{Drug loading efficiency (\%)} = (\text{WL/WF}) \times 100, \quad (2)$$

where W_L, W_T and W_F represent the weight of TA in microcrystals-loaded MSs (mg), total weight of microcrystals-loaded MSs (mg) and feeding weight of TA (mg).

2.6. *In Vitro Drug Release Profiles and Morphological Changes of Microcrystals-Loaded MSs*

In vitro release profiles of TA from the novel MSs were comparatively evaluated with that of a marketed product under accelerated test conditions (45 ± 0.5 °C). To guarantee sink condition during the experiment, 0.5 *w/v* % of SLS and 0.05 *w/v* % of poloxamer 188, were added to 10 mM phosphate buffered saline (pH 7.4). MSs or the marketed product (Triam inj.®, TA 40 mg/mL) containing 20 mg of TA were immersed into 200 mL dissolution medium maintained at 45 ± 0.5 °C and then shaken with an agitation speed of 100 rpm. At predetermined intervals, 1 mL of the release medium was withdrawn and centrifuged at 16,000 *g* for 10 min. The supernatant was diluted two-fold with the mobile phase and TA concentration in the aliquot was determined by HPLC as described above. The equivalent volume of fresh pre-warmed dissolution medium was replenished to maintain a constant medium volume.

The morphological changes of the novel MSs during the in vitro release test were scrutinized by SEM. At predetermined intervals, MSs prepared with the PLGA:PLA ratio of 4:0 (F1), 1:3 (F4), and 0:4 (F5) were withdrawn and centrifuged at 900 *g*. The pelletized MSs were stored at −70 °C for 24 h and lyophilized for 24 h. The appearance of the lyophilized MS samples was observed by SEM with the same procedure described earlier.

2.7. *In Vivo Systemic Exposure and Joint Retention of TA after IA Injection in Rats*

2.7.1. Animals and Experimental Protocols

In Vivo pharmacokinetic studies were performed after approval from the Institutional Animal Care and Use Committee (IACUC) of Seoul National University Bundang Hospital (approval number: BA1608-206/050-01, date of approval: August 9, 2016). Six-week-old male Sprague-Dawley rats (250 ± 20 g) were acquired from Samtako (Kyungki-do, Korea). Four or five rats were housed in each cage and kept in a temperature- and relative humidity-controlled room (23 ± 1 °C and 50 ± 5%, respectively) with a 12-h light-dark cycle. During the acclimatization period, rats were allowed free access to tap water and standardized chow.

After at least three days of the acclimatization period, rats were divided into three groups (n = 9 per group) by a stratified randomization scheme for similar body weights groups. The hair on both hind knee joints was removed using hair removal cream. Prior to IA injection, spray-dried MS (F4 and F8) were re-dispersed in the sterile diluent composed of 0.66 *w/v* % sodium chloride, 0.63 *w/v* % carboxymethylcellulose sodium, and 0.04 *w/v* % polysorbate 80 at the same drug concentration (2.5 mg/mL as TA). Each group received 50 μL of the marketed product, F4, and F8, respectively, using an insulin syringe (31 G) in both knee joints, to administer 125 μg of TA per knee. At the predetermined time, blood samples (approximately 0.2 mL) were collected from the submandibular vein using a 26 G heparinized syringe. Blood samples were centrifuged at 16,000 *g* for 10 min. The obtained plasma samples were then stored at −70 °C until being analyzed by LC-MS/MS assay.

Apart from the systemic exposure evaluation, knee samples were collected to estimate the level of TA in joint tissues. After 3, 7, 21, 28, and 42 days of the IA injection in both knees, two animals from each group were sacrificed, and both knees were removed using bone cutters. After removing any residual substances and adhered tissues, knees were accurately weighed and stored at −70 °C until LC-MS/MS analysis.

2.7.2. LC-MS/MS Analysis of TA Concentration in Plasma and Knee Tissues

The TA concentrations in rat plasma or joint tissue were determined using the LC–MS/MS assay previously reported [23]. In brief, thawed plasma (100 μL) was mixed with 900 μL of methanol and vigorously vortexed for 10 min, to precipitate protein. After centrifuging at 16,000 g, a supernatant

(10 µL) was analyzed through an LC-MS/MS system (Model LC-20 Prominence HPLC, Shimadzu and Model API 2000, AB/SCIEX, Foster City, CA, USA). In the case of articular samples, the frozen knee tissues were thawed and immersed in 2 mL of ACN and shaken overnight to extract TA from the tissue. The extracted solution was centrifuged at 16,000 *g* for 5 min and the supernatant was injected into the LC-MS/MS system. The transitions of 435.1/415.0/15 precursor ion (*m/z*)/product ion (*m/z*)/collision energy (V) were then monitored for TA. Data acquisition/analyses were conducted using Analyst® version 1.5.2 software (ABSciex, Concord, ON, Canada). The assay was validated thoroughly and showed acceptable precision and accuracy, with a lower limit of quantification of 0.2 ng/mL in both rat plasma and knee tissue extract.

2.7.3. Pharmacokinetic Parameters from TA Concentration Profile in Plasma

Pharmacokinetic parameters such as area under the plasma concentration versus time curve ($AUC_{0–7days}$), maximum plasma concentration (C_{max}), time needed to reach the maximum plasma concentration (T_{max}), and terminal half-life ($T_{1/2}$) in plasma were calculated using the linear trapezoidal rule in the BA Calc 2007 pharmacokinetic analysis program (Korea Food & Drug Administration, Seoul, Korea).

2.8. Physicochemical Stability of TA Microcrystals-Loaded MSs

The long-term storage stability of the novel MSs was evaluated in terms of drug crystallinity, drug content, and in vitro release profile. TA microcrystals-loaded MS (F4) power was placed into the scintillation vial and was stored in the chamber maintained at 25 °C and 60% R.H. After 12 months of storage, the drug crystallinity, drug content, and in vitro release profile were evaluated with the same method as previously described.

2.9. Statistical Analysis

Each experiment was performed at least thrice and the data are presented as the mean ± SD. Statistical significance was determined using a one-way analysis of variance (ANOVA) test and was considered to be significant at $p < 0.05$ unless otherwise indicated.

3. Results and Discussion

3.1. Formulation and Physical Characteristics of TA Microcrystals

Various stabilizers were screened to micronize TA powder in the organic solvent using a probe type ultra-sonicator (Table 1). ACN was employed as the vehicle as it exhibited low solubility for the steroidal compound (<1 mg/mL), and high solvation capacity for PLGA and PLA polymers [18]. When the surface stabilizer was not included in the vehicle, TA powder was not uniformly dispersed in the medium after the homogenization process, rather forming large precipitates. The addition of steric stabilizers, such as polysorbate 20, polysorbate 80, PEG 4000, poloxamer 188, and cholesterol could not provide a uniform dispersion of the split TA microcrystals, forming drug aggregates within 24 h. On the other hand, when BKC was included in the organic vehicle at a concentration of 0.1 to 0.5 *w/v* %, TA microcrystals with a median size below 2.1 µm were shaped with re-dispersibility in the organic solvent (Table 1 and Figure 1A). When the concentration of the cationic surfactant was less than 0.05 *w/v* %, it could not afford the re-dispersibility of TA microcrystals. Thus, BKC at the concentration of 0.1 *w/v* % was employed for further preparation of TA microcrystal suspension in ACN.

The morphological feature of the TA microcrystals stabilized by 0.1 *w/v* % BKC was observed by FE-SEM. TA raw material showed characteristic crystal forms, such as hexahedron, octahedron, and dodecahedrons, with different sizes in the range from 2 to 20 µm (Figure 1C). Whereas, the crystal size was markedly decreased to 1–3 µm by the ultra-sonication process (Figure 1D), coinciding with the crystal size as determined by Mastersizer ($d_{0.5}$, 1.7 µm). In spite of crystal size reduction, no noticeable change to the shape was observed in the TA microcrystals. The crystalline state of TA microcrystals was further evaluated by comparing the X-ray diffraction spectrum of TA microcrystals with that of the raw material (Figure 1B). The spectrum of TA microcrystals was identical to that of drug powder, exhibiting distinctive diffraction peaks at 2θ equal to 9.9°, 14.5°, 17.6°, and 24.7°. On the other hand, the cationic surfactant showed no distinctive diffraction peaks over the 2θ range of 5–35°. Taken together, we concluded that TA powder was effectively micronized to 1–3 μm, with no crystalline changes during the ultra-sonication process.

Table 1. Effects of kinds of stabilizers on size, homogeneity, and dispersibility of TA microcrystals in ACN.

Stabilizer (*w/v* %) [1]	Crystal Size ($d_{0.5}$, μm) [2,3]	Homogeneity (SPAN) [2,4]	Dispersibility [5]
- [6]	7.21 ± 1.02	2.14 ± 0.13	Aggregated
Polysorbate 20 0.5%	6.63 ± 0.82	2.11 ± 0.07	Aggregated
Polysorbate 80 0.5%	7.35 ± 0.91	1.99 ± 0.04	Aggregated
Span 20 0.5%	17.7 ± 6.27	1.81 ± 0.11	Aggregated
PEG 4000 0.5%	7.15 ± 3.07	2.10 ± 0.88	Aggregated
Poloxamer 188 0.5%	5.44 ± 1.84	2.00 ± 0.52	Aggregated
Cholesterol 0.5%	8.72 ± 3.49	1.81 ± 0.60	Aggregated
BKC 0.5%	2.11 ± 0.05	1.32 ± 0.03	Re-dispersible
BKC 0.2%	1.94 ± 0.06	1.20 ± 0.03	Re-dispersible
BKC 0.1%	1.73 ± 0.02	1.21 ± 0.01	Re-dispersible
BKC 0.05%	4.75 ± 0.23	1.99 ± 0.02	Aggregated

[1] Weight per volume concentration in ACN. [2] Expressed as mean ± SD (n = 3). [3] Indicates the volume weighted diameter below which 50% of the total particle. [4] Calculated by dividing the difference between $d_{0.9}$ and $d_{0.1}$ by $d_{0.5}$. $d_{0.9}$ and $d_{0.1}$ indicate the volume weighted diameters below which 90% and 10% of the total particle, respectively. [5] Visually evaluated after 24 h storage at room temperature. [6] Indicates without stabilizer.

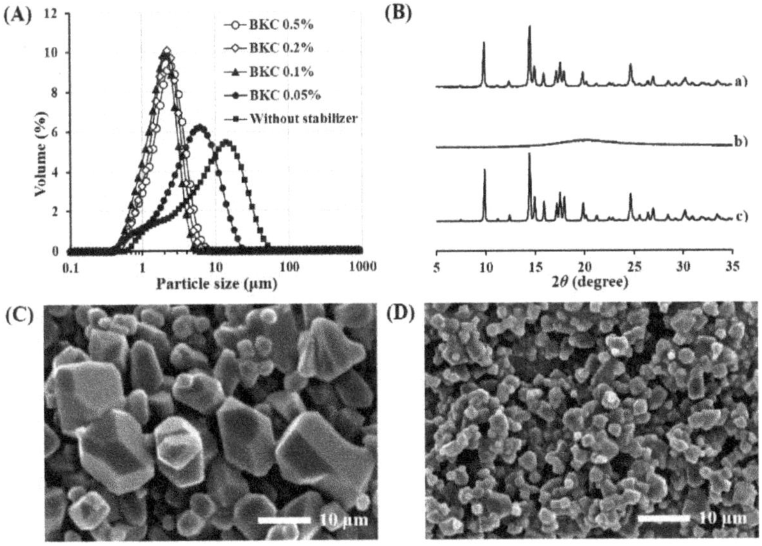

Figure 1. Morphological and physical characteristics of TA microcrystals. (**A**) Size distribution of the drug microcrystals stabilized by BKC, (**B**) XRD patterns of (a) TA raw material, (b) BKC powder, and (c) TA microcrystals, SEM images of (**C**) TA raw material, and (**D**) TA microcrystals stabilized by BKC at the concentration of 0.1 *w/v* %.

3.2. Formulation and Physical Characteristics of TA Microcrystals-Loaded MSs

Various TA microcrystals (1.7 μm median size) or intact TA powder (7.2 μm)-embedded MSs were fabricated using the spray-drying technique, and these particle size, homogeneity, drug loading amount, and loading efficiency are represented in Table 2. The median particle size of MSs prepared ranged from 15.8 to 18.8 μm, with a narrow size distribution possessing a SPAN value below 2.1. The formulation variables, such as TA crystal size, the ratio of PLGA and PLA polymers, and the ratio of drug to polymer, did not cause marked differences in size and homogeneity of TA-loaded MSs (Table 2). The particle size of the novel MS was considered to be suitable for IA prolonged delivery, preventing the trans-synovial efflux of injected TA microcrystals [17,24,25]. There was also no remarkable difference in the drug loading efficiency in the polymeric MSs between the formulas, exhibiting more than 90% drug loading efficiency in all formulations. The absence of the external phase during the spray-drying process might prevent distribution and/or diffusion of TA microcrystals during the external phase, and thus promote TA crystals to be located in the polymeric matrix after solvent evaporation, irrespective of composition variables. On the other hand, the loading amount of TA in MS was adjusted from 0.09 to 0.31 *w/w*, by controlling the drug to polymer weight ratio from 1:2 to 1:10.

Table 2. Compositions and physicochemical characteristics of TA microcrystals-loaded MSs.

	Compositions			Characteristics			
	TA crystal Size (μm) [1,5]	PLGA:PLA Ratio (w:w)	Drug:polymer Ratio (w:w)	Particle Size (μm) [1,5]	SPAN [2,5]	Loading Amount [3,5]	Loading Efficiency (%) [4,5]
F0	7.21 ± 1.02	1:3	1:5	15.9 ± 0.95	2.03 ± 0.04	0.16 ± 0.02	97.2 ± 2.35
F1	1.73 ± 0.02	4:0	1:5	16.2 ± 1.35	1.89 ± 0.05	0.15 ± 0.01	93.5 ± 0.87
F2	1.73 ± 0.02	3:1	1:5	17.2 ± 0.43	1.89 ± 0.02	0.15 ± 0.04	92.7 ± 5.76
F3	1.73 ± 0.02	2:2	1:5	16.1 ± 0.56	2.06 ± 0.02	0.14 ± 0.01	90.9 ± 1.01
F4	1.73 ± 0.02	1:3	1:5	16.9 ± 0.35	1.80 ± 0.01	0.16 ± 0.01	98.1 ± 1.84
F5	1.73 ± 0.02	0:4	1:5	15.9 ± 0.06	2.11 ± 0.01	0.15 ± 0.02	96.3 ± 2.48
F6	1.73 ± 0.02	1:3	1:2	18.9 ± 0.25	1.70 ± 0.01	0.31 ± 0.00	96.0 ± 0.18
F7	1.73 ± 0.02	1:3	1:3	15.8 ± 1.11	2.04 ± 0.07	0.22 ± 0.01	93.7 ± 1.33
F8	1.73 ± 0.02	1:3	1:10	16.9 ± 0.42	1.20 ± 0.01	0.09 ± 0.02	99.5 ± 2.55

[1] Presented as $d_{0.5}$ value; the volume weighted diameter below 50% of the total particle. [2] Calculated by dividing the difference between $d_{0.9}$ and $d_{0.1}$ by $d_{0.5}$: $d_{0.9}$ and $d_{0.1}$ by $d_{0.5}$: $d_{0.9}$ and $d_{0.1}$ are the volume weighted diameters below 90% and 10% of the total particle, respectively. [3] Calculated by dividing the weight of TA in microcrystals-loaded MSs by total weight of microcrystals-loaded MSs (mg). [4] Expressed as the percentage (%) after dividing the weight of TA loaded in MSs by total fed weight of TA (mg). [5] Expressed as mean ± SD (n = 3); Note: Lecithin was included in all formulations at the weight ratio of 5 *w/w* % to the total amount of polymers.

The novel TAs-loaded MSs were further characterized in terms of outer and internal structures and drug crystallinity in MS (Figure 2). The MSs (F4) prepared by the spray-drying technique was highly spherical, with a smooth and homogeneous surface (Figure 2A). In the cross-sectional image, the microcrystals showed different textures from the polymeric matrix and were found to be uniformly embedded in the polymeric matrix (Figure 2B). The number of microcrystals loaded per MS was elucidated by translating the loading amount into the number of MS and TA microcrystals. In the process of converting the weight to a number, the volume of single MS and TA microcrystal was calculated with the assumption that the shape of the MS and microcrystal were spherical and cubic, respectively, with both having a density of 1.0. In the MS formulations prepared with the drug to polymer ratio of 1:2 (F6), 1:3 (F7), 1:5 (F4), and 1:10 (F8), the number of microcrystals embedded in each MS was calculated as 215, 94, 82, and 50, respectively. In the hyperspectral image (Figure 2C), TA microcrystals (red color) were observed to be principally located inside the PLGA/PLA MS (yellow color). However, individual microcrystals were not separately spotted in the image, probably because of the low resolution of Cytoviva®. The characteristic peak of TA microcrystals was identically detected in the microcrystals-load MS (Figure 2D), denoting that TA microcrystals stabilized by BKC were successfully incorporated in the MSs, with no crystalline changes during the fabrication process.

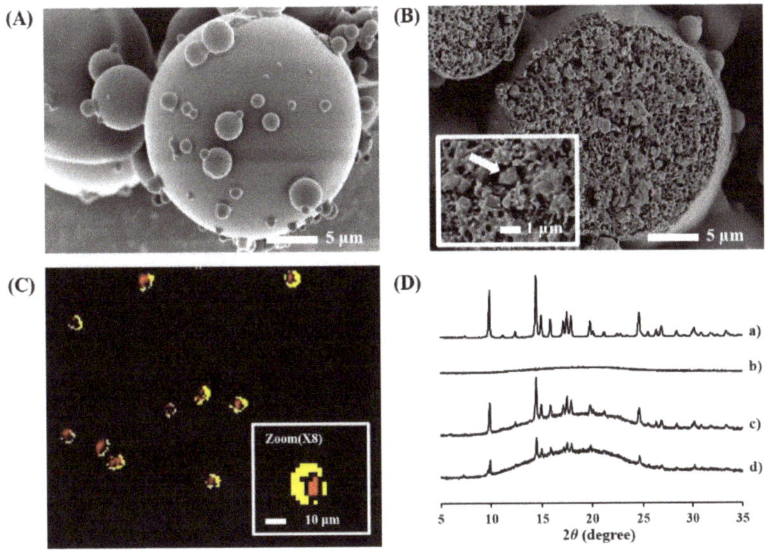

Figure 2. Morphological and physical characteristics of the microcrystals-loaded MSs. Representative micrograph of (**A**) intact and (**B**) cross-sectioned microcrystals-loaded MS (F4), (**C**) hyperspectral image of microcrystals-loaded MSs, and (**D**) XRD patterns of (a) TA microcrystals, (b) blank MS, (c) F4 MS (drug:polymer = 1:5), and (d) F8 MS (1:10); Notes: Inset in (**B**) is ×5000 magnified images and the arrow points to the TA crystal surrounded by the polymeric matrices. In the hyperspectral image (**C**), PLGA/PLA polymers and TA microcrystals are colored as yellow and red, respectively.

3.3. *In Vitro Drug Release and Degradation Profiles of TA Microcrystals-Loaded MSs*

In Vitro drug release profiles from the marketed product, the drug powder- or micronized TA crystals-loaded MSs were evaluated under accelerated dissolution conditions. Although the synovial fluid does not assure the sink condition for the drug dissolution, the in vitro release test under sink conditions was favored for quicker comparison between the formulations. Moreover, as the drug release profiles from PLGA/PLA MSs could be retarded from days to months at body temperature (37 °C), the liberation pattern of the steroidal compound from the MSs was further facilitated by elevating the temperature of the dissolution media (45 °C), promoting the degradation and/or hydrolysis of the biodegradable polymers [26–28]. Actually, the accelerated test at high temperature was reported to be beneficial for faster comparison of release behavior between MS formulas, with high correlation with that obtained at 37 °C [27]. Shen and Burgess (2012) revealed that the time required to reach 100% drug release from the MSs prepared with PLGA polymer with the glass transition temperature (T_g) of 44–48 °C was determined to be 10, 5, 3, and 1.3 days, respectively, at temperatures of 45, 50, 53, and 60 °C. Herein, the temperature of the dissolution medium was set to 45 °C, which did not exceed the T_g values of both polymers (46–52 °C). When the polymeric MSs were exposed to the medium at temperatures above the T_g of the polymer, the drug diffusion coefficient was drastically increased [29], diminishing the difference between the release profiles between the polymeric particulates.

Under the accelerated condition, the marketed product containing 13 µm-sized TA crystals stabilized by polysorbate 20 and sodium carboxymethyl cellulose was rapidly liquefied in the aqueous medium, showing complete drug release within 90 min (Figure 3A). It coincided with a previous report that showed that the TA crystal suspension was readily dissolved in phosphate buffered saline within 2 h [10]. The drug release from the MS with 7.2-µm-sized TA crystals (F0) was not markedly retarded compared to the intact drug crystal, releasing over 95% of TA within 3 h. The incomplete and/or erratic coating thickness of the polymeric layer on the TA crystals probably could not effectively restrain the dissolution and diffusion procedures of the TA crystals into the aqueous media. On the other hand, drug release from the MSs containing smaller TA crystals (median size of 1.7 µm) was markedly impeded compared to the marketed product or 7.2-µm-sized TA crystal-loaded MS, especially as the ratio of PLA increased in the MS (Figure 3A).

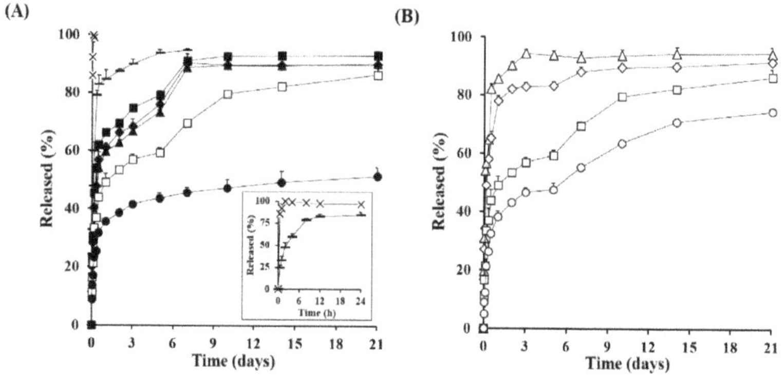

Figure 3. In Vitro release profile of TA from the novel MSs under accelerated conditions (45 °C). (**A**) Drug release profile from the marketed product (×), 7.2 µm-sized TA crystals-loaded MS (F0, –), and 1.7 µm-sized TA crystals-loaded MS prepared with different PLGA:PLA ratios; 4:0 (F1, ■), 3:1 (F2, ♦), 2:2 (F3, ▲), 1:3 (F4, □), and 0:4 (F5, ●) and (**B**) drug release profile from MSs prepared with different drug:polymer ratios; 1:2 (F6, △), 1:3 (F7, ◇), 1:5 (F4, □), and 1:10 (F8, ○); Notes: The inset graph (**A**) is the magnified release profile from the marketed product and 7.2 µm-sized TA crystals-loaded MS. Data are expressed as the mean value (n = 3) and error bars are SDs.

As the glycolic acid has faster hydration/swelling behavior compared to lactic acid [30], MSs prepared with over 50% of PLGA polymers (F1, F2, and F3) showed higher burst release, with over 60% of drug released within 2 days. After the initial burst release, the extent of TA liberated from the F1, F2, and F3 polymers continuously rose, exhibiting over 90% release after 7 days under sink condition. On the other hand, the F4 formula with a PLGA:PLA ratio of 1:3, exhibited a more protracted release profile compared to that of F1, F2, and F3, exhibiting a linear release pattern for 21 days after a 53% initial release in the first 2 days. F5 (PLGA:PLA ratio of 0:4) showed the slowest release profile, displaying only 52% of the accumulated drug release after 21 days. Although there was marked difference in release profile depending on the ratio of PLGA and PLA polymer, the drug release pattern from novel drug microcrystals-loaded MSs were characterized by initial burst release and subsequent slow release profile, which is consistent with the typical release pattern of PLGA/PLA based MPs previous reported [31,32]. In the early phase, TA microcrystals located on or inner compartment near the surface of the MS might be rapidly dissolved by surrounding and/or penetrated aqueous media, and released from polymeric matrix mainly by diffusion mechanism. After initial burst release, the drug release rate tended to be retarded, due to the increased diffusion distance. Afterward, and the remaining steroidal compound in the MSs might be liberated by polymeric degradation and erosion and/or collapse of polymeric MSs.

Different drug release patterns depended on the PLGA:PLA ratio were highly consistent by the morphological changes of TA microcrystals-loaded MSs. As shown in Figure 4, because of the rapid swelling and hydrolysis nature of PLGA polymer, the PLGA MSs (F1) began to collapse and was excavated within three days. Thus, the drug microcrystals embedded in the PLGA MS might be readily exposed to aqueous media, and immediately dissolved under sink condition. On the contrary, because of the greater hydrophobicity of the PLA polymer compared to the PLGA polymer, the hydrolytic degradation of PLA MS (F5, PLGA:PLA = 0:4) progressed slowly. When the MSs were scrutinized at 3 and 7 days, fine pores were formed on the roughed surface and the pore size was gradually enlarged as time elapsed. Nevertheless, the overall globular shape and dimension of MS were retained even at 21 days, supporting the slow and incomplete release profile of TA from the MS (F7). The degradation pattern of F4 prepared with the PLGA:PLA ratio of 1:3 was intermediate between those of MSs prepared with PLGA or PLA polymer F1 and F5, respectively. After surface erosion and pore formation at three days, the MS was then gradually collapsed over 21 days. The drug release rate from F4 was markedly delayed compared to that from F1 but was much faster and higher than that from F5, releasing over 80% of the drug loaded for 21 days. The ratio of PLGA to PLA polymers was fixed to 1:3, expecting the prolonged release pattern for further investigation.

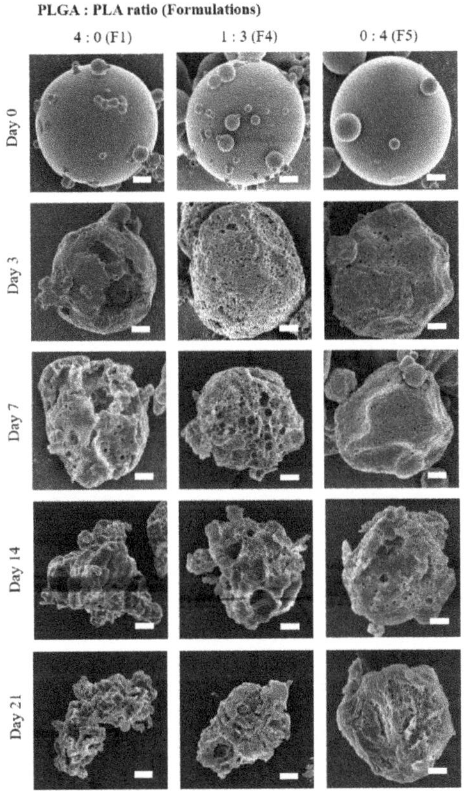

Figure 4. Morphological changes of TA microcrystals-loaded MSs prepared with different PLGA:PLA ratios, 0:4 (F1), 1:3 (F4), 0:4 (F5) under the accelerated release conditions (45 °C); Note: Scale bars in each image indicate 2.0 μm.

The in vitro release profile of TA from the PLGA/PLA MSs prepared with different drug to polymer ratios was further evaluated. As shown in Figure 3B, the initial drug release from the novel MSs were gradually decreased, as the drug to polymer ratio was increased. When the ratio of drug to polymer was 1:2 (F6) or 1:3 (F7), the percentage of drug released for 24 h had reached approximately 86% and 78%, respectively. On the other hand, in the formulations of the drugs:polymer ratio of 1:5 (F4) and 1:10 (F8), the release of the steroidal compound from the MSs were markedly retarded, exhibiting less than 60% of drug release over 5 days. Drug release from the F4 or F8 was steeped after 7 days, probably because of erosion and/or collapse of the MSs, but prolonged for 21 days. Formulas F4 and F8 were further exploited for in vivo pharmacokinetic study in rats, expecting an extended release profile over one month in the knee joint.

3.4. *In Vivo Systemic Exposure and Ioint Retention of TA after IA Injection in Rats*

The systemic exposure and local bioavailability of TA following a single IA injection of the marketed product or the novel MSs (F4 and F8) were evaluated in normal rats. The IA dose of TA treated in all groups was same to 0.25 mg per knee, which was well tolerated in rats [10]. The plasma levels of TA as a function of time following IA injection of the marketed product, F4, and F8 are represented in Figure 5 and the relevant PK parameters are summarized in Table 3. It is recommended that the exposure of the steroidal compound in blood be minimized, as the exogenous corticosteroid can cause Cushing syndrome, incurred impaired wound healing, infection, and muscle weakness [10,33,34]. However, unfortunately, the plasma level of TA was drastically elevated after administration of the marketed product, reaching C_{max} value of 218.7 ng/mL after 3.7 h. This rapid redistribution of TA into the bloodstream is in agreement with earlier reports that intra-articularly injected TA crystals were rapidly absorbed, with a T_{max} value of 4 h in patients with osteoarthritis [6,35,36]. This rapid drug efflux from the knee joint is also correlated with in vitro release profiles, denoting that TA crystalline suspension injected in the joint might be rapidly dissolved and passed out the gap in the synovial membrane. After reaching a C_{max} of 3.7 h, the plasma level of TA sharply decreased below 30 ng/mL after 12 h post-administration of the marketed product.

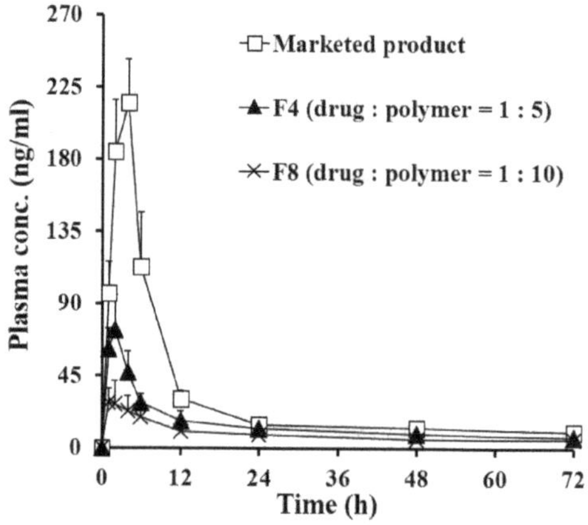

Figure 5. Plasma concentration–time profiles of TA following IA administration of the marketed product, F4 (drug:polymer = 1:5) and F8 (drug:polymer = 1:10) in rats at a dose of 0.5 mg/kg; Note: Each point represents mean ± SD (n = 5).

Table 3. Pharmacokinetic parameters of TA in plasma following IA administration of the marketed product, F4, and F8 in rats.

Parameters	Marketed Product	F4	F8
$AUC_{0\text{–}7day}$ (ng·h/mL)	2787.4 ± 187.4	1500.4 ± 218.9 *	1022.2 ± 161.9 *,**
C_{max} (ng/mL)	218.7 ± 26.6	75.6 ± 17.2 *	32.2 ± 12.0 *,**
T_{max} (h)	3.7 ± 0.8	1.8 ± 0.4 *	1.4 ± 0.5 *
$T_{1/2}$ (h) [1]	5.3 ± 0.1	9.0 ± 0.5 *	13.3 ± 2.7 *,**

[1] Calculated from the plasma concentration–time curve from T_{max} to 24 h after IA injection; Notes: Data are expressed as mean ± SD (n = 5). Statistical analysis was performed using the one-way ANOVA test; * significantly different from the marketed product ($p < 0.05$), ** significantly different from F4 ($p < 0.05$); Abbreviations: $AUC_{0\text{–}7days}$, area under the plasma concentration–time curve until day 7; C_{max}, maximum plasma concentration; T_{max}, time to reach maximum plasma concentration; $T_{1/2}$, elimination half-life of the drug.

In contrast, the systemic exposure of the exogenous corticosteroid was markedly decreased following IA injection of the TA microcrystals-loaded MSs (F4 and F8). The C_{max} values of TA following IA injection of F4 and F8 were determined to 75.6 ng/mL and 32.2 ng/mL, respectively, which were only

34% ($p < 0.05$) and 15% ($p < 0.05$) that of the marketed product. Correspondingly, $AUC_{0–7days}$ values in the F4- and F8-treated groups were drastically decreased to less than 54% ($p < 0.05$) and 37% ($p < 0.05$) of that obtained from the marketed product, respectively. These pharmacokinetic data indicated that the novel MSs remarkably lessened the redistribution of dissolved and/or micronized compound into the bloodstream, prolonging the retention time of TA in the knee. Between the two groups treated with MSs prepared with loading amount of 0.16 (F4) and 0.09 (F8), respectively, the drug exposure to blood was much lowered in the F8-treated group, showing 68% and 42% decreased $AUC_{0-7days}$ and C_{max} values compared to those obtained from the F4-treated group. This pharmacokinetic tendency is explainable with the in vitro release test results, which revealed that the extent of drug released from the polymeric matrix declined as the drug to polymer ratio increased.

The drug remaining in the joint tissue following IA single injection of each formula was further assessed in normal rats (Figure 6). After the IA injection of the marketed product, the drug concentration in the joint at three days post-dosing was only 5.6 μg/g because TA crystals were quickly effluxed from the joint tissue. The percentages of the drug remaining in the joint tissue at 3 and 7 days were calculated to be only 4.5% and 2.4%, respectively. After 21 days, the drug concentration in the joint tissue was below the limit of detection. This result is in line with a previous report that only two of eight patients with osteoarthritis had quantifiable synovial TA concentration at week 6, following IA injection of the marketed product [37]. On the other hand, the novel MS formulations exhibited a markedly profound and prolonged concentration profile in joint tissue compared to the marketed product, exhibiting quantifiable TA concentration over 28 days. Three days after the single administration of F4 or F8, the drug concentration in joints was determined to be 45 μg/g and 67 μg/g, respectively, which is one-third and one-half of the initial dose. In both MSs-treated groups, the TA concentration in the tissue gradually decreased as time elapsed, but approximately 5% of the initial dose was still detected at 28 days. These findings suggested that the retention time of TA in the joint tissue was extended with the sustained-release pattern of the novel MSs.

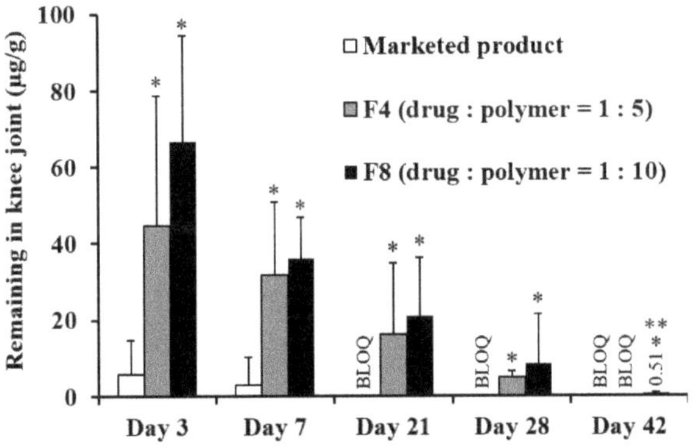

Figure 6. TA remains in rat joint tissue following IA administration of the marketed product, F4, and F8 at a dose of 125 μg of TA per knee; Notes: Vertical bars represent mean ± SD (n = 6). Statistical analysis was performed using the one-way ANOVA test; * significantly different from the marketed product ($p < 0.05$), ** significantly different from F4 ($p < 0.05$). BLOQ value of LC-MS/MS analysis was determined to 0.2 ng/mL.

3.5. Long-Term Stability of TA Microcrystals-Loaded MS

The physicochemical stability of the novel MS was evaluated after 12 weeks of storage under ambient conditions (25 °C, 60% RH). The storage condition of the MSs was set to ambient condition, as there was morphological change above 40 °C, due to softening of the polymer over T_g. At first, the drug crystallinity in the MS was assessed using XRD because a change in drug crystalline nature may occur during storage, affecting the drug chemical stability and release pattern from the MS. Under ambient conditions, the crystallinity of TA microcrystals embedded in the MS was stably maintained over 12 months, with no changes in diffraction pattern (Figure A1). There was also no change in drug content in F8 MS, displaying over 97% drug content after 12 months of storage. The in vitro dissolution pattern was also comparable with that of MSs immediately prepared, exhibiting a sustained-release profile of over 21 days (Figure A1). From these findings, we concluded that the novel MS system was physicochemically stable at least for one year under ambient conditions.

4. Conclusions

A novel parenteral sustained-release system of TA was successfully prepared by micronizing TA powder into 1.7 μm-sized microcrystals, and subsequently embedding into PLGA/PGA polymeric MSs using a spray-drying technique. TA microcrystals were efficiently entrapped into the polymeric MSs, preserving their distinctive crystalline nature. In vitro drug release from the novel MSs was markedly retarded compared to the marketed product and even 7.2 μm-sized TA crystal-loaded MSs, exhibiting a prolonged release profile over 21 days under accelerated conditions (45 °C). In an in vivo pharmacokinetic study in normal rats, the duration that the TA remained in the joint tissue was markedly extended, providing profound drug remains at 28 days following IA single injection. Moreover, TA microcrystals-loaded MSs drastically decreased the systemic exposure of the steroidal compound compared to the marketed product. Thus, the novel IA long-acting system could be a valuable tool, providing both increased drug retention in the knee and diminished systemic exposure of TA following a single administration.

Author Contributions: Conceptualization, M.J.H. and M.J.K.; Data curation, M.J.H., H.T.J. and J.S.P.; Formal analysis, H.R.C., D.Y.K. and Y.S.C.; Funding acquisition, M.J.K.; Investigation, S.H.I., H.T.K. and J.S.P.; Methodology, M.J.H., H.T.J. and J.W.L.; Project administration, Y.W.C., Y.S.C. and M.J.K.; Resources, H.T.J., S.H.I., H.T.K., J.E.L. and J.S.P.; Supervision, Y.W.C., Y.S.C. and M.J.K.; Validation, H.R.C., D.Y.K. and Y.S.C.; Visualization, M.J.H.; Writing—original draft, M.J.H.; Writing—review & editing, Y.W.C., J.W.L., Y.S.C. and M.J.K.

Funding: This research was supported by Basic Science Research Program through the National Research Foundation of Korea (NRF) funded by the Ministry of Science, ICT & Future Planning (NRF-2016R1C1B1010687).

Conflicts of Interest: The authors declare no conflict of interest.

Appendix A

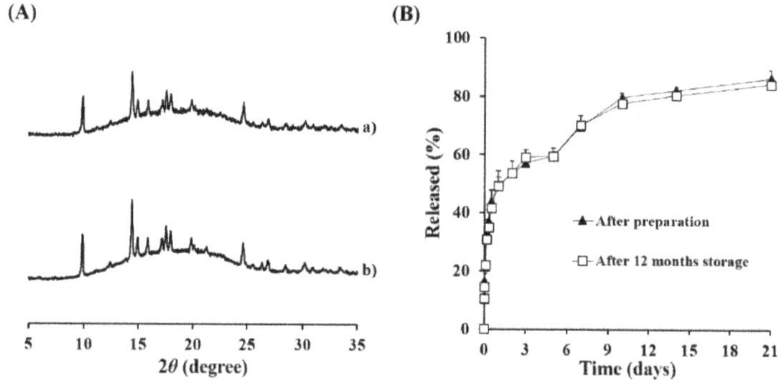

Figure A1. Physicochemical stability of TA microcrystals-loaded MS (F4) after 12 months of storage under ambient conditions. (**A**) XRD patterns of F4 MS (a) after preparation and (b) after 12 months of storage, (**B**) in vitro release profile of F4 after preparation and after 12 months storage under accelerated conditions; Notes: XRD pattern and release profile of F4 MS after preparation are identical to those depicted in Figures 2D and 3B, respectively. Each point in (**B**) in vitro release profile represents mean ± SD (n = 3).

References

1. Gerwin, N.; Hops, C.; Lucke, A. Intraarticular drug delivery in osteoarthritis. *Adv. Drug Deliv. Rev.* **2006**, *58*, 226–242. [CrossRef] [PubMed]
2. Luzardo-Álvarez, A.; Lamela-Gómez, I.; Otero-Espinar, F.; Blanco-Méndez, J. Development, characterization, and in vitro evaluation of resveratrol-loaded poly-(ε-caprolactone) microcapsules prepared by ultrasonic atomization for intra-articular administration. *Pharmaceutics* **2019**, *11*, 249. [CrossRef] [PubMed]
3. Creamer, P. Intra-articular corticosteroid injections in osteoarthritis: Do they work and if so, how? *Ann. Rheum. Dis.* **1997**, *56*, 634–636. [CrossRef] [PubMed]
4. Godwin, M.; Dawes, M. Intra-articular steroid injections for painful knees. Systematic review with meta-analysis. *Can. Fam. Physician* **2004**, *50*, 241–248. [PubMed]
5. Bjordal, J.M.; Johnson, M.I.; Lopes-Martins, R.A.; Bogen, B.; Chow, R.; Ljunggren, A.E. Short-term efficacy of physical interventions in osteoarthritic knee pain. a systematic review and meta-analysis of randomised placebo-controlled trials. *BMC Musculoskelet. Disord.* **2007**, *8*, 51. [CrossRef] [PubMed]
6. Ayhan, E.; Kesmezacar, H.; Akgun, I. Intraarticular injections (corticosteroid, hyaluronic acid, platelet rich plasma) for the knee osteoarthritis. *World J. Orthop.* **2014**, *5*, 351. [CrossRef]
7. Knight, A.D.; Levick, J.R. Morphometry of the ultrastructure of the blood-joint barrier in the rabbit knee. *Q. J. Exp. Physiol.* **1984**, *69*, 271–288. [CrossRef]
8. Ayral, X. Injections in the treatment of osteoarthritis. *Best Pract. Res. Clin. Rheumatol.* **2001**, *15*, 609–626. [CrossRef]
9. Kambhampati, S.P.; Mishra, M.K.; Mastorakos, P.; Oh, Y.; Lutty, G.A.; Kannan, R.M. Intracellular delivery of dendrimer triamcinolone acetonide conjugates into microglial and human retinal pigment epithelial cells. *Eur. J. Pharm. Biopharm.* **2015**, *95*, 239–249. [CrossRef]
10. Rudnik-Jansen, I.; Colen, S.; Berard, J.; Plomp, S.; Que, I.; van Rijen, M.; Woike, N.; Egas, A.; van Osch, G.; van Maarseveen, E.; et al. Prolonged inhibition of inflammation in osteoarthritis by triamcinolone acetonide released from a polyester amide microsphere platform. *J. Control. Release* **2017**, *253*, 64–72. [CrossRef]
11. Elron-Gross, I.; Glucksam, Y.; Margalit, R. Liposomal dexamethasone–diclofenac combinations for local osteoarthritis treatment. *Int. J. Pharm.* **2009**, *376*, 84–91. [CrossRef]
12. Rudnik-Jansen, I.; Woike, N.; de Jong, S.; Versteeg, S.; Kik, M.; Emans, P.; Mihov, G.; Thies, J.; Eijkelkamp, N.; Tryfonidou, M.; et al. Applicability of a modified rat model of acute arthritis for long-term testing of drug delivery systems. *Pharmaceutics* **2019**, *11*, 70. [CrossRef]
13. He, Z.; Wang, B.; Hu, C.; Zhao, J. An overview of hydrogel-based intra-articular drug delivery for the treatment of osteoarthritis. *Colloids Surf. B Biointerfaces* **2017**, *154*, 33–39. [CrossRef]
14. Zhang, Z.; Wei, X.; Gao, J.; Zhao, Y.; Zhao, Y.; Guo, L.; Chen, C.; Duan, Z.; Li, P.; Wei, L.; et al. Intra-articular injection of cross-linked hyaluronic acid-dexamethasone hydrogel attenuates osteoarthritis: An experimental study in a rat model of osteoarthritis. *Int. J. Mol. Sci.* **2016**, *17*, 411. [CrossRef]
15. Ho, M.J.; Kim, S.R.; Choi, Y.W.; Kang, M.J. Recent advances in intra-articular drug delivery systems to extend drug retention in joint. *J. Pharm. Investig.* **2019**, *49*, 9–15. [CrossRef]
16. Evans, C.H.; Kraus, V.B.; Setton, L.A. Progress in intra-articular therapy. *Nat. Rev. Rheumatol.* **2014**, *10*, 11–22. [CrossRef]
17. Bodick, N.; Lufkin, J.; Willwerth, C.; Kumar, A.; Bolognese, J.; Schoonmaker, C.; Ballal, R.; Hunter, D.; Clayman, M. An intra-articular, extended-release formulation of triamcinolone acetonide prolongs and amplifies analgesic effect in patients with osteoarthritis of the knee: a randomized clinical trial. *J. Bone Jt. Surg. Am.* **2015**, *97*, 877–888. [CrossRef]
18. Ho, M.J.; Kim, S.R.; Choi, Y.W.; Kang, M.J. a novel stable crystalline triamcinolone acetonide-loaded PLGA microsphere for prolonged release after intra-articular injection. *Bull. Korean Chem. Soc.* **2016**, *37*, 1496–1500. [CrossRef]
19. Ige, P.P.; Baria, R.K.; Gattani, S.G. Fabrication of fenofibrate nanocrystals by probe sonication method for enhancement of dissolution rate and oral bioavailability. *Colloids Surf. B Biointerfaces* **2013**, *108*, 366–373. [CrossRef]
20. Xiao, C.-D.; Shen, X.-C.; Tao, L. Modified emulsion solvent evaporation method for fabricating core–shell microspheres. *Int. J. Pharm.* **2013**, *452*, 227–232. [CrossRef]

21. Ouyang, H.; Zheng, A.; Heng, P.; Chan, L. Effect of lipid additives and drug on the rheological properties of molten paraffin wax, degree of surface drug coating, and drug release in spray-congealed microparticles. *Pharmaceutics* **2018**, *10*, 75. [CrossRef]
22. Naik, J.B.; Waghulde, M.R. Development of vildagliptin loaded Eudragit® microspheres by screening design: In Vitro evaluation. *J. Pharm. Investig.* **2018**, *48*, 627–637. [CrossRef]
23. César, I.C.; Byrro, R.M.D.; de Santana e Silva Cardoso, F.F.; Mundim, I.M.; de Souza Teixeira, L.; de Sousa, W.C.; Gomes, S.A.; Bellorio, K.B.; Brêtas, J.M.; Pianetti, G.A. Determination of triamcinolone in human plasma by a sensitive HPLC-ESI-MS/MS method: Application for a pharmacokinetic study using nasal spray formulation. *J. Mass Spectrom.* **2011**, *46*, 320–326. [CrossRef]
24. Zulian, F.; Martini, G.; Gobber, D.; Plebani, M.; Zacchello, F.; Manners, P. Triamcinolone acetonide and hexacetonide intra-articular treatment of symmetrical joints in juvenile idiopathic arthritis: A double-blind trial. *Rheumatology* **2004**, *43*, 1288–1291. [CrossRef]
25. Zhang, Z.; Bi, X.; Li, H.; Huang, G. Enhanced targeting efficiency of PLGA microspheres loaded with Lornoxicam for intra-articular administration. *Drug Deliv.* **2011**, *18*, 536–544. [CrossRef]
26. Zolnik, B.S.; Leary, P.E.; Burgess, D.J. Elevated temperature accelerated release testing of PLGA microspheres. *J. Control. Release* **2006**, *112*, 293–300. [CrossRef]
27. Shen, J.; Burgess, D.J. Accelerated in vitro release testing of implantable PLGA microsphere/PVA hydrogel composite coatings. *Int. J. Pharm.* **2012**, *422*, 341–348. [CrossRef]
28. Tomic, I.; Vidis-Millward, A.; Mueller-Zsigmondy, M.; Cardot, J.-M. Setting accelerated dissolution test for PLGA microspheres containing peptide, investigation of critical parameters affecting drug release rate and mechanism. *Int. J. Pharm.* **2016**, *505*, 42–51. [CrossRef]
29. Frank, A.; Rath, S.K.; Venkatraman, S.S. Controlled release from bioerodible polymers: Effect of drug type and polymer composition. *J. Control. Release* **2005**, *102*, 333–344. [CrossRef]
30. Stewart, S.; Domínguez-Robles, J.; Donnelly, R.; Larrañeta, E.; Stewart, S.A.; Domínguez-Robles, J.; Donnelly, R.F.; Larrañeta, E. Implantable polymeric drug delivery devices: Classification, manufacture, materials, and clinical applications. *Polymers* **2018**, *10*, 1379. [CrossRef]
31. Witt, C.; Kissel, T. Morphological characterization of microspheres, films and implants prepared from poly(lactide-*co*-glycolide) and ABA triblock copolymers: Is the erosion controlled by degradation, swelling or diffusion? *Eur. J. Pharm. Biopharm.* **2001**, *51*, 171–181. [CrossRef]
32. Makadia, H.K.; Siegel, S.J.; Makadia, H.K.; Siegel, S.J. Poly lactic-*co*-glycolic acid (PLGA) as biodegradable controlled drug delivery carrier. *Polymers* **2011**, *3*, 1377–1397. [CrossRef]
33. Hopkins, R.L.; Leinung, M.C. Exogenous Cushing's syndrome and glucocorticoid withdrawal. *Endocrinol. Metab. Clin. N. Am.* **2005**, *34*, 371–384. [CrossRef]
34. Wernecke, C.; Braun, H.J.; Dragoo, J.L. The Effect of Intra-articular corticosteroids on articular cartilage. *Orthop. J. Sport. Med.* **2015**, *3*, 232596711558116. [CrossRef]
35. Jüni, P.; Hari, R.; Rutjes, A.W.; Fischer, R.; Silletta, M.G.; Reichenbach, S.; da Costa, B.R. Intra-articular corticosteroid for knee osteoarthritis. *Cochrane Database Syst. Rev.* **2015**, *10*, CD005328. [CrossRef]
36. Hepper, C.T.; Halvorson, J.J.; Duncan, S.T.; Gregory, A.J.M.; Dunn, W.R.; Spindler, K.P. The efficacy and duration of intra-articular corticosteroid injection for knee osteoarthritis: a systematic review of level I studies. *J. Am. Acad. Orthop. Surg.* **2009**, *17*, 638–646. [CrossRef]
37. Kraus, V.B.; Conaghan, P.G.; Aazami, H.A.; Mehra, P.; Kivitz, A.J.; Lufkin, J.; Hauben, J.; Johnson, J.R.; Bodick, N. Synovial and systemic pharmacokinetics (PK) of triamcinolone acetonide (TA) following intra-articular (IA) injection of an extended-release microsphere-based formulation (FX006) or standard crystalline suspension in patients with knee osteoarthritis (OA). *Osteoarthr. Cartil.* **2018**, *26*, 34–42. [CrossRef]

pharmaceutics

MDPI

Article

Formulation, Colloidal Characterization, and In Vitro Biological Effect of BMP-2 Loaded PLGA Nanoparticles for Bone Regeneration

Teresa del Castillo-Santaella [1], Inmaculada Ortega-Oller [2], Miguel Padial-Molina [2], Francisco O'Valle [3], Pablo Galindo-Moreno [2], Ana Belén Jódar-Reyes [1,4] and José Manuel Peula-García [1,5,*]

1 Biocolloid and Fluid Physics Group, Department of Applied Physics, University of Granada, 18071 Granada, Spain
2 Department of Oral Surgery and Implant Dentistry, University of Granada, 18071 Granada, Spain
3 Department of Pathology, School of Medicine & IBIMER, University of Granada, 18071 Granada, Spain
4 Excellence Research Unit "Modeling Nature" (MNat), University of Granada, 18071 Granada, Spain
5 Department of Applied Physics II, University of Malaga, 29071 Malaga, Spain
* Correspondence: jmpeula@uma.es; Tel.: +34-952132722

Received: 20 June 2019; Accepted: 31 July 2019; Published: 3 August 2019

Abstract: Nanoparticles (NPs) based on the polymer poly (lactide-co-glycolide) acid (PLGA) have been widely studied in developing delivery systems for drugs and therapeutic biomolecules, due to the biocompatible and biodegradable properties of the PLGA. In this work, a synthesis method for bone morphogenetic protein (BMP-2)-loaded PLGA NPs was developed and optimized, in order to carry out and control the release of BMP-2, based on the double-emulsion (water/oil/water, W/O/W) solvent evaporation technique. The polymeric surfactant Pluronic F68 was used in the synthesis procedure, as it is known to have an effect on the reduction of the size of the NPs, the enhancement of their stability, and the protection of the encapsulated biomolecule. Spherical solid polymeric NPs were synthesized, showing a reproducible multimodal size distribution, with diameters between 100 and 500 nm. This size range appears to allow the protein to act on the cell surface and at the cytoplasm level. The effect of carrying BMP-2 co-adsorbed with bovine serum albumin on the NP surface was analyzed. The colloidal properties of these systems (morphology by SEM, hydrodynamic size, electrophoretic mobility, temporal stability, protein encapsulation, and short-term release profile) were studied. The effect of both BMP2-loaded NPs on the proliferation, migration, and osteogenic differentiation of mesenchymal stromal cells from human alveolar bone (ABSC) was also analyzed in vitro.

Keywords: BMP-2; PLGA nanoparticles; Pluronic F68

---

1. Introduction

In the context of nanomedicine, tissue regeneration using colloidal micro- and nano-structures having unique size and surface activity has received increasing attention over recent years. Many efforts have been made to improve the engineering of these nano-systems in order to reach a "smart" delivery of bioactive molecules in order to optimize their therapeutic advantages and minimize harmful side effects [1]. With this aim, a broad spectrum of biocompatible nanocarriers has been described, showing properties suitable for different biological and therapeutic applications [2]. Among these varied proposals, polymeric nanosystems represent a major group in which poly lactic-co-glycolic acid (PLGA) is one of the most widely used due to its biocompatibility, biodegradability, and low cytotoxicity, gaining the approval from different drug agencies for human use [3,4].

PLGA-based structures are described as micro- and nanocarriers to deliver a wide variety of active molecules and drugs, synthetic or natural molecules with hydrophilic or hydrophobic properties, and biomolecules from proteins to nucleic acids [5–7]. PLGA micro- and nanosystems can be set up using different formulation techniques, with the possibility of a systemic or local distribution. These systems can be applied not only in tissue regeneration but also in very diverse therapies: Anticancer drug delivery, infections, inflammatory diseases, or gene therapy [3]. Despite this great potential, certain applications, especially in protein encapsulation, are hindered by problems, such as an uncontrolled release profile and protein denaturation [8–11].

The water-in oil-in water (W/O/W) double emulsion method is an "emulsion solvent evaporation" technique frequently used to encapsulate hydrophilic molecules as proteins in PLGA NPs [6,12]. The appropriate choice of organic solvents, the use of polymer-surfactant blends, and the addition of stabilizer-protective agents have proved to be key aspects for optimizing the resulting systems [9,11]. Additionally, a surface specific functionalization can be used to improve their versatility, allowing the chemical surface immobilization of different molecules in order to confer targeting or adhesive properties to these nanocarriers [13].

Within tissue engineering, bone regeneration has a broad range of applications, mostly in the field of dentistry, where PLGA is suggested as a reference polymer to formulate NPs with bone-healing uses [14]. The literature describes the delivery of bioactive molecules, normally growth factors, using polymeric microparticles (MPs) and NPs with PLGA as the main component [13]. Among the bone morphogenetic growth factors, BMP-2 (bone morphogenetic protein 2) has been the most frequently cited, with many examples in which encapsulation or surface adsorption enables adequate entrapment efficiency and diverse release patterns [15–19]. For proteins with a very short half-life, such as BMPs, biodegradable PLGA nanosystems provide protection and optimal dosage for an adequate stimulation of cell differentiation [20,21].

Thus, within this scenario, in the present work, we seek to optimize a nano-particulate system in order to carry out and control the release of BMP-2 using as a starting point the synthesis procedure of a lysozyme-loaded NP system, previously described for the encapsulation of that model protein [11]. Also, to encapsulate BMP-2, we prepared a second system in which this protein was co-adsorbed with bovine serum albumin onto the surface of empty NPs. The size and morphology, the protein encapsulation efficiency, the surface characteristics, and the colloidal and temporal stability were studied to complete the physico-chemical characterization of both NP systems.

The release profile of BMP-2 indicates the potential of a PLGA nanocarrier for bone regeneration and depends heavily on the polymer degradation by hydrolysis [22]. However, over the short term, during which the release does not depend on this chemical degradation, proper control of release is necessary in order to modulate other physical processes. Thus, we focused our release experiments on the short-term using different techniques to compare the two NP samples and establish the corresponding BMP-2 release profiles. Finally, the biological activity (cell migration, proliferation, and osteogenic differentiation) was tested in vitro using mesenchymal stromal cells (MSCs) derived from alveolar bone [23].

2. Materials and Methods

2.1. *Nanoparticle Synthesis*

2.1.1. Formulation

Poly(lactide-co-glycolide) acid (PLGA 50:50) ($[C_2H_2O_2]_x[C_3H_4O_2]_y$), x = 50, y = 50 (Resomer® 503H, (Evonik, Essen, Germany), 32–44 kDa was used as the polymer, and polymeric surfactant Pluronic F68 (Poloxamer 188) (Sigma-Aldrich, St. Louis, MO, USA) as the emulsifier. Their structure, based on a poly(ethylene oxide)-block-poly(propylene oxide)-block-poly(ethylene oxide), is expressed as PEOa-PPOb-PEOa with a = 75 and b = 30. Human recombinant bone morphogenetic protein, rhBMP-2 (Sigma-H4791), was used as therapeutic biomolecule. Water was purified in a Milli-Q

Academic Millipore system. A double-emulsion synthesis method was used following a procedure previously described with slight modifications [11]. In this method, 100 mg of PLGA and 3 mg of deoxycholic acid (DC) were dissolved in a tube containing 1 mL of ethyl acetate (EA) and vortexed. In total, 40 μL of a buffered solution at pH 12.8, with or without rhBMP-2 (200 μg/mL), were added and immediately sonicated (Branson Ultrasonics 450 Analog Sonifier) for 1 min (Duty cycle dial: 20%, Output control dial: 4) with the tube surrounded by ice. This primary W/O emulsion was poured into a plastic tube containing 2 mL of a buffered solution (pH 12) of F68 at 1 mg/mL, and vortexing for 30 s. Then, the tube surrounded by ice was sonicated at the maximum amplitude for the micro tip for 1 min (Output control: 7). This second W/O/W emulsion was poured into a glass containing 10 mL of the buffered F68 solution and kept under magnetic stirring for 2 min. The organic solvent was then rapidly extracted by evaporation under vacuum to a final volume of 8 mL. The resulting empty and BMP-2 encapsulated NP systems were named NP and NP-BMP2, respectively. A detailed scheme of the synthesis procedure, with a yield based on the PLGA component always higher than 85%, is shown in Figure S1 of the Supplementary Materials.

2.1.2. Cleaning and Storage

After the organic solvent evaporation, the sample was centrifuged for 10 min at 20 °C at 12,000 rpm. The supernatant was filtered using Millipore nanofilters, 0.1 μm for measuring the free non-encapsulated protein. The pellet was then resuspended in phosphate buffer (1.15 mM NaH_2PO_4), PB, to a final volume of 4 mL and kept refrigerated at 4 °C. Under these conditions, the systems kept colloidal stability at least for one month.

2.1.3. Protein Loading and Encapsulation Efficiency

The initial protein loading was optimized for the nanoparticle formulation, preserving the final colloidal stability after the evaporation step and taking into account the amounts shown in the literature for this growth factor when encapsulated inside PLGA NPs [24,25]. Thus, we chose 2 μg as the initial total mass of rhBMP-2, which means a relation of 2×10^{-5}% *w*/*w* (rhBMP-2/PLGA). The amount of encapsulated rhBMP-2 was calculated by measuring the difference between the initial added amount, and the free non-encapsulated protein present in the supernatant after the cleaning step, which was tested by a specific enzyme-linked immuno-sorbent assay following the instructions of the manufacturer (ELISA, kit RAB0028 from Sigma-Aldrich, St. Louis, MO, USA). Then, protein-encapsulation efficiency (EE) was calculated as follows:

$$EE = \frac{M_I - M_F}{M_I} \times 100$$

where M_I is the initial total mass of rhBMP-2, and M_F is the total mass of rhBMP-2 in the aqueous supernatant.

2.1.4. Physical Protein Adsorption

Bovine serum albumin (BSA) and rhBMP-2 were coupled on the empty nanoparticle surface by a physical adsorption method. The appropriate volume of an aqueous protein solution containing 0.5 mg of BSA and 2 μg of rhBMP-2 was mixed with 5 mL of acetate buffer (pH 5) containing empty NPs with 12.5 mg of PLGA. This provided a starting amount of proteins corresponding to 0.04% *w*/*w* (protein/PLGA), while the mass relation between proteins was 0.4 *w*/*w* (rhBMP-2/BSA). This solution was incubated at room temperature for 2 h under mechanical stirring. The nanoparticles were separated from the buffer solution by centrifugation, and after the supernatants were filtered (Millipore nanofilters, 0.1 μm), they were qualitatively analyzed by gel electrophoresis while the protein quantification was made by a bicinchoninic acid protein assay (BCA) (Sigma-Aldrich, St. Louis, MO, USA) for BSA and the specific ELISA for rhBMP-2. The nanoparticle pellet was resuspended in phosphate buffer (pH 7.4) and stored at 4 °C. This system was named NP-BSA-BMP2.

2.1.5. Protein Separation by Gel Electrophoresis, SDS-PAGE

The protein-loaded NPs and different supernatants were treated at 90 °C for 10 min in the following buffer: 62.5 mM Tris-HCl (pH 6.8 at 25 °C), 2% (*w*/*v*) sodium dodecyl sulfate (SDS), 10% glycerol, 0.01% (*w*/*v*) bromophenol blue, 40 mM dithiothreitol (DTT). Samples were then separated by size in porous 12% polyacrylamide gel (1D SDS polyacrylamide gel electrophoresis), under the effect of an electric field. The electrophoresis was run under constant voltage (130 V, 45 min) and the gels were stained using a Coomassie Blue solution (0.1% Coomassie Brilliant Blue R-250, 50% methanol and 10% glacial acetic acid) and destained with the same solution lacking the dye.

2.2. Nanoparticle Characterization: Morphology, Size, Concentration, and Electrokinetic Mobility

NPs were imaged by scanning electron microscopy (SEM) with a Zeiss SUPRA 40VP field-emission scanning electron microscope from the Scientific Instrumentation Center of the University of Granada (CIC, UGR).

The hydrodynamic size distribution of the NPs was evaluated by nanoparticle tracking analysis (NTA) with a NanoSight LM10-HS (GB) FT14 (NanoSight, Amesbury, UK) and an sCMOS camera. The particle concentration according to the diameter (size distribution) was calculated as an average of at least three independent size distributions. The total concentration of NPs of each system was determined in order to control the number of particles used in cell experiments. The measurement conditions for all samples were 25 °C, a viscosity of 0.89 cP, a measurement time of 60 s, and a camera gain of 250. The camera shutter was 11 and 15 ms for the empty and BMP-loaded NPs, respectively. The detection threshold was fixed at 5.

The electrophoretic mobility of the NPs was determined using a Zetasizer® NanoZeta ZS device (Malvern Instrument Ltd., Malvern, UK) working at 25 °C with an He-Ne laser of 633 nm, and a 173° scattering angle. Each data point was taken as an average over three independent sample measurements. For each sample, the electrophoretic mobility distribution and the average electrophoretic mobility (μ-average) were determined by the technique of laser Doppler electrophoresis.

2.3. Colloidal and Temporal Stability in Biological Media

The average hydrodynamic diameter and the polydispersity index (PDI) by dynamic light scattering (DLS) of each NP system were measured in different media (phosphate buffer (PB) saline phosphate buffer (PBS), and cell culture medium: Dulbecco's modified Eagle's medium, DMEM (Sigma)). Also, data on temporal stability were gathered by repeating these analyses at different times after synthesis (0, 1, and 5 days) and after 1 month under storage conditions.

In vitro release experiments were conducted as follows: 1 mL of each sample for each incubation time was suspended in PBS at 37 °C. After the corresponding time (24, 48, 96, 168 h), NPs were separated from the supernatant of released proteins by centrifugation for 10 min at 14,000 rpm (10 °C). The NP pellet was suspended in 1 mL of 0.05 M NaOH and stirred for 2 h for a complete polymer degradation. The alkaline protein solution was assayed by BCA and ELISA to quantify the unreleased amount. The protein released was calculated taking into account the total encapsulated amount. All experiments were made in triplicate.

2.4. Cell Interactions

For all biological in vitro studies, a cell population cultured from the maxillary alveolar bone was used. This population was previously characterized and confirmed to present all characteristics of a mesenchymal stromal cell population (MSC) [23]. Cells were taken from healthy human donors after the approval from the Ethics Committee for Human Research from the University of Granada (424/CEIH/2018). Regular Dulbecco's modified Eagle's medium (DMEM) with 1 g/L glucose (DMEM-LG) (Gibco), 10% fetal bovine serum (FBS) (Sigma-Aldrich, St. Louis, MO, USA), 1:100 of non-essential amino acid solution (NEAA) (Gibco), 0.01 μg/mL of basic fibroblast growth factor (bFGF)

(PeproTech, London, UK), 100 U/mL of penicillin/streptomycin, and 0.25 µg/mL of amphotericin B was used as culture medium for all experiments. Cultures were maintained at 37 °C in a 5% CO_2 atmosphere (2000 cells/well). All biological experiments were repeated in triplicate at least 3 times per condition.

2.4.1. Cell Migration

A cell-migration assay was conducted as previously described [26,27]. Briefly, MSCs were distributed on to three wells for each condition and allowed to grow to a cell confluency close to 99%, in 24-wells/plate at 3000 cells/cm^2, and in each well three different scratches were made. Then, cells were starved for 24 h by adding culture medium without serum. A scratch was made using a pipette tip along the diameter of the well. A wash step with PBS was performed to remove the scratched cells. Fresh complete culture media was added and supplemented depending on the assigned group (BMP-2, NP- BMP2, and NP-BSA-BMP2 at 1.25, 2.5, and 5 ng/mL of BMP-2). Afterwards, nine images were taken from the same area in each condition until 48 h later. On these images, the scraped area was measured by ImageJ software (National Institute of Health, Bethesda, MD, USA; http://rsbweb.nih.gov/ij/). The reduction in the scratched area over time was measured considering the area at time 0 as 100% open.

2.4.2. Cell Proliferation

Proliferation was evaluated by a sulphorhodamine (SRB) assay [28]. The assay was conducted by seeding the cells at 1500 cells/cm^2 in a 96-well plate at a confluence not higher than 50%. After cell attachment, the different supplements were added (BMP-2, NP- BMP2, and NP-BSA-BMP2 at 1.25, 2.5, and 5 ng/mL of BMP-2) and the cells were maintained in culture for up to 7 days. At each time point, the cells were washed with 1X PBS and fixed by adding ice-cold 10% trichloroacetic acid for 20 min at 4 °C. Then, the cells were washed 3 times with dH_2O and dried until all time points were collected. Each well received 0.4% SRB in 1% acetic acid for 20 min at room temperature with gentle shaking. The staining was finished by washing each well 3 times with 1% acetic acid and drying it at room temperature for 24 h. The dye was retrieved from the cells by adding 10 mM Tris Base at pH 10.5 and gently shaking for 10 min. The solution recovered was then distributed in a 96-well plate and the optical absorbance was read at 492 nm.

2.4.3. Osteogenic Differentiation

Osteogenic differentiation was evaluated by adding osteogenic media to the cell culture in combination with free BMP-2, NP-BMP2, and NP-BSA-BMP2 at the highest dosages used in previous experiments. Cells were seeded at 3000 cells/cm^2 and cultured to reach an 85% to 90% confluency. This was followed by the addition of induction media containing 10 mM of β-glycerophosphate (Fluka, 50020), 0.1 µM of dexamethasone (Sigma-Aldrich, D2915) and 0.05 mM of L-ascorbic acid (Sigma-Aldrich, A8960). Cell cultures were maintained for 7 days to analyze early activity. At day 7, cells were collected in 1 mL of TRIzol®. Then, RNA was extracted and converted to cDNA. Alkaline phosphatase (ALP) was then evaluated, expression being calculated relative to glyceraldehyde-3-phosphate dehydrogenase protein (GAPDH) by the $2^{-\Delta\Delta Ct}$ method. These procedures were conducted as described elsewhere [23]. Forward and reverse primer sequences were AGCTCATTTCCTGGTATGACAAC and TTACTCCTTGGAGGCCATGTG for GAPDH, and TCCAGGGATAAAGCAGGTCTTG and CTTTCTCTTTCTCTGGCACTAAGG for ALP.

2.4.4. Statistical Evaluation

Cell migration and proliferation were evaluated by ANOVA followed by Tukey multiple comparisons test for pairwise analysis. Comparison between the levels of ALP at 4 vs. 7 days were analyzed by paired Student's *t* test. In all cases, a *p* value lower than 0.05 was established as statistical significance.

3. Results and Discussion

3.1. Nanoparticle Formulation

Double emulsion-solvent evaporation has been described as a robust and frequently used method to produce biomolecule-loaded PLGA NPs [6,12,13,29]. A formulation previously optimized by our group enabled the preservation of the biological activity of encapsulated biomolecules using a slightly aggressive organic solvent. Moreover, deoxycholic acid has been used in the first step of the formulation in order to improve the colloidal stability of NPs and, simultaneously, to obtain NP surfaces enriched with carboxylic groups, improving their versatility and allowing a subsequent chemical immobilization of different specific ligands [30]. By means of this improved formulation, in the present work, we developed empty nanoparticles (NPs) or nanoparticles encapsulating rhBMP-2 (NP-BMP2). A schematic description of the synthesis procedure is shown in Figure S1 of the Supplementary Data. For NP-BMP2, we achieved a protein-encapsulation efficiency (EE) of 97 ± 2%. This result is consistent with the literature in which several authors have reported similarly high values encapsulating this protein inside PLGA nano- and microparticles [31,32]. Our formulation has several factors leading to this very high EE value: The low protein/polymer relation in mass [33], the affinity of rhBMP-2 to an unspecific interaction with hydrophobic surfaces [31], or the addition of stabilizers (poloxamer) in the second step of the double-emulsion procedure [13]. The absence of rhBMP-2 in the supernatant resulting from the centrifugation step in the cleaning process was verified by ELISA and SDS-PAGE, in which a clear band corresponding to 14 kD of rhBMP-2 polypeptidic chains is shown for lane A in Figure 1, corresponding to NP-BMP2. The mass of protein encapsulated, around 2 µg, is similar to that of different PLGA micro- and nanosystems described in the literature [18,34,35]. Taking into account the storage conditions for our samples, this corresponds to 500 ng/mL, which represents a sufficient concentration for practical applications since this growth factor shows in vitro biological activities at very low dosages (5–20 ng/mL) [13].

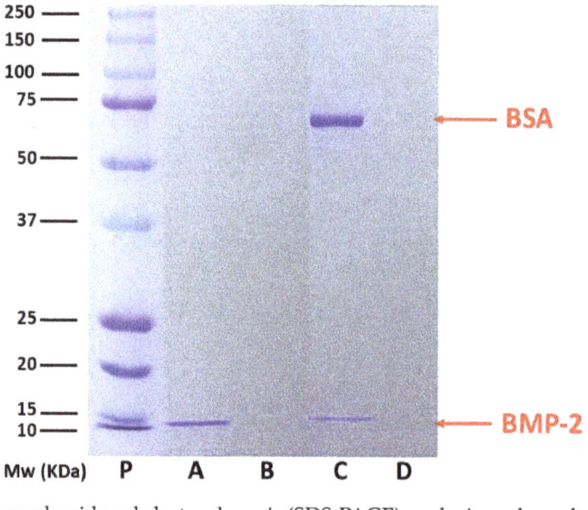

Figure 1. SDS polyacrylamide gel electrophoresis (SDS-PAGE) analysis under reducing conditions of solid PLGA Nanoparticles (PLGA NPs) and liquid (supernatant) fractions of different NP systems. Lane **P**: Protein standards; lane **A**: NP-BMP2 (bone morphogenetic protein); lane **B**: supernatant of NP-BMP2 after synthesis and encapsulation of rhBMP-2; lane **C**: NP after physical adsorption of BSA/rhBMP-2; lane **D**: supernatant after physical adsorption of BSA(bovine serum albumin)/rhBMP-2 on NP system.

On the other hand, a second nanosystem resulted, modifying the way in which rhBMP-2 is incorporated in the nanocarrier. There are several examples of surface adsorption of different growth factors in micro- and nanoparticles [35–37], and surface immobilization over the encapsulation has recently been proposed as a way to modulate the later release of biomolecules. This process, which depends on the slow diffusion of biomolecules through the polymeric matrix, is consequently highly influenced by the protein–polymer interaction [38,39] and polymer degradation [3,6]. Thus, this new focus on the use of PLGA NPs for biomolecule delivery was explored by immobilizing the protein rhBMP-2 on the surface of empty NPs by means of simple physical adsorption. This process is known to be governed by electrostatic and hydrophobic interactions between protein molecules and NP surfaces [40].

For this, the surface-charged groups, the hydrophilicity, the net charge of the protein molecules, and the characteristics of the adsorption medium are the reference parameters. Thus, we designed a co-adsorption experiment in which a mixture of rhBMP-2 and BSA (0.4% *w*/*w*, rhBMP-2/BSA) interact simultaneously with the PLGA NP surface. Albumins are routinely used as protective proteins when growth factors are incorporated in PLGA NPs [13,19]. Moreover, a surface distribution of BSA molecules can improve the colloidal stability of NPs at physiological pH due to their net negative charge under these conditions [41]. Figure S2 from Supplementary Materials shows a scheme of the co-adsorption process. The adsorption efficiency is higher than 95% and in SDS-PAGE from Figure 1, two bands characteristic of both proteins can be seen in lane C, corresponding to the NP-BSA-BMP2 nanosystem. However, lane D, corresponding to the run of the supernatant from the centrifugation of the nanosystem after adsorption processes, shows the absence of any protein. This result is fully explained by taking into account the pH of the medium (pH 5.0), near the isoelectric point of BSA, where the adsorption of this protein onto negatively charged nanoparticles presents a maximum [40,42]. The immobilization of rhBMP-2 on the negatively charged surface of NPs proves they are electrostatically favored due to the positive net charge of this protein at acid and neutral pH.

3.2. *Nanoparticle Characterization*

3.2.1. Nanoparticle Size

SEM and STEM micrographs (Figure 2) show that the samples consist of spherical particles of different diameters (between 150 and 450 nm), a range similar to that found in a previous work in which NPs were loaded with lysozyme following a similar synthesis protocol [11]. In that work, the DLS technique failed to provide a reliable size distribution. Therefore, the NTA technique was directly used to determine the hydrodynamic size of the BMP2-loaded NPs (see NTA video in the Supplementary Material).

The size distributions for empty (NP) and BMP-loaded NPs (NP-BMP2) from NTA (Figure 3 and videos S1, S2) were consistent with the SEM images. Particles with diameters between 100 and 500 nm were found to have the highest particle concentration at around 200 nm. The loading with BMP had an effect on the size distribution, leading to more defined peaks. These measurements enabled us to determine the concentration of particles in the measured sample: $6.88 \pm 0.09 \times 10^8$ pp/mL and $5.19 \pm 0.12 \times 10^8$ pp/mL for NP and NP-BMP2 nanosystems, respectively. These values were used (by taking into account the corresponding dilution) to control the number of particles added in the cell experiments.

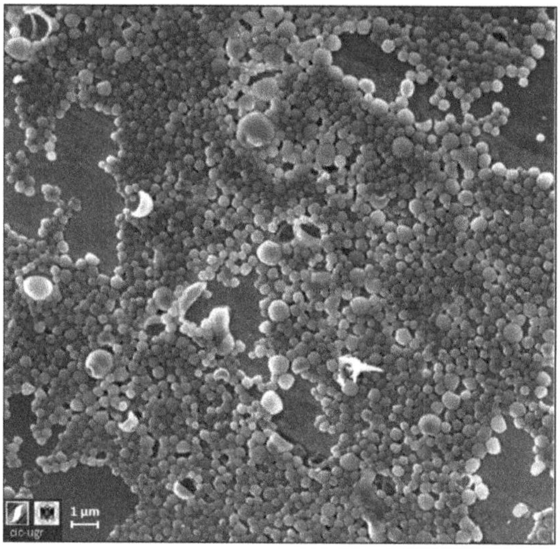

Figure 2. Scanning electron microscopy (SEM) micrograph of rhBMP-2-loaded nanoparticles (NP-BMP2).

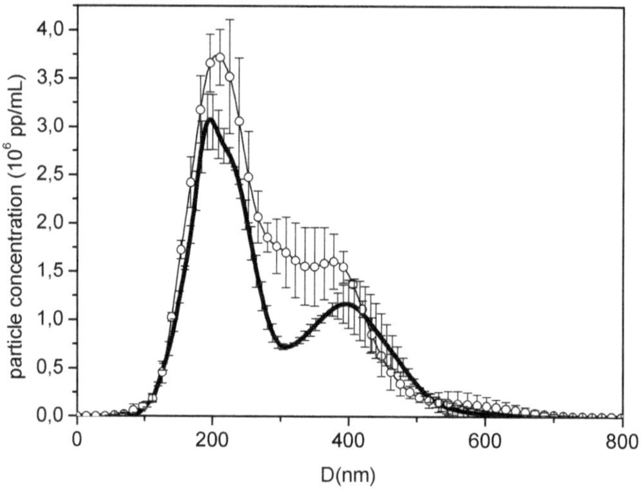

Figure 3. Hydrodynamic diameter distribution of NP (circles) and NP-BMP2 (thick black line) measured at pH 7.0 (phosphate buffer) by nanoparticle tracking analysis (NTA).

3.2.2. Electrokinetic Mobility and Colloidal Stability

The surface charge of nanoparticles can be analyzed using an electrokinetic study by measuring the electrophoretic mobility (μ_e) under different conditions. Figure 4 shows the μ_e and zeta potential values for the three nanosystems: NP, NP-BMP2, and NP-BSA-BMP2, at low ionic strength and different pH values. The electric surface charge of NPs resides in the carboxylic groups of the uncapped PLGA and deoxycholic acid molecules. These functionalized groups are additionally useful due to the possibility of a chemical surface vectorization in order to develop directed delivery nanocarriers [43]. It was previously confirmed that protonation of these acidic surface groups at pH values under their pKa value was tightly correlated with a loss of surface charge and consequently a reduction (in absolute value) of the electrophoretic mobility of the colloidal system [44,45]. Usually, when colloidal particles are coated by protein molecules, the μ_e values change markedly compared with the same bare surfaces and are influenced by the electrical charge of the adsorbed protein molecules [46,47]. The electrokinetic behavior of the NP-BMP2 system remains similar to that of NP, and encapsulation of rhBMP-2 does not affect the surface charge distribution. A similar result was reported by d'Angelo et al. on encapsulating different growth factors in PLGA-poloxamer blend nanoparticles in the same proportion *w/w* of protein/polymer [24]. This may be due to the low amount of encapsulated protein and its distribution in the inner part of the NPs (far from the surface). In our system, this internal distribution may be favored by the encapsulating conditions where the basic pH (pH 12.0) of the water phase containing rhBMP-2 allows a negative charge of these protein molecules, thereby preventing their electrostatic specific interaction with acidic groups of the NPs.

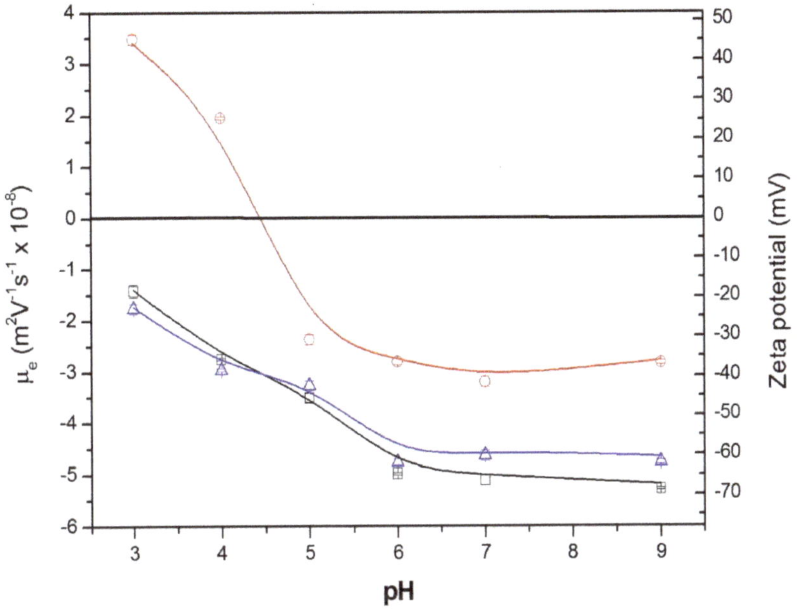

Figure 4. Electrophoretic mobility and zeta potential vs. pH in buffered media of low salinity (ionic strength equal to 0.002 M) for the different nanosystems: (black square) NP; (blue triangle) NP-BMP2; (red circle) NP-BSA-BMP2.

The electrokinetic distribution for the NP-BSA-BMP2 system radically changes. As previously shown, the very high adsorption efficiency leads to NPs with both proteins adsorbed around their surface. This situation is closely correlated with the μ_e values from Figure 4. Taking into account the *w/w* relation between adsorbed proteins (250 times higher for BSA), albumin molecules modulate the behavior at pH values below their isoelectric point (pI 4.7), where the positive net charge of BSA masks the original surface charge of NPs and even changes their original values to positive ones. This is a typical result found for this protein-covering colloidal particles [42,48]. At neutral and basic pH values, BSA molecules have a negative net charge, and the slight decrease in the absolute μ_e values could be due to the reduction of the negative net surface charge of NPs, which may be shielded, at least in a small part, by the positive charge of rhBMP-2 molecules under their basic isoelectric point (pI 9.0).

The colloidal stability for the different nanosystems (NP, NP-BMP2, and NP-BSA-BMP2) was determined by analyzing the size distributions in various media (PB, PBS, and DMEM) at different times after synthesis (0, 1, and 5 days). Size distributions similar to the original ones were found for the two formulations, NP and NP-BMP2, in all the media analyzed. This result was similar to that previously found for these types of NPs encapsulating lysozyme [11], in which the combination of electrostatic and steric interactions generated by surface chemical groups of NPs confer the stability mechanism that prevents colloidal aggregation [33]. The decrease of the absolute value of the zeta potential for the NP-BSA-BMP2 system as a consequence of surface protein distribution does not affect its colloidal stability. This system also maintains the same size distribution in the different media. It is commonly accepted that a zeta potential higher than +30 or −30 mV will give rise to a stable colloidal system [49] and the zeta potential value for NP-BSA-BMP2 is above −30 mV. Colloidal stability in PBS and DMEM, typically used media for the development of scaffold or cell interactions, respectively, assures the potential use of these nanosystems for in vitro or in vivo living environments. Additionally, these systems maintained their size under storage in PB, at 4 °C for at least 1 month (data not shown), showing this to be an adequate medium for sample storage.

3.2.3. Protein Release

One of the main problems for micro- or nanosystems of PLGA drug delivery is to find the appropriate release pattern for encapsulated/attached protein molecules. A wide spectrum of formulations modulates this property by the use of different types of synthesis processes, PLGA polymers, co-polymers, and stabilizers [3,13]. An adequate limitation and control in the burst release is critical for BMPs in order to ensure long-term continuous release that, favored by the polymer degradation, provides better in vivo action in driving bone and cartilage regeneration [20]. Therefore, we previously developed a dual PLGA nanosystem for controlled short-term release, where protein diffusion and protein–polymer interaction are the main factors governing this process [11].

In the present work, NP-BMP2 and NP-BSA-BMP2 nanosystems represent two different ways in which rhBMP-2 was incorporated into the nanocarrier. Figure 5A shows the cumulative release of both proteins, rhBMP-2 and BSA, for different systems as a function of time in a short-term period (7 days). The encapsulated rhBMP-2 protein reaches an amount released of around 30% of the initial encapsulated one while adsorbed rhBMP-2, despite its surface distribution, is three times lower. However, BSA shows released amounts up to 80% of the initial adsorbed ones. In all cases, error bars correspond to the standard deviations from three independent experiments. Under these conditions, the growth factor encapsulated in NP-BMP2 presents a release pattern similar to that previously found with the same formulation but using lysozyme as the protein [11]. Poloxamer in the water phase of the synthesis process can be key in modulating both specific and unspecific interfacial protein interactions [50]. Thus, the relation between protein–polymer interaction and protein diffusion appears to be well balanced, preventing an excessive initial burst and simultaneously maintaining the needed protein flux to release around a third of the encapsulated rhBMP-2 in 7 days. Although an excessive initial burst has been widely reported for PLGA NPs related with protein molecules close to the surface [6], this situation did not appear for the NP-BMP2 system, this being consistent with the electrokinetic behavior that did not show the presence of protein near surface. The literature offers some examples with reduced short-term release of BMP-2 using more hydrophilic PLGA-PEG co-polymers [16] or a different synthesis process [25].

The release performance for the NP-BSA-BMP2 system, also shown in Figure 5A, presents notable differences. The electrokinetic profile has previously justified the surface location of BSA and rhBMP-2 on the surface, which could lead to a fast release of both proteins. However, results from Figure 5A,B show this trend only for the BSA protein that is released from NPs, with about 20% of the initial amount remaining after seven days. However, up to 90% of the initial load of rhBMP-2 protein, unlike BSA, remains attached to the surface. The NP surface with hydrophilic groups form poloxamer molecules and a negative charge due to the abundant presence of carboxylic groups (end-groups of PLGA and deoxycholic acid molecules) favor a desorption process for BSA, whose molecules have a negative charge under release conditions (physiological pH). This agrees with the results of other authors who, even after encapsulating BSA in PLGA-poloxamer blend NPs, achieved a fast burst release of above

40% to 50% of the initial protein amount [33]. Moreover, the co-encapsulation of albumins with growth factors could strongly affect its release profile, causing an initial burst [21,24]. Otherwise, the specific electrostatic attraction between positive rhBMP-2 molecules and negative surface groups slows down the short time release of this protein. This result is in agreement with the low release of adsorbed BMP previously found using PLGA micro- and nanoparticles with uncapped acid end groups [38,51]. Thus, the combination of different methods for trapping BMP-2 into and around NPs shows up the possibility of attaining a properly controlled release, balancing the interactions between polymers, stabilizers, and protein.

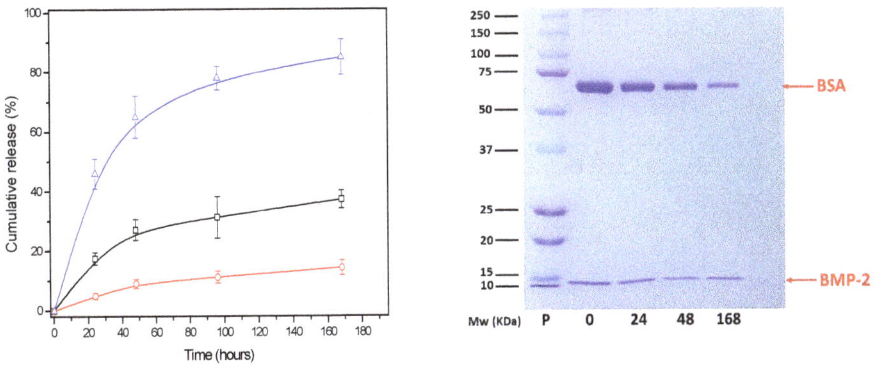

(A) **(B)**

Figure 5. (**A**) Cumulative release of rhBMP-2 for NP-BMP2 (black square) and NP-BSA-BMP2 (red circle) systems; and cumulative release of BSA for NP-BSA-BMP2 (blue triangle) system, incubated for different times at 37 °C in saline phosphate buffer (pH 7.4). (**B**) SDS-PAGE analysis under reducing conditions of solid fraction of NP-BSA-BMP2 after release at different times where the number of each lane corresponds to the time in hours.

3.3. Biological Activity and Interactions

3.3.1. Cell Migration

Cell migration is the first and necessary step in tissue regeneration [52]. Thus, a regenerative agent must accelerate cell migration or, at least, not interfere with it. In the present study, we found no differences between the groups, doses, and control in terms of closure of a scratched area (ANOVA with Tukey multiple comparisons test) (Figure 6). In contrast to our findings, previously published data suggests a positive effect of BMP-2 on cell migration [53,54]. However, in those studies, the doses applied, and the cell types were different than in the current experiments. We used lower doses of BMP-2 in order to test whether, even at low dosages, BMP-2 could still provide benefits if protected in a nanoparticle system. As mentioned, we demonstrated no negative effect of the system on cell migration. Our results nonetheless support the idea that BMP-2 activity is mediated by the activation of the phosphoinositide 3-kinase (PI3K) pathway, a common group of signaling molecules that participate in several process with BMP-2 and other molecules [26,54]. It should also be mentioned that the timeframe of a migration assay is short. Thus, the potential advantages of a controlled-release system as the one under study might be limited. That is, the release of BMP-2 from the nanoparticles, as demonstrated in Figure 5, is limited to the first 48 h. Thus, a sustained positive effect on migration activity over time could be hypothesized.

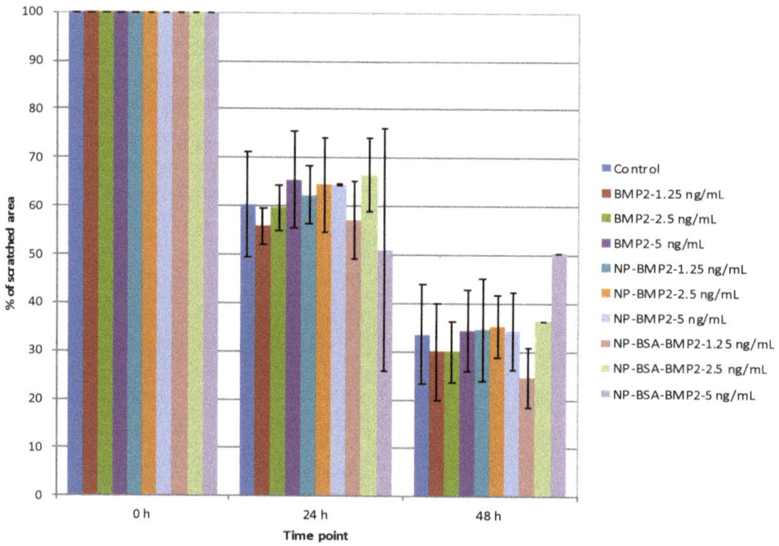

Figure 6. Migration assay. Percentage of scratched area closure at 24 and 48 h on different groups and doses.

3.3.2. Cell Proliferation

Proliferation is another of the cell activities required for tissue regeneration. However, this property must be balanced with both migration and differentiation, and not all three characteristics increase at the same time and with the same ratios [55]. In fact, reportedly, when a dose of BMP-2 induces higher proliferation, it decreases differentiation [56]. This property has been extensively analyzed but discrepancies can still be detected in the literature. Therefore, Kim et al. analyzed different doses of BMP-2 and its effect on cell proliferation and apoptosis. It was confirmed in vitro that high doses, but still lower than those used clinically, reduce cell proliferation and increase apoptosis [57]. This should be avoided. We have found that although free BMP-2 does not induce higher proliferation than the control at any of the doses applied nor time points (ANOVA with Tukey multiple comparisons test), the same amount of BMP-2 encapsulated or adsorbed onto PLGA nanoparticles boosts proliferation, this being statistically significant when using a dose of 2.5 ng/mL or higher (ANOVA with Tukey multiple comparisons test) (Figure 7). These dosages are still lower than those suggested in previous studies. Apart from that difference, a positive effect on proliferation was still achieved. Moreover, following the release pattern from Figure 5, more BMP-2 is expected to be released over time beyond the 7-day time frame. Thus, a sustained induction effect could be expected as well until full confluency of the cell culture.

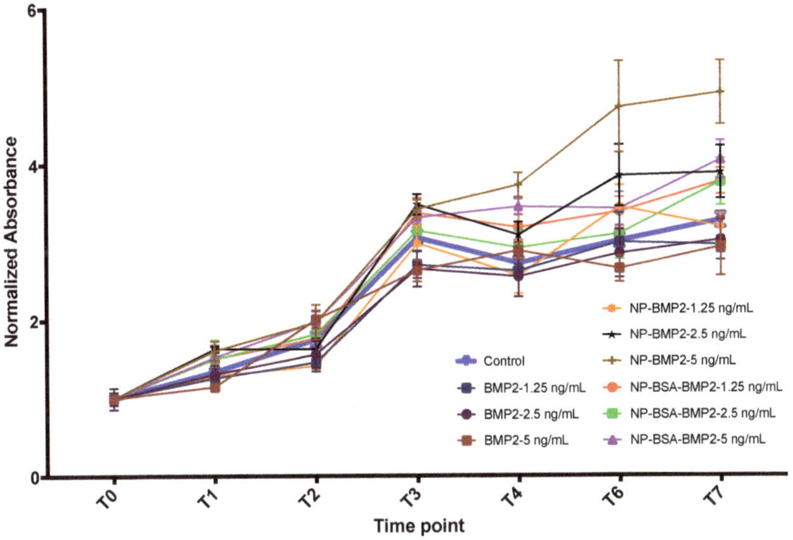

Figure 7. Proliferation of human mesenchymal stromal cells (MSCs) as measured by sulphorhodamine (SRB) absorbance. Results were normalized to T0 in each group.

3.3.3. Osteogenic Differentiation

It has been confirmed that cell differentiation induced by BMP-2 needs the presence of permissive osteoinductive components. Particularly, β-glycerophosphate has been shown to exert a synergistic effect with BMP-2 in inducing cell differentiation [56]. Thus, to test for osteogenic differentiation, we analyzed the expression of ALP mRNA. Maximum ALP activity was found to occur 10 days after stimulation with PLGA-based microparticles containing BMP-2 in co-encapsulation with human serum albumin [16]. Although other tests could have been used to reinforce our findings, ALP is known to modulate the deposition of mineralized nodules, thus indicating osteoblastic activity. For all of this, we supplemented the differentiation media with β-glycerophosphate and either free BMP-2, NP-BMP2, or NP-BSA-BMP2 for 4 and 7 days so that we could capture the early dynamics of the expression of the gene. In our study, we identified an increase in the expression of ALP in all groups from day 4 to day 7 (Figure 8). Although ALP at day 7 in the BMP-2 group appears to be higher than for the other two groups, the change did not prove significant. In fact, differences between groups were not statistically significant within any time period. Noteworthy though, the increase was not significant within the BMP-2 group ($p = 0.141$, Student's t test), but it was significant within the other two groups ($p = 0.025$ and $p = 0.003$; NP-BMP2 and NP-BSA-BMP2 groups, respectively). This, again, could be taken as a confirmation of the sustained release of the protein from the nanoparticle system beyond the earlier time points.

This and both the migration and proliferation studies described below lead us to confirm that the system proposed can maintain a proper release of BMP-2 over time, sustaining a positive effect on cell migration and proliferation with initial reduced doses of BMP-2. The fact that the excessive initial burst is prevented is important for the application of this nanotechnology in bone regeneration, as in dentistry. In this way, the negative effects of initial high doses of BMP-2 are avoided at the same time as the molecule is protected from denaturalization inside the NP. Thus, the regenerator effects are maintained over time.

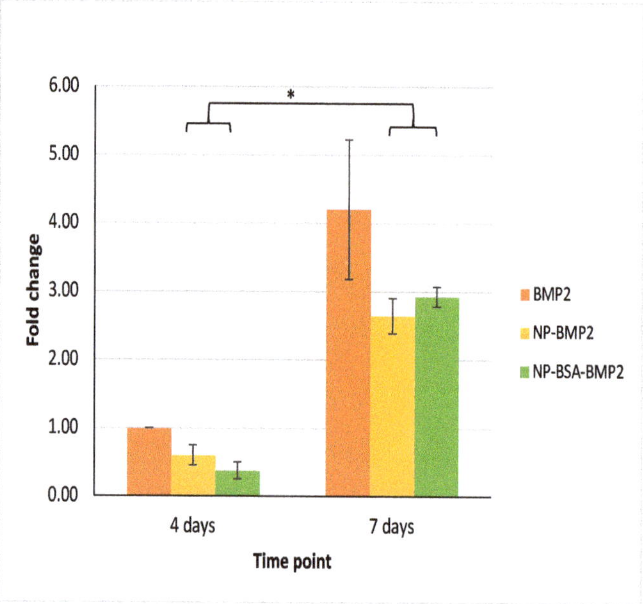

Figure 8. Relative fold change in the expression of ALP mRNA (control group: BMP2 at 4 days). * = Statistical significance of the comparison over time ($p = 0.025$ and $p = 0.003$, Student's *t* test; NP-BMP2 and NP-BSA-BMP2 groups).

4. Conclusions

In this work, a delivery PLGA-nanosystem previously developed for model proteins was chosen as the reference system to carry and deliver the growth factor BMP-2. This NP system, with a dual size distribution, was developed following a double-emulsion formulation in which the process and the components used were optimized to reach the appropriate colloidal and biological behavior. Encapsulation and adsorption are two different processes to load BMP-2 in PLGA NPs. Both were tested to elucidate the factors controlling them and their influence in the physico-chemical and biological properties of nanosystems. We verified that protein–polymer specific interactions have a major role in the way that protein molecules are carried and delivered from NPs. In vitro experiments showed that BMP-2-loaded PLGA NPs are the nanocarriers with the best release profile over the short-term without an initial burst and with moderate and sustained release of active protein before the onset of polymer degradation. Therefore, the biological activity is positive with no negative interaction with migration or proliferation but rather the induction of cell differentiation through the expression of ALP.

Supplementary Materials: The following are available online at http://www.mdpi.com/1999-4923/11/8/388/s1, Figure S1. Scheme of the formulation of NP-BMP2; Figure S2: Scheme of the protein adsorption process for NP-BSA-BMP2; Video S1. NTA experiments for NP-BMP2; Video S2. NTA experiments for empty NPs.

Author Contributions: Conceptualization, J.M.P.-G. and P.G.-M.; methodology, J.M.P.-G., A.B.J.-R. and M.P.-M.; investigation, T.d.C.-S., I.O.-O., J.M.P.-G., A.B.J.-R. and M.P.-M.; resources, A.B.J.-R., P.G.-M., F.O.-R.; writing—original draft preparation, J.M.P.-G. and M.P.-M.; writing—review and editing, J.M.P.-G., M.P.-M., A.B.J.-R., T.d.C.-S.; supervision, J.M.P.-G., P.G.-M. and F.O.-R.; funding acquisition, A.B.J.-R. and P.G.-M.

Funding: This research was funded by the Consejería de Economía, Innovación, Ciencia y Empleo de la Junta de Andalucía (Spain) through research groups FQM-115 and CTS-1028, by the following research project: MAT2013-43922-R—European FEDER support included—(MICINN, Spain) and by MIS Ibérica S.L.

Acknowledgments: The authors wish to express their appreciation for the technical support to D. Darío Abril-García.

Conflicts of Interest: The authors declare no conflict of interest.

References

1. Van Rijt, S.; Habibovic, P. Enhancing regenerative approaches with nanoparticles. *J. R. Soc. Interface* **2017**, *14*. [CrossRef]
2. Kumar, B.; Jalodia, K.; Kumar, P.; Gautam, H.K. Recent advances in nanoparticle-mediated drug delivery. *J. Drug Deliv. Sci. Technol.* **2017**, *41*, 260–268. [CrossRef]
3. Mir, M.; Ahmed, N.; Rehman, A.U.R. Recent applications of PLGA based nanostructures in drug delivery. *Colloids Surf. B Biointerfaces* **2017**, *159*, 217–231. [CrossRef] [PubMed]
4. Jana, S.; Jana, S. Natural polymeric biodegradable nanoblend for macromolecules delivery. In *Recent Developments in Polymer Macro, Micro and Nano Blends*; Woodhead Publishing: Cambridge, UK, 2017; pp. 289–312. ISBN 9780081004081.
5. Danhier, F.; Ansorena, E.; Silva, J.M.; Coco, R.; Le Breton, A.; Préat, V. PLGA-based nanoparticles: An overview of biomedical applications. *J. Control. Release* **2012**, *161*, 505–522. [CrossRef] [PubMed]
6. Ding, D.; Zhu, Q. Recent advances of PLGA micro/nanoparticles for the delivery of biomacromolecular therapeutics. *Mater. Sci. Eng. C* **2018**, *92*, 1041–1060. [CrossRef] [PubMed]
7. Arias, J.L.; Unciti-Broceta, J.D.; Maceira, J.; del Castillo, T.; Hernández-Quero, J.; Magez, S.; Soriano, M.; García-Salcedo, J.A. Nanobody conjugated PLGA nanoparticles for active targeting of African Trypanosomiasis. *J. Control. Release* **2015**, *197*, 190–198. [CrossRef]
8. Giteau, A.; Venier-Julienne, M.C.; Aubert-Pouëssel, A.; Benoit, J.P. How to achieve sustained and complete protein release from PLGA-based microparticles? *Int. J. Pharm.* **2008**, *350*, 14–26. [CrossRef]
9. Fredenberg, S.; Wahlgren, M.; Reslow, M.; Axelsson, A. The mechanisms of drug release in poly(lactic-co-glycolic acid)-based drug delivery systems—A review. *Int. J. Pharm.* **2011**, *415*, 34–52. [CrossRef]
10. White, L.J.; Kirby, G.T.S.; Cox, H.C.; Qodratnama, R.; Qutachi, O.; Rose, F.R.A.J.; Shakesheff, K.M. Accelerating protein release from microparticles for regenerative medicine applications. *Mater. Sci. Eng. C* **2013**, *33*, 2578–2583. [CrossRef]
11. Ortega-Oller, I.; del Castillo-Santaella, T.; Padial-Molina, M.; Galindo-Moreno, P.; Jódar-Reyes, A.B.; Peula-García, J.M. Dual delivery nanosystem for biomolecules. Formulation, characterization, and in vitro release. *Colloids Surf. B Biointerfaces* **2017**, *159*, 586–595. [CrossRef]
12. McClements, D.J. Encapsulation, protection, and delivery of bioactive proteins and peptides using nanoparticle and microparticle systems: A review. *Adv. Colloid Interface Sci.* **2018**, *253*, 1–22. [CrossRef]
13. Ortega-Oller, I.; Padial-Molina, M.; Galindo-Moreno, P.; O'Valle, F.; Jódar-Reyes, A.B.; Peula-García, J.M. Bone Regeneration from PLGA Micro-Nanoparticles. *BioMed Res. Int.* **2015**, *2015*, 1–18. [CrossRef] [PubMed]
14. Bapat, R.A.; Joshi, C.P.; Bapat, P.; Chaubal, T.V.; Pandurangappa, R.; Jnanendrappa, N.; Gorain, B.; Khurana, S.; Kesharwani, P. The use of nanoparticles as biomaterials in dentistry. *Drug Discov. Today* **2019**, *24*, 85–98. [CrossRef]
15. Ji, Y.; Xu, G.P.; Zhang, Z.P.; Xia, J.J.; Yan, J.L.; Pan, S.H. BMP-2/PLGA delayed-release microspheres composite graft, selection of bone particulate diameters, and prevention of aseptic inflammation for bone tissue engineering. *Ann. BioMed. Eng.* **2010**, *38*, 632–639. [CrossRef] [PubMed]
16. Kirby, G.T.S.; White, L.J.; Rahman, C.V.; Cox, H.C.; Qutachi, O.; Rose, F.R.A.J.; Hutmacher, D.W.; Shakesheff, K.M.; Woodruff, M.A. PLGA-Based Microparticles for the Sustained Release of BMP-2. *Polymers* **2011**, *3*, 571–586. [CrossRef]
17. Qutachi, O.; Shakesheff, K.M.; Buttery, L.D.K. Delivery of definable number of drug or growth factor loaded poly(dl-lactic acid-co-glycolic acid) microparticles within human embryonic stem cell derived aggregates. *J. Control. Release* **2013**, *168*, 18–27. [CrossRef] [PubMed]
18. Wang, Y.; Wei, Y.; Zhang, X.; Xu, M.; Liu, F.; Ma, Q.; Cai, Q.; Deng, X. PLGA/PDLLA core-shell submicron spheres sequential release system: Preparation, characterization and promotion of bone regeneration in vitro and in vivo. *Chem. Eng. J.* **2015**, *273*, 490–501. [CrossRef]

19. Zhang, H.-X.; Zhang, X.-P.; Xiao, G.-Y.; Hou, -Y.; Cheng, L.; Si, M.; Wang, S.-S.; Li, Y.-H.; Nie, L. In vitro and in vivo evaluation of calcium phosphate composite scaffolds containing BMP-VEGF loaded PLGA microspheres for the treatment of avascular necrosis of the femoral head. *Mater. Sci. Eng. C* **2016**, *60*, 298–307. [CrossRef]
20. Begam, H.; Nandi, S.K.; Kundu, B.; Chanda, A. Strategies for delivering bone morphogenetic protein for bone healing. *Mater. Sci. Eng. C* **2017**, *70*, 856–869. [CrossRef]
21. Balmayor, E.R.; Feichtinger, G.A.; Azevedo, H.S.; Van Griensven, M.; Reis, R.L. Starch-poly-ε-caprolactone microparticles reduce the needed amount of BMP-2. *Clin. Orthop. Relat. Res.* **2009**, *467*, 3138–3148. [CrossRef]
22. Xu, Y.; Kim, C.S.; Saylor, D.M.; Koo, D. Polymer degradation and drug delivery in PLGA-based drug–polymer applications: A review of experiments and theories. *J. BioMed. Mater. Res. Part B Appl. Biomater.* **2017**, *105*, 1692–1716. [CrossRef] [PubMed]
23. Padial-Molina, M.; de Buitrago, J.G.; Sainz-Urruela, R.; Abril-Garcia, D.; Anderson, P.; O'Valle, F.; Galindo-Moreno, P. Expression of Musashi-1 during osteogenic differentiation of oral MSC: An in vitro study. *Int. J. Mol. Sci.* **2019**, *20*, 2171. [CrossRef] [PubMed]
24. D'Angelo, I.; Garcia-Fuentes, M.; Parajó, Y.; Welle, A.; Vántus, T.; Horváth, A.; Bökönyi, G.; Kéri, G.; Alonso, M.J. Nanoparticles based on PLGA:poloxamer blends for the delivery of proangiogenic growth factors. *Mol. Pharm.* **2010**, *7*, 1724–1733. [CrossRef] [PubMed]
25. Chang, H.-C.; Yang, C.; Feng, F.; Lin, F.-H.; Wang, C.-H.; Chang, P.-C. Bone morphogenetic protein-2 loaded poly(D,L-lactide-co-glycolide) microspheres enhance osteogenic potential of gelatin/hydroxyapatite/β-tricalcium phosphate cryogel composite for alveolar ridge augmentation. *J. Formos. Med. Assoc.* **2017**, *116*, 973–981. [CrossRef] [PubMed]
26. Padial-Molina, M.; Volk, S.L.; Rios, H.F. Periostin increases migration and proliferation of human periodontal ligament fibroblasts challenged by tumor necrosis factor -α and Porphyromonas gingivalis lipopolysaccharides. *J. Periodontal Res.* **2014**, *49*, 405–414. [CrossRef] [PubMed]
27. Liang, C.-C.; Park, A.Y.; Guan, J.-L. In vitro scratch assay: A convenient and inexpensive method for analysis of cell migration in vitro. *Nat. Protoc.* **2007**, *2*, 329–333. [CrossRef]
28. Houghton, P.; Fang, R.; Techatanawat, I.; Steventon, G.; Hylands, P.J.; Lee, C.C. The sulphorhodamine (SRB) assay and other approaches to testing plant extracts and derived compounds for activities related to reputed anticancer activity. *Methods* **2007**, *42*, 377–387. [CrossRef]
29. Iqbal, M.; Zafar, N.; Fessi, H.; Elaissari, A. Double emulsion solvent evaporation techniques used for drug encapsulation. *Int. J. Pharm.* **2015**, *496*, 173–190. [CrossRef]
30. Sánchez-Moreno, P.; Ortega-Vinuesa, J.L.; Boulaiz, H.; Marchal, J.A.; Peula-García, J.M. Synthesis and characterization of lipid immuno-nanocapsules for directed drug delivery: Selective antitumor activity against HER2 positive breast-cancer cells. *Biomacromolecules* **2013**, *14*, 4248–4259. [CrossRef]
31. Lochmann, A.; Nitzsche, H.; von Einem, S.; Schwarz, E.; Mäder, K. The influence of covalently linked and free polyethylene glycol on the structural and release properties of rhBMP-2 loaded microspheres. *J. Control. Release* **2010**, *147*, 92–100. [CrossRef]
32. Kempen, D.H.R.; Lu, L.; Hefferan, T.E.; Creemers, L.B.; Maran, A.; Classic, K.L.; Dhert, W.J.A.; Yaszemski, M.J. Retention of in vitro and in vivo BMP-2 bioactivities in sustained delivery vehicles for bone tissue engineering. *Biomaterials* **2008**, *29*, 3245–3252. [CrossRef]
33. Santander-Ortega, M.J.; Csaba, N.; González, L.; Bastos-González, D.; Ortega-Vinuesa, J.L.; Alonso, M.J. Protein-loaded PLGA–PEO blend nanoparticles: Encapsulation, release and degradation characteristics. *Colloid Polym. Sci.* **2010**, *288*, 141–150. [CrossRef]
34. Chung, Y.I.; Ahn, K.M.; Jeon, S.H.; Lee, S.Y.; Lee, J.H.; Tae, G. Enhanced bone regeneration with BMP-2 loaded functional nanoparticle-hydrogel complex. *J. Control. Release* **2007**, *121*, 91–99. [CrossRef]
35. La, W.-G.; Kang, S.-W.; Yang, H.S.; Bhang, S.H.; Lee, S.H.; Park, J.-H.; Kim, B.-S. The Efficacy of Bone Morphogenetic Protein-2 Depends on Its Mode of Delivery. *Artif. Organs* **2010**, *34*, 1150–1153. [CrossRef]
36. Fu, Y.; Du, L.; Wang, Q.; Liao, W.; Jin, Y.; Dong, A.; Chen, C.; Li, Z. In vitro sustained release of recombinant human bone morphogenetic protein-2 microspheres embedded in thermosensitive hydrogels. *Die Pharm.* **2012**, *67*, 299–303. [CrossRef]

37. Rahman, C.V.; Ben-David, D.; Dhillon, A.; Kuhn, G.; Gould, T.W.A.; Müller, R.; Rose, F.R.A.J.; Shakesheff, K.M.; Livne, E. Controlled release of BMP-2 from a sintered polymer scaffold enhances bone repair in a mouse calvarial defect model. *J. Tissue Eng. Regen. Med.* **2014**, *8*, 59–66. [CrossRef]
38. Pakulska, M.M.; Elliott Donaghue, I.; Obermeyer, J.M.; Tuladhar, a.; McLaughlin, C.K.; Shendruk, T.N.; Shoichet, M.S. Encapsulation-free controlled release: Electrostatic adsorption eliminates the need for protein encapsulation in PLGA nanoparticles. *Sci. Adv.* **2016**, *2*, e1600519. [CrossRef]
39. Fu, C.; Yang, X.; Tan, S.; Song, L. Enhancing Cell Proliferation and Osteogenic Differentiation of MC3T3-E1 Pre-osteoblasts by BMP-2 Delivery in Graphene Oxide-Incorporated PLGA/HA Biodegradable Microcarriers. *Sci. Rep.* **2017**, *7*, 12549. [CrossRef]
40. Peula, J.M.; de las Nieves, F.J. Adsorption of monomeric bovine serum albumin on sulfonated polystyrene model colloids 1. Adsorption isotherms and effect of the surface charge density. *Colloids Surf. A Physicochem. Eng. Asp.* **1993**, *77*, 199–208. [CrossRef]
41. Peula, J.M.; de las Nieves, F.J. Adsorption of monomeric bovine serum albumin on sulfonated polystyrene model colloids 3. Colloidal stability of latex—Protein complexes. *Colloids Surf. A Physicochem. Eng. Asp.* **1994**, *90*, 55–62. [CrossRef]
42. Peula, J.M.; Hidalgo-Alvarez, R.; De Las Nieves, F.J. Coadsorption of IgG and BSA onto sulfonated polystyrene latex: I. Sequential and competitive coadsorption isotherms. *J. Biomater. Sci. Polym. Ed.* **1996**, *7*, 231–240. [CrossRef]
43. Siafaka, P.I.; Üstündağ Okur, N.; Karavas, E.; Bikiaris, D.N. Surface modified multifunctional and stimuli responsive nanoparticles for drug targeting: Current status and uses. *Int. J. Mol. Sci.* **2016**, *17*, 1440. [CrossRef]
44. Peula-García, J.M.; Hidalgo-Alvarez, R.; De Las Nieves, F.J. Colloid stability and electrokinetic characterization of polymer colloids prepared by different methods. *Colloids Surf. A Physicochem. Eng. Asp.* **1997**, *127*, 19–24. [CrossRef]
45. Santander-Ortega, M.J.; Lozano-López, M.V.; Bastos-González, D.; Peula-García, J.M.; Ortega-Vinuesa, J.L. Novel core-shell lipid-chitosan and lipid-poloxamer nanocapsules: Stability by hydration forces. *Colloid Polym. Sci.* **2010**, *288*, 159–172. [CrossRef]
46. Peula-Garcia, J.M.; Hidaldo-Alvarez, R.; De las Nieves, F.J. Protein co-adsorption on different polystyrene latexes: Electrokinetic characterization and colloidal stability. *Colloid Polym. Sci.* **1997**, *275*, 198–202. [CrossRef]
47. Santander-Ortega, M.J.; Bastos-González, D.; Ortega-Vinuesa, J.L. Electrophoretic mobility and colloidal stability of PLGA particles coated with IgG. *Colloids Surf. B Biointerfaces* **2007**, *60*, 80–88. [CrossRef]
48. Peula, J.M.; Callejas, J.; de las NIeves, F.J. Adsorption of Monomeric Bovine Serum Albumin on Sulfonated Polystyrene Model Colloids. II. Electrokinetic Characterization of Latex-Protein Complexes. In *Surface Properties of Biomaterials*; Butterworth and Heinemann: Oxford, UK, 1994; pp. 61–69.
49. Sun, D. Effect of Zeta Potential and Particle Size on the Stability of SiO_2 Nanospheres as Carrier for Ultrasound Imaging Contrast Agents. *Int. J. Electrochem. Sci.* **2016**, 8520–8529. [CrossRef]
50. del Castillo-Santaella, T.; Peula-García, J.M.; Maldonado-Valderrama, J.; Jódar-Reyes, A.B. Interaction of surfactant and protein at the O/W interface and its effect on colloidal and biological properties of polymeric nanocarriers. *Colloids Surf. B Biointerfaces* **2019**, *173*, 295–302. [CrossRef]
51. Schrier, J.A.; DeLuca, P.P. Porous bone morphogenetic protein-2 microspheres: Polymer binding and in vitro release. *AAPS PharmSciTech* **2001**, *2*, 66–72. [CrossRef]
52. Padial-Molina, M.; O'Valle, F.; Lanis, A.; Mesa, F.; Dohan Ehrenfest, D.M.; Wang, H.-L.; Galindo-Moreno, P. Clinical application of mesenchymal stem cells and novel supportive therapies for oral bone regeneration. *BioMed Res. Int.* **2015**, *2015*. [CrossRef]
53. Inai, K.; Norris, R.A.; Hoffman, S.; Markwald, R.R.; Sugi, Y. BMP-2 induces cell migration and periostin expression during atrioventricular valvulogenesis. *Dev. Biol.* **2008**, *315*, 383–396. [CrossRef]
54. Gamell, C.; Osses, N.; Bartrons, R.; Rückle, T.; Camps, M.; Rosa, J.L.; Ventura, F.; Imamura, T. BMP2 induction of actin cytoskeleton reorganization and cell migration requires PI3-kinase and Cdc42 activity. *J. Cell Sci.* **2008**, *121*, 3960–3970. [CrossRef]
55. Friedrichs, M.; Wirsdöerfer, F.; Flohé, S.B.; Schneider, S.; Wuelling, M.; Vortkamp, A. BMP signaling balances proliferation and differentiation of muscle satellite cell descendants. *BMC Cell Biol.* **2011**, *12*, 26. [CrossRef]

56. Hrubi, E.; Imre, L.; Robaszkiewicz, A.; Virág, L.; Kerényi, F.; Nagy, K.; Varga, G.; Jenei, A.; Hegedüs, C. Diverse effect of BMP-2 homodimer on mesenchymal progenitors of different origin. *Hum. Cell* **2018**, *31*, 139–148. [CrossRef]
57. Kim, H.K.W.; Oxendine, I.; Kamiya, N. High-concentration of BMP2 reduces cell proliferation and increases apoptosis via DKK1 and SOST in human primary periosteal cells. *Bone* **2013**, *54*, 141–150. [CrossRef]

pharmaceutics

MDPI

Article

Biocompatible PLGA-Mesoporous Silicon Microspheres for the Controlled Release of BMP-2 for Bone Augmentation

Silvia Minardi [1,†], Joseph S. Fernandez-Moure [2,†], Dongmei Fan [1], Matthew B. Murphy [1], Iman K. Yazdi [1], Xuewu Liu [3], Bradley K. Weiner [1,4,*] and Ennio Tasciotti [1]

1. Center for Biomimetic Medicine, Houston Methodist Research Institute, Houston, TX 77030, USA; silvia.minardi@northwestern.edu (S.M.); dongmeifan@gmail.com (D.F.); matthewmurphy21@gmail.com (M.B.M.); ikyazdi@houstonmethodist.org (I.K.Y.); etasciotti@houstonmethodist.org (E.T.)
2. Division of Trauma, Acute and Surgical Critical Care, Department of Surgery, Duke University, Durham, NC 27517, USA; joseph.fernandezmoure@duke.edu
3. Department of Nanomedicine, Houston Methodist Research Institute, Houston, TX 77030, USA; xliu@houstonmethodist.org
4. Department of Orthopedic Surgery, Houston Methodist Hospital, Houston, TX 77030, USA

* Correspondence: BKWeiner@houstonmethodist.org; Tel.: +713-441-9000

† Both authors contributed equally as first authors.

Received: 14 December 2019; Accepted: 28 January 2020; Published: 1 February 2020

Abstract: Bone morphogenetic protein-2 (BMP-2) has been demonstrated to be one of the most vital osteogenic factors for bone augmentation. However, its uncontrolled administration has been associated with catastrophic side effects, which compromised its clinical use. To overcome these limitations, we aimed at developing a safer controlled and sustained release of BMP-2, utilizing poly(lactic-*co*-glycolic acid)-multistage vector composite microspheres (PLGA-MSV). The loading and release of BMP-2 from PLGA-MSV and its osteogenic potential in vitro and in vivo was evaluated. BMP-2 in vitro release kinetics was assessed by ELISA assay. It was found that PLGA-MSV achieved a longer and sustained release of BMP-2. Cell cytotoxicity and differentiation were evaluated in vitro by MTT and alkaline phosphatase (ALP) activity assays, respectively, with rat mesenchymal stem cells. The MTT results confirmed that PLGA-MSVs were not toxic to cells. ALP test demonstrated that the bioactivity of BMP-2 released from the PLGA-MSV was preserved, as it allowed for the osteogenic differentiation of rat mesenchymal stem cells, in vitro. The biocompatible, biodegradable, and osteogenic PLGA-MSVs system could be an ideal candidate for the safe use of BMP-2 in orthopedic tissue engineering applications.

Keywords: BMP-2; silicon; microsphere; PLGA; bone regeneration; controlled release

---

1. Introduction

Autologous bone grafting remains the gold standard for spinal fusion, traumatic non-union and total hip arthroplasty complicated by osteolysis [1–3], yet it comes with morbidity (separate incision, graft site pain, potential infection, etc.) and might provide insufficient volumes of bone for complex or multilevel reconstruction [4]. Bone morphogenetic protein-2 (BMP-2) is a transforming growth factor known to play a key role in the development and repair of bone and cartilage [5]. Initially, it appeared to provide an ideal solution [6–8] to enhance bone growth but as its clinical use has expanded, multiple complications associated with BMP-2 use have come to light including local wound problems, chemical radiculitis, bony overgrowth into the canal or foramen, osteoclast activation with associated bony resorption and device displacement, and, possibly, cancer when used at very high doses in an off-label manner [1,9–11]. It is thought these complications were associated with the uncontrolled release and systemic distribution of supraphysiologic doses of this potent growth factor [1,12]. One promising way to avoid these complications is via the controlled local delivery of very small but effective doses using the combination of growth factors with controlled drug delivery vehicles [13–17].

To date, however, systems designed to provide this controlled release have been limited by: (a) a burst release phenomena which leads to similar supraphysiologic dosing, inefficiently sustained dosing, and uncontrolled delivery as that seen with delivery systems currently used clinically; (b) the inability to preserve the quaternary structure of drugs following release from the delivery system; and (c) delivery system (polymer) degradation byproducts that have a secondary negative impact on the structure of the drugs released [16,18]. Therefore, a new carrier system capable of sustained, regulated, local release of small but effective doses that do not impact BMP-2 functionality are needed to allow the avoidance of the biological complications associated with burst supraphysiologic dosing [16].

With the development of nanomaterials, several types of particles, both nano and micro, have been used as growth factor delivery carriers [19]. Of these, nanoporous silicon has emerged as a material uniquely capable of the preservation of protein stability and function with predictable degradation properties in physiologic fluids and systems [16,17,19–22]. The breakdown product, orthosilic acid, has also been shown to stimulate mineralization by osteogenic cells while retaining the ability to buffer the breakdown products of coating polymer (poly(lactic-*co*-glycolic acid), PLGA) [19,23].

Recently, by integrating the drug preserving and encapsulating capabilities of nanoporous silicon with the further controlled release capabilities achieved by polymer encapsulation (using PLGA), we optimized a double controlled delivery system for the controlled and sustained temporospatial release of growth factors [17]. The platforms consisted of the mesoporous silicon-based multistage vector (MSV), encapsulated within a PLGA microparticle (PLGA-MSV). In our previous study, we demonstrated that PLGA-MSV is able to efficiently load a growth factor (i.e., PDGF-BB) and release it in a controlled fashion in vivo, with a significant reduction of the initial burst release, while preserving its functionality (i.e., inducing vascularization) [17].

The aim of the current study was to optimize the release of BMP-2 through the PLGA-MSV delivery system and assess its effectiveness in inducing osteogenesis in vitro.

2. Materials and Methods

2.1. Preparation of PLGA-MSV Microspheres

The PLGA-MSV microspheres were fabricated by a modified S/O/W emulsion method as in our previously published studies [7]. Briefly, PLGA (Sigma Aldrich, St. Louis, MO, USA was dissolved in dichloromethane (DCM; Fisher Scientific, Loughborough, UK) to form PLGA/DCM organic phase solution (10% and 20% *w*/*v*). BMP-2 loaded particles (8×10^7) were suspended into 1 mL of PLGA/DCM solutions (10% and 20% *w*/*v* respectively) by vortex mixing and sonication for 2 min. The organic phase containing the MSV particles was transferred into 3 mL of PVA (2.5% *w*/*v*) solution and emulsified for 1 min by vortex mixing. The primary emulsion was then gradually dispersed into 50 mL of PVA solution (0.5% *w*/*v*). The resulting suspension was stirred continually for 2 h under a biochemical hood, and the DCM evaporated rapidly during the stirring process. PLGA-MSV microspheres were washed with deionized water 3 times and lyophilized overnight. The freeze-dried BMP-2 loaded PLGA-MSV microspheres were then stored at −80 °C.

2.2. Characterization of PLGA-MSV Microspheres

The morphology of the microspheres was characterized by scanning electron microscope (SEM; Nova NanoSEM 230, FEI, Lincoln, NE, USA) and confocal microscope (Nikon A1 laser confocal microscope). The samples were sputter coated with 8 nm of platinum (Pt; Cressington sputter coater 208 HR System, Ted Pella, Inc., Watford, UK) and examined by SEM under a voltage of 3 kV, spot size 3.0, and a working distance of 5 mm.

2.3. Loading of BMP-2 into Nanoporous Silicon Particles (MSV)

Two hundred microliters of BMP-2 growth factor solution (Peprotech) was added into 8×10^7 oxidized nanoporous silicon particles in an Eppendorf tube. The suspension was mixed throughout by vortex mixing and sonication. The tube was gently rotated on a rotator at room temperature for 2 h to allow the adsorption of BMP-2 into the MSV particles. The BMP-2 loaded particles were then spun down by centrifugation (Sorvall Legend X1R Centrifuge, Thermo Scientific, Waltham, MA, USA) (4500 rpm for 5 min), lyophilized overnight, and stored at −80 °C for future use. The concentrations of the BMP-2 loading solution and the supernatant were measured by Elisa assay to determine the amount of BMP-2 loaded into the MSV particles.

2.4. Evaluation of Growth Factor (BMP-2) In Vitro Release

The BMP-2 loaded PLGA-MSV microspheres (10% and 20% *w*/*v*) containing 8×10^7 of MSV particles were dispersed into 0.5 mL of 1% BSA solution at 37 °C. The BMP-2 loaded PLGA microspheres (10% and 20% *w*/*v*) were used as a control. At predetermined time intervals, the suspension was spun down at 4500 rpm for 5 min and 0.5 mL of each supernatant was collected, and replaced with 0.5 mL of fresh 1% BSA solution. The amount of BMP-2 released from BMP-2 loaded PLGA-MSV microspheres was detected using an enzyme-linked immunosorbent assay kit (BMP-2 ELISA, R&D Systems, Minneapolis, MN, USA).

2.5. Cell Isolation and Culture

The study protocol and all operations were reviewed and approved by the Houston Methodist Research Institute's Institutional Animal Care and Use Committee (IACUC, protocol IS00003525, 18 August 2010). All investigators complied with the National Research Council's Guide for the Care and Use of Laboratory Animals. Male Sprague Dawley rats (N = 10) with an average weight of 310 g were used in the study. The rodents underwent a mandatory 48 h acclimation time prior to any surgical procedures and were housed in pairs with ad libitum water and chow until the study period began. Bone marrow stromal cells (BM-MSCs) were isolated from male Sprague Dawley rats as previously described. Briefly, femora and tibiae bones were removed from male Sprague Dawley rats (100–125 g) sacrificed by CO_2 overdose under isoflurane anesthesia. Bones were cleaned of connective tissues, ligaments, and muscle by scalpel, washed thoroughly in phosphate-buffered saline (PBS, Invitrogen, Carlsbad, CA) containing 2% penicillin and streptomycin (P/S, Invitrogen). The proximal and distal ends of each bone were removed and the marrow was gently flushed out with PBS containing 1% fetal bovine serum (FBS, HyClone, ThermoFisher 124 Scientific, Logan, UT, USA) and 1% P/S. The diaphysis regions were crushed using a mortar 125 and pestle while submerged in PBS washes until the PBS appeared clear, indicating complete removal of the remaining marrow and perivascular cells. The total BM fraction was kept on ice for up to 1 h prior to further purification. The total BM cell fractions was counted by hemocytometer and then purified to mononuclear cells by centrifugation on Ficoll (150× *g* for 30 min without brake). The mononuclear BM populations were counted, resuspended in standard media (alpha-MEM (αMEM, Invitrogen) with 20% FBS, 1% P/S, 1% sodium pyruvate and 1% GlutaMAX (Invitrogen)). Cells were seeded in T75 flasks at a density of approximately 105 cells/cm^2 and cultured in hypoxic conditions (5% O_2, 5% CO_2) to maintain their multipotency. Upon reaching 80% confluency, cells were passaged and split 1:4 in new flasks.

Cells were cultured in α-minimum essential medium (αMEM; Invitrogen) supplemented with 20% (*v*/*v*) defined fetal calf serum (Invitrogen), 1%, L-glutamine (Invitrogen), 1% sodium pyruvate (Invitrogen), 100 U/mL penicillin, and 100 μg/mL streptomycin (Invitrogen) as the standard growth media. Osteogenic growth media included 10 mM β-glycerophosphate, 0.1 mM ascorbate-2-phosphate, and 100 nM dexamethasone. Cells were maintained at 37 °C in a humidified 5% CO_2 atmosphere. Cell culture media was changed every 3 days.

2.6. Cell Metabolic Activity—MTT Assay

MTT assay of BM-MSC treated with PLGA-MSV microspheres was performed to quantify metabolic activity [8]. Two thousand and five hundred BM-MSCs were seeded and cultured in a 24-cell culture well plate in the presence of the PLGA and PLGA-MSV microspheres (BM-MSCs:Particles, 1:5). Cells only were used as a control. MTT (3-(4,5-Dimethylthiazol-2-yl)-2,5-diphenyltetrazolium bromide) assay was performed on day 1, 4, and 7. Cell culture media was removed from cell culture wells and 500 μL of MTT working solution (0.5 mg/mL) were added into the wells. The cells were incubated in the MTT working solution at 37 °C for 4 h. The solution was removed from the cell culture wells and replaced with 500 μL of dimethyl sulfoxide (DMSO; Sigma Aldrich). The cells were incubated with DMSO at room temperature for 30 min. The solutions were transferred to a 96 well plate and the absorbance of the colored solutions was quantified by a spectrophotometer (Synergy H4 Hybrid Reader, BioTek, Winooski, VT, USA) at 570 nm. DMSO was used as blank. Cells only and PLGA particle wells were used as controls.

2.7. Alkaline Phosphatase (ALP) Activity Assay

Osteogenic differentiation was measured by ALP activity, a biomarker of osteoblastic differentiation. The assay was carried out according to manufacturer guidelines for the spectrophotometric procedure using ALP reagent (Vector Laboratories, Inc. Burlingame, CA, USA). The BM-MSCs were plated into a 24 well cell culture plate at a density of 2500 cells per well with the PLGA-MSV microspheres (20% *w*/*v*) loaded with BMP-2 was used as the experiential group. Cells alone, cells cultured with BMP-2, and empty PLGA-MSV microspheres without BMP-2 (20% *w*/*v*) were used as controls. Cell culture media was changed every 3 days. Cells were cultured in α-minimum essential medium (αMEM; Invitrogen) supplemented with 10% (*v*/*v*) fetal calf serum (Invitrogen), 100 U/mL penicillin, and 100 μg/mL streptomycin (Invitrogen) as the standard growth media. Culture conditions were 37 °C in a humidified 5% CO_2 atmosphere. The osteogenic growth media included 10 mM β-glycerophosphate, 0.1 mM ascorbate-2-phosphate, and 100 nM dexamethasone. Cell culture media was replenished twice weekly. Cells were cultured in standard media until 60% confluence and then switched to osteogenic media. ALP assays were performed at week 1, 2, and 3. The medium was aspirated and 1 mL of PBS was added into each well to wash the cells. Cells were washed 3 times with PBS, and fixed in 10% buffered formalin for 15 min. Cells were then washed twice in deionized (DI) water and covered in ALP stain made fresh. ALP staining stock solution was made from ALP substrate kit III (Vector Laboratories. Inc., Burlingame, CA, USA) into 5 mL of 100 mM tris-HCl (pH = 8.2) solution. Absorbance of cells after staining was used to quantify ALP activity. The experiments were performed in triplicate.

2.8. Von Kossa Co-Staining

Cell cultures were co-stained for calcium-triphosphate mineral deposition by Von Kossa staining. Co-staining of the cell culture was performed to simultaneous both quantify ALP activity and provide a qualitative assessment of mineral deposition. Following ALP staining, cells were washed twice in DI water and soaked in in 1% aqueous silver nitrate ($AgNO_3$) and placed under ultraviolet (UV) light for 60 min and then rinsed with DI water. To remove unreacted silver, 5% sodium thiosulfate was added for 5 min, removed and cells rinsed in DI water. Following this cell nuclei were counterstained with Nuclear Fast Red (Sigma Aldrich) for 5 min, rinsed in DI water, and serially dehydrated prior to characterization.

3. Results

3.1. PLGA-MSV Characterization

The PLGA-MSVs were characterized by SEM-energy-dispersive X-ray (EDX) and confocal microscopy. SEM images show that the spherical-shaped PLGA-MSVs with smooth surfaces had a wide distribution from a few microns to approximately 50 μm, with an average diameter of 23 ± 3

($n = 50$; Figure 1A). EDX spectrum showed the presence of a Si peak (Figure 1B), which indicates the presence of MSV particles inside of the PLGA microspheres. The full encapsulation of MSV was further confirmed by optical microscopy, where hemispherical MSVs appeared yellow in color (Figure 1C). MSVs were also loaded with a reporter protein (FITC-BSA) and imaged by confocal microscopy. Imaging clearly demonstrated that the FITC-BSA loaded MSVs (in green) were fully encapsulated (Figure 1D–F).

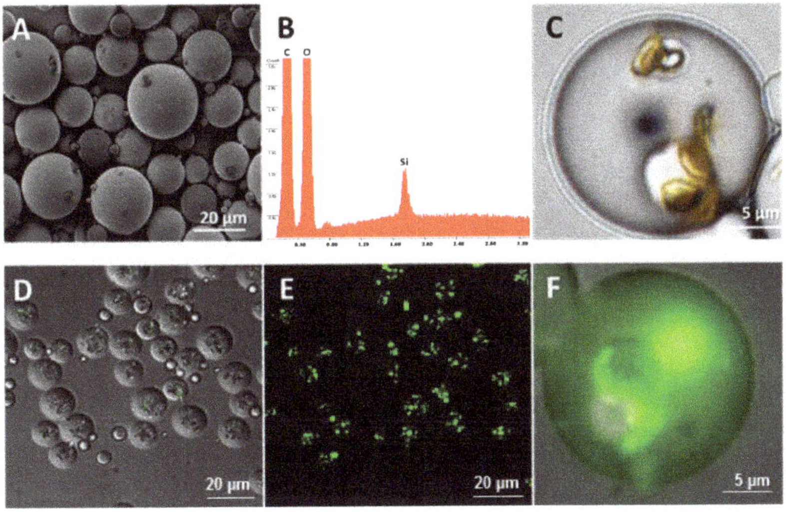

Figure 1. SEM and confocal images of PLGA-multistage vector (MSV). Representative SEM image of PLGA-MSV (**A**). EDX spectrum of PLGA-MSV microspheres showing the presence of the silicon peak corresponding to MSV (**B**). Confocal laser microscopy Z-stack of PLGA-MSVs loaded with FITC-BSA (in green), showing that MSVs were fully encapsulated into the PLGA microsphere (**D**,**E**). Close up of a PLGA-MSV (**C**,**F**).

3.2. PLGA-MSV Loading with BMP-2 and In Vitro Release

Recombinant human BMP-2 was used in this study. BMP-2 is a 29 kDa protein with an isoelectric point of 8.21 (Peprotech). The BMP-2 loaded PLGA-MSV microspheres (with a PLGA coating of 10% and 20% *w*/*v*) were used for in vitro sustained release studies. The loading efficiency of BMP-2 into MSV particles is shown in Figure 2A. Consistent with mass transport theory, the greater number of particles added to the solution of BMP-2, the greater amount of loading achieved and thus a higher loading efficiency. The in vitro release profiles of BMP-2 from different types of microspheres were monitored for 40 days. A delayed burst release over three days was demonstrated by 10% *w*/*v* PLGA-MSV microspheres (Figure 2B). When monitored to 40 days 90% of release was achieved within 10 days and sustained release equilibrium near 24 days (Figure 2C). In contrast, BMP-2 release from 20% *w*/*v* PLGA-MSV microspheres showed a more linear-like release with a more linear early release profile (Figure 2B). This slower rate of BMP-2 release continued until equilibrium was reached at 41 days (Figure 2C).

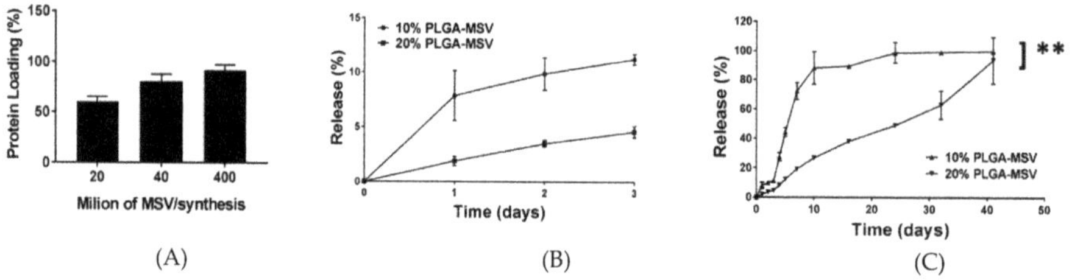

Figure 2. The in vitro loading and release profiles of BMP-2 using PLGA-MSV microspheres (10%, 20% *w*/*v*; $n = 3$). (**A**) The loading profiles of BMP-2 into MSV particles with fixed BMP-2 concentrations and mass, and varied MSV microparticle number. Cumulative release profile of BMP-2 from different 10% and 20% PLGA-MSV in the first three days (**B**) and over 41 days, (**C**). Values are reported as mean ± standard deviation. A value of $p < 0.05$ was considered statistically significant: ** $p < 0.01$ (calculated by *t* test analysis).

3.3. Cell Metabolic Activity—MTT Assay

Cell metabolic activity in the presence of the PLGA and PLGA-MSV microspheres was analyzed using MTT assay. Figure 3 shows the results of MTT assay for BM-MSCs cultured with PLGA-MSV microspheres at 1, 4, and 7 days. Metabolic activity among cells alone (control) and cells with the PLGA-MSV microspheres shared the similar trend and there was no significant difference among of the three groups at 1, 4, or 7 days. Cells in each group achieved a comparable metabolic activity and reached equilibrium at each time point.

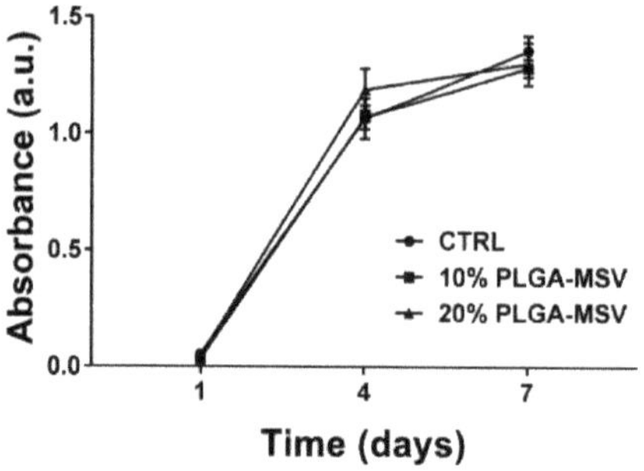

Figure 3. MTT assay for rat bone marrow stromal cells (BM-MSCs) cultured with 10% PLGA-MSV and 20% PLGA-MSV microspheres over 7 days ($n = 3$). Cell viability is reported as absorbance at 490 nm. The cell metabolic activity of BM-MSCs in the presence of both formulations of PLGA-MSV was comparable to that of untreated cells (CTRL). Values are reported as mean ± standard deviation. No statistical significance was found among selected experimental conditions.

3.4. PLGA-MSV-BMP2 Enhances Alkaline Phosphatase Activity (ALP)

ALP is synthesized by osteoblasts and is a biomarker of presumed ECM production during BM-MSC differentiation [24]. As shown in Figure 4, at weeks 1 and 2, BMP-2 loaded 20% PLGA-MSV(PLGA-MSV-BMP2) microspheres showed higher ALP activities compared to cells alone and cells treated with BMP-2 (BMP2). At week 2, cells treated with PLGA-MSV-BMP2 showed the highest ALP activity during the duration of the investigation. The ALP activities decreased to a fairly low level for all controls and experimental groups after 3 weeks. Although, interestingly, the ALP activity of the control remained elevated at this time point.

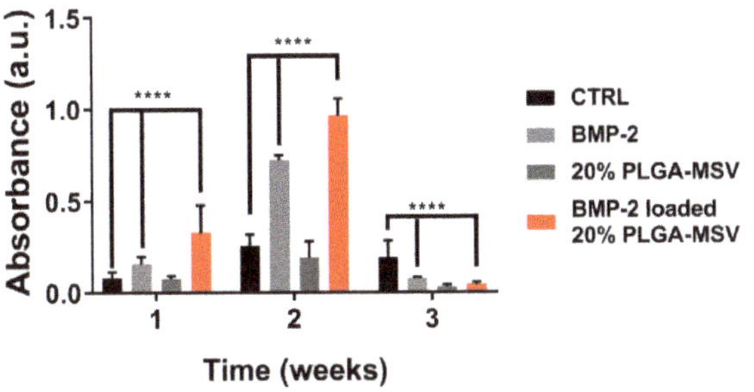

Figure 4. ALP activity of untreated BM-MSCs (CTRL), BM-MSCs treated with soluble BMP-2 or 20% PLGA-MSV loaded or non-loaded with BMP-2 over 3 weeks (mean ± SD, $n = 3$). Values are reported as mean ± standard deviation. Data was analyzed through 2-way ANOVA. A value of * $p < 0.05$ was considered statistically significant (**** $p < 0.0001$).

3.5. *ALP Staining and Von Kossa Staining*

All groups were co-stained for ALP and calcium using standard ALP staining and Von Kossa stain. ALP staining data of BM-MSC treated with BMP2 and PLGA-MSV-BMP2 correlated with ALP quantitative activity data. BMP2 and PLGA-MSV-BMP2 showed qualitatively increased staining at 1 and 2 weeks while untreated cells showed increased ALP staining at 3 weeks (Figure 5).

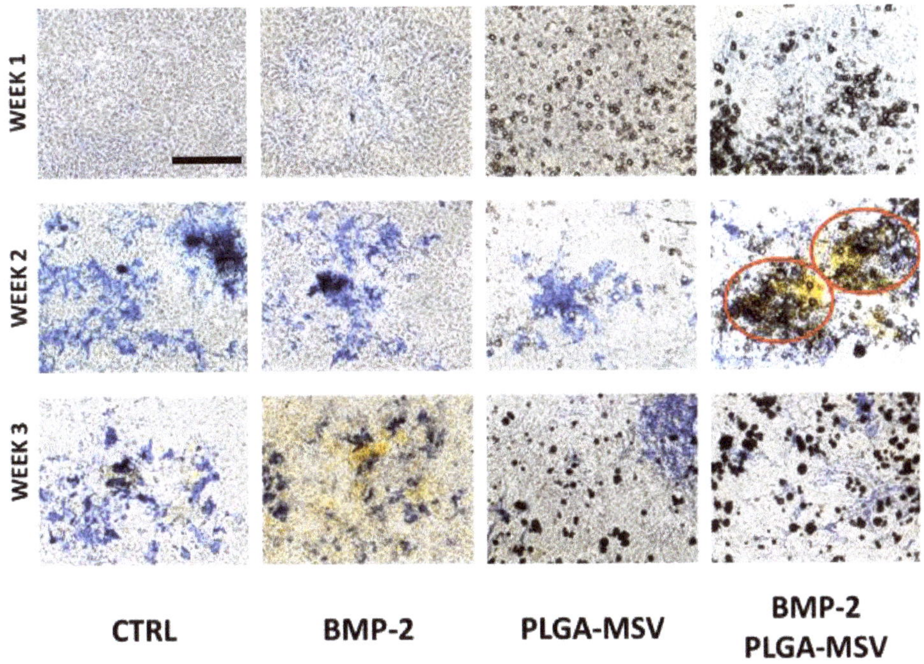

Figure 5. ALP and Von Kossa co-staining of BM-MSCs and BM-MSCs with experimental groups. ALP activities increased from week 1 to week 2 and then decreased from week 2 to week 3 in BMP2 and BMP2-PLGA-MSV. The cells in the presence of the PLGA-MSV-BMP-2 microspheres at week 2 showed the strongest ALP activity (red circles) among all of the control and experimental groups over the 3-week culture period. Von Kossa staining progressively increased in all groups with BMP2 and BMP2-PLGA-MSV showing the most staining at week 3 (Scale bar: 60 μm).

Von Kossa staining was performed to assess functional activity of BM-MSC osteoinduction. Calcium deposition was delayed compared to ALP staining. Calcium (black stain) is seen in the PLGA-MSV-BMP2 group as early as one week and continues to increase throughout the 3 weeks having the most deposition of all groups. In the control and BMP2 group calcium begins to be seen at 2 weeks yet the deposition of calcium only continues to increase in the BMP2 group. PLGA-MSV did not demonstrate appreciable deposition until week 3 (Figure 5).

4. Discussion

This study reports the development and investigation of a PLGA based protein delivery system, PLGA-MSV microspheres, for controlled protein release to stimulate cell differentiation. Herein, we studied the loading and release of BMP-2 from MSV, the relationship of PLGA:BMP concentration on BMP-2 release, and the effect of long-term constant BMP-2 stimulation on BM-MSCs. BMP-2 was successfully encapsulated into the microsphere system and the performance was tested in vitro. The experiments show that: (1) BMP/PLGA-MSV microspheres had the ability to release small but effective doses of BMP-2 for 40 days in a controlled, linear fashion; (2) the release profile of the BMP/PLGA-MSV microspheres was dependent on the PLGA coating of 10% or 20%; (3) the PLGA-MSV system did not impact cell metabolic activity; and (4) BMP-2 released from PLGA-MSV microspheres was capable of osteoinduction of BM-MSCs.

Previously, published reports with the clinical and pre-clinical use of BMP-2 have demonstrated bone augmentation, but the growth occurred in an uncontrolled and non-physiologic fashion. Partly due to the short half-life and need for supraphysiologic dosing, this resulted in heterotopic bone formation. Clinically this has translated in to the development of multiple complications [1,10]. Currently formulated BMP devices must be used at high concentrations resulting in problems with protein release timing stability, and need for accessory factors [11,12]. Additionally, the bone regenerative process under these conditions is very different from the normal physiologic processes in normal bone healing in which multiple BMPs and other growth factors are present in overlapping temporal patterns [13,14]. Here we reported a biomolecular release vehicle capable of controlling release kinetics of an encapsulated molecule for localized protein delivery, in this case BMP2. This phenomenon is affected by the percentage of PLGA used indicating an added level of spatiotemporal control of the system. Increasing PLGA percentage slowed release kinetics from a burst release to a more constant release and could theoretically lead to more prolonged and slowed release kinetics [25,26]. The slower release kinetics may be more biologically accurate and induce a more sustained response in contrast to the massive burst release kinetics often seen from PLGA alone [27]. Here we demonstrated this system is capable of long-term delivery as well as in vitro osteoinduction.

As many of the issues attributed to the use of growth factors in regenerative medicine have been centered on the lack of controlled release, this model presents a system for localized delivery within a clinically relevant scenario such as a healing fracture. MSCs are the major regenerative cell involved in the fracture healing process [28]. Their stimulation is paramount for the successful regeneration of tissue and organized bone formation. Without their coordinated efforts through properly induced pathways, dysregulated and disorganized growth occurs leading to weakened fracture callus formation and recurrent injuries [15,16]. Previously, investigators have demonstrated the efficacy of BMP2 release on the activity of MSCs [17]. Here we show that normal cellular function is not inhibited by the PLGA-MSV system and that sustained BMP-2 release results in MSC osteoinduction and increased ALP activity compared to BMP-2 alone. Cells can proliferate normally in the presence of the PLGA-MSV microspheres, which, indicates that these materials were nontoxic to cells and compatible with normal cell function. When loaded with BMP, the PLGA-MSV system was used to stimulate BM-MSC in vitro. ALP enzyme activity was increased in the PLGA-MSV-BMP-2 group and this correlated with ALP staining. These findings translated to a persistent and increasing deposition of calcium seen in the Von Kossa co-stain. The delayed deposition of calcium is consistent with normal biological function and an expected outcome [29]. The PLGA-MSV-BMP system induced early and robust ALP activity at one week compared to BMP2 alone. At this same time point calcium is already detected in the Von Kossa stain of the PLGA-MSV-BMP2 group. This suggests that the slowed release may in fact have greater biological activity compared to high dose free BMP. This localized effect may provide insights into the further concomitant use of MSCs and PLGA-MSV vectors for regenerative purposes in other applications.

There are several limitations with the current study. First the lack of free PLGA-BMP2 makes direct comparison difficult. There have been several studies looking at the use of PLGA as a protein carrier and more specifically a BMP2 carrier. These studies reported the release kinetics of BMP2 from PLGA and its impact in vitro on osteoinduction. Given the historical findings with PLGA alone as the carrier we chose to focus on the use of MSV encapsulated in PLGA. Second, while ALP activity, staining and calcium deposition all form part of the evaluation of BM-MSC osteoinduction further studies in the transcriptional changes and more quantitative histological assessments could be performed. This was out of the scope of this initial work and is the focus of subsequent works in progress. Third, an additional comparison group looking into the effects of PLGA-MSV ratio on bioactivity toward BM-MSC is needed. While there were significant differences between the release profiles between 10% and 20% PLGA-MSV-BMP2 no biologic data was obtained. This is currently a subject of current investigation. Lastly, this is only an in vitro study with one cell line. The biologic milieu of the in vivo micro-environment is composed of a complex interplay of signaling and cell types. While our in vitro findings are promising they warrant analysis in an in vivo fracture model to assess the efficacy of localized controlled release.

5. Conclusions

In conclusion, this biocompatible, biodegradable, and osteogenic PLGA-MSV microsphere system holds promise as a candidate for the delivery of safe, local, small but effective doses of bioactive proteins for pharmaceutical induction of osteoregeneration. The local and long-term delivery of proteins and other bioactive molecules for orthopedic applications may avoid the complications associated with burst release and supraphysiologic dosing of BMP-2 currently afforded by available delivery systems.

Author Contributions: Conceptualization, S.M., E.T., D.F.; methodology, S.M., D.F., M.B.M., I.K.Y., X.L.; data curation and analysis: S.M., D.F., M.B.M., I.K.Y.; writing—original draft preparation, S.M., D.F, J.S.F.-M. writing—review and editing, S.M., J.S.F.-M.; supervision, B.K.W., E.T.; project administration, B.K.W., E.T.; funding acquisition, B.K.W., E.T. All authors have read and agreed to the published version of the manuscript.

Funding: This research was funded by Defense Advances Research Projects Agency (DARPA), grants W81XWH-14-1-0600, Log #SC130156; W81XWH-15-1-0718, Log #4170002.

Conflicts of Interest: The authors declare no conflict of interest.

References

1. Minardi, S.; Corradetti, B.; Taraballi, F.; Sandri, M.; Van Eps, J.; Cabrera, F.J.; Weiner, B.K.; Tampieri, A.; Tasciotti, E. Evaluation of the Osteoinductive Potential of a Bio-Inspired Scaffold Mimicking the Osteogenic Niche for Bone Augmentation. *Biomaterials* **2015**, *62*, 128–137. [CrossRef] [PubMed]
2. Xenakis, T.; Koukoubis, T.; Hantes, K.; Varytimidis, S.; Soucacos, P.N. Bone Grafting in Total Hip Arthroplasty for Insufficient Acetabulum. *Acta Orthop Scand.* **1997**, *68* (Suppl. 275), 33–37. [CrossRef] [PubMed]
3. Azi, M.L.; Aprato, A.; Santi, I.; Kfuri, M.; Masse, A.; Joeris, A. Autologous Bone Graft in the Treatment of Post-Traumatic Bone Defects: A Systematic Review and Meta-Analysis. *Bmc Musculoskelet Disord* **2016**, *17*, 465. [CrossRef] [PubMed]
4. Dimitriou, R.; Mataliotakis, G.I.; Angoules, A.G.; Kanakaris, N.K.; Giannoudis, P.V. Complications Following Autologous Bone Graft Harvesting from the Iliac Crest and Using the Ria: A Systematic Review. *Injury* **2011**, *42*, S3–S15. [CrossRef] [PubMed]
5. Poon, B.; Kha, T.; Tran, S.; Dass, C.R. Bone Morphogenetic Protein-2 and Bone Therapy: Successes and Pitfalls. *J. Pharm Pharm.* **2016**, *68*, 139–147. [CrossRef]

6. Mistry, A.S.; Antonios, G.M. Tissue Engineering Strategies for Bone Regeneration. In *Regenerative Medicine Ii*; Springer: Berlin/Heidelberg, Germany, 2005; pp. 1–22.
7. Geiger, M.; Li, R.H.; Friess, W. Collagen Sponges for Bone Regeneration with Rhbmp-2. *Adv. Drug Deliv. Rev.* **2003**, *55*, 1613–1629. [CrossRef]
8. Minardi, S.; Taraballi, F.; Cabrera, F.J.; Van Eps, J.; Wang, X.; Gazze, S.A.; Fernandez-Mourev, J.S.; Tampieri, A.; Francis, L.; Weiner, B.K. Biomimetic Hydroxyapatite/Collagen Composite Drives Bone Niche Recapitulation in a Rabbit Orthotopic Model. *Mater. Today Bio.* **2019**, *2*, 100005. [CrossRef]
9. Chrastil, J.; Low, J.B.; Whang, P.G.; Patel, A.A. Complications Associated with the Use of the Recombinant Human Bone Morphogenetic Proteins for Posterior Interbody Fusions of the Lumbar Spine. *Spine* **2013**, *38*, E1020–E1027. [CrossRef]
10. James, A.W.; LaChaud, G.; Shen, J.; Asatrian, G.; Nguyen, V.; Zhang, X.; Ting, K.; Soo, C. A Review of the Clinical Side Effects of Bone Morphogenetic Protein-2. *Tissue Eng. Part. B: Rev.* **2016**, *22*, 284–297. [CrossRef]
11. Liu, Y.; Schouten, C.; Boerman, O.; Wu, G.; Jansen, J.A.; Hunziker, E.B. The Kinetics and Mechanism of Bmp-2 Release from Calcium Phosphate-Based Implant-Coatings. *J. Biomed. Mater. Res. Part. A* **2018**, *106*, 2363–2371. [CrossRef]
12. Schmidt-Bleek, K.; Willie, B.M.; Schwabe, P.; Seemann, P.; Duda, G.N. Bmps in Bone Regeneration: Less Is More Effective, a Paradigm-Shift. *Cytokine Growth Factor Rev.* **2016**, *27*, 141–148. [CrossRef] [PubMed]
13. Zhang, W.; Zhang, Z.; Zhang, Y. The Application of Carbon Nanotubes in Target Drug Delivery Systems for Cancer Therapies. *Nanoscale Res. Lett.* **2011**, *6*, 555. [CrossRef] [PubMed]
14. Poojari, R.; Srivastava, R. Composite Alginate Microspheres as the Next-Generation Egg-Box Carriers for Biomacromolecules Delivery. *Expert Opin. Drug Deliv.* **2013**, *10*, 1061–1076. [CrossRef] [PubMed]
15. Webber, M.J.; Matson, J.B.; Tamboli, V.K.; Stupp, S.I. Controlled Release of Dexamethasone from Peptide Nanofiber Gels to Modulate Inflammatory Response. *Biomaterials* **2012**, *33*, 6823–6832. [CrossRef] [PubMed]
16. Minardi, S.; Pandolfi, L.; Taraballi, F.; De Rosa, E.; Yazdi, I.K.; Liu, X.; Ferrari, M.; Tasciotti, E. PLGA-Mesoporous Silicon Microspheres for the in Vivo Controlled Temporospatial Delivery of Proteins. *Acs Appl. Mater. Interfaces* **2015**, *7*, 16364–16373. [CrossRef]
17. Minardi, S.; Pandolfi, L.; Taraballi, F.; Wang, X.; De Rosa, E.; Mills, Z.D.; Liu, X.; Ferrari, M.; Tasciotti, E. Enhancing Vascularization through the Controlled Release of Platelet-Derived Growth Factor-Bb. *Acs Appl. Mater. Interfaces* **2017**, *9*, 14566–14575. [CrossRef]
18. Minardi, S.; Taraballi, F.; Pandolfi, L.; Tasciotti, E. Patterning Biomaterials for the Spatiotemporal Delivery of Bioactive Molecules. *Front. Bioeng. Biotechnol.* **2016**, *4*, 45. [CrossRef]
19. Fan, D.; De Rosa, E.; Murphy, M.B.; Peng, Y.; Smid, C.A.; Chiappini, C.; Liu, X.; Simmons, P.; Weiner, B.K.; Ferrari, M. Mesoporous Silicon-Plga Composite Microspheres for the Double Controlled Release of Biomolecules for Orthopedic Tissue Engineering. *Adv. Funct. Mater.* **2012**, *22*, 282–293. [CrossRef]
20. De Rosa, E.; Chiappini, C.; Fan, D.; Liu, X.; Ferrari, M.; Tasciotti, E. Agarose Surface Coating Influences Intracellular Accumulation and Enhances Payload Stability of a Nano-Delivery System. *Pharm. Res.* **2011**, *28*, 1520–1530. [CrossRef]
21. Martinez, J.O.; Chiappini, C.; Ziemys, A.; Faust, A.M.; Kojic, M.; Liu, X.; Ferrari, M.; Tasciotti, E. Engineering Multi-Stage Nanovectors for Controlled Degradation and Tunable Release Kinetics. *Biomaterials* **2013**, *34*, 8469–8477. [CrossRef]
22. Martinez, J.O.; Evangelopoulos, M.; Chiappini, C.; Liu, X.; Ferrari, M.; Tasciotti, E. Degradation and Biocompatibility of Multistage Nanovectors in Physiological Systems. *J. Biomed. Mater. Res. A* **2014**, *102*, 3540–3549. [CrossRef]
23. Reffitt, D.M.; Ogston, N.; Jugdaohsingh, R.; Cheung, H.F.; Evans, B.A.; Thompson, R.P.; Powell, J.J.; Hampson, G.N. Orthosilicic Acid Stimulates Collagen Type 1 Synthesis and Osteoblastic Differentiation in Human Osteoblast-Like Cells in Vitro. *Bone* **2003**, *32*, 127–135. [CrossRef]
24. Hanna, H.; Mir, L.M.; Andre, F.M. In Vitro Osteoblastic Differentiation of Mesenchymal Stem Cells Generates Cell Layers with Distinct Properties. *Stem Cell Res.* **2018**, *9*, 203. [CrossRef] [PubMed]
25. Makadia, H.K.; Siegel, S.J. Poly Lactic-*co*-Glycolic Acid (PLGA) as Biodegradable Controlled Drug Delivery Carrier. *Polymers (Basel)* **2011**, *3*, 1377–1397. [CrossRef] [PubMed]
26. Clark, A.; Milbrandt, T.A.; Hilt, J.Z.; Puleo, D.A. Mechanical Properties and Dual Drug Delivery Application of Poly(Lactic-*co*-Glycolic Acid) Scaffolds Fabricated with a Poly(β-Amino Ester) Porogen. *Acta Biomater.* **2014**, *10*, 2125–2132. [CrossRef]

27. Beachler, D.C.; Yanik, E.L.; Martin, B.I.; Pfeiffer, R.M.; Mirza, S.K.; Deyo, R.A.; Engels, E.A. Bone Morphogenetic Protein Use and Cancer Risk among Patients Undergoing Lumbar Arthrodesis: A Case-Cohort Study Using the Seer-Medicare Database. *J. Bone Jt. Surg Am.* **2016**, *98*, 1064–1072. [CrossRef]
28. Lin, W.; Xu, L.; Zwingenberger, S.; Gibon, E.; Goodman, S.B.; Li, G. Mesenchymal Stem Cells Homing to Improve Bone Healing. *J. Orthop Transl.* **2017**, *9*, 19–27. [CrossRef]
29. Borras, T.; Comes, N. Evidence for a Calcification Process in the Trabecular Meshwork. *Exp. Eye Res.* **2009**, *88*, 738–746. [CrossRef]

pharmaceutics

Article

PLGA-BMP-2 and PLA-17β-Estradiol Microspheres Reinforcing a Composite Hydrogel for Bone Regeneration in Osteoporosis

Patricia García-García [1], Ricardo Reyes [2,3], Elisabet Segredo-Morales [1], Edgar Pérez-Herrero [1,2], Araceli Delgado [1,2,*] and Carmen Évora [1,2,*]

1 Department of Chemical Engineering and Pharmaceutical Technology, University of La Laguna, 38206 La Laguna, Spain; patg0991@gmail.com (P.G.-G); esegredm@ull.edu.es (E.S.-M.); eperezhe@ull.edu.es (E.P.-H.)
2 Institute of Biomedical Technologies (ITB), University of La Laguna, 38206 La Laguna, Spain; rreyesro@ull.edu.es
3 Department of Biochemistry, Microbiology, Cell Biology and Genetics, University of La Laguna, 38206 La Laguna, Spain
* Correspondence: adelgado@ull.edu.es (A.D.); cevora@ull.edu.es (C.É.)

Received: 23 October 2019; Accepted: 29 November 2019; Published: 3 December 2019

Abstract: The controlled release of active substances—bone morphogenetic protein 2 (BMP-2) and 17β-estradiol—is one of the main aspects to be taken into account to successfully regenerate a tissue defect. In this study, BMP-2- and 17β-estradiol-loaded microspheres were combined in a sandwich-like system formed by a hydrogel core composed of chitosan (CHT) collagen, 2-hidroxipropil γ-ciclodextrin (HP-γ-CD), nanoparticles of hydroxyapatite (nano-HAP), and an electrospun mesh shell prepared with two external electrospinning films for the regeneration of a critical bone defect in osteoporotic rats. Microspheres were made with poly-lactide-*co*-glycolide (PLGA) to encapsulate BMP-2, whereas the different formulations of 17β-estradiol were prepared with poly-lactic acid (PLA) and PLGA. The in vitro and in vivo BMP-2 delivered from the system fitted a biphasic profile. Although the in vivo burst effect was higher than in vitro the second phases (lasted up to 6 weeks) were parallel, the release rate ranged between 55 and 70 ng/day. The in vitro release kinetics of the 17β-estradiol dissolved in the polymeric matrix of the microspheres depended on the partition coefficient. The 17β-estradiol was slowly released from the core system using an aqueous release medium ($D_{eff} = 5.58{\cdot}10^{-16} \pm 9.81{\cdot}10^{-17} m^2 s^{-1}$) and very fast in MeOH-water (50:50). The hydrogel core system was injectable, and approximately 83% of the loaded dose is uniformly discharged through a 20G needle. The system placed in the defect was easily adapted to the defect shape and after 12 weeks approximately 50% of the defect was refilled by new tissue. None differences were observed between the osteoporotic and non-osteoporotic groups. Despite the role of 17β-estradiol on the bone remodeling process, the obtained results in this study suggest that the observed regeneration was only due to the controlled rate released of BMP-2 from the PLGA microspheres.

Keywords: BMP-2-microspheres; hydrogel system; 17-βestradiol release; bone regeneration; osteoporosis; poly-lactide-*co*-glycolide; polylactic acid

---

1. Introduction

Regeneration of bone critical defects is still a challenge in the orthopedic field. Local treatment with bone morphogenetic protein (BMP-2) incorporated in different biomaterial scaffolds has demonstrated to be efficient to induce new bone formation for critical bone defect in several animal models. Nowadays, collagen sponges loaded with recombinant BMP-2 are clinically available as bone graft substitutes for the treatment of nonunion and critical-sized bone defects. Although BMP-2 is a potent osteogenic agent, a controlled release profile is required for safety and efficacy. Thus, the scaffold to fill the bone defect should not only be designed to act as support and guide for tissue growth, but also to control the release rate of active substances. Although many materials and structures have been proposed to construct these scaffolds, the control of the release rate has not always been taken into account. In fact, in some cases the protein is incorporated in the material by incubation and, unless a material-protein interaction occurs, a burst release at an early period would be expected. Consequently, a high dose of BMP-2 would be in blood circulation, leading to a high risk of side effects and, at the same time, a significant loss of the protein in the site of action. In addition, some authors showed that BMPs in these cases might also stimulate bone resorption due to the high dose of BMP-2 associated to uncontrolled release [1]. To minimize the osteoclastic effect of BMPs, some authors proposed the addition of an anti-catabolic agent. The most frequently studied combination has been BMP-2 with bisphosphonates as anti-resorption agents such as alendronate [2,3] and zolendronate [4]. The results indicated good bone regeneration with improved bone quality and mineralization in different localizations compared to BMP-2 alone [5,6].

Among the several natural polymeric scaffolds prepared for bone regeneration, chitosan is a biomaterial frequently used for this purpose. In a recent extensive review [7] based on chitosan (CHT) applied in bone tissue regeneration, different advantageous aspects were showed and discussed such as biocompatibility, capacity for BMPs sustained release, improvement of cell proliferation, and increase of in vitro and in vivo differentiation and mineralization. However, bone graft materials to simulate bone structure, e.g., collagen, the major ECM of bone tissue, and hydroxyapatite (HAP), a mineral component of the bone, have also been widely studied [6,8–14].

Although the aforementioned studies indicated positive results, the mentioned strategies applied in osteoporosis (OP) conditions have not always been effective [15]. According to recent reports, the prolonged subcutaneous administration of alendronate and the low level of estrogen in OP alters the evolution of calvarial bone repair due to estrogen, Transforming Growth Factor beta 1 (TGF-β1), and α-estrogen receptor (α-ER) interaction [16]. Menopausal women are the population most affected by this disease due to estrogen deficiency. Previous reports revealed that local implantation of scaffolds loaded with combinations of BMP-2 and 17β-estradiol formulated in microspheres of polylactic acid (PLA) or PLGA, in rat calvaria critical defects increased the bone repair in OP rats, but the new bone that refilled the defect was less mineralized compared to non-OP groups [17,18].

As some biomaterials may promote bone regeneration, controlled release of the active substances is required for efficient and safe bone regeneration [19,20]. In this study, we propose a loaded BMP-2 and 17β-estradiol sandwich-like system, comprising two polymeric external films and a core of a biocomposite hydrogel containing microspheres, to provide sustained release of active substances. The core system composed of CHT, collagen, HAP nanoparticles (nano-HAP), polyethylene glycol (PEG-400) and 2-Hidroxipropil γ-Ciclodextrin HP-γ-CD was previously characterized in terms of composition, rheological behavior and mass-transfer using RITC-dextran as macromolecule model and 17β-estradiol in microspheres [21]. In the present study, we aim to study the influence of the release rate of 17β-estradiol on the osteogenic effect induced by BMP-2 released from PLGA microspheres within core system after sandwich-like system implantation in a OP rats critical size defect. Therefore, 17β-estradiol was incorporated into the system in 3 forms: free and dispersed in the core, encapsulated in microspheres prepared with a mixture of PLA and PLGA dispersed within the hydrogel and lastly encapsulated in the PLGA films shell prepared by electrospinning technique.

2. Materials and Method

2.1. Materials

PLGA 75:25 (Resomer® RG755-S), PLGA 50:50 (Resomer® RG504), PLGA 85:15 (Resomer® RG858-S), and PLA (Resomer® RG203-S) were supplied by Evonic Industries (Darmstadt, Germany). Chitosan

(Protasan® UP-CL-213) was purchased from NovaMatrix (Sandvika, Norway). 2-Hidroxipropil γ-Ciclodextrin (CAVASOL® W8 HP), was supplied by Wacker Chemical (Burghausen, Germany). The bovine collagen type I was purchased from CellSystems Biotechnologie (Vertrieb GmbH, Germany). Riboflavin (RB), Poly(ethylene glycol) 400, Poly(vinyl alcohol) (PVA, Mw 33–70 kDa; 87–90% hydrolyzed), 17β-estradiol, and all the other reagents were purchased from Sigma-Aldrich, (St. Louis, MO, USA). The recombinant human bone morphogenetic protein 2 (BMP-2) was bought from Biomedal Life Sciences (Sevilla, Spain). Citrate-coated carbonated apatite nanoparticles (nano-HAP) were kindly donated (Jaime Gómez-Morales, PhD, Laboratory of Crystallographic Studies, CSIC, Granada, Spain).

2.2. Microspheres Preparation and Characterization

The BMP-2 microspheres were prepared by the double emulsion method (w/o/w) previously described [22]. Briefly, 200 μL of an aqueous solution (0.2% PVA) of BMP-2 (260 μg) was emulsified with 1 mL of a PLGA mixture (150 mg) of RG504 and RG858 [4:1] in methylene chloride (DCM) by vortexing 1 min (position 10, Genie® Industries 2, Sciencies Industries Inc. USA). Then, this emulsion was poured into 10 mL of 0.2% PVA solution vortexed 15 s, then poured into 100 mL of 0.1% PVA and kept under magnetic stirring for 1 h for solvent evaporation.

The 17β-estradiol microspheres were prepared by a modified solvent evaporation method previously described [17]. Briefly a mixture of 17β-estradiol (4 mg), PLA-S RG203-S (160 mg) and PLGA RG858 (40 mg) dissolved in 0.6 mL of DCM:Methanol (DCM:MeOH) (80:20) was emulsified with 4 mL of 1% PVA aqueous solution by vortexing 1 min (position 10), and then added to 100 mL of 0.16% PVA solution, under magnetic stirring for organic solvent evaporation (1 h).

Both type of microspheres were collected by filtration (Pall Corporation, pore size 45 μm, Sigma-Aldrich, USA), lyophilized, and stored at 4 °C until use.

Microspheres were characterized in terms of size (Mastersizer 2000, Malver Instruments, Malvern, UK) and morphology (SEM, Jeol JSM-6300, Tokyo, Japan). To determine the BMP-2 encapsulation efficiency and to carry out the BMP-2 release assays, some batches were prepared with ^{125}IBMP-2. The BMP-2 was labeled with ^{125}INa (Perkin-Elmer) by the iodogen method [23]. The content of 17β-estradiol in the microspheres was determined spectrophotometrically at $\lambda = 280$ nm previous dissolution in a mix of DCM:MeOH (80:20).

To determine the solubility of 17β-estradiol in the polymer matrix of the microspheres, differential scanning calorimetry (DSC 025, TA Instruments, New Castle, DE, USA) was performed. 17β-estradiol and lyophilized microspheres were analyzed after drying in an oven at 37 °C overnight. In addition, samples of polymer blends (RG 203-S and RG 858, 4:1) and samples of the polymer blend with excess 17β-estradiol (8.5%) were dissolved in DCM:MeOH (80:20) and maintained in a hood for 24 h. Then, samples were placed 48 h more in a vacuum desiccator to complete the evaporation of the organic solvent. The analysis of all samples was performed with the same thermal program in two thermal cycles under a nitrogen atmosphere (50 mL/min). In the first cycle, temperature was increased to 40 °C (10 °C/min) and then cooled to −20 °C (5 °C/min) to avoid possible water interference. Once the samples were stabilized, they underwent a final heating cycle from −20 °C to 270 °C (10 °C/min).

2.3. Fabrication and Characterization of the Film

The film was fabricated by a previously described electrospinning method [24]. Briefly, 7 mg of 17β-estradiol and 300 mg of a mixture of PLGAs, RG755-S and RG858 [4:1] were dissolved in 2 mL of hexafluoroisopropanol (Sigma-Aldrich, Steinheim, Germany) and electrospun at 7kV; flow rate of 3.0 mL/h and 10 cm of distance from the collector.

The film quality was checked in terms of porosity, thickness and fiber diameter using helium pycnometer (AccuPyc 1330, Micromeritics, Norcross, GA, USA), stereo microscopy (Leica M205C, Leica Las, v3 sofware), and SEM (Jeol, JSM-6300, Tokyo, Japan), respectively.

2.4. Core System Preparation and Characterization

To prepare the core system, approximately 20 mg of microspheres were dispersed in 50 μL of the hydrogel composed by a mixture of collagen type I (5 mg/mL), HP-γ-CD (34 mg/mL), RB (0.4 mg/mL), CHT (5 mg/mL), PEG-400 (150 mg/mL), and 5 mg of nano-HAP. Then, the hydrogel was cross-linked with 5% *w/w* TPP sterile aqueous solution (0.5 μL/μL of hydrogel) and visible light blue at 468 nm (Dental device) for 3 min [21]. The dose of BMP-2 was 6 μg in microspheres and the total 17β-estradiol dose was 200 μg in three different forms: electrospun films, microspheres, or dispersed into the gel.

The quality of the core system was checked according to its rheological characteristics and porosity previously described [21]. In addition, water uptake and mass loss assays were carried out by incubation of aliquots of 300 μL of the core system in 5 mL of sterile MilliQ water (37 °C) under orbital agitation (25 rpm). At specific times, six samples were withdrawn, we then removed excess water, weighed, and freeze-dried the samples. Then, three samples were visualized by SEM (Jeol JSM-6300) to see the evolution of the internal structure after incubation. The other three samples were used to record the dried weight and calculate the percentage of mass loss and water uptake, applying Equations (1) and (2), respectively, where W_0 is the initial weight of the sample and W_w and W_d are the weights of the wet and dried sample, respectively, at the different times tested.

$$Mass\ loss(\%) = \frac{(W_0 - W_d)}{W_0} \times 100 \tag{1}$$

$$Water\ uptake(\%) = \frac{(W_w - W_d)}{W_d} \times 100 \tag{2}$$

To test the syringeability of the core system two syringes of 1 mL were loaded with a suspension of microspheres of 17β-estradiol in the hydrogel, up to 0.5 mL. Then, 4 doses of 50 μL each were unloaded from both syringes assayed for fluidity through a 20G needle and dose uniformity. For this, the discharged samples were lyophilized, and the 17β-estradiol content evaluated by spectrophotometry at 280 nm, after dissolution in DCM: MeOH (80:20).

2.5. In Vitro Release Assays

BMP-2 in vitro release assays were carried out by incubating an amount of ^{125}I-BMP-2 microspheres and an amount of core system with the same amount of ^{125}I-BMP-2 microspheres in sterile MilliQ water at 37 °C and 25 rpm. The amount of BMP-2 released was calculated by measuring the radioactivity of supernatant samples with a gamma counter (Cobra® II, Packard).

The in vitro release of 17β-estradiol from the different formulations (dispersed in the core system, microspheres, microspheres incorporated to the core system, and electrospun films) was carried out at 37 °C and 25 rpm using two release media: an aqueous solution of sodium lauryl sulfate (SLS) 1% [25] and MeOH:water (50:50) [26,27]. The released 17β-estradiol was measured in the supernatant using the spectrophotometric method. The effective diffusion coefficient, D_{eff} in the matrix of the microspheres and the mass transfer coefficient of the drug in the boundary layer h, were calculated according to the non-steady-state Fick law, as previously described in detail [21]. Whether or not the released fraction of 17β-estradiol from the microspheres dispersed in the core system was analyzed, and Equations (3)–(6) were applied for D_{eff} and h calculation.

$$\frac{M_t}{M_\infty} = 1 - \sum_{n=1}^{\infty} \frac{6\,L^2}{\beta_n^2\left(\beta_n^2 + L^2 - L\right)} exp\left(-\frac{\beta_n^2}{R^2}\,D_{eff}\,t\right) \tag{3}$$

where M_t and M_∞ are the total mass of drug released to the media at time t and at the end of the experiment, respectively. The $\beta_n s$ are the infinite roots (eigenvalues) of the Equation (4):

$$\beta_n \cot \beta_n + L - 1 = 0 \tag{4}$$

L is the dimensionless mass transfer Biot number Equation (5):

$$L = \frac{h\,R}{D_{eff}} \quad (5)$$

For large values of L, the roots of Equation (4) are multiples of the number *pi* and Equation (3) can be simplified in the Equation (6), that is, a simplified solution of non-steady-state Fick law:

$$\frac{M_t}{M_\infty} = 1 - \frac{6}{\pi^2} \sum_{n=1}^{\infty} \frac{1}{n^2} exp\left(-\frac{n^2\,\pi^2}{R^2}\, D_{eff}\, t \right) \quad (6)$$

As stated previously, to minimize the residual sum of squares, "genetic algorithms" already implanted in R software (R Foundation for Statistical Computing, version 3.6.1., 2019, Vienna, Austria) were used [21].

2.6. Animal Experiments

All animal experiments were carried out in conformity with the European Directive (2010/63UE) on Care and Use in Experimental Procedures. Furthermore, the animal protocols were approved on 5 November 2014 by the Ethics Committee for Animal Cares of the University of La Laguna (CEIBA) with identification code CEIBA2014-0128. All surgical procedures were made under aseptic conditions.

2.6.1. Animal Models

Forty female adult Sprague-Dawley rats approximately 12 weeks old, weighing 200–250 g, were divided in 4 groups of 10 each. The experimental osteoporosis was induced in 3 groups by three different protocols, OVX, chronic administration of DEX and OD. The forth group was the sham, non-osteoporotic control group (non-OP). The bilateral ovariectomy was carried out under isoflurane anesthesia, via dorsal approach to the animals of OVX and OD groups. Analgesia consisted in buprenorphine administered subcutaneously (0.05 mg/kg) before surgery and paracetamol (70 mg/100 mL) in the water, for 3 days post-surgery. The DEX group received 0.3 mg/kg body weight of dexamethasone-21-isonicotinate (Deyanil retard, Fatro Ibérica, Barcelona, Spain) administered subcutaneously once ievery two weeks [28] up to the time of euthanasia. Then, two weeks after the ovariectomy, the rats of group OD were chronically treated with DEX as the DEX group. The 40 rats were sacrificed after 12 weeks and the calvaria and femurs were extracted to be histologically analyzed. The results of these analyzes were used to evaluate the 3 protocols tested to induce OP.

2.6.2. Animals Groups

Fifty female Sprague-Dawley rats (12 weeks old), weighing 200–250 g, were divided into 2 groups of 25 each: OP and non-OP. The rats of the OP group were ovariectomized and the rats of the non-OP group underwent similar surgery but the ovaries were not resected. Twelve weeks post-surgery, 8 mm critical size cranial defects were created surgically with a trephine burr in the rats under isoflurane and the systems were placed into the defects [20]. Analgesia treatment was administrated.

Female rats were divided into 5 groups of 10 rats each—5 OP and 5 non-OP—and the applied regenerative treatment is reflected in Table 1. The implantation of the systems was carried out following a two steps procedure. First, a layer of film (bottom film), previously soaked in the blood produced during the surgery, was placed in the defect then 50 μL of the hydrogel mixed with the microspheres and partially cross-linked with UV light, was discharged. Second, the hydrogel was completely cross-linked by dripping 25 μL of sodium tripolyphosphate (TPP) forming the core system, after 5 min, a second layer of film (soaked in blood) was placed on the top, like a sandwich, and the wound was then closed.

Table 1. Experimental groups to evaluate regenerative efficiency.

Group Designations	Treatment
Blank I (B)	System loaded with blank microspheres and blank films
Blank II (B-HAP)	System loaded with blank microspheres and 5 mg of nano-HAP and blank films
BMP+EF	System loaded with 6 μg of BMP-2 in microspheres and 200 μg of 17β-estradiol in the 2 films
BMP+EMs	System loaded with 6 μg of BMP-2 in microspheres and 200 μg of 17β-estradiol in microspheres, blank films
BMP+ED	System loaded with 6 μg of BMP-2 and 200 μg of 17β-estradiol dispersed, blank films

2.6.3. ^{125}I-BMP-2 in Vivo Release Assay

The BMP-2 release kinetics was monitored periodically by measuring the remaining ^{125}I-BMP-2 at the rat calvarial defect site (n = 5) using an external probe-type gamma counter (Captus ®, Capintec Inc., Ramsey, NJ, USA), as previously described and validated [29].

2.7. Rat Mesenchymal Stem Cells (rMSCs) Osteogenic Differentiation

The rMSCs were obtained by centrifugal isolation as previously described [30] from the bone marrow of the femur of OVX female Sprague-Dawley rats. Briefly, the cells were resuspended in high glucose DMEM (HyClone® Utah) supplemented with 10% fetal bovine serum (Biowest, South America Origin), 1% penicillin–streptomycin (PAA, Pasching, Austria), and 2 mM L-Glutamine stable (Biowest, France) (Complete Culture Medium, CCM). Then, cells were cultured in flasks of 75 cm^2 and subcultured by incubating at 37°C and 5% CO_2. The culture medium was changed every 2–3 days.

To test the osteogenic differentiation, 50,000 cells (passage 2) in 20 μL of CCM were added over aliquots of 300 μL of the core system (hydrogel with microspheres) with and without nano-HAP and incubated at 37 °C and 5% CO_2 for 1.5 h for cell adhesion. The homogeneous cell distribution was checked by light microscopy. Afterwards, 500 μL of CCM were added to each well, after 3 days incubation the medium was changed to CCM supplemented with 10 mM β-glycerol phosphate, 10^{-7} M dexamethasone and 50 μM ascorbate-2-phosphate. At 7, 14, and 21 days of culture, three wells of each time point were washed (2 times) with Hank's balanced salt solution (HBSS 1x) and cooled at 4 °C. Then, 500 μL of 0.1 M buffer Tris-HCl, 0.1M NaCl, and 0.05 M $MgCl_2$ (pH = 9.2–9.5) containing Nitro blue tetrazolium chloride (NBT, Roche Diagnostics, Mannheim, Germany) and 5-Bromo-4-chloro-3-indolyl phosphate (BCIP, Roche Diagnostics, Mannheim, Germany) were added and incubated at 37 °C and 5% CO_2 under soft agitation for 1.5 h. Then, the NBT/BCIP was removed and the cells were fixed with a solution of 3.7–4% p-formaldehyde buffered to pH = 7.0 (Panreac®, Barcelona, Spain) during 30 min. After this, the formaldehyde was removed, and the wells were washed 3 times with HBBS 1x. Immediately after this, cells were visualized by stereo microscopy (Leica M205C, Leica Las, v3 software). In addition, samples were dehydrated in a graded series of ethanol before being embedded in Paraplast® and microtome (Shandon Finesse 325, Thermo Fisher Scientific, Madrid, Spain) sections were observed by light microscopy (LEICA DM 4000B, Barcelona, Spain).

2.8. Histology, Immunohistochemical, and Histomorphometrically Evaluation

First, to check the osteoporotic-like condition, 12 weeks after the 12 rats undergone the different protocols were sacrificed and the femurs and calvaria were analyzed. The femurs and calvaria were fixed (4% paraformaldehyde solution), decalcified in Histofix® Decalcifier (Panreac, Barcelona, Spain) and prepared for histological analysis as previously described [20,22].

Bone morphology was analyzed by hematoxylin–erythrosin staining. The histomorphometric analyze was carried out in femurs by measuring the following parameters; thickness of the cortical bone (Ct.Wi) and number (Tb.N), width (Tb.Wi) and separation (Tb.Sp) of the trabeculae in cancellous bone. In the calvaria bone, the histomorphometric analysis was carried out by measuring the following parameters, cortical bone thickness (CBT) and intercortical space thickness (IST) occupied by trabecular bone in transversal sections of calvaria.

To determine the capacity of the bone active substances, so as to regenerate the critical size defect practiced in the calvaria of the rats, samples of the 10 groups of 5 rats each were examined.

Samples were processed as previously described [22]. New bone formation was identified by hematoxylin–erythrosin staining. Bone mineralization was assessed with VOF trichrome stain, in which red and brown staining indicates advanced mineralization, whereas less mineralized, newly formed bone stains blue [31]. Sections were analyzed by light microscopy (LEICA DM 4000B, Barcelona, Spain). Computer-based image analysis software (Leica Q-win V3 Pro-Image Analysis System, Barcelona, Spain) was used to evaluate all sections. A region of interest (ROI) within the defect (50 mm^2) for quantitative evaluation of new bone formation was defined. New bone formation was expressed as a percentage of repair with respect to the original defect area within the ROI. From the total bone repair, the areas of mature bone (MB) and immature bone (IB) were determined, and the MB/IB ratio for each experimental group as well as between non-osteoporotic and osteoporotic-like animals was calculated.

For immunohistochemical analysis, sections were deparaffined and rehydrated in Tris-buffered saline (TBS) (pH 7.4, 0.01 M Trizma base, 0.04 M Tris hydrochloride, 0.15 M NaCl), which was used for all further incubations and rinse steps. Sections were incubated in citrate buffer (pH 6) at 90 °C for antigen retrieval, followed by incubation in 0.3% hydrogen peroxide in TBS buffer for 20 min. After a rinse step, sections were blocked with 2% FBS in TBS–0.2% Triton X-100 (blocking buffer). The indirect immunohistochemical procedure was carried out by incubating the sections with osteocalcin (OCN) polyclonal antiserum (1/100) (Millipore, Barcelona, Spain) in blocking buffer overnight at 4 °C. Sections were rinsed three times, then incubated with biotin-SP-conjugated donkey anti-rabbit F(ab) fragment (1/200) (Millipore, Barcelona, Spain) in blocking buffer for 1 h followed, after another rinse step, by incubation in peroxidase-conjugated streptavidin (1/300) (Millipore, Barcelona, Spain) for 1 h. Peroxidase activity was revealed in Tris–HCl buffer (0.05 M, pH 7.6) containing 0.05% of 3,3′-diaminobenzidine tetrahydrochloride (Sigma, Poole, UK) and 0.004% hydrogen peroxide. Reaction specificity was confirmed by replacing the specific antiserum with normal serum or by pre-adsorption of the specific antiserum with the corresponding antigen.

OCN staining was evaluated using computer-based image analysis software (ImageJ, NIH, Bethesda, MD, USA). OCN staining was measured by applying a fixed threshold to select for positive staining within the ROI. Positive pixel areas were divided by the total surface size (mm^2) of the ROI. Values were normalized to those measured from blank scaffolds and are reported as relative staining intensities.

Statistical analysis was performed with SPSS.25 software. We compared the distinct treatments by means of a one-way analysis of variance (ANOVA) with a Tukey multiple comparison post-test. Significance was set at $p < 0.05$. Results are expressed as means ± SD.

3. Results

3.1. Sandwich-Like System Characterization

The characteristics of the microspheres, electrospun film, and core system are shown in Table 2.

Table 2. Characteristics of the component of the sandwich-like system. Microspheres: size and encapsulation efficiency. Electrospun film: thickness and porosity of the film and average diameter of the fibers). Core system: porosity freshly prepared and lyophilized and water uptake and mass loss after incubation in MilliQ water, 37 °C, and 25 rpm.

Microspheres	**Size (μm)**			**E.E. (%)**
17β-estradiol	101.4 (10% < 29.73, 90% < 198.31)			83.5 ± 1.84
BMP-2	112.1 (105 < 60.3, 905 < 174.8)			71 ± 7
Film	Thickness (μm)	Fiber diameter (μm)		Porosity (%)
	63.4 ± 4.3	1.2 ± 0.26		71.9 ± 0.41
Core system (Hydrogel + microspheres)	Porosity (%)	Incubation time	Water uptake (%)	Mass loss (%)
	72	7 days	135.9 ± 2.6	18.85 ± 7
		28 days	138.9 ± 11.0	29.19 ± 4.88

SEM image of the microspheres is shown in Figure 1A and the differential scanning calorimetry thermograms of the 17β-estradiol microspheres components are plotted in Figure 2. The glass transition temperature (Tg) of the mixture of polymers RG 203-S and RG 858 [4:1] was located at 52–58 °C, in the temperature range of the PLA (RG203-S) and PLGA (RG858). The DSC analysis of pure 17β-estradiol showed three endothermic peaks (Figure 2A), the first two at 118.1 °C and 174.4 °C, previously attributed to the partial and complete loss of hydrogen-bound water and reticular water, respectively. The third at 179.4 °C corresponds to the melting point [32]. This last peak, characteristic of the crystalline structure of 17β-estradiol, was not detected in the spectrum of the polymers and 17β-estradiol blend or in the thermogram of the microspheres (Figure 2B). These results indicated that 17β-estradiol was dissolved in the polymer by at least 8.5%.

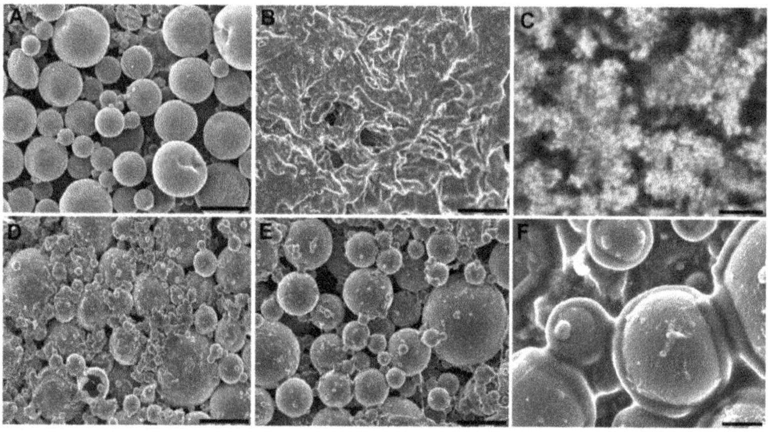

Figure 1. SEM images (**A**) microspheres, (**B**) hydrogel, (**C**) hydrogel high magnification detail, (**D**) core system freshly prepared, (**E**) internal structure of the core system after 4 weeks incubation in water at 37 °C and 25 rpm, and (**F**) high magnification detail of Figure 1E. Scale bars: (**A–E**) 100 μm, (**C**) 1 μm, (**F**) 20 μm.

The integrity of the system was assayed throughout the 4-week test duration. The SEM images of the internal structure of the hydrogel (Figure 1B,C), the freshly prepared core system (Figure 1D), and after 4 weeks incubation showed that the microspheres were homogeneously dispersed in the hydrogel and are trapped in the core system during incubation (Figure 1E,F). The core system absorbed a significant amount of water during the first days of incubation which was maintained over time.

Contrarily, the system lost little mass: less than 20% during the first week and approximately 35% after 4 weeks (Table 2). In addition, the core system flew well through the 20 G needle and the average dose discharged was 83.5 ± 6% of the loaded dose.

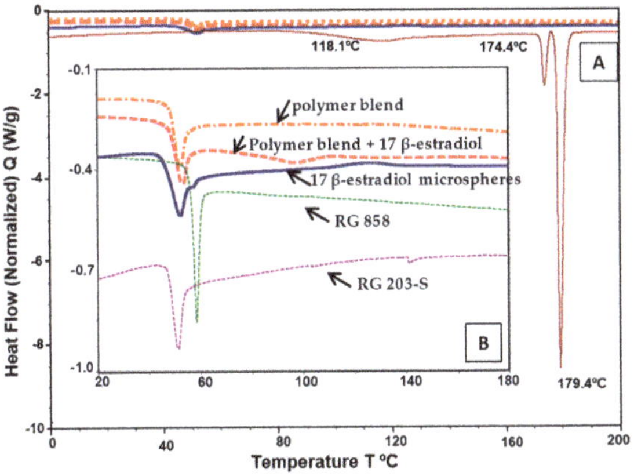

Figure 2. Differential scanning calorimetry thermograms. (**A**) Curve of pure 17β-estradiol. (**B**) Curves of PLA (RG203-S); PLGA (RG858), polymer blend (RG203-S:RG858, [4:1]), and polymer blend with 8.5% of 17β-estradiol, previously dissolved in DCM:MeOH (80:20) and the curve of the microspheres of 17β-estradiol (**B**).

3.2. *Osteogenic Differentiation*

The alkaline phosphatase positive (ALP+) cell count, in the hydrogels pre-seeded with rMSCs and cultured in the osteogenic differentiation culture medium, showed a discrete and progressive increase in the number of cells between 7 and 21 days of culture in the hydrogels without and with nano-HAP; the number being significantly higher in those containing nano-HAP (Figure 3). Likewise, qualitatively, the cells presented, in the scaffolds with nano-HAP, greater intensity of color, suggesting greater ALP activity (Figure 3) than without nano-HAP.

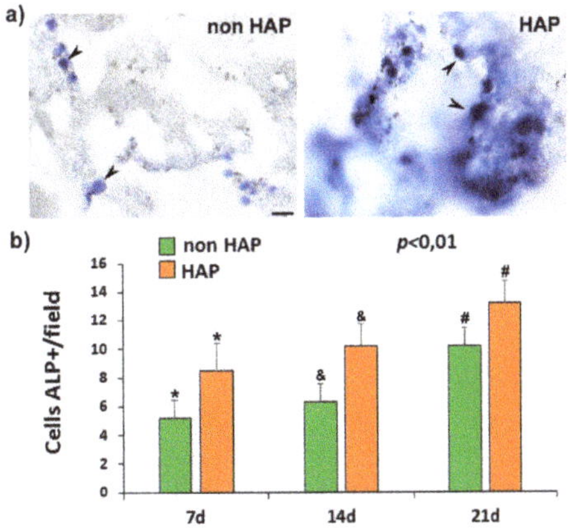

Figure 3. Alkaline phosphatase (ALP) activity in rMSCs cultures. (**a**) Representative images from hydrogels 20 days after cultured showing the AP-positive (ALP+) cells' morphology (arrowheads) in hydrogel without (non-HAP) and with nanohydroxyapatite (HAP). (**b**) Graphic showing the number of APL+ cells/ microscopic field at different time points of analyses (7, 14, and 21 days) after culture in each system. Scale bar = 20 μm. The identical symbol on different bars indicates significant differences.

3.3. *Release Profiles of* ^{125}*I-BMP-2 and 17β-Estradiol.*

Although the hydrogel provoked a strong reduction of the burst effect, the in vitro release of BMP-2 from the microspheres and from the core system showed a biphasic profile. During the first 24 h, approximately 7% of the protein was release from the core system versus 27% from the microspheres directly dispersed in the medium. Afterward, the release rate was kept in the range of 60 to 55 ng/day (Figure 4). The in vivo release profile was also biphasic, with a first phase that lasted up to 7 days whereas approximately 50% of the protein released. Then, the release rate was reduced to 70 ng/day, which is slightly higher than the in vitro rate.

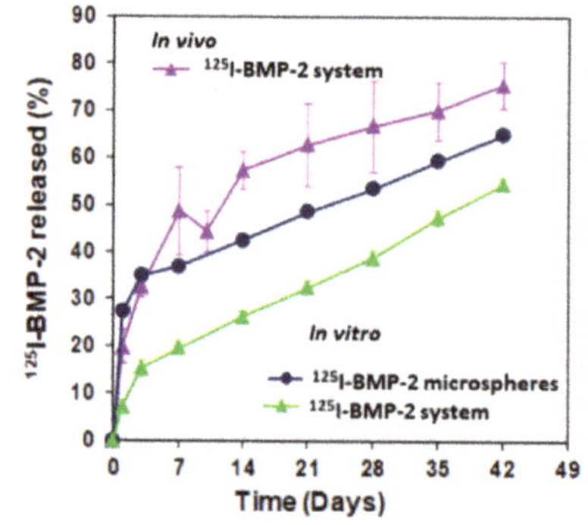

Figure 4. BMP-2 release assays. In vivo release profile of ^{125}I-BMP-2 from PLGA-microspheres in the system after implantation in the rat calvaria defect (n = 5). In vitro release of ^{125}I-BMP-2 (incubation in water at 37 °C and 25 rpm) from PLGA-microspheres and from the PLGA microspheres dispersed in the system.

The in vitro release rate of 17β-estradiol in the aqueous medium depended on the formulation (Figure 5B); 100% and 70% of the 17β-estradiol dispersed in the hydrogel and in the electrospun film were detected in the medium after 4 weeks incubation, respectively. The release profile was characterized by a high burst effect; approximately half of the dose was released during the first day. By contrast, the 17β-estradiol release rate was extremely slow from the microspheres alone and from the microspheres included in the hydrogel (core system). Both release profiles were similar: less than 20% was delivered in 4 weeks. However, in the MeOH: water (50:50) medium there were not differences in the transfer profiles. The presence of MeOH modifies the solubility of 17β-estradiol, showing a strong burst effect that varied in a range between 70 and 85% in the first 24 h (Figure 5A).

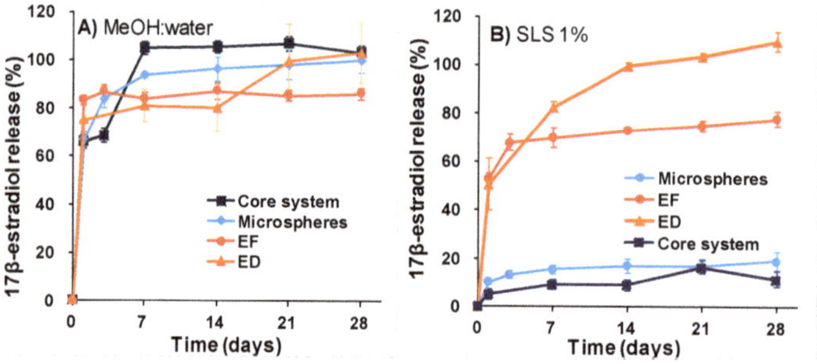

Figure 5. In vitro 17-β-estradiol release profiles in different release media at 37 °C and 25 rpm. (**A**) In MeOH:water (50:50). (**B**) In aqueous solution of SLS 1%. (ED) 17β-estradiol dispersed in the core system; (EF) 17β-estradiol in the electrospinned film; (microspheres) 17β-estradiol pre-encapsulated in microspheres and (core system) 17β-estradiol microspheres in the hydrogel.

The estimated values of D_{eff} and h for 17β-estradiol release in the different media are showed in Table 3. Although the value of D_{eff} of 17β-estradiol in SLS was significantly lower compared to the MeOH:water (50:50), there were not differences for h regardless of the release media used. h is the contribution of hydrogel, as part of the boundary layer, to the whole mass transfer process, and the value of this coefficient should not change by varying the release media. The values of R^2 (Table 3), together with the comparison of experimental and predicted values of the released fractions shows a good fit of the data to the proposed model for both release media.

Table 3. Estimated values of effective diffusion coefficient (D_{eff}) and mass transfer coefficient (h) for 17β-estradiol release in different media applying Equations (3)–(6).

Release Media	D_{eff} (m²/s)	h (m/s)	R^2 (%) Value
MeOH:water (50:50)	$2.28 \cdot 10^{-15} \pm 5.00 \cdot 10^{-17}$	$7.56 \cdot 10^{-10} \pm 2.89 \cdot 10^{-10}$	95.47 ± 0.07
SLS 1%	$5.58 \cdot 10^{-16} \pm 9.81 \cdot 10^{-17}$	$4.01 \cdot 10^{-10} \pm 4.94 \cdot 10^{-10}$	92.49 ± 2.71

3.4. Histology, Immunohistochemical, and Histomorphometrically Evaluation

3.4.1. Osteoporotic Model

The osteoporotic model was assessed in both long bone (femur) and flat bone (calvaria). The histological analysis of calvaria showed evident changes in the structure and microarchitecture of the bone among the different experimental groups. Although the non-OP animals showed a normal bone structure in cortical bone (CB) and trabecular bone (TB), in the intercortical space (ICS), for the different OP models (DEX, OVX, and OD (Figure 6a), it was observed a progressive decrease in cortical bone thickness (CBT) and an increase in the intercortical space thickness (IST), being the group OD the one that presented greater alteration of the tissue bone structure (Figure 6b).

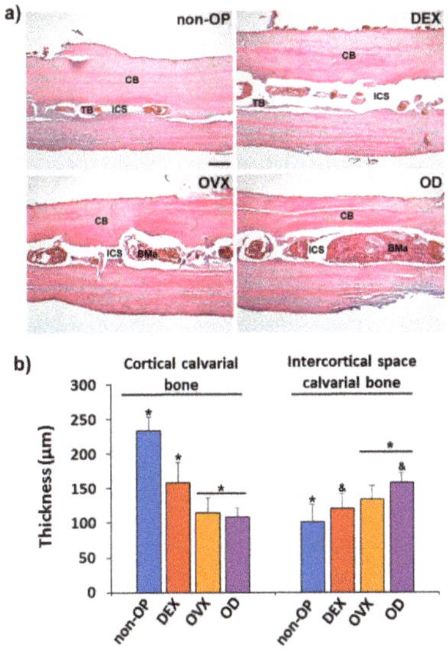

Figure 6. Validation of the OP model in calvarial bone. (**a**) Representative images in transversal section of calvaria in non-osteoporotic animals (non-OP) and in each of the different experimental models of osteoporosis tested showing the bone structure in each one. (**b**) Histomorphometric analysis of the cortical bone thickness and intercortical space thickness evaluated in calvaria in the different models of osteoporosis. CB: cortical bone; BMa: bone marrow; ICS: intercortical space; TB: trabecular bone. Scale bar = 100 μm. The identical symbol on different bars indicates significant differences.

The histological analysis of the femurs showed evident changes in the structure and microarchitecture of the bone among the different experimental groups. Although the non-OP animals showed a normal bone structure, in the different OP models (DEX, OVX and OD), structural changes were observed, both at the level of cortical and cancellous bone, showing a less compact bone and with a more porous structure (Figure 7a). The histomorphometric analysis revealed differences in the parameters measured in cancellous bone (Tb.N, Tb.Wi., and Tb.Sp.), with a significant reduction in all of them, in OVX and OD compared to the non-OP and DEX animals (Figure 7b–d). The cortical bone thickness (Ct.Wi.), although it showed a slight reduction in the groups (OVX and OD) with respect to the non-OP and DEX groups, was not significant (Figure 7e).

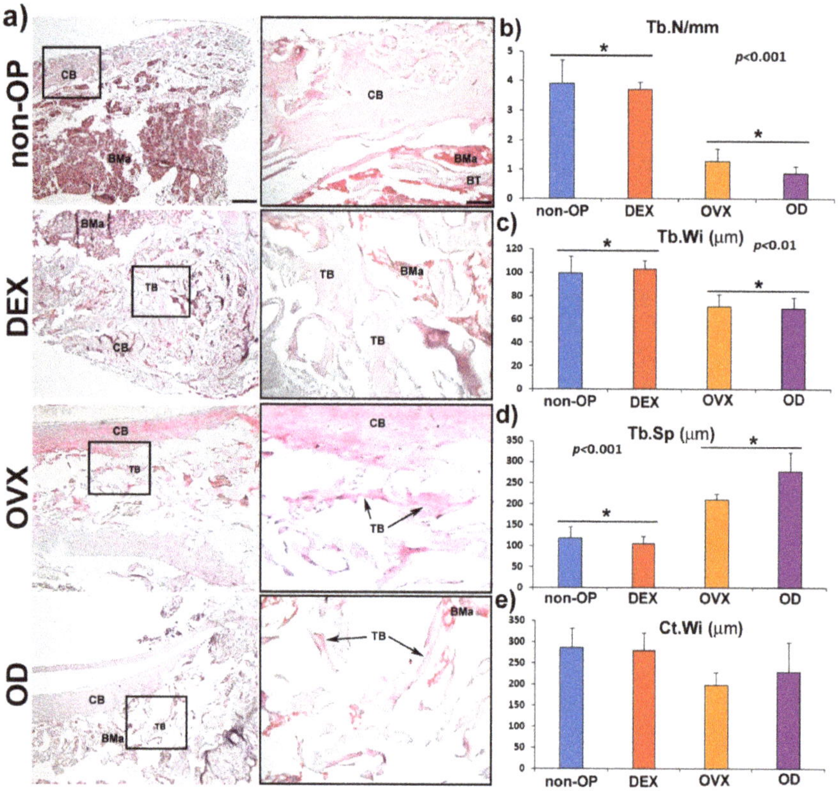

Figure 7. Validation of the OP model in long bone (femur). (**a**) The left column shows representative panoramic images in longitudinal section of rat femur in non-osteoporotic animals (non-OP) and in each of the different experimental models of osteoporosis tested. The right column shows high magnification images of the distal portion of the femur, showing differences in the microarchitecture of the bone in the different models. The column on the left shows detail of the boxed areas in which the structural characteristics of the compact and trabecular bone can be observed in each of the models. Histomorphometric analysis of the different parameters evaluated in femur in the different models of osteoporosis (**b**), Tb. N (mm), (**c**) Tb. Wi (μm), (**d**) Tb. Sp. (μm), and (**e**) Ct. Wi (μm). CB: cortical bone; TB: Trabecular bone; BMa: bone marrow. Scale bars: Left column 250 μm. Right column 50 μm. The identical symbol on different bars indicates significant differences.

3.4.2. Calvarial Critical Size Defect

The histological analysis at the level of the calvarial defect showed a few new bone formations in the blank groups (B and B HAP), being limited to the margins of the defect in both, non-OP, and OP groups (Figure 8a). The groups implanted with BMP-2 + 17β-estradiol in the three different formulations showed a greater area of newly formed bone in the defect area, not only in the margins, but also in inner zone of the defect (Figure 8a). The newly formed bone in the different experimental groups of non-OP and OP animals showed a normal morphology and VOF staining, revealing significant areas of mineralization, slightly higher in the groups of non-OP animals (Figure 8a).

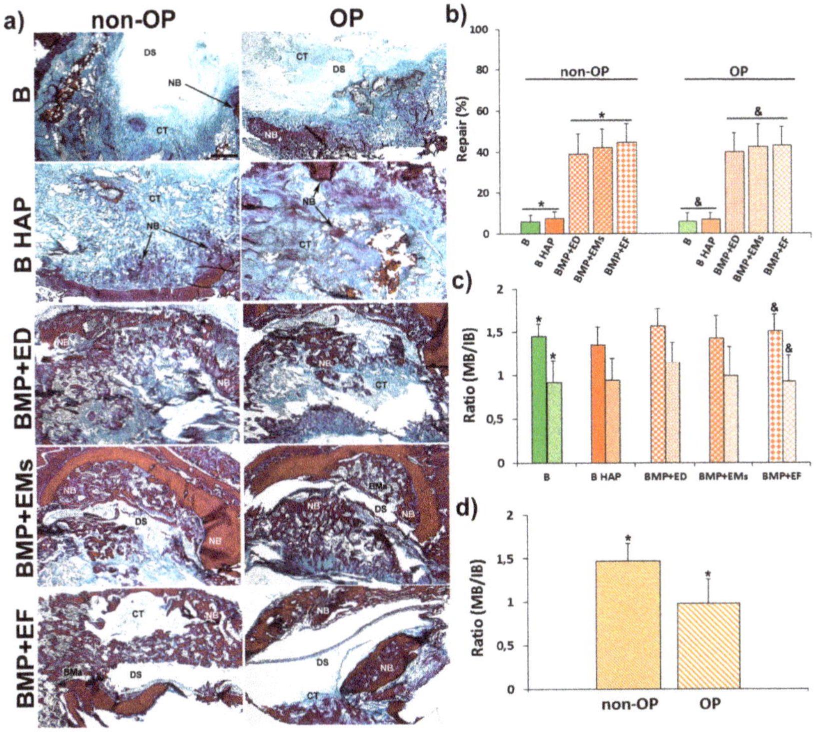

Figure 8. Repair process in calvarial defect. (**a**) Representative images in horizontal section of calvarial critical size defects in non-OP and OP rats showing the repair response at the defect level in the different experimental groups 12 weeks post-implantation. (**b**) Histomorphometrical analysis comparing of the degrees of repair (%) among the different experimental groups in non-OP and OP rats 12 weeks post-implantation. (**c**) Histomorphometric analysis showing the ratio between mature bone and immature bone (MB/IB) among the different experimental groups (**d**) and between non-OP and OP rats, estimated using VOF staining. Bars represent means ± SD (n = 4). The identical letter on different bars indicates significant differences. BMa: bone marrow; CT: connective tissue; NB: newly formed bone; DS: defect site. Scale bar = 1 mm. The identical symbol on different bars indicates significant differences.

The histomorphometric analysis showed little repair response in the blanks groups (B and B HAP) of non-OP and OP animals, with repair percent between 6 and 8%. The groups implanted with BMP-2 + 17β-estradiol in the three different formulations, on the contrary, showed a significantly higher repair response of 38–45%, with no differences being observed between non- OP and OP animals (Figure 8b).

The histomorphometric analysis of mature and immature bone showed a higher quantity of mature bone, and therefore with a greater degree of mineralization, in the experimental groups of non-OP with respect to OP animals. The ratio between mature and immature bone (MB/IB) showed individually higher values in all non-OP with respect to OP groups as well as on the whole, with values of 1.47 in non-OP animals and 0.99 in OP (Figure 8c,d).

The immunohistochemical analysis of osteocalcin (OCN), a marker of late osteogenesis and mineralization, showed a low immunoreaction in the blank groups (B and B HAP) in both non-OP and OP animals, with no differences between them (Figure 9a). In the groups implanted with BMP-2 + 17β-estradiol in the three different formulations, the immunoreaction was higher and more intense with respect to the blank groups, in this case being slightly higher in the non-OP animals (Figure 9a).

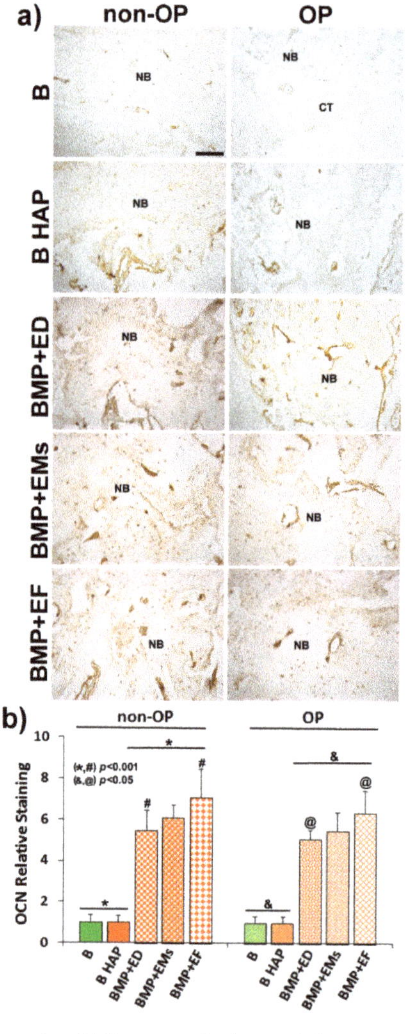

Figure 9. OCN relative expression. (**a**) Representative images in horizontal section of calvarial critical size defects in non-OP and OP rats showing OCN immunoreactivity in the different experimental groups 12 weeks post-implantation. (**b**) Histomorphometric analysis showing the relative staining values for OCN-ir. Bars represent means ± SD (n = 4). The identical letter on different bars indicates significant differences. CT: connective tissue; NB: newly formed bone. Scale bar = 100 μm. The identical symbol on different bars indicates significant differences.

The histomorphometric analysis confirmed the histological data, showing slightly higher relative staining values in the BMP-2 +17β-estradiol groups in non-OP animals (Figure 9b).

4. Discussion

In the present study, a BMP-2-17β-estradiol hydrogel system, with porosity of approximately 72%, was evaluated for regeneration of a critical size defect in rat calvaria. Although the system had already been characterized in terms of rheological behavior, porosity, interactions between components, mass transfer parameters, and cell viability, here the good injectability of the system was showed, and its characterization has been completed by testing the water uptake and mass loss as well as differentiation studies in cultures of osteoporotic rMSCs. In vitro release profiles of 17β-estradiol in two media and the in vitro and in vivo release profile of BMP-2 were also analyzed.

The osteogenic differentiation of osteoporotic rMSCs seeded on the system was assessed in order to test the effect of the incorporation of the nano-HAP on the cell behavior. The results showed greater ALP activity and a greater number of differentiated cells in the systems with nano-HAP. Therefore, nano-HAP systems were subsequently used in the in vivo experiments.

First, a histological evaluation of the femur and calvaria of rats suffering the three treatments for osteoporosis induction was carried out. OP is a systemic bone disease characterized by the increase of bone porosity, loss of bone mass and changes in the microstructure of the skeletal. Consequently, the OP population has an increased risk of fracture.

Despite the high number of OP studies and the several publications dedicated to tissue repair in non-OP specimens, very few reports devoted to bone defect regeneration in OP have been published. As OP might be primary post-menopausal or secondary, due to corticoid chronic administration, three animal models were used for OP induction: OVX, chronic glucocorticoid treatment, and the combination of both. In previous reports, a combination OD rat model was used. However, the high deterioration observed in the animals, the risk of induced additional disorders on the skeletal and the fact that bone loss reverses after corticoid stop [33] justify the bone histological study of the different treatment, in order to simplify the model and improve animal welfare. In general, OP condition is established through the analysis of long bones and lumbar spine but few data of the effect on the calvaria of the animals used as OP models are available [34]. Most of the publications on the regeneration of calvaria critical size defect in OP animals do not report the effects of OP in the calvaria [35,36]. In the present study, the data comparing the response of the femurs and calvarias to the three treatments revealed that the effect of OVX was similar to OD combination, consequently the fabricated system was tested in OVX rats.

As some authors have observed a delay in bone consolidation of OVX rats [37] and as this combination of BMP-2 and 17β-estradiol, formulated in microspheres, when applied to a critical calvaria defect, improved bone healing in OP rats, but the new bone was less mineralized [17,18], we tried to prolong the release of active substances to cover this delay. The drugs were incorporated to the hydrogel system pre-encapsulated in microspheres for prolonged controlled release. To reduce the release rate of the active substances the microspheres were prepared with a mixture of polymers, 25% of the RG 858 was incorporated to the RG 504 for BMP-2 microspheres as well as to the RG203-S for 17β-estradiol microspheres. RG 858 is a PLGA 85:15, of high molecular weight with a degradation rate lower than that of RG 504 and RG203-S. The BMP-2 release profiles showed a two-phase behavior with a weak burst effect that coincides with the period in which the system loses mass and uptakes a high amount of water. However, the burst effect of BMP-2 that can be seen from the microspheres was damped by the hydrogel, probably due to the interaction of the protein with the chemical groups of the HAP [6,13]. Afterwards, the second phases were practically parallel, which indicated a mass transfer process controlled by the access of water inside the microspheres, dissolution and diffusion of the protein throughout the porous of the polymeric matrix.

The release profile of 17β-estradiol as liposoluble drug in a MeOH:water (60:40) release medium was previously characterized [21]. By contrast, here two release media were used, MeOH:water (50:50) and an aqueous solution of SLS, because we suspected that the solvent affected the release rate of the lipophilic substance. In fact, the release of 17β-estradiol in MeOH:water was fast regardless of the formulations. However, the in vitro release of 17β-estradiol was formulation-dependent when an aqueous medium was used; a high decrease of the release rate of the drug from the microspheres was observed. As DSC results indicated, 17β-estradiol formed a solid solution in the microspheres, which indicates that the release process takes place by molecular diffusion of 17β-estradiol within the microspheres, governed by the partition coefficient, and consequently the aqueous medium dissolved it very slowly. In addition, the higher estimated values of D_{eff} in the MeOH:water compared with the aqueous medium, confirmed the dependency on the release media. The calorimetric analysis of the electrospun sheet was not carried out because the amount of 17β-estradiol would have had to be increased to be detected and its characteristics would have been modified. Although 17β-estradiol is expected to also be dissolved in the polymer of the film, the large specific surface area that microfibers expose to the medium causes the drug to release rapidly. Similarly, 17β-estradiol dispersed in the hydrogel was very little retained. Obviously, neither media are physiological, but it seems more correct to use an aqueous medium to predict release in vivo. Although, with reservations, one would also expect a slightly faster release rate in vivo as the biological components present in the tissue could accelerate the drug release from the system.

Despite the beneficial role of the nano-HAP controlling the BMP-2 burst effect as well as its positive effect on the proliferation and osteogenic differentiation of rMSC, which justifies the use of nano-HAP in the system, the reparative effect of the blank scaffolds with and without nano-HAP was not enough to be considered useful. Unlike that observed in this study, other authors showed better bone repair in different bone defects practiced in osteoporotic goats implanted with a system of type I collagen containing nano-HAP than without [38]. By contrast, another study [6] found, as in the present study, that the use of nano-HAP and calcium sulfate bone substitute scaffolds in rat critical calvaria defect showed no effect on repair and mineralization at 8 and 12 weeks with respect to the empty defect. In both studies, the systems were loaded with BMP-2 combined with 17β-estradiol or zolendronic acid [6] that might abolish the effect of the HAP observed in vitro.

Although previously our group reported [17,18] a better result in bone regeneration of OP animals combining BMP-2 and 17β-estradiol, in this study, the repair effect observed has been similar to that observed in non-OP groups. The three combinations of BMP-2 with 17β-estradiol in each of the three formulations used showed the same effect at 12 weeks. However, the ratios of mature and immature bone in normal and osteoporotic animals showed significant differences, indicating that the quality of the repaired bone, at least after 12 weeks, was better in normal animals.

Although these results coincide with previous work, in the present study it seems that the mineralization of the bone formed slightly improved in the OP animals. The difference in the relative osteocalcin expression was not statistically significant. These results might suggest that a longer release of BMP-2 together with the composition of the system, presence of collagen and nanoHAP favor the mineralization process. In any case, it would be necessary in future to conduct studies aimed at discerning the role of each of these components in the process. However, we have not been able to reproduce the positive effect of 17β-estradiol combined with BMP-2 in a hydrogel composed of Pluronic, Tetronic and, cyclodextrin, with other scaffolds of different composition [18]. In addition, according to the present study, the fact that the different release profiles of 17β-estradiol had no effect on the repair of the defect indicates that 17β-estradiol, when applied locally, and regardless of the release rate (available dose) and of the obvious role that it plays on bone remodeling, does not justify its inclusion as active substance in the repair of bone defects neither in normal animals nor in osteoporotic ones. Therefore, new strategies and alternative drugs are currently being designed trying to accelerate mineralization of new bone in OP groups.

5. Conclusions

The prepared hydrogel system resulted to be easily injectable and solidified fast due to crosslinking of collagen and chitosan chains. The system helps control the burst effect of BMP-2 pre-encapsulated in PLGA microspheres, probably due to the nano-HAP. Release of 17β-estradiol from PLA-PLGA microspheres was more complex and is governed by the partition coefficient of the drug which is in solid dissolution in the microspheres. The system was biocompatible both in vivo and in vitro. However, the regenerative effect detected in the critical bone defect of both OP and non -OP rats was mainly due to the osteogenic effect of BMP-2 released in a controlled rate for 6 weeks. A delay in the mineralization of the new bone which fills the defect in OP animals was observed. 17β-estradiol released from different formulations and included in the system does not improve bone repair.

Author Contributions: Conceptualization, E.P.-H., A.D., and C.É.; methodology, E.P.-H., A.D., C.É., R.R., E.S.-M., P.G.-G; validation, R.R.; formal analysis, E.P.-H., A.D., C.É., R.R., P.G.-G.; investigation, E.P.-H., A.D., C.É., R.R. and P.G.-G.; resources, A.D. and C.É.; writing—original draft preparation, R.R., A.D., and C.É.; writing—review and editing, A.D. and C.É.; supervision, A.D. and C.É.; funding acquisition, A.D. and C.E. All authors have made a substantial contribution to the work.

Funding: This research and the APC was funded by Ministry of Science and Technology (MAT2014-55657-R).

Conflicts of Interest: The authors declare no conflicts of interest.

References

1. Murphy, C.M.; Schindeler, A.; Gleeson, J.P.; Yu, N.Y.; Cantrill, L.C.; Mikulec, K.; Peacock, L.; O'Brien, F.J.; Little, D.G. A collagen–hydroxyapatite scaffold allows for binding and co-delivery of recombinant bone morphogenetic proteins and bisphosphonates. *Acta Biomater.* **2014**, *10*, 2250–2258. [CrossRef] [PubMed]
2. Cho, T.H.; Kim, I.S.; Lee, B.; Park, S.N.; Ko, J.H.; Hwang, S.J. Early and Market Enhancement of New Bone Quality by alendronate-Loaded Collagen Sponge Combined with Bone Morphogenetic Protein-2 at High Dose: A Long-Term Study in Calvarial Defects in Rat Model. *Tissue Eng. Part A* **2017**, *23*, 1343–1360. [CrossRef] [PubMed]
3. Kim, H.C.; Son, J.M.; Kim, C.J.; Yoon, S.Y.; Kim, I.R.; Park, B.S.; Shin, S.H. Effect of bisphosphonate and recombinant human bone morphogenetic protein 2 on bone healing of rat calvarial defects. *Maxillofac. Plast. Reconstr. Surg.* **2015**, *37*, 16. [CrossRef] [PubMed]
4. Horstmann, P.F.; Raina, D.B.; Isaksson, H.; Hettwer, W.; Lidgren, L.; Petersen, M.M.; Tägil, M. Composite Biomaterial as a Carrier for Bone-Active Substances for Metaphyseal Tibial Bone Defect Reconstruction in Rats. *Tissue Eng. Part A* **2017**, *23*, 1403–1412. [CrossRef]
5. Raina, D.B.; Isaksson, H.; Teotia, A.K.; Lidgren, L.; Tägil, M.; Kumar, A. Biocomposite macroporous cryogels as potential carrier scaffolds for bone active agents augmenting bone regeneration. *J. Control Release* **2016**, *10*, 365–378. [CrossRef]
6. Teotia, A.K.; Raina, D.B.; Singh, C.; Sinha, N.; Isaksson, H.; Tägil, M.; Lidgren, L.; Kumar, A. Nano-Hydroxyapatite Bone Substitute Functionalized with Bone Active molecules for Enhanced Cranial Bone Regeneration. *Appl. Mater. Interfaces* **2017**, *1*, 6816–6828. [CrossRef]
7. Venkatesan, J.; Anil, S.; Kim, S.K.; Shim, M.S. Chitosan as a vehicle for growth factor delivery: Various preparations and their applications in bone tissue regeneration. *Int. J. Biol. Macromol.* **2017**, *104*, 1383–1397. [CrossRef]
8. Ma, X.; He, Z.; Han, F.; Zhong, Z.; Chen, L.; Li, B. Preparation of collagen/hydroxyapatite/alendronate hybrid hydrogels as potential scaffolds for bone regeneration. *Colloids Surf. B Biointerfaces* **2016**, *1*, 81–87. [CrossRef]
9. Cholas, R.; Kunjalukkal Padmanabhan, S.; Gervaso, F.; Udayan, G.; Monaco, G.; Sannino, A.; Licciulli, A. Scaffolds for bone regeneration made of hydroxyapatite microspheres in a collagen matrix. *Mater. Sci. Eng. C Mater. Biol. Appl.* **2016**, *63*, 499–505. [CrossRef]
10. Zhang, P.; Hong, Z.; Yu, T.; Chen, X.; Jing, X. In vivo mineralization and osteogenesis of nanocomposite scaffold f poly(lactide-co-glycolide) and hydroxyapatite surface-grafted wit poly(L-lactide). *Biomaterials* **2009**, *30*, 58–70. [CrossRef]

11. Zhang, B.; Zhang, P.B.; Wang, Z.L.; Lyu, Z.W.; Wu, H. Tissue-engineered composite scaffold of poly(lactide-co-glycolide) and hydroxyapatite nanoparticles seeded with autologous mesenchymal stem cells for bone regeneration. *J. Zhejiang Univ. Sci. B* **2017**, *18*, 963–976. [CrossRef] [PubMed]
12. Gupta, V.; Lyne, D.V.; Barragan, M.; Berkland, C.J.; Detamora, M.S. Microsphere-Based Scaffolds Encapsulating Tricalcium Phosphate and Hydroxyapatite for Bone Regeneration. *J. Mater. Sci. Mater. Med.* **2016**, *27*, 121. [CrossRef] [PubMed]
13. Quinlan, E.; López-Noriega, A.; Thompson, E.; Kelly, H.M.; Cryan, S.A.; O'Brien, F.J. Development of collagen-hydroxyapatite scaffolds incorporating PLGA and alginate microparticles for the controlled delivery of rhBMP-2 for bone tissue engineering. *J. Control. Release* **2015**, *28*, 71–79. [CrossRef] [PubMed]
14. Wang, X.; Wu, X.; Xing, H.; Zhang, G.; Shi, Q.; E, L.; Liu, N.; Yang, T.; Wang, D.; Qi, F.; et al. Porous Nanohydroxyapatite/Collagen Scaffolds Loading Insulin PLGA Particles for Restoration of Critical Size Bone Defect. *ACS Appl. Mater. Interfaces* **2017**, *5*, 11380–11391. [CrossRef] [PubMed]
15. Segredo-Morales, E.; García-García, P.; Évora, C.; Delgado, A. BMP delivery systems for bone regeneration: Healthy vs osteoporotic population. Review. *J. Drug Deliv. Sci. Tec.* **2017**, *42*, 107–118. [CrossRef]
16. Giovanini, A.F.; de Sousa Passoni, G.N.; Göhringer, I.; Deliberador, T.M.; Zielak, J.C.; Storrer, C.L.M.; Costa-Casagrande, T.A.; Scariot, R. Prolonged use of alendronate alters the biology of cranial repair in estrogen-deficients rats' associated simultaneous immunohistochemical expression of TGF-β1 + α-ER+, and BMPR1B. *Clin. Oral Investig.* **2018**, *22*, 1959–1971. [CrossRef]
17. Segredo-Morales, E.; Reyes, R.; Arnau, M.R.; Delgado, A.; Évora, C. In situ gel-forming system for dual BMP-2 and 1717β-estradiol controlled release for bone regeneration in osteoporotic rats. *Drug Deliv. Transl. Res.* **2018**, *8*, 1103–1113. [CrossRef]
18. Segredo-Morales, E.; García-García, P.; Reyes, R.; Pérez-Herrero, E.; Delgado, A.; Évora, C. Bone regeneration in osteoporosis by delivery BMP-2 and PRGF from tetronic-alginate composite thermogel. *Int. J. Pharm.* **2018**, *30*, 160–168. [CrossRef]
19. Seo, B.B.; Koh, J.T.; Song, S.C. Tuning physical properties and BMP-2 release rates injectable hydrogel systems for an optimal bone regeneration effect. *Biomaterials* **2017**, *122*, 91–104. [CrossRef]
20. Rodríguez-Évora, M.; Delgado, A.; Reyes, R.; Hernández-Daranas, A.; Soriano, I.; San Román, J.; Évora, C. Osteogenic effect of local, long versus short term BMP-2 delivery from a novel SPU-PLGA-βTCP concentric system in a critical size defect in rats. *Eur. J. Pharm. Sci.* **2013**, *16*, 873–884. [CrossRef]
21. Pérez-Herrero, E.; García-García, P.; Gómez-Morales, J.; Llabrés, M.; Delgado, A.; Évora, C. New injectable two-step forming hydrogel for delivery of bioactive substances in tissue regeneration. *Regen. Biomater.* **2019**, *6*, 149–162. [CrossRef] [PubMed]
22. Hernández, A.; Reyes, R.; Sánchez, E.; Rodríguez-Évora, M.; Delgado, A.; Évora, C. In vivo osteogenic response to different ratios of BMP-2 and VEGF released from a biodegradable porous system. *J. Biomed. Mater. Res.* **2012**, *100*, 2382–2391. [CrossRef] [PubMed]
23. Fraker, P.J.; Speck, J.C. Proteion and cell membrane iodinations with a spraringly soluble chloroamide, 1,3,4,6-tetrachloro-3a,6a-diphenylglycoluril. *Biochem. Biophys. Res. Commum.* **1978**, *80*, 849–857. [CrossRef]
24. García-García, P.; Reyes, R.; Pérez-Herrero, E.; Arnau, M.R.; Évora, C.; Delgado, A. Alginate-hydrogel versus alginate-solid system. Efficacy in bone regeneration in osteoporosis. *Eur. J. Pharm. Biopharm.*. Manuscript submitted.
25. *Pharmaceutical Compounding–Estradiol Tablets*; The United States Pharmacopeial Convention: Baltimore, MD, USA, 2014.
26. Birnbaum, D.T.; Kosmala, J.D.; Henthorn, D.B.; Brannon-Peppas, L. Controlled release of b-estradiol from PLAGA microparticles:The effect of organic phase solvent on encapsulation and release. *J. Control Release* **2000**, *65*, 375–387. [CrossRef]
27. Zaghloul, A.A. *β*-Estradiol biodegradable microspheres: Effect of formulation parameters on encapsulation efficiency and in vitro release. *Pharmazie* **2006**, *61*, 775–779.
28. Govindarajan, P.; Khassawna, T.; Kampschulte, M.; Böcker, W.; Huerter, B.; Dúrselen, L.; Faulenbach, M.; Heiss, C. Implications of combined ovariectomy and glucocorticoid (dexamethasone) treatment on mineral, microarchitectural, biomechanical and matrix properties of rat bone. *Int. J. Exp. Pathol.* **2013**, *94*, 387–398. [CrossRef]

29. Delgado, J.; Évora, C.; Sánchez, E.; Baro, M.; Delgado, A. Validation of a method for non-invasive in vivo measurement of growth factor release from a local delivery system in bone. *J. Control Release* **2006**, *114*, 223–229. [CrossRef]
30. Dobson, K.R.; Reading, L.; Haberey, M.; Marine, X.; Scutt, A. Centrifugal isolation of bone marrow from bone: An improved method for the recovery and quantitation of bone marrow osteoprogenitor cells from rat tibiae and femurae. *Celcif. Tissue Int.* **1999**, *65*, 411–413. [CrossRef]
31. Martínez-Sanz, E.; Ossipov, D.A.; Hilborn, J.; Larsson, S.; Jonsson, K.B.; Varghese, O.P. Bone reservoir: Injectable hyaluronic acid hydrogel for minimal invasive bone Augmentation. *J. Control Release* **2011**, *152*, 230–240. [CrossRef]
32. Turek, A.; Olakowska, E.; Borecka, A.; Janeczek, H.; Sobota, M.; Jaworska, J.; Kacmarczyk, B.; Jarzabek, B.; Gruchlik, A.; Libera, M.; et al. Shape-Memory Terpolymer Rods with 17-17β-estradiol for the Treatment of Neurodegenerative Diseases: An In Vitro and In Vivo Study. *Pharm. Res.* **2016**, *33*, 2967–2978. [CrossRef] [PubMed]
33. Zhang, Z.; Ren, H.; Shen, G.; Qiu, T.; Liang, D.; Yang, Z.; Yao, Z.; Tang, J.; Jiang, X.; Wei, Q. Animal models for glucocorticoid-induced postmenopausal osteoporosis: An updated review. *Biomed. Pharmacother.* **2016**, *84*, 438–446. [CrossRef] [PubMed]
34. Calciolari, E.; Mardas, N.; Dereka, X.; Kostomitsopoulos, N.; Petrie, A.; Donos, N. The effect of experimental osteoporosis on bone regeneration: Part 1, histology findings. *Clin. Oral. Impl.* **2017**, *28*, 101–110. [CrossRef] [PubMed]
35. Durão, S.F.; Gomes, P.S.; Colaço, B.J.; Silva, J.C.; Fonseca, H.M.; Duarte, J.R.; Felino, A.C.; Fernandes, M.H. The biomaterial-mediated healing of critical size bone defects in the ovariectomized rat. *Osteoporos Int.* **2014**, *25*, 1535–1545.
36. Engler-Pinto, A.; Siéssere, S.; Calefi, A.; Oliveira, L.; Ervolino, E.; Souza, S.; Furlaneto, F.; Messora, M.R. Effects of leukocyte- and platelet-rich fibrin associated or not with bovine bone graft on the healing of bone defects in rats with osteoporosis induced by ovariectomy. *Clin. Oral Impl. Res.* **2019**, *30*, 962–976. [CrossRef]
37. Cortet, B. Bone repair in osteoporosic bone: Postmenopausal and cortione-induced osteoporosis. *Osteoporos Int.* **2011**, *22*, 2007–2010. [CrossRef]
38. Alt, V.; Cheung, W.H.; Chow, S.K.; Thormann, U.; Cheung, E.N.; Lips, K.S.; Shenettler, R.; Leung, K.S. Bone formation and degradation behavior of nanocrystalline hydroxyapatite with or without collagen-type 1 in osteoporotic bone defects—An experimental study in osteoporotic goats. *Injury* **2016**, *47*, 58–65. [CrossRef]

pharmaceutics

MDPI

Article

Pharmacokinetic Profile and Anti-Adhesive Effect of Oxaliplatin-PLGA Microparticle-Loaded Hydrogels in Rats for Colorectal Cancer Treatment

Sharif Md Abuzar [1,2], Jun-Hyun Ahn [1,2], Kyung Su Park [3], Eun Jung Park [4,*], Seung Hyuk Baik [4,*] and Sung-Joo Hwang [1,2,*]

1. College of Pharmacy, Yonsei University, 85 Songdogwahak-ro, Yeonsu-gu, Incheon 21983, Korea
2. Yonsei Institute of Pharmaceutical Sciences, Yonsei University, 85 Songdogwahak-ro, Yeonsu-gu, Incheon 21983, Korea
3. Advanced Analysis Center, Korea Institute of Science and Technology, Hwarang-ro, Seongbuk-gu, Seoul 02792, Korea
4. Division of Colon and Rectal Surgery, Department of Surgery, Gangnam Severance Hospital, Yonsei University College of Medicine, Seoul 06273, Korea

* Correspondence: camp79@yuhs.ac (E.J.P.); whitenoja@yuhs.ac (S.H.B.); sjh11@yonsei.ac.kr (S.-J.H.); Tel.: +82-32-749-25-4518 (S.-J.H.); Fax: +82-32-749-4105 (S.-J.H.)

Received: 14 June 2019; Accepted: 1 August 2019; Published: 5 August 2019

Abstract: Colorectal cancer (CRC) is one of the most malignant and fatal cancers worldwide. Although cytoreductive surgery combined with chemotherapy is considered a promising therapy, peritoneal adhesion causes further complications after surgery. In this study, oxaliplatin-loaded Poly-(D,L-lactide-co-glycolide) (PLGA) microparticles were prepared using a double emulsion method and loaded into hyaluronic acid (HA)- and carboxymethyl cellulose sodium (CMCNa)-based cross-linked (HC) hydrogels. From characterization and evaluation study PLGA microparticles showed smaller particle size with higher entrapment efficiency, approximately 1100.4 ± 257.7 nm and 77.9 ± 2.8%, respectively. In addition, microparticle-loaded hydrogels showed more sustained drug release compared to the unloaded microparticles. Moreover, in an in vivo pharmacokinetic study after intraperitoneal administration in rats, a significant improvement in the bioavailability and the mean residence time of the microparticle-loaded hydrogels was observed. In HC21 hydrogels, $AUC_{0–48h}$, C_{max}, and T_{max} were 16012.12 ± 188.75 ng·h/mL, 528.75 ± 144.50 ng/mL, and 1.5 h, respectively. Furthermore, experimental observation revealed that the hydrogel samples effectively protected injured tissues from peritoneal adhesion. Therefore, the results of the current pharmacokinetic study together with our previous report of the in vivo anti-adhesion efficacy of HC hydrogels demonstrated that the PLGA microparticle-loaded hydrogels offer novel therapeutic strategy for CRC treatment.

Keywords: oxaliplatin; PLGA; hydrogel; intra-abdominal anti-adhesion barrier; colorectal cancer

1. Introduction

Colorectal cancer (CRC) is one of the most malignant cancers worldwide. Although the incidence of CRC is low, the number of new cases that are malignant and fatal is still the highest among both men and women. In addition, peritoneal carcinomatosis is one of the numerous manifestations of CRC identified at the first diagnosis in more than 10% of CRC patients, and it is extremely fatal, with median survival of approximately 6 months [1,2]. The treatment strategies for both CRC and peritoneal carcinomatosis patients depend on the type and stage, and optimal therapy comprises cytoreductive surgery combined with chemotherapy [3].

Peritoneal adhesion is one of the most common postoperative complications associated with cytoreductive surgery. Previous studies indicated their occurrence in more than 50% cases, and they have even higher recurrence rates (85–93%) [4,5]. As cytoreductive surgery and intraperitoneal chemotherapy are closely related to the completeness of cytoreduction, problems associated with cytoreductive surgery are still a big challenge in medical science. Therefore, hydrogel-based anti-adhesion barriers came into the spotlight to reduce adhesion by mechanical separation of injured tissue surfaces during peritoneal repair after surgery [6].

Oxaliplatin (Figure 1) is a third-generation, platinum-based, systemic chemotherapeutic agent for CRC [7], and it is expected to be similarly effective in peritoneal carcinomatosis. Therefore, it is currently being used as a part of the standard chemotherapy regimen, FOLFOX (oxaliplatin with 5-fluorouracil and leucovorin) [8], for the clinical treatment of metastatic CRC [9]. However, a recent study conducted in a murine model confirmed that intraperitoneally administered oxaliplatin enhanced peritoneal tissue concentration while reducing its systemic absorption, suggesting a possible decrease in toxicity associated with systemic chemotherapy [10]. Therefore, the new approach comprising cytoreductive surgery followed by intraperitoneal oxaliplatin delivery has resulted in significant improvement in the disease states of CRC and peritoneal carcinomatosis [11].

Biodegradable microparticles have been widely investigated for the controlled delivery of chemotherapeutic agents [12,13]. In addition, neither an initial burst release, nor the presence of a lag phase are desirable for the chemotherapeutic agents as it can be associated with adverse effects. Poly-(D,L-lactide-*co*-glycolide) (PLGA) is an unique copolymer approved by the regulatory agencies for the manufacturing of bioresorbable surgical sutures. Owing to its biodegradability and biocompatibility, PLGA become the first line materials for the production of injectable microparticle based controlled release system [14]. A recent report of metformin/irinotecan-loaded nanoparticles have shown an initial 2 h burst release, however, increase the drug retention time in tumor and increase drug circulation [15]. Moreover, alternative reports of paclitaxel-loaded PLGA microparticles have been evaluated for cancer therapy. These models revealed protection of the therapeutic payload from premature burst release, and enable the sustained release of paclitaxel [16,17].

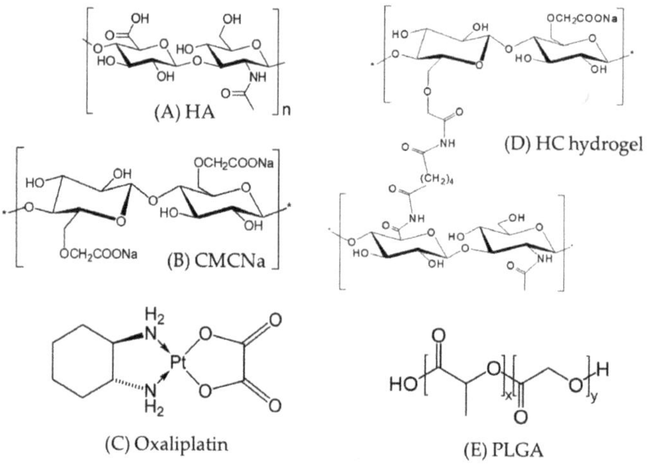

Figure 1. Chemical structures of Poly-(D,L-lactide-*co*-glycolide) (PLGA) microparticles and hyaluronic acid (HA) and carboxymethyl cellulose sodium (CMCNa)-based cross-linked (HC) hydrogel components (**A**) HA; (**B**) CMCNa; (**C**) HC hydrogel; (**D**) Oxaliplatin; and (**E**) PLGA.

Recently we succeeded in synthesizing novel hyaluronic acid (HA) and carboxymethyl cellulose sodium (CMCNa)-based cross-linked (HC) hydrogels loaded with oxaliplatin [18]. These novel HC hydrogels significantly prevented intraperitoneal adhesion and offered the highest anti-adhesion barrier in an in vivo rat model. However, pharmacokinetic evaluation of oxaliplatin-loaded HC hydrogels was not carried out. Several reports of microparticles-loaded hydrogels in various drug delivery system showed controlled drug release for over a prolonged period [19,20]. Therefore, to provide an adequate anti-adhesion barrier effect after cytoreductive surgery and deliver intraperitoneal chemotherapy, oxaliplatin-loaded PLGA microparticles were prepared, characterized, and loaded into HC hydrogels. In addition, an in vitro oxaliplatin release study of PLGA microparticle-loaded HC hydrogels in comparison with that of commercially available Guardix-Sol® hydrogel was carried out. Furthermore, in vivo pharmacokinetic analysis was carried out to demonstrate the efficacy of intraperitoneal chemotherapy with oxaliplatin-PLGA microparticles loaded into HC hydrogels.

2. Materials and Methods

2.1. Materials and Animals

Oxaliplatin was a gift from Boryung Pharm (Ansan, Korea). Resomer® RG 502 H (Poly-(D,L-lactide-*co*-glycolide)) PLGA, molecular weight (MW) approximately 7000–17,000, was purchased from Evonic Ind., (Darmstadt, Germany). Hyaluronic acid (HA, MW 1000 kDa) was a gift from Huons (Seongnam, Korea). Poly-(vinyl alcohol) (PVA, MW approximately 89,000–98,000), carboxymethyl cellulose sodium (CMCNa, MW approximately 700 kDa), and adipic acid dihydrazide (ADH) were obtained from Sigma Aldrich (St. Louis, MO, USA). 1-Ethyl-3-(3-(dimethylaminopropyl) carbodiimide (EDC) was purchased from Tokyo Chemical (Tokyo, Japan). All other chemicals were of reagent grade, and Milli-Q® water (Millipore1, Molsheim, France) was used throughout the study.

Male Sprague-Dawley (SD) rats were purchased from YoungBio (Seongnam, Republic of Korea). The animals were housed in a semi-specific pathogen free facility using standard cages at 19 ± 1 °C and 50 ± 5% relative humidity, with a 12-h light-dark cycle. The rats were fed a standard diet, provided with purified water, and allowed to move freely. All experiments were approved by the Institutional Animal Care and Use Committee (IACUC-201809-788-01) at Yonsei University, Seoul, Korea, and were performed according to IACUC guidelines.

2.2. Preparation of Oxaliplatin-PLGA Microparticles

Oxaliplatin-PLGA microparticles were prepared using a double emulsion method [21]. Briefly, oxaliplatin (20 mg) was dissolved in 5 mL of deionized water (first aqueous solution, W_1) containing 0.5% (*w*/*v*) PVA. The resulting oxaliplatin containing W_1 was added dropwise to 12.5 mL DCM, containing 1 g of PLGA 502 H, using a homogenizer ULTRA-TURRAX® (IKA-WERKE GMBH & Co. KG, Staufen, Germany) at 12,000 rpm for 2 min. This primary emulsion (W_1/O) was slowly added to 100 mL of 1% (*w*/*v*) PVA solution (second aqueous solution, W_2) and emulsified at 400 rpm using magnetic stirrer for 3 h while the DCM was allowed to evaporate completely under vacuum. The resultant microparticles were collected, washed using distilled water, centrifuged, and freeze-dried using an ilShinBioBase (Seoul, Korea) freeze-drier under vacuum (5 mTorr). Samples were pre-frozen for 2 h at −60 ± 1.0 °C prior to final drying at −88 ± 1.0 °C for 2 days. The collected freeze-dried microparticles were stored at 4 °C and further evaluated by scanning electronic microscopy (SEM) to characterize the morphological features.

2.3. Synthesis of HA-CMCNa Cross-Linked Hydrogels

Hydrogels were synthesized using variable ratios of HA and CMCNa (Figure 1) by chemical cross-linking, as described by our previous study [18]. Briefly, HA and CMCNa at ratios of 1:2 and 2:1 (*w*/*w*) were dissolved in deionized water with continuous stirring. After complete dissolution, 2 mM/L of ADH was added to the mixture, and the pH was adjusted to 4.75 ± 0.05 using 0.1 M HCl. After vortex mixing for 1 min, 2 mM/L EDC was added, and the resulting mixture was stirred thoroughly at room temperature (25 ± 1.0 °C) for at least 12 h to allow for complete cross-linking. The reaction was maintained at pH 4.75 ± 0.05 by the addition of 0.1 M HCl. Finally, the cross-linking reaction was terminated by elevating the pH of the mixture to 7.0 ± 0.05 via the slow addition of 0.1 M NaOH. The hydrogel samples were purified as previously described by Luo et al. [22]. Briefly, the hydrogel samples were dispensed into a dialysis bag (Mw cut-off 18,000 Da) previously treated with a 70% ethanol (EtOH) solution and dried. The bag was submerged and dialyzed for 12 h in 0.1 M NaCl followed by alternating solutions of 25% EtOH for 6 h and DW for 6 h. The remaining hydrogels in the dialysis bag were rinsed using EtOH, and then centrifuged at 2000 rpm for 10 min to remove the remaining ADH. Finally, the precipitated hydrogels (HC12 and HC21) were collected and stored until further experiments.

2.4. Preparation of Oxaliplatin-PLGA Microparticle-Loaded Hydrogel

Guardix-Sol® is a commercial anti-adhesive hydrogel used for the prevention of post-operative adhesion. Above, we synthesized HC12 and HC21 hydrogel composites with variable weight ratios of HA and CMCNa. Next, oxaliplatin-PLGA microparticles were dispersed gently in the hydrogel (HC12, HC21, and Guardix-Sol®). Briefly, precisely weighed oxaliplatin-PLGA microparticles were mixed with 10 mL of hydrogels at room temperature (25 ± 1.0 °C) with continuous stirring for 30 min. The final concentration of oxaliplatin was 2 mg/ mL of hydrogel. After complete dissolution was visually observed, samples were evaluated for in vitro and in vivo release of oxaliplatin.

2.5. Characterization of Oxaliplatin-PLGA Microparticles

2.5.1. Morphology

The morphology of oxaliplatin-PLGA microparticles was evaluated using SEM (JSM-6700F, JEOL, Tokyo, Japan). Briefly, a small amount of powder was sprinkled onto double-sided adhesive tape attached to an aluminum stub and was sputter-coated with gold under vacuum. Photographs were taken at 5× magnification with an accelerating voltage of 1–5 kV to reveal the surface characteristics of the particles.

2.5.2. Particle Size Analysis

The mean particle size and distribution of oxaliplatin-PLGA microparticles were analyzed by dynamic light scattering (DLS) using an electrophoretic light scattering spectrophotometer (ELS-Z, Otsuka Electronics, Hirakata, Japan).

2.5.3. Rheological Measurements

The rheological measurements for HC12, HC21, and Guardix-Sol® hydrogels were performed using a Brookfield rheometer (Brookfield Digital Rheometer Model DV-III, DV3T™ Rheometer, Middleboro, MA, USA) equipped with a Peltier system for temperature control. Precisely, about 0.5 g of sample was applied to the plate and allowed to equilibrate. Measurements were performed at 37 ± 0.5 °C with shear rates ranging from 200–500 s^{-1}. Before each measurement, the samples were allowed to rest for 5 min at 37 ± 0.5 °C. Results were analyzed with Brookfield software (Firmware version 1.2.2-9).

2.5.4. Encapsulation Efficiency

Freeze-dried oxaliplatin-PLGA microparticles (eq. 2 mg of oxaliplatin) were dissolved in 5 mL of DCM. After 5 min of vortexing, particles were allowed to dissolve properly, and the tube was gently swung in incubator at 37 ± 0.5 °C for 1 h. Then, 5 mL of DW was added to the tube and vortexed vigorously for 1 h. The suspension was centrifuged at 12,000 rpm for 5 min to precipitate PLGA. The upper aqueous phase was collected, and the concentration of oxaliplatin was determined by HPLC. Encapsulation efficiency (EE %) was calculated using the following Equation (1):

$$\text{EE }\% = \text{M}_{\text{actual oxaliplatin}}/\text{M}_{\text{theoretical oxaliplatin}} \times 100 \quad (1)$$

2.6. *In Vitro Oxaliplatin Release from PLGA Microparticles Loaded into Hydrogels*

To evaluate the in vitro oxaliplatin release rate from oxaliplatin-PLGA microparticles, an amount equivalent to 2 mg of oxaliplatin was weighed and suspended in 1 mL of HPLC grade water. The suspension was transferred into dialysis bags (MWCO 14,000) and then submerged in 20 mL of double distilled water in capped 50 mL Falcon® tube. Samples were shaken at 30 rpm at the predetermined time points (1, 2, 3, 4, 6, 8, and 12 h); 1 mL of sample was collected from the medium and 1 mL of pre-warmed medium was immediately added to the tubes. Experiments were performed in triplicate (n = 3), and the collected samples were analyzed with HPLC after necessary dilution. The release profile was expressed as the ratio of cumulative oxaliplatin release to initial oxaliplatin loading versus time, using the following Equation (2):

$$\text{Percent of oxaliplatin release} = M_t/M_0 \times 100 \quad (2)$$

For the in vitro oxaliplatin release from the PLGA microparticle-loaded hydrogels (from Section 2.4.), a similar method was used. Briefly, 1 mL PLGA microparticle-loaded hydrogel (HC12, HC21, or Guardix-Sol®) was transferred to one end-closed dialysis bags (MWCO 14,000). The other end of the bag was closed properly, and the samples were submerged in 20 mL of double distilled water in capped 50 mL Falcon® tube. Samples were shaken at 30 rpm at similar time points (1, 2, 3, 4, 6, 8, and 12 h). Subsequently, 1 mL of sample was collected from the medium and 1 mL of pre-warmed medium was immediately added to the tubes. Experiments were performed in triplicate (n = 3), and the collected samples were analyzed with HPLC after necessary dilution with the mobile phase. The release profile was expressed as cumulative oxaliplatin release calculated using Equation (2).

2.7. *In Vivo Oxaliplatin Release in SD Rat's Intraperitoneal Cavity*

In vivo oxaliplatin release from the prepared hydrogels was evaluated in the intraperitoneal cavity of SD rats, and oxaliplatin solution was also evaluated for comparison. Briefly, a total of fifteen male SD rats, aged 4–6 weeks, were randomly divided into three groups (n = 5, per group) and were anesthetized with isoflurane. Intra-abdominal adhesions were induced to mimic the postoperative surgical conditions, as described previously [23]. The peritoneum was exposed by a 5-cm ventral midline incision. The left abdominal sidewall was scraped with a 1 cm^2 piece of 100-grit sandpaper 200 times. In group 1, oxaliplatin solution was introduced, whereas, group 2 and 3 were exposed to oxaliplatin powder and oxaliplatin-PLGA microparticles, loaded into HC21 hydrogel, respectively. All the rats from each group were administered oxaliplatin in the form of a solution or loaded into hydrogel at a dose level of 5 mg/kg body weight [24]. The abdominal layers and skin incision were then completely closed, and each rat was kept separately in single cases. All surgeries were performed by the same individual. At predetermined time interval (0.5, 1, 1.5, 2, 3, 5, 8, 14, 24, and 48 h), approximately 1-mL blood samples were collected from the conjunctiva using a capillary tube and kept in an ice bath.

Blood samples collected from the rat's conjunctiva were centrifuged at 10,000 rpm (9425× *g*) for 10 min at 4 °C. The supernatant plasma was obtained and stored at −80 °C until analyzed. The frozen plasma samples were thawed, and approximately 200 ± 5 mg was weighed. The samples were oxidized into metals and organic materials by treating with 6 mL nitric acid using a microwave sample pre-treatment machine equipped with platinum (Pt) sensor (Microwave Reaction System, Multi-wave PRO, Anton Paar, Graz, Austria). Samples were heated at predefined temperature and pressure (200 ± 0.5 °C and 4.0 MPa) for 1 h and diluted up to 30 mL with deionized water, and the concentrations of Pt were analyzed using ICP-MS (NexION 300 D, PerkinElmer, Waltham, MA, USA). Pharmacokinetic parameters $AUC_{0\text{–}48h}$ (area under curve), C_{max} (peak concentration), and T_{max} (time to peak concentration) were calculated using noncompartmental analysis.

After the last blood sample collection at 48 h, rats were housed individually in standard case with a 12-h light-dark cycle. The rats were fed a standard diet, provided with purified water, and were allowed to move freely. After 10 days, they were sacrificed, and the peritoneum was opened. For all rats, adhesion type and extent were assessed, and photographs were taken.

2.8. *High-Performance Liquid Chromatography (HPLC) Analysis*

In vitro oxaliplatin release and % EE were determined by using HPLC Agilent 1200 Infinity Series HPLC system (Agilent Technologies, Waldbronn, Germany). Briefly, XTerra™ RPC_{18} column (particle size 5 μm, inside diameter 4.6 mm, and length 250 mm; Waters Corporation, Milford, MA, USA) at 210 nm using an HPLC-UV spectrometer (Agilent 1290 infinity). The mobile phase was a 20:80 mixture of ACN and deionized water and was set to a flow rate of 0.8 mL/min (Model 1260 Quat Pump VL). The samples were diluted as necessary, and 20 μL of each sample was injected using an autosampler (Model 1260 ALS).2.9. Statistical Analysis

In vitro percent (%) cumulative oxaliplatin release and in vivo pharmacokinetic study data are expressed as the means ± standard deviations. The *t*-test or two-sided RM ANOVA and Bonferroni test were applied to the analyses of the differences between the groups. A p value < 0.05 was considered as a statistically significant difference.

3. Results and Discussion

3.1. *Characterization of PLGA Microparticles*

PLGA-based microparticles were prepared using the double emulsion method. The microparticles were characterized for size, morphology, and encapsulation efficiency. As shown in Table 1, oxaliplatin-loaded PLGA microparticles had a particle size of more than 1 μm diameter, with a small standard deviation. Encapsulation efficiency was 77.9 ± 2.8%, which is excellent for a hydrophilic drug like oxaliplatin. Although the encapsulation of hydrophilic drugs in PLGA microparticles is challenging due to the partitioning of weakly associated drugs from the oil phases to the external water phase, the double emulsion technique is the most commonly used method for encapsulating hydrophilic drugs. In addition, the SEM image shown in Figure 2A reveals the morphology of the microparticles. Uniform size and spherical particles were observed, which is in agreement with the particle size results. To fulfill the aims of this study, PLGA microparticles were further loaded into HC hydrogels and the in vitro release and in vivo pharmacokinetic studies were conducted.

Table 1. Compositions and characterizations of Poly-(D,L-lactide-*co*-glycolide) (PLGA) microparticles and hyaluronic acid (HA) and carboxymethyl cellulose sodium (CMCNa)-based cross-linked (HC) hydrogels.

Particles	Method	Weight Ratio				Particle Size (nm) (Mean ± SD)	Encapsulation Efficiency (%) (Mean ± SD)
		Oxaliplatin	PLGA 502 H	HA	CMCNa		
Oxaliplatin-PLGA Microparticles	Double Emulsion	1	50	-	-	1100.4 ± 257.7	77.9 ± 2.8
HC12 Hydrogel	Synthesis by cross-linking reaction	-	-	1	2	-	-
HC21 Hydrogel		-	-	2	1	-	-

3.2. *Preparation and Characterization of HC Hydrogels*

Cross-linked HC hydrogels were prepared using variable weight ratios of HA and CMCNa. HC hydrogel synthesis was carried out based on our previous description [18]. In the presence of ADH (as a nucleophile), EDC (a water-soluble carbodiimide) linked both HA and CMCNa molecules with its amine group. EDC is not incorporated into the final product, but converts into a non-toxic water-soluble urea derivative, which is removed by dialysis. Moreover, CH_2COO-anions provided by CMCNa react with H^+ ions, leading to carboxyl group formation and facilitation of cross-linking. Therefore, the reaction in the presence of EDC is pH-dependent and was performed at pH 4.75 ± 0.05.

To evaluate the rheology of the HC hydrogel along with the commercial product Guardix-Sol®, a Brookfield Digital Rheometer was used. All measurements were performed at 37 ± 0.5 °C, with shear rates ranging from 200 to 500 s^{-1}. All hydrogels followed non-Newtonian shear thinning (pseudo) plastic flow behavior (Figure 2B). Decreasing viscosity with increasing shear rates was observed, which demonstrated viscosity-dependent shear rates [25].

In addition, variable ratios of the composition (HA and CMCNa) alter the viscosity of the HC hydrogels. The order could be written as follows: HC21 > HC12, as markedly increased viscosity was observed with HA. The highest viscosity was observed with the HC21 formulation, whereas the lowest viscosity was observed with HC12. The viscosity of Guardix-Sol® was measured for comparison with those of the synthesized HC hydrogels, and it showed moderate viscosity. The rheology results correlated with the mechanical characteristics of the polymer, in which viscosity scaled linearly with molecular weight. The mechanical properties (entanglement phenomenon) of the polymers increased with increasing molecular weight. Finally, the hydrogels were loaded with oxaliplatin-PLGA microparticles and their in vitro and in vivo characteristics were evaluated.

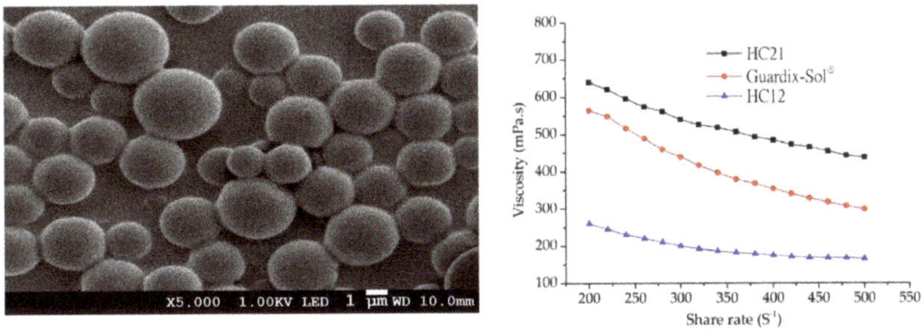

(A) (B)

Figure 2. (**A**) SEM image of oxaliplatin-PLGA microparticles (at X5.000 magnification); (**B**) rheology properties of hydrogel samples (■) HC21, (●) Guardix-Sol®, and (▲) HC12 (measurements were performed at 37 ± 0.5 °C).

3.3. *In Vitro Oxaliplatin Release Study*

An in vitro oxaliplatin release study was conducted on oxaliplatin-PLGA microparticles generated using the double emulsion method. In addition, oxaliplatin-PLGA microparticles and oxaliplatin powder were dispersed into hydrogels (HC12, HC21, and Guardix-Sol®), and an in vitro oxaliplatin release study was conducted. Plots of cumulative oxaliplatin release (%) versus time (h) are shown in Figure 3A–C. In vitro release profile for oxaliplatin-PLGA microparticles (red closed circle; (●) showed immediate release. With an initial burst release, about 63.2% oxaliplatin was released at 4 h. Oxaliplatin release from PLGA microparticles is a combination of diffusion and bioerosion mechanism [26]. During the diffusion stage, oxaliplatin release occurs by diffusion through aqueous channels. The release could also be triggered by the external pores on the surface or through leaching of the oxaliplatin at or near the surface, leading to an initial burst release. Besides, the standard deviation values of the results of the in vitro release study for the hydrogels were low, which indicate excellent reproducibility of the release behavior. However, when oxaliplatin-PLGA microparticles were loaded into three different hydrogel systems, HC12, HC21, and Guardix-Sol®, slight initial burst release was observed. Although the rate of burst release was almost the same for the three systems, HC12, with the lowest viscosity, showed the maximum release. In contrast, the higher the viscosity of the hydrogel system, the lower the burst and sustained release observed. This may be due to low leaching of the drugs and high protection of the microparticles by the hydrogels.

Table 2 demonstrated the kinetic models adopted for evaluation. Where M_t was the cumulative drug released at time t and M was the initial drug present in the PLGA microparticles-loaded hydrogels. k_1, k_H, and k_{KP} are the first order, Higuchi, and Korsmeyer-Peppas release constant, respectively.

M_t/M_∞ was the fraction of drug released at time t, and n was the diffusional release exponent symbolic of the release mechanism. The rate of oxaliplatin release from the HC12, Guardix-Sol® and HC21-loaded PLGA microparticles were slower and constant compared with the oxaliplatin powder-loaded hydrogels, as indicated by the K_H values of 18.9686, 17.5626, and 20.6559% $h^{1/2}$, respectively. Release constants were determined using the slope of the appropriate plots, and the regression coefficient (R^2) was obtained through linear regression analysis (Table 2). The coefficient of determination (R^2) was used as an indicator of curve fit for each of the considered models [27]. The regression coefficients R^2, from the Korsmeyer-Peppas plots were 0.9342, 0.9554, and 0.9366 for HC12, Guardix-Sol® and HC21, respectively, thus the log of cumulative drug release was proportional to the log of time. The best linear fits were observed for the PLGA microparticles-loaded hydrogels using both the Higuchi and Korsmeyer-Peppas models, suggesting that oxaliplatin release from the PLGA microparticles-loaded hydrogels were diffusive process. Oxaliplatin gradually dissolved into the fluid within the swelled hydrogels, then slowly diffused from polymeric networks of the hydrogels.

Table 2. Release rate constants and regression coefficient R^2 obtained from drug release profile based on kinetic equations.

Equations	Loaded in HC12 Hydrogel		Loaded in Guardix-Sol® Hydrogel		Loaded in HC21 Hydrogel		PLGA Microparticle
	Oxaliplatin Powder	PLGA Microparticle	Oxaliplatin Powder	PLGA Microparticle	Oxaliplatin Powder	PLGA Microparticle	
[a] Higuchi model: k_H (% $h^{1/2}$)	28.5970	18.9686	22.1635	17.5626	24.3401	20.6559	20.0882
R^2	0.6738	0.9611	0.7431	0.9783	0.7746	0.9485	0.8026
[b] First—Order model: k_1 (h^{-1})	0.0302	0.0421	0.0263	0.0469	0.0379	0.0522	0.0219
R^2	0.1639	0.6646	0.2327	0.7224	0.2640	0.6451	0.4145
[c] Korsmeyer—Peppas model: k_{KP}	0.4517	0.5033	0.3750	0.5475	0.5352	0.6308	0.2884
R^2	0.5495	0.9342	0.6099	0.9554	0.6512	0.9366	0.7767

[a] Higuchi model: ($Mt = k_H \times t^{1/2}$); [b] First—Order model: ($\ln M_t = \ln M + k_1 \times t$); [c] Korsmeyer—Peppas model: ($Mt/M\infty = k_{KP} \times t^n$).

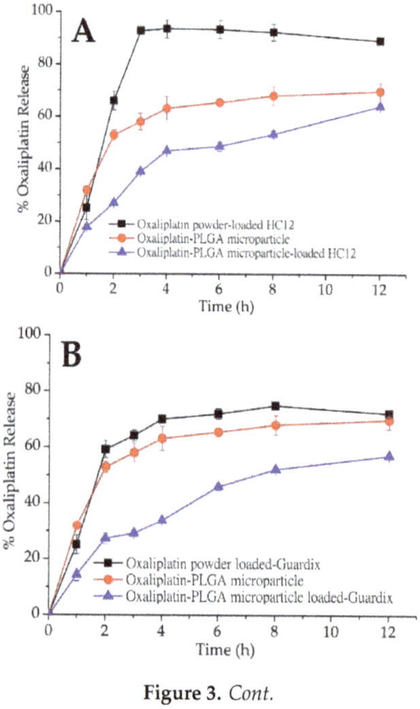

Figure 3. *Cont.*

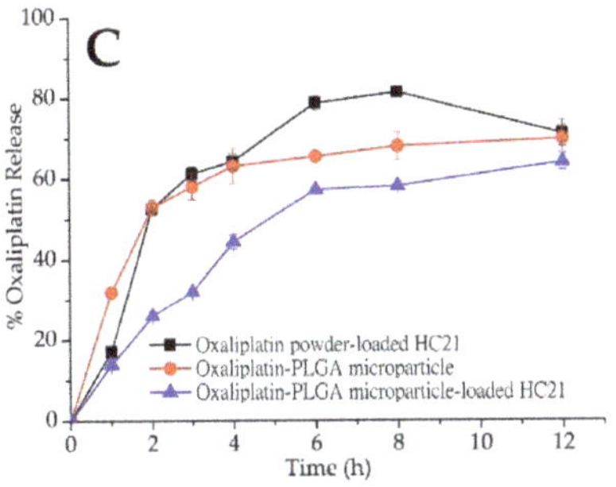

Figure 3. In vitro oxaliplatin release from PLGA microparticles (red, ●). Oxaliplatin release profile from oxaliplatin powder and oxaliplatin-PLGA microparticles-loaded in (**A**) HC12; (**B**) Guardix-Sol®; and (**C**) HC21 hydrogel. Mean ± S.D. (n = 3).

Although hydrogels are swellable materials with high water content, which readily allow the release or leaching of hydrophilic drugs through the channels, high viscosity might hinder this process. Therefore, release may be completed through bioerosion. In this study, three different hydrogel systems—HC12, HC21, and Guardix-Sol®, were loaded with oxaliplatin powder and oxaliplatin-PLGA microparticles. The HC12 hydrogel system (low viscosity) was loaded with oxaliplatin powder and more than 90% oxaliplatin release occurred within 3 h, followed by saturated release behavior up to 12 h. However, oxaliplatin-PLGA microparticles loaded into the HC12 hydrogel showed a slower release pattern than that of the oxaliplatin powder. Approximately 50% oxaliplatin was released after 4 h, followed by 64.2% release up to the end of the experiment ($p < 0.05$). Due to the less hydrophilicity of the PLGA copolymers, they absorb less water, and subsequently degrade more slowly. Therefore, drug release from the PLGA microparticles was delayed. In the case of Guardix-Sol®, with moderate viscosity, a sustained release profile was observed for the oxaliplatin-PLGA microparticles. During the first 2 h of the release study, less than 30% of oxaliplatin was released; however, 59.25% of oxaliplatin was released after the same amount of time when loaded with oxaliplatin powder ($p < 0.05$).

In addition, the HC21 hydrogel system (high viscosity) showed a similar sustained release profile when loaded with oxaliplatin-PLGA microparticles. In all three systems, more rapid oxaliplatin release was observed when oxaliplatin was loaded as a powder compared to that when loaded as microparticles. The release rate was faster than those of both oxaliplatin-PLGA microparticles and oxaliplatin-PLGA microparticles-loaded hydrogels. Additionally, both Guardix-Sol® and HC21 showed similar sustained release patterns up to 12 h. Although PLGA microparticles were generated using the well-known double emulsion method, further studies are necessary to establish process parameters and extensive characterization of the particles.

3.4. *Pharmacokinetics in Rats*

The bioavailability of oxaliplatin-PLGA microparticles loaded into the HC21 hydrogel was evaluated in rats. Figure 4 shows the mean concentration–time profiles of oxaliplatin in rats after a single dose (5 mg/kg) of oxaliplatin solution, oxaliplatin-PLGA microparticles, and oxaliplatin powder loaded into the HC21 hydrogel. The samples were introduced through a midline incision to open the peritoneum immediately after introducing intra-abdominal adhesions by multiple scraping to mimic postoperative surgical conditions. Intraperitoneal absorption of oxaliplatin solution was obviously higher than that of the hydrogel loaded samples (oxaliplatin-PLGA microparticles and oxaliplatin powder). For oxaliplatin solution $AUC_{0\text{–}48h}$, C_{max}, and T_{max} were 8181.51 ± 176.89 ng·h/mL, 2265.28 ± 192.51 ng/mL, and 1.0 h, respectively. In the case of oxaliplatin powder loaded into the HC21 hydrogel, $AUC_{0\text{–}48h}$, C_{max}, and T_{max} were 16,571.37 ± 139.13 ng·h/mL, 690.63 ± 140.54 ng/mL, and 1.5 h, respectively. Intraperitoneal bioavailability of oxaliplatin was significantly increased by up to 2-fold in the HC21 hydrogel (Table 3). In particular, 1.9-fold higher bioavailability was observed in the case of oxaliplatin-PLGA microparticles loaded into the HC21 hydrogel compared to that of the oxaliplatin solution. AUC_{0-48h}, C_{max}, and T_{max} were 16,012.12 ± 188.75 ng·h/mL, 528.75 ± 144.50 ng/mL, and 1.5 h, respectively. Moreover, mean residence time (MRT) was increased by 2.7-fold for oxaliplatin-PLGA microparticles when loaded into the HC21 hydrogels ($p < 0.05$). When the oxaliplatin-PLGA microparticles were loaded into the HC21 hydrogel, oxaliplatin release rate decreased due to the multi-layered encapsulation, which was observed from the in vitro release study. Therefore, in the high, moderate, and low viscous hydrogels (HC21, Guardix-Sol®, and HC12, respectively), oxaliplatin release from the PLGA microparticles, followed by release from the hydrogels, was observed in a sustained manner (Figure 3). Furthermore, using sustained-release hydrogels allows a longer residence time of the component at the application site (intraperitoneal after cytoreductive surgery), which offers additional benefits through the prevention of intra-abdominal adhesion and improvement in the delivery of intraperitoneal oxaliplatin from the PLGA microparticles.

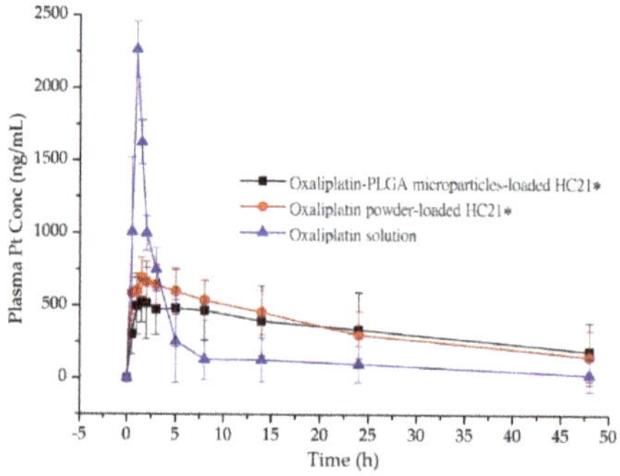

Figure 4. Plasma concentration-time profile of Pt in rats after intra-peritoneal administration of oxaliplatin powder and oxaliplatin-PLGA microparticles-loaded in HC21 hydrogel, and oxaliplatin solution. Data are expressed as the Mean ± standard deviation (n = 5). *, $p < 0.05$ represents a significant difference, two-sided RM ANOVA and Bonferroni test.

Table 3. In vivo pharmacokinetic parameters after intra-peritoneal administration to rats (n = 5).

Formulations	Pharmacokinetic Parameters			
	C_{max} (ng/mL)	T_{max} (h)	AUC_{0-48h} (ng·h/mL)	MRT (h)
Oxaliplatin solution	2265.28 ± 192.51	1	8181.51 ± 176.89	4.36 ± 0.65
Oxaliplatin powder loaded HC21	690.63 ± 140.54	1.5	16571.37 ± 139.13	10.29 ± 0.33
Oxaliplatin-PLGA microparticle loaded HC21	528.75 ± 144.50	1.5	16012.12 ± 188.75	12.02 ± 0.41

3.5. *Intraperitoneal Anti-Adhesion Effect*

An extensive investigation of the intraperitoneal anti-adhesion efficacy of HC hydrogels was conducted in our previous study, after introducing peritoneal injury to generate peritoneal adhesion [18]. HC hydrogel groups (treated) showed the highest anti-adhesion barrier compared to the control group (non-treated). In the present study, the pharmacokinetic parameters of the PLGA microparticles loaded into the HC21 hydrogel were evaluated and compared with those of oxaliplatin powder loaded into hydrogels. Artificial injury on the left abdominal sidewall was introduced to mimic surgical conditions and observe the anti-adhesion barrier effect of hydrogel samples followed by the intraperitoneal delivery of oxaliplatin. The method was adopted from experienced surgeons certainly feasible in human administration after cytoreductive surgery in colorectal cancer treatment. Intraperitoneal anti-adhesion effect was observed and recorded photographically. Figure 5 shows a photographical representation of adhesion characteristics of the study groups.

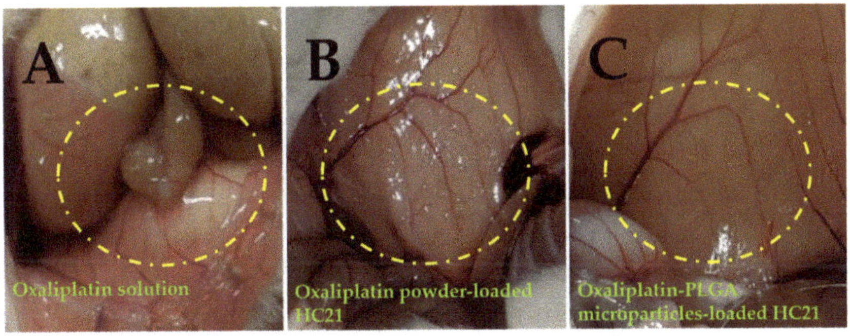

Figure 5. Photographical expression of in vivo anti-adhesion efficacy observed in rats after introducing intra-peritoneal injury. (**A**) oxaliplatin solution; (**B**) oxaliplatin powder-loaded HC21 hydrogel; and (**C**) oxaliplatin-PLGA microparticles-loaded HC21 hydrogel.

HC21 hydrogel samples exhibited an efficient anti-adhesion barrier effect towards the injury. Although hydrogels were not intended to cure the injury, long resident time hydrogels significantly prevented adhesion along with rapid recovery of the injury, possibly due to mechanical separation from the injury site [28]. As reported previously, the washing away or displacement of the oxaliplatin solution from the application site allows less barrier or protection to the injury site [29]. Therefore, dense adhesion on the abdominal wall was observed, and, although current study was not intended to be carried out extensively, adhesion scoring and extent (as disclosed in a previous report) and photographic representation demonstrated the superiority of the hydrogels in terms of the anti-adhesion barrier. In the hydrogel-loaded samples, no or minor adhesion was observed in the treated rat abdomen (Figure 5B,C). Although two different samples were evaluated after loaded in HC21 hydrogel both oxaliplatin powder and oxaliplatin-PLGA microparticles loaded hydrogels showed significant improvement as anti-adhesion barrier. Subsequently, both hydrogel systems showed rapid wound recovery effect may be due to mechanical separation- could be seen from the blood spot in Figure 5C. Instead, oxaliplatin solution presents no barrier to the injure site and dense abdominal adhesion was observed. The HC21 hydrogel-loaded sample showed enhanced anti-adhesion protection to the injury site, suggesting its potential therapeutic application in CRC patients who have undergone cytoreductive surgery. Although the current study evaluated the pharmacokinetic profile of PLGA microparticle-loaded HC hydrogels, a further pharmacodynamics study on a CRC model is required to evaluate the composition of the hydrogels.

4. Conclusions

Oxaliplatin-PLGA microparticles were loaded into hydrogels to improve intraperitoneal chemotherapy with an anti-adhesion barrier effect. In an in vitro release study, we observed sustained oxaliplatin release from PLGA microparticles loaded into hydrogels in comparison with the unloaded sample. Moreover, in the in vivo pharmacokinetics study on rats, PLGA microparticle-loaded hydrogels showed higher bioavailability compared to oxaliplatin solution. Furthermore, visual observations revealed that the HC hydrogels were effective as an intra-abdominal anti-adhesion barrier. Therefore, intraperitoneal delivery of PLGA microparticle-loaded hydrogels, which have an effective intra-abdominal anti-adhesion barrier, is expected to provide an optimized alternative therapy for CRC. Earlier reports of cytotoxicity study for PLGA (RG502H) microparticles exhibited non-cytotoxic at increasing concentration [30]. Additionally, both HA and CMCNa are considered as inert and non-cytotoxic. However, cytotoxicity profile for the novel oxaliplatin-PLGA microparticles-loaded HC hydrogels were yet to revealed. As the investigation undertaken in current study was preliminary, pharmacokinetic and anti-adhesion efficacy of this multi-complex drug delivery system, further studies are required to address cytotoxicity, cell viability, toxicity in liver function, and efficacy in CRC model in vivo.

Author Contributions: Conceptualization, S.M.A., J.-H.A.; Analytical Support, S.M.A., K.S.P.; Methodology, J.-H.A., E.J.P.; Writing—Original Draft Preparation, S.M.A.; Writing—Review and Editing, J.-H.A., E.J.P., S.-J.H., S.H.B.; Supervision, S.H.B., S.-J.H.; Project Administrator, S.-J.H.

Funding: This study was supported by a grant from Basic Science Research Program through the National Research Foundation of Korea (NRF) funded by the Ministry of Science and ICT & Future Planning, Republic of Korea (NRF-2017R1A2B2011520 and NRF-2019R1F1A1056350), and University-Centered Labs-2018R1A6A1A03023718).

Acknowledgments: The authors would like to acknowledge Yonsei Center for Research Facilities (YCRF, Seoul, Korea), Yonsei University for writing assistance.

Conflicts of Interest: The authors declare no conflict of interest.

References

1. Verwaal, V.J.; van Ruth, S.; Witkamp, A.; Boot, H.; van Slooten, G.; Zoetmulder, F.A.N. Long-term survival of peritoneal carcinomatosis of colorectal origin. *Ann. Surg. Oncol.* **2005**, *12*, 65–71. [CrossRef] [PubMed]
2. Koppe, M.J.; Boerman, O.C.; Oyen, W.J.G.; Bleichrodt, R.P. Peritoneal carcinomatosis of colorectal origin—Incidence and current treatment strategies. *Ann. Surg.* **2006**, *243*, 212–222. [CrossRef] [PubMed]
3. Verwaal, V.J.; van Ruth, S.; de Bree, E.; van Slooten, G.W.; van Tinteren, H.; Boot, H.; Zoetmulder, F.A.N. Randomized trial of cytoreduction and hyperthermic intraperitoneal chemotherapy versus systemic chemotherapy and palliative surgery in patients with peritoneal carcinomatosis of colorectal cancer. *J. Clin. Oncol.* **2003**, *21*, 3737–3743. [CrossRef] [PubMed]
4. Diamond, M.P.; Freeman, M.L. Clinical implications of postsurgical adhesions. *Hum. Reprod. Update* **2001**, *7*, 567–576. [CrossRef] [PubMed]
5. Dizerega, G.S. Contemporary Adhesion Prevention. *Fertil. Steril.* **1994**, *61*, 219–235. [CrossRef]
6. Oh, A. Trends of anti-adhesion adjuvant—Review. *Biomater. Res.* **2013**, *17*, 138–145.
7. Lim, W.Q.; Phua, S.Z.F.; Chen, H.Z.; Zhao, Y.L. An oxaliplatin(iv) prodrug-based supramolecular self-delivery nanocarrier for targeted colorectal cancer treatment. *Chem. Commun.* **2018**, *54*, 12762–12765. [CrossRef]
8. Serrano, D.R.; Hernandez, L.; Fleire, L.; Gonzalez-Alvarez, I.; Montoya, A.; Ballesteros, M.P.; Dea-Ayuela, M.A.; Miro, G.; Bolas-Fernandez, F.; Torrado, J.J. Hemolytic and pharmacokinetic studies of liposomal and particulate amphotericin B formulations. *Int. J. Pharm.* **2013**, *447*, 38–46. [CrossRef]
9. Song, N.; Pogue-Geile, K.L.; Gavin, P.G.; Yothers, G.; Kim, S.R.; Johnson, N.L.; Lipchik, C.; Allegra, C.J.; Petrelli, N.J.; O'Connell, M.J.; et al. Clinical Outcome From Oxaliplatin Treatment in Stage II/III Colon Cancer According to Intrinsic Subtypes Secondary Analysis of NSABP C-O7/NRG Oncology Randomized Clinical Trial. *JAMA Oncol.* **2016**, *2*, 1162–1169. [CrossRef]
10. Piche, N.; Leblond, F.A.; Sideris, L.; Pichette, V.; Drolet, P.; Fortier, L.P.; Mitchell, A.; Dube, P. Rationale for Heating Oxaliplatin for the Intraperitoneal Treatment of Peritoneal Carcinomatosis A Study of the Effect of Heat on Intraperitoneal Oxaliplatin Using a Murine Model. *Ann. Surg.* **2011**, *254*, 138–144. [CrossRef]
11. Elias, D.; Raynard, B.; Farkhondeh, F.; Goere, D.; Rouquie, D.; Ciuchendea, R.; Pocard, M.; Ducreux, M. Peritoneal carcinomatosis of colorectal origin—Long-term results of intraperitoneal chemohyperthermia with oxaliplatin following complete cytoreductive surgery. *Gastroen. Clin. Biol.* **2006**, *30*, 1200–1204. [CrossRef]
12. Makkouk, A.; Joshi, V.B.; Wongrakpanich, A.; Lemke, C.D.; Gross, B.P.; Salem, A.K.; Weiner, G.J. Biodegradable Microparticles Loaded with Doxorubicin and CpG ODN for In Situ Immunization Against Cancer. *AAPS J.* **2015**, *17*, 184–193. [CrossRef]
13. Ramasamy, T.; Ruttala, H.B.; Gupta, B.; Poudel, B.K.; Choi, H.G.; Yong, C.S.; Kim, J.O. Smart chemistry-based nanosized drug delivery systems for systemic applications: A comprehensive review. *J. Control. Release* **2017**, *258*, 226–253. [CrossRef]
14. Blasi, P. Poly(lactic acid)/poly(lactic-co-glycolic acid)-based microparticles: An overview. *J. Pharm. Investig.* **2019**, *49*, 337–346. [CrossRef]
15. Taghizadehghalehjoughi, A.; Hacimuftuoglu, A.; Cetin, M.; Ugur, A.B.; Galateanu, B.; Mezhuev, Y.; Okkay, U.; Taspinar, N.; Taspinar, M.; Uyanik, A.; et al. Effect of metformin/irinotecan-loaded poly-lactic-co-glycolic acid nanoparticles on glioblastoma: In vitro and in vivo studies. *Nanomedicine* **2018**, *13*, 1595–1606. [CrossRef]

16. Palileo, A.; Munoz-Sagastibelza, M.; Martello-Rooney, L. Treatment with paclitaxel causes upregulation in resistance protein tubulin beta III in a type 2 human endometrial cancer cell line. *Gynecol. Oncol.* **2019**, *154*, e15. [CrossRef]
17. Dwivedi, P.; Han, S.Y.; Mangrio, F.; Fan, R.; Dwivedi, M.; Zhu, Z.A.; Huang, F.S.; Wu, Q.; Khatik, R.; Cohn, D.E.; et al. Engineered multifunctional biodegradable hybrid microparticles for paclitaxel delivery in cancer therapy. *Mater. Sci. Eng. C-Mater.* **2019**, *102*, 113–123. [CrossRef]
18. Lee, J.E.; Abuzar, S.M.; Seo, Y.; Han, H.; Jeon, Y.; Park, E.J.; Baik, S.H.; Hwang, S.-J. Oxaliplatin-loaded chemically cross-linked hydrogels for prevention of postoperative abdominal adhesion and colorectal cancer therapy. *Int. J. Pharm.* **2019**, *565*, 50–58. [CrossRef]
19. Naghizadeh, Z.; Karkhaneh, A.; Khojasteh, A. Simultaneous release of melatonin and methylprednisolone from an injectable in situ self-crosslinked hydrogel/microparticle system for cartilage tissue engineering. *J. Biomed. Mater. Res. A* **2018**, *106*, 1932–1940. [CrossRef]
20. Khaing, Z.Z.; Agrawal, N.K.; Park, J.H.; Xin, S.J.; Plumton, G.C.; Lee, K.H.; Huang, Y.J.; Niemerski, A.L.; Schmidt, C.E.; Grau, J.W. Localized and sustained release of brain-derived neurotrophic factor from injectable hydrogel/microparticle composites fosters spinal learning after spinal cord injury. *J. Mater. Chem. B* **2016**, *4*, 7560–7571. [CrossRef]
21. Jain, R.A. The manufacturing techniques of various drug loaded biodegradable poly(lactide-co-glycolide) (PLGA) devices. *Biomaterials* **2000**, *21*, 2475–2490. [CrossRef]
22. Liu, L.; Liu, D.R.; Wang, M.; Du, G.C.; Chen, J. Preparation and characterization of sponge-like composites by cross-linking hyaluronic acid and carboxymethylcellulose sodium with adipic dihydrazide. *Eur. Polym. J.* **2007**, *43*, 2672–2681. [CrossRef]
23. Marshall, C.D.; Hu, M.S.; Leavitt, T.; Barnes, L.A.; Cheung, A.T.; Malhotra, S.; Lorenz, H.P.; Longaker, M.T. Creation of Abdominal Adhesions in Mice. *J. Vis. Exp.* **2016**, *144*. [CrossRef]
24. Graham, M.A.; Lockwood, G.F.; Greenslade, D.; Brienza, S.; Bayssas, M.; Gamelin, E. Clinical pharmacokinetics of oxaliplatin: A critical review. *Clin. Cancer Res.* **2000**, *6*, 1205–1218.
25. Milas, M.; Rinaudo, M.; Roure, I.; Al-Assaf, S.; Phillips, G.O.; Williams, P.A. Comparative rheological behavior of hyaluronan from bacterial and animal sources with cross-linked hyaluronan (hylan) in aqueous solution. *Biopolymers* **2001**, *59*, 191–204. [CrossRef]
26. Bittner, B.; Witt, C.; Mader, K.; Kissel, T. Degradation and protein release properties of microspheres prepared from biodegradable poly(lactide-co-glycolide) and ABA triblock copolymers: Influence of buffer media on polymer erosion and bovine serum albumin release. *J. Control. Release* **1999**, *60*, 297–309. [CrossRef]
27. Rehman, F.; Volpe, P.L.O.; Airoldi, C. The applicability of ordered mesoporous SBA-15 and its hydrophobic glutaraldehyde-bridge derivative to improve ibuprofen-loading in releasing system. *Colloid Surf. B* **2014**, *119*, 82–89. [CrossRef]
28. Peppas, N.A.; Bures, P.; Leobandung, W.; Ichikawa, H. Hydrogels in pharmaceutical formulations. *Eur. J. Pharm. Biopharm.* **2000**, *50*, 27–46. [CrossRef]
29. Chen, C.H.; Chen, S.H.; Mao, S.H.; Tsai, M.J.; Chou, P.Y.; Liao, C.H.; Chen, J.P. Injectable thermosensitive hydrogel containing hyaluronic acid and chitosan as a barrier for prevention of postoperative peritoneal adhesion. *Carbohydr. Polym.* **2017**, *173*, 721–731. [CrossRef]
30. Kasturi, S.P.; Qin, H.; Thomson, K.S.; El-Bereir, S.; Cha, S.C.; Neelapu, S.; Kwak, L.W.; Roy, K. Prophylactic anti-tumor effects in a B cell lymphoma model with DNA vaccines delivered on polyethylenimine (PEI) functionalized PLGA microparticles. *J. Control. Release* **2006**, *113*, 261–270. [CrossRef]

pharmaceutics

MDPI

Article

Design and Evaluation of a Poly(Lactide-*co*-Glycolide)-Based In Situ Film-Forming System for Topical Delivery of Trolamine Salicylate

Yujin Kim [1], Moritz Beck-Broichsitter [2] and Ajay K. Banga [1,*]

1 Centre for Drug Delivery Research, Department of Pharmaceutical Sciences, College of Pharmacy, Mercer University, Atlanta, GA 30341, USA

2 MilliporeSigma a Business of Merck KGaA, Frankfurter Strasse 250, 64293 Darmstadt, Germany

* Correspondence: Banga_AK@mercer.edu; Tel.: +678-547-6243

Received: 5 June 2019; Accepted: 5 August 2019; Published: 12 August 2019

Abstract: Trolamine salicylate (TS) is a topical anti-inflammatory analgesic used to treat small joint pain. The topical route is preferred over the oral one owing to gastrointestinal side effects. In this study, a poly(lactide-*co*-glycolide) (PLGA)-based in situ bio-adhesive film-forming system for the transdermal delivery of TS was designed and evaluated. Therefore, varying amounts (0%, 5%, 10%, 20%, and 25% (*w*/*w*)) of PLGA (EXPANSORB® DLG 50-2A, 50-5A, 50-8A, and 75-5A), ethyl 2-cyanoacrylate, poly (ethylene glycol) 400, and 1% of TS were dissolved together in acetone to form the bio-adhesive polymeric solution. In vitro drug permeation studies were performed on a vertical Franz diffusion cell and dermatomed porcine ear skin to evaluate the distinct formulations. The bio-adhesive polymeric solutions were prepared successfully and formed a thin film upon application in situ. A significantly higher amount of TS was delivered from a formulation containing 20% PLGA (45 ± 4 μg/cm^2) and compared to PLGA-free counterpart (0.6 ± 0.2 μg/cm^2). Furthermore, the addition of PLGA to the polymer film facilitated an early onset of TS delivery across dermatomed porcine skin. The optimized formulation also enhanced the delivery of TS into and across the skin.

Keywords: NSAIDs; PLGA; polymeric film; topical drug delivery; trolamine salicylate

1. Introduction

Topical drug delivery has many advantages over systemic drug administration, such as minimizing side effects, bypassing first-pass metabolism, and ensuring better patient compliance [1]. However, dermal drug delivery is usually limited to certain drug molecules, owing to the well-known barrier function of the skin. The outermost layer of the skin, the stratum corneum, generally only allows permeation of molecules with a molecular weight <500 Da, a log$P$ between 1 and 3, and a melting point of <250 °C [2].

Among the available topical formulations, non-steroidal anti-inflammatory drugs (NSAIDs) constitute approximately 18 world-wide marketed molecules (some examples include diclofenac, methyl salicylate, salicylic acid, and ketoprofen) [3]. As an example, topical salicylates are known to be absorbed by the dermal tissue and have been reported to be effective in relieving local pain [4]. Trolamine salicylate (TS), a derivative of salicylic acid offers many benefits over other topical analgesics, including a lack of distinct odor, low systemic absorption upon dermal or topical administration, and low skin irritation [5]. Although topical salicylates are widely used, compliance is often an issue, owing to the frequent dosing regimen (three to four times per day).

The development of topical formulations has made an important contribution to medical practice. To deliver active pharmaceutical ingredients into or through the skin, various types of formulations are known, such as gels and emulsions. The type of vehicle chosen depends on the properties of the drug and the intended target area. In addition, the hydrophilicity and lipophilicity of the drug must be compatible with the vehicle. "Conventional" vehicles are inelegant and have the drawback of poor control of both the amount of drug applied and the area of skin exposed [6]. As a result, the use of topical formulations to deliver molecules to systemic circulation is less than ideal, thus resulting in substantial variability in the extent and duration of drug effects [7]. To overcome the shortcomings of skin drug delivery, the former Alza Corporation pioneered a transdermal patch in which system design and explicit control of the surface area led to improved passive drug delivery to the systemic circulation at a predetermined rate. Transdermal patches are recommended to be applied to a flat surface area to ensure adhesion to skin over a prolonged period of time. However, in the case of arthritis and other joint pain, the use of a patch is not suitable because of the uneven surface at the application sites. Pain associated with arthritis requires drug application at the pain region, which is usually difficult to cover with conventional patches [8]. In addition, clinical trials have shown that the topical application of NSAIDs delivers drugs at a higher local concentration to the tissue, resulting in better pain management [9].

Bioadhesive in situ film-forming systems provide many advantages such as higher dosing flexibility, extended release properties, higher patient compliance, improved cosmetic appearance, and less chance for loss of formulation by rubbing [10].

Poly(lactide-*co*-glycolide) (PLGA) is one of the most promising polymers used for the fabrication of drug delivery devices and tissue engineering applications [11]. PLGA is biocompatible and biodegradable, and its properties such as the erosion profile and mechanical strength can be controlled as needed. Furthermore, PLGA can be engineered to control the drug release behavior by changing the polymer molecular weight and the molar ratio of lactide to glycolide [12–14]. However, a detailed characterization of such systems is required to prevent potential dose dumping, and an inconsistent drug release profile [15].

In this study, a PLGA-based in situ bioadhesive film-forming system for the transdermal delivery of TS was designed and evaluated for drug permeation across dermatomed porcine ear skin. Different types of plasticizer and solvents were screened to optimize the film forming behavior. The effects of different ingredients were studied on the permeation profile of TS. The optimized formulations were developed by finding a proper ratio between polymer, adhesive, and plasticizer and showed enhancement of TS delivery into and across the skin.

2. Materials and Methods

2.1. Materials

All polymers (Table 1) were kindly donated by Merck KGaA (Darmstadt, Germany). Trolamine salicylate and ethyl-2-cyanoacrylate were purchased from Sigma (St. Louis, MO, USA). Glycerol and Poly(ethylene glycol) 400 was purchased from VWR (Richmond, BC, Canada) and Acros organics (Morris Plains, NJ, USA), respectively. Propylene glycol was obtained from EKI Industries (Joliet, IL, USA). Porcine ear skin was procured from a local slaughterhouse. Phosphate-buffered saline (PBS; 10 mM, pH 7.4), acetonitrile, ethyl acetate, and acetone were purchased from Fisher Scientific (Waltham, MA, USA).

Table 1. Properties of the utilized PLGA polymers.

PLGA Polymer	Lactide/Glycolide Molar Ratio	End Group	Molecular Weight (Indicative Range, kDa)
EXPANSORB® DLG 50-2A	1:1	–COOH	5–20
EXPANSORB® DLG 50-5A	1:1	–COOH	42–65
EXPANSORB® DLG 50-8A	1:1	–COOH	80–130
EXPANSORB® DLG 75-5A	3:1	–COOH	37–84

2.2. Development of Film-Forming Polymeric Solutions

2.2.1. Screening of Components

To choose a compatible solvent for PLGA, 1 g of PLGA and 0.25 g of cyanoacrylate (CA) were dissolved together in 1 g of four different solvents (ethanol, ethyl acetate, propylene glycols, and acetone) and left overnight mixing. To select a proper plasticizer, we added a few drops (0.25 g) of one of the three different plasticizers namely glycerol, poly(ethylene glycol) 400 (PEG 400), and propylene glycol to the polymeric mixture to increase the flexibility of the film. Each formulation (6.4 μL) was then applied on the dermatomed porcine ear skin, and the film forming behavior was investigated.

2.2.2. Optimization of Formulations

Optimized film-forming solutions were prepared by the addition of the PLGA and CA to the solvent. After obtaining a clear solution, the plasticizer and the drug were added. The solutions were left overnight mixing in glass vials to allow the drug to dissolve completely at room temperature. A detailed composition of each formulation can be found in Table 2.

Table 2. Composition of the distinct formulations (% (*w*/*w*)).

	Different Types of PLGA				Different Amounts of PEG 400				Different Concentrations of PLGA				
	F1	F2	F3	F4	F1′	F5	F6	F7	F1″	F8	F9	F10	F11
Code	50-2A	50-5A	50-8A	75-5A	4:1:1	4:1:0	4:1:2	4:1:3	20%	0%	5%	10%	25%
Polymer	50-2A	50-5A	50-8A	75-5A	50-2A	50-2A	50-2A	50-2A	50-2A	50-2A	50-2A	50-2A	50-2A
PLGA	20	20	20	20	20	24	17	15	20	0	5	10	25
Cyanoacrylate	5	5	5	5	5	6	4.25	3.75	5	15	12.5	10	2.5
Plasticizer													
PEG 400	5	5	5	5	5	0	8.75	11.25	5	15	12.5	10	2.5
Solvent													
Acetone	69	69	69	69	69	69	69	69	69	69	69	69	69
Drug													
Trolamine Salicylate	1	1	1	1	1	1	1	1	1	1	1	1	1

2.2.3. Effect of Plasticizer PEG 400, Types of PLGA, and Various Concentration of PLGA for Topical Delivery of TS

PLGA, 50-2A, was used to test the effects of plasticizer on the permeation profiles of TS. The ratio of PLGA to CA was kept at four to one. Then, the proportion of the plasticizer was increased from 0 to 3. The highest amount PEG 400 was added to F7, at 11.25 g (11.25% (*w*/*w*)), then reduced to 8.75 g (F6; 8.75% (*w*/*w*)), 5 g (F′1; 5% (*w*/*w*)), and finally to no PEG 400 in F5 (Table 2).

For the comparison of different types of PLGA, the ratio of each component was kept at 4:1:1 (PLGA:cyanoacrylate:PEG 400). To investigate the effects of different amounts of PLGA in TS permeation profiles, we incorporated different amounts of PLGA into formulations. PLGA 50-2A was used in all formulations, and the ratio of PEG 400 and cyanoacrylate was kept at 1:1. To increase the amount of PLGA, the amount of cyanoacrylate was decreased from 25 to 0 in increments of 5.

2.3. Evaluation of the Formulations

2.3.1. Crystallization Study and Solvent Evaporation Study

To observe the crystallization of the drug in the films, we placed 6.4 μL of formulations on a microscopic slide and evaluated the samples under a microscope after the evaporation of acetone. To study the extent and time of solvent evaporation, 6.4 μL of F1 formulation was placed on a microscopic slide using a positive displacement pipette. The slides were left in a convection oven (32 °C) for a pre-determined time and weighed at 2 min, 1 h and 24 h to calculate the percentage evaporation of the solvent over time.

2.3.2. In Vitro TS Release Study

The release profiles of TS from the formulations F1, F2, F3, and F4 (different PLGA types) were determined to study the impact of the polymer properties on the drug release kinetics. Polymer films containing 64 μg of TS (equal to a surface area of 0.64 cm^2) was added to the bottom of a glass vial. Then, 5 mL of PBS were added and left under constant shaking at 150 rpm and 30 °C. Samples (300 μL) were withdrawn from the vials at predetermined time points, and the same amount of fresh receptor solution was replaced. The samples were analyzed by high-performance liquid chromatography (HPLC) as outlined below.

2.3.3. In Vitro Permeation Study

Skin Preparation

The outer region of full-thickness ear skin was removed carefully with a scalpel. The skin was then washed with PBS, dried, wrapped with Parafilm, and stored at −80 °C until use. The porcine ear skin was dermatomed with a Dermatome 75 mm (Nouvag AG, Goldach, Switzerland) prior to use. The average thickness of the dermatomed skin pieces was 0.50 ± 0.02 mm.

Evaluation of Skin Integrity

Before conducting a permeation study, we evaluated the integrity of the skin membrane by measuring the skin resistance value. The skin was mounted on a vertical Franz diffusion cell with the stratum corneum side facing up, and then 300 μL of PBS was added to the donor compartment. A silver electrode and silver chloride electrode were placed in the receptor and the donor chambers, respectively, without touching the skin membrane. Two electrodes were connected to a digital multimeter (34410A $6\frac{1}{2}$ digit multimeter; Agilent Technologies, Santa Clara, CA, USA) and waveform generator (Agilent 33220A, 20 MHz function/arbitrary waveform generator) [16]. The resistance of the employed skin (R_s) was calculated according to the following formula:

$$R_s = V_s R_L/(V_0 - V_s) \quad (1)$$

where R_L and V_0 were 100 kΩ and 100 mV, respectively. Skin pieces with a resistance lower than 10 KΩ were discarded.

Next, permeation studies were performed on a jacketed Franz diffusion cell with a diffusion area of 0.64 cm^2 (PermeGear, Bethlehem, PA, USA) on dermatomed porcine ear skin. PBS was used as the receptor solution (5 mL). The skin was clamped between the donor and receptor chambers of a vertical diffusion cell with the stratum corneum side in contact with the donor solution. The distinct formulations (64 μg of TS per 0.64 cm^2) was added in the donor compartment. The temperature of the receptor medium was maintained at 37 °C, and the skin surface temperature was about 32 °C. The amount of drug diffused over the dermatomed porcine ear skin was determined by removal of aliquots of 300 μL at pre-determined time points over 72 h from the receptor compartment. The volume was immediately replaced with the same amount of fresh buffer. The samples were analyzed by HPLC as outlined below. A skin extraction study was carried out to determine the drug amount that had penetrated into the skin. After 72 h, the skin samples were removed from Franz diffusion cell and the residual polymer film was removed with D-squame tape. The skin surface was wiped off with two cotton buds dipped in receptor solution. Then, the skin was dried with two cotton buds. The utilized tape and the four cotton buds were pooled and extracted with 30 mL of receptor solution. After the cleaning procedure, the epidermis and dermis were separated. The tissue was minced manually and added to 1 mL of receptor solution. Samples were shaken for 4 h followed by filtration using a 0.45 μm membrane filter and analyzed by HPLC.

2.4. Quantitative Analysis

TS was quantified by HPLC analysis. A Waters Alliance HPLC system (2795 Separating Module; Waters Co., Milford, MA, USA) equipped with a Photodiode detector (Waters 2475) and a Kinetex EVO C18 column (5 μm, 100 Å, 150 × 4.6 mm; Phenomenex, Torrance, CA, USA) was used. The mobile phase consisted of a 70/30 (*v*/*v*) mixture of acetonitrile and 0.1% of trifluoroacetic acid in distilled water. The injection volume and the flow rate were set to 20 μL and 1.0 mL/min, respectively, and the detection wavelength of TS was 304 nm. The reversed-phase HPLC method provided a linear range of 0.1–100 μg/mL ($R^2 = 0.999$).

2.5. Statistical Analyses

All results are reported as the mean with the standard error of the mean (SE) from at least three replicates. Statistical calculations were performed with GraphPad Prism Version 8.0 (GraphPad Software, Inc., San Diego, CA, USA). One-Way ANOVA followed by Tukey HSD post hoc test was applied to compare the results of different groups. Statistically significant differences were denoted by $p < 0.05$.

3. Results

3.1. In Situ Film-Forming Behavior

PLGA and CA did not dissolve completely in propylene glycol and ethanol. Complete dissolution was achieved in ethyl acetate and acetone. However, ethyl acetate did not evaporate upon application and the obtained polymer films were difficult to handle. Therefore, acetone was chosen as the solvent for the system. A successful film formation behavior on the dermatomed porcine ear skin was observed with the plasticizer PEG 400. Upon administration, the solvent evaporated instantaneously and left a thin and transparent film layer behind.

3.2. Crystallization Study and Solvent Evaporation Study

Formulation (F1–F11) was observed under the light microscope after complete evaporation of the solvent. No drug crystals were observed in the polymer films during the slide crystallization study, indicating the solubility of TS in the film matrix after evaporation of the organic solvent (Figure 1). After 2 min in the convection oven, the result demonstrated 75.3 ± 4.1% ($n = 3$) solvent evaporation, and it did not further evaporate for 24 h.

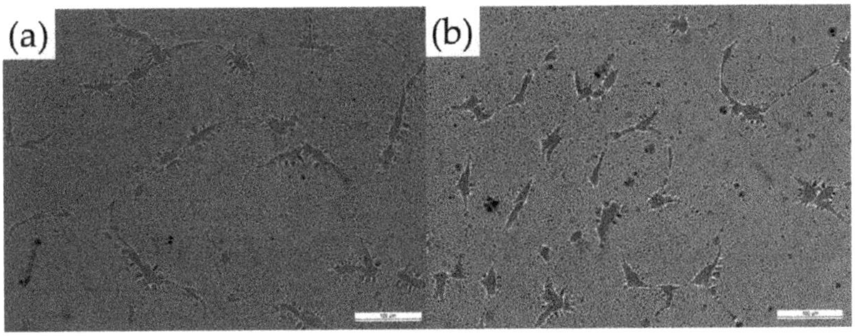

Figure 1. Microscopic images of the 20% PLGA formulation (F1) on glass sides (**a**) without and (**b**) with added drug (10× magnification) did not show drug crystallization after complete evaporation of the solvent.

3.3. *In Vitro Release Study for Different Types of PLGA*

In the first hour, F3 and F4 released 34% and 38%, respectively. In contrast, a smaller amount was released after 1 h with F1 and F2 (20% and 13%, respectively). After 72 h, the drug release from F1, F2, and F3 was completed (>90%), whereas only 68% of TS was released from F4 (Figure 2).

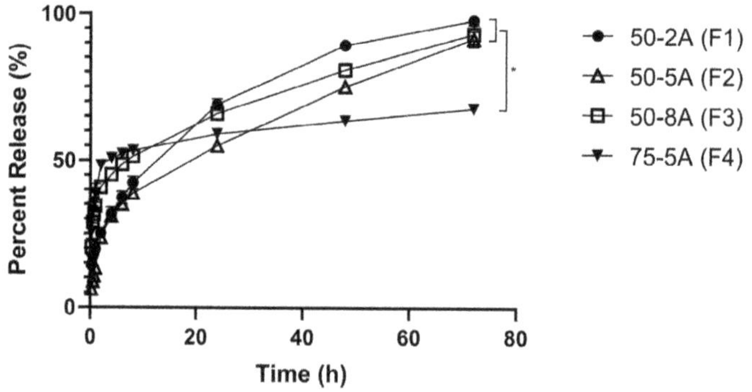

Figure 2. In vitro drug release profile from different types of PLGA (n = 6).

3.4. *Effects of Each Ingredient and Permeation Study*

3.4.1. Effects of Plasticizer

Four formulations with different amounts of PEG 400 were prepared, and a thin film was formed upon application on the skin. The amount of TS delivered in the receptor solution after 72 h was found to be 16 ± 3 μg/cm^2 and 24 ± 3 μg/cm^2 in F5 and F7, respectively. A significantly higher amount of TS was delivered to the receptor with F1′ (47 ± 8 μg/cm^2) and F6 (47 ± 5 μg/cm^2). Moreover, formulation without any plasticizer (F5) delivered only 2.7 ± 0.3 μg/cm^2, the least amount of drug into the skin (Figure 3). The highest amount of PEG 400 facilitated the largest amount of TS delivery into the skin (individual values in Table 3). Because there was no significant difference between the F1′ and F6 groups in the amount of drug in the receptor, the F1′ ratio was chosen to carry out permeation studies to investigate the effects of different types of PLGA.

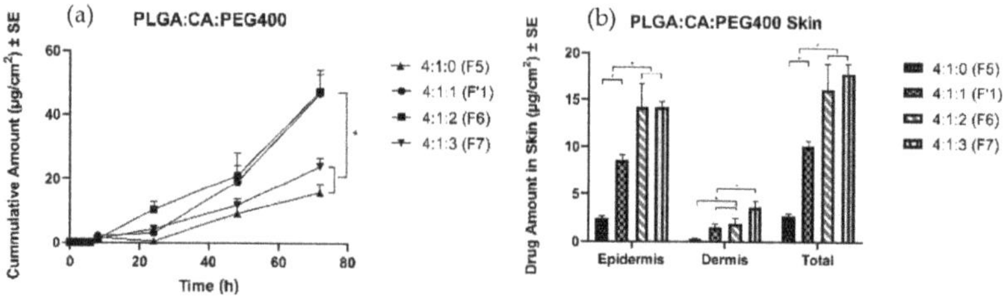

Figure 3. Permeation profiles of TS through porcine ear skin to study the effects of plasticizer. The group is representative of the ratio between PLGA:CA:PEG 400. (**a**) Cumulative amount (**b**) Average amount in the epidermis, dermis, and total skin.

Table 3. Amounts of TS extracted after 72 h in the epidermis and dermis and the total amount of TS for all groups.

	Groups	Epidermis (µg/cm²)	Dermis (µg/cm²)	Total (µg/cm²)
Different types of PLGA	50-2A (F1)	10.4 ± 1.4	1.5 ± 0.4	11.9 ± 1.7
	50-5A (F2)	12.6 ± 2.5	1.3 ± 0.4	13.9 ± 2.4
	50-8A (F3)	8.6 ± 1.3	1.1 ± 0.4	9.8 ± 1.5
	75-5A (F4)	8.7 ± 0.5	1.0 ± 0.2	9.7 ± 0.7
Effect of plasticizer	4:1:1 (F1′)	8.6 ± 0.6	1.6 ± 0.3	10.1 ± 0.6
	4:1:0 (F5)	2.5 ± 0.2	0.3 ± 0.1	2.7 ± 0.3
	4:1:2 (F6)	14.3 ± 2.5	1.9 ± 0.5	16.2 ± 2.8
	4:1:3 (F7)	14.2 ± 0.7	3.7 ± 0.6	17.9 ± 1.0
Different amount of PLGA	20% (F1″)	9.1 ± 2.0	18.2 ± 4.0	27.4 ± 6.0
	0% (F8)	6.3 ± 0.6	0.40 ± 0.1	6.7 ± 0.5
	5% (F9)	9.9 ± 1.0	0.62 ± 0.2	10.5 ± 1.2
	10% (F10)	11.5 ± 1.5	22.9 ± 3.0	34.4 ± 4.5
	25% (F11)	9.5 ± 0.8	1.9 ± 0.4	11.4 ± 1.1

3.4.2. Effects of Different Types of PLGA

All formulations were able to form a polymeric solution, and then a thin layer of film was formed upon application on the skin. After 72 h, all three formulations with the one to one ratio (G/L) delivered 17 ± 3, 17 ± 3, and 19 ± 3 µg/cm^2, respectively, and no significant difference was observed in the cumulative amount of TS in the receptor. In contrast, F4 delivered 11 ± 1 µg/cm^2, a significantly lower amount than was observed for the other three groups. There was no significant difference in the drug amount delivered into the skin (Figure 4).

3.4.3. Effects of Concentration of PLGA

All formulations were able to form a polymeric solution and formed a thin layer of film after complete evaporation of the solvent. The highest amount of TS (45 ± 4 µg/cm^2) was delivered with 20% PLGA (F1″), followed by the formulation with 25% of PLGA (F11) (21 ± 4 µg/cm^2). The formulations with 0%, 5%, and 10% PLGA delivered a significantly lower amount of TS, at 0.6 ± 0.2, 1.6 ± 0.3, and 5.6 ± 0.7 µg/cm^2, respectively (Figure 5). The formulation with 10% (F10) and 20% (F1″) PLGA delivered significantly higher amounts into the skin. However, the 0%, 5%, and 25% PLGA formulations delivered a lower amount of TS into the skin (values in Table 3).

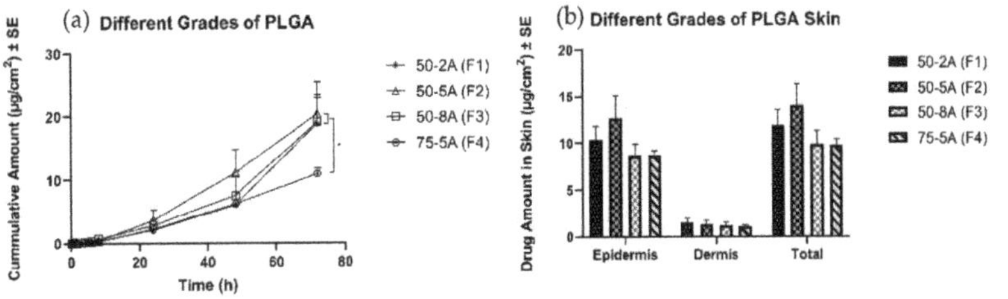

Figure 4. Permeation profiles of TS through porcine ear skin for the different types of PLGA. The group is representative of different types of PLGA: (**a**) Cumulative amount (**b**) Average amount in the epidermis, dermis, and total skin.

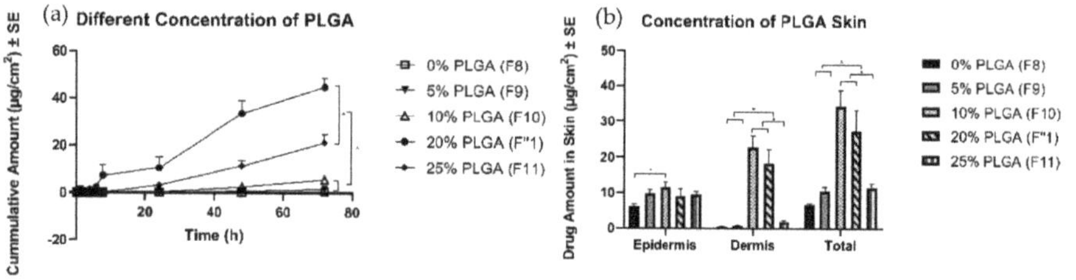

Figure 5. Permeation profiles of TS through porcine ear skin for different concentrations of PLGA. The group is representative of different concentrations of PLGA: (**a**) Cumulative amount (**b**) Average amount in the epidermis, dermis, and total skin.

4. Discussion

4.1. Parameters for the Development of Film-Forming Polymeric Solutions

Limitations of conventional formulations for topical drug delivery include poor adherence to the skin, poor permeability, and a low compliance rate [17]. Bioadhesive film, because it is an intermediate between transdermal patches and semisolid dosage forms, has the advantages of both systems, such as transparency, no stickiness, convenience, less frequent dosing, sustained drug release, and resistance to wiping off. In addition, incorporating cosmetic or therapeutic agents is more convenient with this system than conventional topical drug delivery systems, owing to lower potential loss of the formulation by rubbing.

First, the most important parameters for polymeric solutions are the polymers. Suitable polymers for successful film forming systems may possess transparency, flexibility, and drug encapsulation ability at moderate temperature. To develop a suitable formulation to decrease the frequency of application of topical analgesics, common film-forming polymers, such as hydroxypropyl, methylcellulose, polyvinyl pyrrolidine, and acrylate copolymer, were searched in the literature. However, many of the polymers were hydrophilic and might not provide water resistance in daily life. PLGA was chosen to be incorporated in the polymeric solution because of its many benefits, such as its water resistance, nonirritating and nonallergic qualities, and capability for drug incorporation [15]. To provide an efficient local drug delivery system, Eskandari et al. have investigated the use of butyl-2-cyanoacrylate. To ensure good adhesion of the film to the skin, CA, which is compatible with PLGA, was chosen. CA gets stiffened by polymerization process in the presence of moisture [18–20]. When the polymeric film is applied on the skin, the film is formed upon polymerization of PLGA and CA due to the moisture of the skin. Polymerized CA is an excellent polymer candidate for drug delivery because it is a biodegradable, hemostatic, nonallergenic tissue adhesive with local antibacterial properties, and it is inexpensive and widely available [18,19,21]. In addition, it is commonly used and is acceptable to the public, owing to its use in liquid bandage formulations [22]. Currently, CA is used in surgical and clinical adhesives in various surgical procedures such as the treatment of arteriovenous malformation, retinal ruptures, and skin graft placements [23,24]. Eskandari et al. have shown that CA can successfully incorporate drugs for slow release of antibiotics to specific areas without causing an inflammatory response [25].

Solvents play an important role in film formation. The solvent used in the film forming system is responsible for drug-polymer solubility and also affects drug permeation. Some common solvents used in polymeric solutions are ethanol, ethyl acetate, propylene glycols, and acetone [26]. The combination of the PLGA and the adhesive was tested for dissolution in different solvents such as acetone, ethyl acetate, and ethanol. All the ingredients were dissolved readily in acetone and formed a film upon application to the skin. Acetone is approved in use of topical formulation per the inactive ingredient list of FDA [27]. Also, a toxicology report by US. Department of Health and Human Services reported that dermal exposure to acetone did not affect the human health negatively [28].

Apart from the polymer and solvent, other excipients such as plasticizers must be added into the formulation. Plasticizers are low molecular weight resins that may interact with polymer chains and affect polymer-polymer bonding [29]. These interactions may affect film flexibility as well as the permeability toward drug substances. The commonly used plasticizers in polymeric solutions include fatty acid esters, glycol derivatives, phthalate esters, and phosphate esters [30]. The plasticizer used should be compatible with the polymers and should have low skin permeability. Hence, choosing a proper plasticizer is important for the polymeric matrix. In addition, determining the right amount of plasticizer is critical for film-forming polymeric solutions. After the selection of two polymers to form the polymeric solution, proper plasticizers were screened for tolerance to mechanical stress, which may be exerted on the formed film by the movement of the skin. In selecting an appropriate plasticizer, compatibility with the polymers and plasticization efficiency were considered. After application on dermatomed porcine skin, the acetone evaporated, and a thin transparent film was formed in situ. Based on the results, PEG 400 was incorporated as a plasticizer to enhance the mechanical properties and flexibility of the film.

4.2. Characterization Methods for Film-Forming Polymeric Solutions

In a transdermal delivery system, drug solubility in the polymer matrix system is a critical factor that may affect the rate and extent of drug permeation [31]. If a drug is not soluble in a polymer matrix, it will become supersaturated and hence unstable. Moreover, when a drug is successfully incorporated in a polymeric matrix, many studies have reported that the stability of the drug improves [32,33]. PLGA formulations are widely used in transdermal and dermal delivery and have a safety profile showing a lack of skin irritation and uniform drug dosing [15,34].

The drug was successfully dissolved in the polymeric matrix, and the absence of crystallization, which might hinder drug permeation into the skin, was verified. In patch development, to improve the permeation and release profiles of drugs, the investigation of the supersaturated state is an important step. A possible strategy to study the solubility of a drug is to disperse it uniformly into a polymeric matrix to prevent drug crystallization [35]. Different techniques have been described to determine the solubility of drugs in polymer matrixes [31,36,37]. Such solubility testing can be time consuming and difficult. Among different techniques, Jain et al. have compared two different methods to predict drug solubility in the polymeric matrix: differential scanning calorimetry and slide crystallization study [31]. In that study, the author found that the experimental solubility potential and the theoretical values from a solubility calculator were similar. In addition, low drug solubility in the drug-polymer matrix was shown to negatively affect drug release. The author concluded that simple slide crystallization was able to predict the saturation solubility of a drug in the polymer. In the present study, after complete evaporation of solvents from our drug-polymeric matrix, crystallization of the drug was not observed under a light microscope, and we confirmed the drug solubility and stability in the system. The complete TS solubility in the polymer matrix system after evaporation of acetone was studied to ascertain that the stability and solubility of TS was not a factor affecting the rate and extent of drug permeation.

The amount of residual solvent was also measured and calculated. Instant solvent evaporation and complete dryness of the formulation was observed after 2 min. As a result, approximately 25% of the residual solvent was left on the application site. Per the International Conference on Harmonisation residual solvent guideline, solvents are classified in the three categories which set limits depending on the toxicity data for each solvent. Acetone is categorized as a Class III solvent, which has a low toxic potential and is safe to use in human [38]. Furthermore, Baino et al. reported that the drug concentration increases upon evaporation of the solvent, which may result in a greater driving force to deliver the drug into the system [39]. However, this variable was eliminated by keeping the solid content of all the formulations constant. Thus, residual solvent was not a factor that led to differences in the permeation profile of TS from the formulations tested.

4.3. *In Vitro Release Study for Different Types of PLGA*

In this study, the release profiles of TS from different PLGA polymers were determined to study the effects of the PLGA parameters on the drug release kinetics. In terms of developing a polymeric system, knowing the release mechanisms and the physicochemical properties is important. The two main release mechanisms related to drug release from a PLGA-based system are degradation and diffusion. Often, the release rate is initially determined by diffusion followed by degradation controlled at the final stage [40]. PLGA systems commonly release the drug in a bi-phasic or a tri-phasic pattern. In this study, for higher molecular weight PLGA polymers, we observed a bi-phasic profile of drug release, with a burst release followed by relatively slow diffusion. Lower molecular weight PLGA groups showed a monophasic release profile, which is preferable for drug delivery systems because it follows a zero-order release profile [41]. The formulation with higher molecular weight compounds 50-8A and 75-5A released almost 40% after 1 h, thus indicating dose dumping, whereas the release profiles for 50-2A and 50-5A showed a gradual release until 72 h. In this study, the film was submerged in the receptor solution, which was more hydrophilic than the polymeric solution. Because the drug was hydrophilic, it may be preferentially located in the receptor solution over the PLGA system. Thus, the higher molecular weight, and hence more lipophilic PLGA, resulted in a more rapid release of the drug than was observed in the groups with lower molecular weight. Based on the result of the release study, 50-2A was chosen for the final formulation to avoid dose dumping effect as well as provide a gradual release of TS until 72 h.

4.4. *In Vitro Permeation Study*

In vitro permeation studies, using the vertical Franz diffusion cell model, have been commonly used to evaluate drug delivery into and across the skin, and is well established as a reliable tool [42,43]. In addition, FDA recommends and endorses the use of an in vitro permeation test to evaluate topical products. FDA guidelines have included the merits of this approach and have established a strong correlation between the in vitro permeation test (IVPT) and in vivo bioavailability data with narrow inter- and intra-variability between data [44–46].

4.4.1. Effects of Plasticizers

Plasticizers are important components of film forming systems because they improve the appearance of the film, prevent film cracking, increase film flexibility, and confer desired mechanical properties [47]. By selecting an appropriate plasticizer and optimizing its concentration in the formulation, the release rate of a therapeutic compound can be controlled. To observe the effects of the plasticizer, we kept the formulation ratio between the polymer and the adhesive at 4 to 1 (PLGA:CA), and the same amounts of other components were included, i.e., solvents and drug concentrations. The ratio of plasticizer was increased from 0 to 4 (0%, 5%, 8.75%, and 11.25% *w*/*w*). The polymeric films without any plasticizer and 11.25% *w*/*w* plasticizer delivered a significantly lower amount of TS after 72 h, than did those with 5 and 8.75% *w*/*w* PEG 400. The films with a plasticizer concentration of 8.75% *w*/*w* did not significantly improve the diffusion of the drug in the receptor (as compared with the results with 5% *w*/*w* plasticizer concentration). However, the amount of PEG 400 showed a linear relationship up to 8.75% *w*/*w* with regarding TS delivery into the skin. The formulation with 11.25% *w*/*w* did not show a significant increase over 8.75% *w*/*w* regarding TS delivery into the skin. Hence, 5% *w*/*w* was considered the optimum concentration for the plasticizer. These results may have been due to a polymer-plasticizer interaction that affected the release of the drug from the system. Barhate et al. have studied the carvedilol permeation profile with and without PEG 400 as a plasticizer and have found that the incorporation of PEG 400 increases the film flexibility and permeation rate [48]. The results suggested that too little or too much plasticizer in the polymeric solution can negatively affect the permeation profile of TS. In formulating a polymeric solution, minor variations might be acceptable; however, major changes in the composition should be carefully considered, because they might unfavorably affect the properties of the film forming system and consequently the mechanical or cosmetic performance in drug delivery into the skin.

4.4.2. Effects of Different Types of PLGA

PLGA is one of the most commonly used biodegradable polymers in the field of biomedical devices because of its favorable degradation characteristics and the possibility for sustained drug delivery. It can be engineered to control drug release by changing the polymer molecular weight and ratio of lactide to glycolide. Commercially available PLGA combinations are 50:50, 65:35, 75:25, and 85:15 (LA:GA) in different molecular weights [49]. These physical properties affect the solubility, glass transition temperature, and inherent viscosity, thus resulting in tensile strength and polymer chain flexibility. These parameters help delivery systems or medical devices achieve the desired dosage and release interval. However, evaluation of the system is required to minimize the potential toxicity from dose dumping, inconsistent release, and drug-polymer interactions [50]. In this study, four different types of PLGA were chosen and incorporated in the polymeric solution. Three formulations were 50:50 (LA:GA) with different molecular weights. One of the formulations was 75:25 (LA:GA). There was no significant difference among formulation groups with the same ratio of LA:GA, but the group with a higher proportion of lactide delivered less TS after 72 h. Moreover, the amount of drug delivered into the skin did not show significant differences across all four groups. PGA is a hydrophilic and highly crystalline polymer with a relatively fast degradation rate. Although PGA and PLA are similar structurally, they exhibit different physicochemical properties because of the presence of a methyl group on the alpha carbon. Delayed release of the drug from the higher ratio of LA delivered fewer drugs to the receptor at the end of the study. However, the same ratio with different molecular weights did not show a difference in drug delivery to the receptor and skin. Farahani has investigated the degradation mechanism of PLGA (50:50) and has reported that up to eight weeks are required for complete degradation [51]. In this study, drug diffusion was found to be important factor affecting in vitro release of the drug from the formulation; hence the rate of diffusion influenced drug delivery into and across the skin. Three formulations with 50:50 (LA:GA, different molecular weights) did not result in a difference in permeation profile of TS into and across skin; however, the formulation with 75:25 (LA:GA) delivered a significantly lower amount of TS due to its slower diffusion rate.

4.4.3. Effects of Different Amounts of PLGA

To test the effects of the amounts of PLGA in TS permeation profiles, we varied the amount of PLGA without changing the ratio between CA and plasticizer. The amount of PLGA was increased from 0% to 25% *w*/*w*. Formulations with 20% and 25% PLGA delivered significantly higher amounts than the other groups. Formulations with lower amounts of PLGA required more CA. CA has been used as a synthetic adhesive for tissue adhesive application since the 1980s. CA polymerizes rapidly in the presence of weak basic conditions, such as in water. The interaction between skin and the polymer results in the impressive adhesive strength of CA. Moreover, CA is superior to rival polymers in terms of its strong wet adhesion, drug incorporation, and rapid curing [52–55]. The trend in permeation profile was observed because the drug's affinity toward CA was greater than that toward PLGA. Although CA has good adhesion to skin, the rapid polymerization of CA is associated with heat dissipation at the application site, and the brittleness can be problematic [56]. PLGA content was increased while the amount of CA was reduced proportionally to maintain a similar solid content of all formulation. The cumulative drug amount linearly increased with an increase in the amount of PLGA in the formulation. In the optimized formulation, 20% *w*/*w* PLGA was added, resulting in the highest delivery after 72 h without compromising its adhesive property. In this study, an in-situ film forming system was developed by minimizing the CA component and maximizing the amount of PLGA in solution, thus enhancing the delivery of TS into and across the skin.

5. Conclusions

Film forming solutions were successfully formulated with polymers from different chemical groups such as PLGA and polymerized CA. These formulations contained a combination of polymers, a volatile solvent, and other optional excipients such as plasticizers, and fixed concentrations were used for all excipients involved. The optimized formulation also enhanced the delivery of TS into and across the skin.

Author Contributions: Y.K. performed the experiments and drafted the manuscript. A.K.B. supervised the project. A.K.B. and M.B.-B. reviewed the entire contents of the manuscript.

Funding: This project was funded by Merck KGaA, Frankfurter Str. 250, 64293 Darmstadt, Germany, grant number [UR1710050]. The life science business of Merck KGaA, Darmstadt, Germany operates as MilliporeSigma in the U.S. and Canada. Product designations of Merck KGaA, Darmstadt, Germany, and third parties may appear in this material. For details on the ownership of mentioned trademarks, please refer to publicly available resources such as tmdn.org.

Conflicts of Interest: The founding sponsors had no role in the design of the study, writing of manuscript, or the decision to publish the results but provided scientific edits to the manuscript draft which were not related to the collection, analyses, or interpretation of data.

References

1. Singh, G. Recent considerations in nonsteroidal anti-inflammatory drug gastropathy. *Am. J. Med.* **1998**, *105*, 31S–38S. [CrossRef]
2. Banga, A.K. *Transdermal and Intradermal Delivery of Therapeutic Agents: Application of Physical Technologies*; CRC Press: Boca Raton, FL, USA, 2011.
3. Trottet, L.; Maibach, H. *Dermal Drug Selection and Development*; Springer: Berlin, Germany, 2017.
4. Cross, S.E.; Anderson, C.; Roberts, M.S. Topical penetration of commercial salicylate esters and salts using human isolated skin and clinical microdialysis studies. *Br. J. Clin. Pharmacol.* **1998**, *46*, 29–35. [CrossRef] [PubMed]
5. Barkin, R.L. The pharmacology of topical analgesics. *Postgrad. Med.* **2013**, *125*, 7–18. [CrossRef] [PubMed]
6. Alkilani, A.Z.; McCrudden, M.T.; Donnelly, R.F. Transdermal drug delivery: Innovative pharmaceutical developments based on disruption of the barrier properties of the stratum corneum. *Pharmaceutics* **2015**, *7*, 438–470. [CrossRef] [PubMed]
7. Wiedersberg, S.; Guy, R.H. Transdermal drug delivery: 30+ years of war and still fighting! *J. Control. Release* **2014**, *190*, 150–156. [CrossRef] [PubMed]
8. Haroutiunian, S.; Drennan, D.A.; Lipman, A.G. Topical NSAID therapy for musculoskeletal pain. *Pain Med.* **2010**, *11*, 535–549. [CrossRef] [PubMed]
9. Cagnie, B.; Vinck, E.; Rimbaut, S.; Vanderstraeten, G. Phonophoresis versus topical application of ketoprofen: comparison between tissue and plasma levels. *Phys. Ther.* **2003**, *83*, 707–712. [PubMed]
10. Schroeder, I.Z.; Franke, P.; Schaefer, U.F.; Lehr, C.-M. Development and characterization of film forming polymeric solutions for skin drug delivery. *Eur. J. Pharm. Biopharm.* **2007**, *65*, 111–121. [CrossRef] [PubMed]
11. Lü, J.-M.; Wang, X.; Marin-Muller, C.; Wang, H.; Lin, P.H.; Yao, Q.; Chen, C. Current advances in research and clinical applications of PLGA-based nanotechnology. *Expert Rev. Mol. Diagn.* **2009**, *9*, 325–341. [CrossRef] [PubMed]
12. Mohamed, F.; van der Walle, C.F. Engineering biodegradable polyester particles with specific drug targeting and drug release properties. *J. Pharm. Sci.* **2008**, *97*, 71–87. [CrossRef]
13. Allison, S.D. Effect of structural relaxation on the preparation and drug release behavior of poly (lactic-*co*-glycolic) acid microparticle drug delivery systems. *J. Pharm. Sci.* **2008**, *97*, 2022–2035. [CrossRef] [PubMed]
14. Mundargi, R.C.; Babu, V.R.; Rangaswamy, V.; Patel, P.; Aminabhavi, T.M. Nano/micro technologies for delivering macromolecular therapeutics using poly (D, L-lactide-*co*-glycolide) and its derivatives. *J. Control. Release* **2008**, *125*, 193–209. [CrossRef] [PubMed]
15. Makadia, H.K.; Siegel, S.J. Poly Lactic-*co*-Glycolic Acid (PLGA) as Biodegradable Controlled Drug Delivery Carrier. *Polymers* **2011**, *3*, 1377–1397. [CrossRef] [PubMed]

16. Bakshi, P.; Jiang, Y.; Nakata, T.; Akaki, J.; Matsuoka, N.; Banga, A.K. Formulation development and characterization of nanoemulsion-based formulation for topical delivery of heparinoid. *J. Pharm. Sci.* **2018**, *107*, 2883–2890. [CrossRef] [PubMed]
17. Kathe, K.; Kathpalia, H. Film forming systems for topical and transdermal drug delivery. *Asian J. Pharm. Sci.* **2017**, *12*, 487–497. [CrossRef]
18. Amarante, M.; Constantinescu, M.A.; O'Connor, D.; Yaremchuk, M.J. Cyanoacrylate fixation of the craniofacial skeleton: an experimental study. *Plast. Reconstr. Surg.* **1995**, *95*, 639–646. [CrossRef]
19. Bonutti, P.M.; Weiker, G.G.; Andrish, J.T. Isobutyl cyanoacrylate as a soft tissue adhesive. An in vitro study in the rabbit Achilles tendon. *Clin. Orthop. Relat. Res.* **1988**, *229*, 241–248.
20. Ambrose, C.G.; Clyburn, T.A.; Louden, K.; Joseph, J.; Wright, J.; Gulati, P.; Gogola, G.R.; Mikos, A.G. Effective treatment of osteomyelitis with biodegradable microspheres in a rabbit model. *Clin. Orthop. Relat. Res.* **2004**, *421*, 293–299. [CrossRef]
21. Cheski, P.J.; Matthews, T.W. Endoscopic reduction and internal cyanoacrylate fixation of the zygoma. *J. Otolaryngol.* **1997**, *26*, 2.
22. Meskin, S.W.; Ritterband, D.C.; Shapiro, D.E.; Kusmierczyk, J.; Schneider, S.S.; Seedor, J.A.; Koplin, R.S. Liquid bandage (2-octyl cyanoacrylate) as a temporary wound barrier in clear corneal cataract surgery. *Ophthalmology* **2005**, *112*, 2015–2021. [CrossRef]
23. Shermak, M.A.; Wong, L.; Inoue, N.; Crain, B.J.; Im, M.J.; Chao, E.; Manson, P.N. Fixation of the craniofacial skeleton with butyl-2-cyanoacrylate and its effects on histotoxicity and healing. *Plast. Reconstr. Surg.* **1998**, *102*, 309–318. [CrossRef] [PubMed]
24. Trott, A.T. Cyanoacrylate tissue adhesives: an advance in wound care. *JAMA* **1997**, *277*, 1559–1560. [CrossRef] [PubMed]
25. Eskandari, M.M.; Ozturk, O.G.; Eskandari, H.G.; Balli, E.; Yilmaz, C. Cyanoacrylate adhesive provides efficient local drug delivery. *Clin. Orthop. Relat. Res.* **2006**, *451*, 242–250. [CrossRef] [PubMed]
26. Williams, A. Chemical penetration enhancement, possibilities and problems. *Dermal Absorpt. Toxic. Assess.* **1998**, *12*, 645–651.
27. US Food and Drug Administration. *Inactive Ingredient Search for Approved Drug Products*; FDA Database: Silver Spring, MD, USA, 2017.
28. Hansen, H.; Wilbur, S.B. *Toxicological Profile for Acetone*; Agency for Toxic Substances and Disease Registry: Atlanta, GA, USA, 1994.
29. Gal, A.; Nussinovitch, A. Plasticizers in the manufacture of novel skin-bioadhesive patches. *Int. J. Pharm.* **2009**, *370*, 103–109. [CrossRef]
30. Bharkatiya, M.; Nema, R.; Bhatnagar, M. Designing and characterization of drug free patches for transdermal application. *Int. J. Pharm. Sci. Drug Res.* **2010**, *2*, 35–39.
31. Jain, P.; Banga, A.K. Induction and inhibition of crystallization in drug-in-adhesive-type transdermal patches. *Pharm. Res.* **2013**, *30*, 562–571. [CrossRef]
32. Liechty, W.B.; Kryscio, D.R.; Slaughter, B.V.; Peppas, N.A. Polymers for drug delivery systems. *Annu. Rev. Chem. Biomol. Eng.* **2010**, *1*, 149–173. [CrossRef]
33. Duncan, R. Polymer conjugates as anticancer nanomedicines. *Nat. Rev. Cancer* **2006**, *6*, 688. [CrossRef]
34. Naves, L.; Dhand, C.; Almeida, L.; Rajamani, L.; Ramakrishna, S.; Soares, G. Poly (lactic-*co*-glycolic) acid drug delivery systems through transdermal pathway: An overview. *Prog. Biomater.* **2017**, *6*, 1–11. [CrossRef]
35. Marsac, P.J.; Shamblin, S.L.; Taylor, L.S. Theoretical and practical approaches for prediction of drug–polymer miscibility and solubility. *Pharm. Res.* **2006**, *23*, 2417. [CrossRef]
36. Mahieu, A.l.; Willart, J.-F.o.; Dudognon, E.; Danède, F.; Descamps, M. A new protocol to determine the solubility of drugs into polymer matrixes. *Mol. Pharm.* **2013**, *10*, 560–566. [CrossRef]
37. Kim, J.-H.; Choi, H.-K. Effect of additives on the crystallization and the permeation of ketoprofen from adhesive matrix. *Int. J. Pharm.* **2002**, *236*, 81–85. [CrossRef]
38. Guideline, I.H.T. Impurities: Guideline for residual solvents Q3C (R5). *Curr. Step* **2005**, *4*, 1–25.
39. Baino, F. Biomaterials and implants for orbital floor repair. *Acta Biomater.* **2011**, *7*, 3248–3266. [CrossRef]
40. D'Souza, S.S.; Faraj, J.A.; DeLuca, P.P. A model-dependent approach to correlate accelerated with real-time release from biodegradable microspheres. *Aaps Pharmscitech* **2005**, *6*, E553–E564. [CrossRef]
41. Fredenberg, S.; Wahlgren, M.; Reslow, M.; Axelsson, A. The mechanisms of drug release in poly (lactic-*co*-glycolic acid)-based drug delivery systems—A review. *Int. J. Pharm.* **2011**, *415*, 34–52. [CrossRef]

42. Salamanca, C.; Barrera-Ocampo, A.; Lasso, J.; Camacho, N.; Yarce, C. Franz diffusion cell approach for pre-formulation characterisation of ketoprofen semi-solid dosage forms. *Pharmaceutics* **2018**, *10*, 148. [CrossRef]
43. Abd, E.; Yousef, S.A.; Pastore, M.N.; Telaprolu, K.; Mohammed, Y.H.; Namjoshi, S.; Grice, J.E.; Roberts, M.S. Skin models for the testing of transdermal drugs. *Clin. Pharmacol. Adv. Appl.* **2016**, *8*, 163. [CrossRef]
44. Miranda, M.; Sousa, J.J.; Veiga, F.; Cardoso, C.; Vitorino, C. Bioequivalence of topical generic products. Part 1: Where are we now? *Eur. J. Pharm. Sci.* **2018**, *123*, 260–267. [CrossRef]
45. Roberts, M. Correlation of Physicochemical Characteristics and In Vitro Permeation Test (IVPT) Results for Acyclovir and Metronidazole Topical Products. Available online: https://www.fda.gov/media/110256/download (accessed on 31 July 2019).
46. U.S. Food and Drug Administration. FY2016 Regulatory Science Report: Topical Dermatological Drug Products. Available online: https://www.fda.gov/industry/generic-drug-user-fee-amendments/fy2016-regulatory-science-report-topical-dermatological-drug-products (accessed on 31 July 2019).
47. Wypych, G. *Handbook of Plasticizers*; ChemTec Publishing: Toronto, ON, Canada, 2004.
48. Barhate, S.D.; Patel, M.; Sharma, A.S.; Nerkar, P.; Shankhpal, G. Formulation and evaluation of transdermal drug delivery system of carvedilol. *J. Pharm. Res.* **2009**, *2*, 663–665.
49. Chereddy, K.K.; Payen, V.L.; Préat, V. PLGA: From a classic drug carrier to a novel therapeutic activity contributor. *J. Control. Release* **2018**, *289*, 10–13. [CrossRef]
50. Martanto, W.; Davis, S.P.; Holiday, N.R.; Wang, J.; Gill, H.S.; Prausnitz, M.R. Transdermal delivery of insulin using microneedles in vivo. *Pharm. Res.* **2004**, *21*, 947–952. [CrossRef]
51. Farahani, T.D.; Entezami, A.A.; Mobedi, H.; Abtahi, M. Degradation of poly (D, L-lactide-*co*-glycolide) 50: 50 implant in aqueous medium. *Iran. Polym. J.* **2005**, *14*, 753–763.
52. Bré, L.P.; Zheng, Y.; Pêgo, A.P.; Wang, W. Taking tissue adhesives to the future: from traditional synthetic to new biomimetic approaches. *Biomater. Sci.* **2013**, *1*, 239–253. [CrossRef]
53. Leonard, F.; Kulkarni, R.; Brandes, G.; Nelson, J.; Cameron, J.J. Synthesis and degradation of poly (alkyl α-cyanoacrylates). *J. Appl. Polym. Sci.* **1966**, *10*, 259–272. [CrossRef]
54. Petersen, B.; Barkun, A.; Carpenter, S.; Chotiprasidhi, P.; Chuttani, R.; Silverman, W.; Hussain, N.; Liu, J.; Taitelbaum, G.; Ginsberg, G.G. Tissue adhesives and fibrin glues: November 2003. *Gastrointest. Endosc.* **2004**, *60*, 327–333. [CrossRef]
55. Kull, S.; Martinelli, I.; Briganti, E.; Losi, P.; Spiller, D.; Tonlorenzi, S.; Soldani, G. Glubran2 surgical glue: in vitro evaluation of adhesive and mechanical properties. *J. Surg. Res.* **2009**, *157*, e15–e21. [CrossRef]
56. Bhagat, V.; Becker, M.L. Degradable adhesives for surgery and tissue engineering. *Biomacromolecules* **2017**, *18*, 3009–3039. [CrossRef]

pharmaceutics

MDPI

Article

Photodynamic Therapy of Ovarian Carcinoma Cells with Curcumin-Loaded Biodegradable Polymeric Nanoparticles

Lili Duse [1,†], Michael Rene Agel [1,†], Shashank Reddy Pinnapireddy [1], Jens Schäfer [1], Mohammed A. Selo [2,3], Carsten Ehrhardt [2] and Udo Bakowsky [1,*]

1 Department of Pharmaceutics and Biopharmaceutics, University of Marburg, Robert-Koch-Str. 4, 35037 Marburg, Germany; lili.duse@pharmazie.uni-marburg.de (L.D.); michael.agel@pharmazie.uni-marburg.de (M.R.A.); shashank.pinnapireddy@pharmazie.uni-marburg.de (S.R.P.); j.schaefer@staff.uni-marburg.de (J.S.)
2 School of Pharmacy and Pharmaceutical Sciences and Biomedical Sciences Institute, Trinity College Dublin, Dublin 2, Ireland; mohammeda.mohsin@uokufa.edu.iq (M.A.S.); ehrhardc@tcd.ie (C.E.)
3 Faculty of Pharmacy, University of Kufa, 31001 Kufa, Iraq
* Correspondence: ubakowsky@aol.com; Tel.: +49-6421-282-5884
† Both authors contributed equally to this work.

Received: 20 May 2019; Accepted: 13 June 2019; Published: 15 June 2019

Abstract: Accumulation of photosensitisers in photodynamic therapy in healthy tissues is often the cause of unwanted side effects. Using nanoparticles, improved bioavailability and site-specific drug uptake can be achieved. In this study, curcumin, a natural product with anticancer properties, albeit with poor aqueous solubility, was encapsulated in biodegradable polymeric poly(lactic-*co*-glycolic acid) (PLGA) nanoparticles (CUR-NP). Dynamic light scattering, laser Doppler anemometry and atomic force microscopy were used to characterise the formulations. Using haemolysis, serum stability and activated partial thromboplastin time tests, the biocompatibility of CUR-NP was assessed. Particle uptake and accumulation were determined by confocal laser scanning microscopy. Therapeutic efficacy of the formulation was tested in SK-OV-3 human ovarian adenocarcinoma cells post low level LED irradiation by determining the generation of reactive oxygen species and cytotoxicity. Pharmacologic inhibitors of cellular uptake pathways were used to identify the particle uptake mechanism. CUR-NP exhibited better physicochemical properties such as stability in the presence of light and improved serum stability compared to free curcumin. In addition, the novel nanoformulation facilitated the use of higher amounts of curcumin and showed strong apoptotic effects on tumour cells.

Keywords: PLGA; nanoscaled drug delivery; LED; cancer; serum stability; reactive oxygen species; cellular uptake

1. Introduction

Light has long been used in the treatment of diseases as different as psoriasis, rickets or cancer [1]. With an alarming mortality rate, cancer is one of the most prevalent causes of death worldwide [2]. Hence, tremendous efforts have been made to develop novel safe, selective and effective cancer therapies. One such treatment, photodynamic therapy, exemplifies a novel minimally invasive tumour targeting therapy, in which tumour tissues can be selectively destroyed by three main mechanisms [3,4]. In the first case there is a combination of three individually inert entities viz. a photoactive drug molecule (photosensitiser; PS), oxygen and a light source of a specific wavelength, which upon combining, activate the photosensitiser and exhibit toxicity towards cancer cells and tissues [5,6]. This process of photosensitisation hugely relies upon the presence of molecular oxygen and forms the basis for a successful photodynamic therapy (PDT) [7]. After absorbing energy from the light source, the PS interacts with molecular oxygen via energy transfer process or electron transfer process and generates reactive oxygen species (ROS), which oxidise cellular and subcellular organelles to induce either apoptosis or necrosis leading to cell death [8,9]. Since cytotoxic ROS occurs only after irradiation of the PS, tumour tissues can be targeted with high precision. Due to its high reactivity and short half-life, ROS generation only targets cells lying in close proximity of the site of irradiation [1]. The half-life of singlet oxygen in biological systems is ~0.01–0.04 μs, with a restricted site of action of a radius spanning 0.01–0.02 μm [5,10]. PDT can also damage the tumour-associated vasculature, leading to tumour infarction. Finally, PDT can activate the immune response against tumour cells. All three mentioned mechanisms can also influence each other and act synergistically. PDT has an added advantage of almost no or minimal effect towards healthy tissues, making this therapy site specific [11]. It could also be combined with other therapies or immune-stimulatory agents like microbial adjuvants or cytokine for T-cell therapy and adoptive cellular therapy, respectively [1]. Another advantage of PDT is its effectiveness against otherwise chemoresistant cell types [12]. Furthermore, the photosensitiser can be administered by various means, such as systemically, locally or topically [1].

For the success of any PDT, it is important to choose an optimal PS. Despite many studies performed using different PS, only a few have reached the stage of advanced human clinical trials or even U.S. Food and Drug Administration (FDA) approval for clinical use [13]. A very promising photosensitiser for photodynamic treatment of tumours is curcumin. Curcumin (diferuloylmethane), is a naturally occurring yellowish polyphenol extract from the rhizomes of turmeric (*Curcuma longa*), which is cultivated widely in south and southeast tropical Asia [14]. Curcumin is the most active component of turmeric, making up approx. 0.5–3.14% of the dry weight of turmeric powder. Curcumin is used for a variety of therapeutic activities against many different diseases and conditions such as skin diseases, pulmonary and gastrointestinal systems, aches, pains, wounds, sprains, and liver disorders [15].

Offering a variety of potential applications, curcumin has been commonly used as a food-colouring agent (E100) as well as in pharmaceutical research due to its anti-inflammatory, anti-oxidative, anti-carcinogenic and anti-microbial effects [16,17]. It is a proven anticancer agent and therefore a perfect choice for a photosensitiser [18]. However, curcumin exhibits a very low level of bioavailability, since it is a highly lipophilic substance (solubility 11 ng/mL in phosphate buffer saline, PBS; pH 5) [19]. Systemically absorbed curcumin is prone to an extensive first-pass metabolism and undergoes a fast metabolic reduction with biliary excretion [20]. In 1978, Wahlstrom and Blennow observed the uptake of curcumin using Sprague–Dawley rats. After oral administration of 1 g/kg curcumin only negligible amounts of curcumin could be found in blood plasma [21].

These limitations can be overcome by the use of polymeric nanoparticles such as carrier systems, wherein the active substance is protected from degradation in the physiological environment and the bio-membrane permeability and cellular uptake of highly hydrophobic molecules are facilitated [17]. Several studies have revealed a remarkable enhancement of curcumin's phototoxicity through nanoparticle encapsulation [22–26]. One such polymer is PLGA, which is an FDA and European Medicines Agency (EMA) approved biocompatible and biodegradable polymer [27,28]. In the body, PLGA undergoes hydrolysis wherein the endogenous metabolite monomers lactic acid and glycolic acid are and easily metabolised by the body (to water and carbon dioxide) via Krebs cycle [27,29,30]. The degradation time of PLGA and the release of drug from the nanoparticles mainly depends on the nature of copolymer composition [31]. In this study, PLGA 50:50 was chosen due to its ability to hydrolyse much faster than other types of PLGA with a half-life (50% loss of molecular weight) of 15 days for microspheres [32,33].

Another important aspect in PDT is the choice of the irradiation device. There is an array of radiation devices available for PDT such as lasers or lamps [13]. In the current study, with an aim to make PDT more cost effective, efficient and safe, we have utilised a custom manufactured prototype low level light emitting diode (LED) array for the irradiation of PS [34]. LEDs are energy efficient, generate less thermal energy in the form of heat and are versatile in terms of structural form. They have the ability to emit light over a wide area with a relatively homogenous output, which enables treatment of larger lesions in fewer therapy sessions. Using a combination of curcumin loaded PLGA nanoparticles and LED-based PDT, superficial tumours (melanomas and lymphomas) and accessible adenocarcinomas (such as ovarian and cervical) could be treated effectively. The PLGA nanoparticles can be administered intratumourally or could be surface modified to enable systemic application.

In the present study, we have exploited the combination of biodegradable PLGA nanoparticles and LED-based PDT using curcumin as a photosensitiser. The nanoparticles have been characterised for their size and surface charge using Dynamic light scattering (DLS) and laser Doppler anemometry (LDA), respectively. Structural morphology was analysed using atomic force microscopy (AFM) and transmission electron microscopy (TEM). The photo-destructive effects have been evaluated in vitro by cytotoxicity assays and confocal laser scanning microscopy (CLSM). The cellular uptake in SK-OV-3 cells was also analysed using three different uptake inhibitors. Irradiation experiments were carried out in SK-OV-3 tumour cells and the oxidative stress induced during the irradiation was analysed by determining the inhibition of ROS. Furthermore, haemolysis assays using fresh blood, serum stability using serum and activated partial thromboplastin time tests (aPTT) using plasma were used to demonstrate the biocompatibility of the curcumin-loaded polymeric nanoparticles.

2. Materials and Methods

2.1. Materials

Curcumin (95% purity) was obtained from Alfa Aesar (Ward Hill, MA, USA); PLGA (Resomer RG 503 H) was supplied by Evonik (Essen, Germany); poly(vinyl alcohol) (PVA, Mowiol 4-88) was purchased from Kuraray (Hattersheim, Germany); polysorbate 80 (Tween 80), 3-(4,5-dimethylthiazol-2-yl)-2,5-diphenyltetrazolium bromide (MTT), 2′,7′-dichlorofluorescin diacetate (DCFDA) and tert-butyl hydroperoxide (TBHP) were obtained from Sigma-Aldrich Chemie (Taufkirchen, Germany). Inhibitors of cellular uptake, viz. Filipin III, chlorpromazine and dynasore were also obtained from Sigma Aldrich. Ultrapure water, generated by a PURELAB flex 4 device (ELGA LabWater, High Wycombe, UK) was used for all experiments. For cell culture studies, ultrapure water was additionally autoclaved and filter-sterilised using 0.2 µm polyethersulphone membrane filters (Sarstedt, Nümbrecht, Germany) prior to use. HPLC-grade ethyl acetate (VWR, Darmstadt, Germany) was used to prepare the organic phases. All other chemicals used were of analytical grade. All buffers used in this study were prepared in the laboratory, unless stated otherwise.

2.2. Light Source

The LED device used in this study was custom made by Lumundus (Eisenach, Germany) to be usable with microtiter plates. The device is equipped with an array of two different LEDs, capable of emitting light at wavelengths of 457 nm and 620 nm, respectively. Irradiation time, current (i.e., 20, 40, 60, 80, and 100 mA) and wavelength settings are adjustable as per the energy requirement. Radiation intensity was calculated based on the current and irradiation time.

2.3. Cell culture

The human ovarian adenocarcinoma cell line SK-OV-3 was procured from American Type Culture Collection (ATCC, Manassas, VA, USA). SK-OV-3 cells were cultivated in Iscove's modified Dulbecco's medium (Capricorn Scientific, Ebsdorfergrund, Germany) supplemented with 10% foetal bovine serum (Sigma-Aldrich, Taufkirchen, Germany) at 37 °C and 7% CO_2 in humidified atmosphere. The medium was replaced every other day and the cells were sub-cultured upon reaching 80% confluency.

2.4. Formulation of Nanoparticles

Curcumin-loaded nanoparticles (CUR-NP) and unloaded PLGA nanoparticles (NP) were prepared from PLGA by the emulsion–diffusion–evaporation technique as previously described by Kumar et al. [35]. In a pilot study, the method was optimised by varying formulation parameters (e.g., concentration of PLGA, PVA and curcumin; homogenisation speed and time). Briefly, 200 mg of PLGA were dissolved in 5 mL ethyl acetate at room temperature. Curcumin stock solution was prepared by dissolving curcumin in ethyl acetate (2 mg/mL). Equal volumes of PLGA and curcumin stock solutions were mixed and filtered through a 0.45 μm nylon syringe filter (Pall Corporation, New York, NY, USA). The organic solution was added dropwise to an aqueous solution of 2% PVA (*w*/*w*) and the resulting emulsion was stirred in a sealed tube for about 3 h, before homogenising for 10 min at 13,400 rpm using an Ultra-Turrax T25 homogeniser (IKA-Werke, Staufen, Germany). Nanoprecipitation was induced by adding water at a constant rate of 120 mL/h using a syringe pump (Perfusor, B. Braun, Melsungen, Germany). Finally, the organic solvent was evaporated by stirring overnight at room temperature in a dark environment. The nanosuspension was adjusted to a final volume of 50 mL and a concentration of 0.1 mg/mL curcumin. Unloaded nanoparticles were prepared by the same protocol, except that pure ethyl acetate was used instead of the curcumin stock solution. To remove agglomerates, the nanoformulations were centrifuged at 2000× *g* for 45 s using an Eppendorf Centrifuge 5418 (Eppendorf, Hamburg, Germany). The resulting pellet was discarded, and the supernatant was washed three times with water to separate the nanospheres from non-encapsulated curcumin. Between each washing step, centrifugation at 16,000× *g* for 45 min was performed. For prolonged storage, the CUR-NP were freeze-dried using an Alpha 1-4LSC lyophiliser (Martin Christ Gefriertrocknungsanlagen, Osterode am Harz, Germany) using 0.5% PVA as a cryoprotectant. Afterwards the nanoformulations were stored at 4 °C protected from light.

2.5. Dynamic Light Scattering and Laser Doppler Anemometry

To determine particle size distribution and zeta (ζ)-potential of the particles, DLS and LDA were used, respectively (Zetasizer Nano ZS, Malvern Instruments, Malvern, UK). A viscosity of 0.88 mPa × s and a refractive index of 1.33 of water at 25 °C were assumed for data interpretation. Measurement position and laser attenuation were automatically adjusted by the instrument. The instrument performs 15 size runs per measurement with each lasting 10 s. For ζ-potential measurements, the instrument automatically performs 15–100 runs per measurement, depending upon the sample. All samples were diluted with filtered PBS buffer (1:100) [36]. Data from at least three independent experiments were measured for both DLS and LDA analysis.

2.6. Morphological Characterisation

To investigate the morphology of CUR-NP and to confirm size data obtained by DLS measurements, AFM was performed using a Nanowizard 3®(JPK Instruments, Berlin, Germany) and a Digital Nanoscope IV Bioscope (Veeco Instruments, Santa Barbara, CA, USA), respectively. Formulations were diluted 1:100 with water and 20 μL of the diluted sample were placed onto a silica wafer or an untreated microscopic glass slide. The samples were left to settle onto the surface for a few minutes and the remaining fluid was absorbed by a lint-free wipe (Kimtech Precision Wipes, Kimberly-Clark, Fullerton, CA, USA). Measurements were performed in tapping mode, in which the cantilever oscillated with determined amplitude close to its resonance frequency, with scan rates from 0.5 to 1 Hz. A HQ:NSC16/AL_BS (Anfatec Instruments, Oelsnitz, Germany) cantilever was used [37]. Data were processed using JPKSPM data processing software (v. 5.1.8, JPK instruments).

For the TEM analysis, the nanoparticle suspension was applied onto 300-mesh copper grids. Samples were then negative stained thrice with 2% uranyl acetate (pH 4.2), which was alternated by washing steps with water. The samples were then allowed to dry overnight before being examined under the TEM (LEO 912 AB, Carl Zeiss, Jena, Germany) [22].

2.7. Yield, Encapsulation Efficiency and Loading Capacity

The freeze-dried nanoparticles were weighted, and the percentage yield was calculated using the following equation

$$\%\text{Yield} = \frac{\text{Dry weight of nanoparticles}}{\text{Weight of drug and polymer used for NP preparation}} \times 100 \quad (1)$$

The encapsulation efficiency (%EE) was determined by extracting curcumin from nanospheres that were previously washed and separated from free curcumin, as previously described in Section 2.4. The nanosuspension was mixed with an equal volume of acetonitrile (a solvent for PLGA and curcumin) and sonicated for 20 min. The absorbance of this solution was measured spectrophotometrically at 425 nm using a Multiskan GO micro plate reader (Thermo Fischer Scientific, Waltham, MA, USA), and the amount of drug quantified using a calibration curve recorded with known curcumin concentrations. For determination of the total drug content, nanoparticles without any washing procedure were measured in the same way. The %EE (*w*/*w*) was calculated as follows:

$$\%\text{EE} = \frac{\text{Drug}_{\text{encapsulated}}}{\text{Drug}_{\text{total}}} \times 100 \quad (2)$$

In addition, the loading capacity (LC) was calculated using the equation:

$$\%\text{LC} = \frac{\text{Drug}_{\text{encapsulated}}}{\text{Dry weight of nanoparticles}} \times 100 \quad (3)$$

2.8. In Vitro Drug Release

For determination of drug release, 10 mg lyophilised nanoparticles were redispersed in 30 mL PBS buffer (pH 5.5 and pH 7.4) containing 1% polysorbate 80 (*v*/*v*) to assure sink conditions. This dispersion was stored under absence of light in a shaking incubator (TH 15/KS 15, Edmund Bühler, Bodelshausen, Germany) at 37 °C under slight agitation (50 rpm). For the following seven days, 500 μL samples were drawn and replaced with fresh medium. The samples were centrifuged for 10 min at 16,000× *g* (Centrifuge 5418, Eppendorf AG) and the amount of curcumin present in the supernatant was determined spectrophotometrically at 425 nm, using Multiskan GO micro plate reader (Thermo Fischer Scientific).

2.9. Serum Stability

To study the stability of the formulations in physiologically relevant conditions, 2 mL serum was diluted with 20 mM HEPES (pH 7.4) to make a 60% serum solution. Curcumin-loaded PLGA nanoparticles were added in a ratio of 1:5 (*v*/*v*). The mixture was incubated at 37 °C in a shaking incubator at 100 rpm for 24 h. Samples were further diluted to a ratio of 1:20 (*v*/*v*) with 20 mM HEPES (pH 7.4) prior to DLS and LDA analyses at different time points. All measurements were carried out in triplicates [38].

2.10. In Vitro Irradiation Experiments

Cells were seeded onto in 96-well plates (10,000 cells/0.35 cm^2 (per well); Nunclon Delta, Thermo Fisher Scientific, Waltham, MA, USA) and were allowed to adhere overnight. Various concentrations of CUR-NP and free curcumin (dissolved in DMSO) were added to the wells and incubated for 4 h. Irradiation was performed at a wavelength of 457 nm with a radiation fluence of 8.6 J/cm^2 and the plates were incubated overnight. Dark (un-irradiated) plates were used as control [39]. Subsequently, the medium was replaced with fresh medium containing MTT dye and cells were incubated for 4 h. Absorbance of the resulting formazan crystals dissolved in DMSO was measured using a plate reader

(FLUOstar Optima; BMG Labtech, Offenburg, Germany) at 570 nm. Viability of untreated cells was considered as 100%.

2.11. Cellular Uptake Studies

To determine the cellular uptake mechanism, SK-OV-3 cells were seeded at a density of 10,000 cells/0.35 cm^2 (per well) in 96-well plates (Nunclon Delta, Nunc/Thermo Fischer Scientific, Waltham, MA, USA) and incubated overnight. The next day, cells were washed with PBS containing Ca^{2+} and Mg^{2+} (pH 7.4) and were incubated for 30 min with pharmacological inhibitors of different endocytic mechanisms viz. dynasore (80 μM), chlorpromazine (14 μM) and Filipin III (8 mM) [40]. After incubation with the uptake inhibitors, cells were washed again with PBS and exposed to 50 μM CUR-NP or free curcumin (dissolved in DMSO) for 4 h. After irradiation at 457 nm with a radiation fluence of 8.6 J/cm^2, cells were incubated with MTT and the absorbance from resulting formazan crystals (dissolved in DMSO) was determined at 570 nm as described above.

2.12. Reactive Oxygen Species

ROS were determined using 2′,7′-dichlorofluorescin diacetate (DCFDA, Abcam, Cambridge, UK) according to the supplier's protocol with slight modifications [34]. Briefly, SK-OV-3 cells were seeded in 96-well plates as mentioned above. The next day, cells were incubated with CUR-NP for 4 h. Tert-butyl hydroperoxide (TBHP, 50 μM) was used as positive control. Cells were subsequently washed using PBS containing Ca^{2+} and Mg^{2+} (pH 7.4) and supplemented with fresh medium. Irradiation was carried out at 457 nm with a radiation fluence of 8.6 J/cm^2. The cells were washed again with PBS and incubated with culture medium (IMDM without phenol red) containing 25 μM DCFDA for 1 h. Cell culture lysis reagent (Promega, Mannheim, Germany) was used to lyse the cell and fluorescence was recorded using a FLUOstar Optima plate reader (λ_{ex} 480 nm/λ_{em} 520 nm).

2.13. Intracellular Visualisation of PDT

Cells were seeded onto 12-well plates (90,000 cells/3.5 cm^2 (per well); Nunclon Delta) containing cover slips (15 mm in diameter). The plates were incubated for 24 h, before being used for the experiments. CUR-NP suspension was added dropwise to each well. After 4 h, the supernatant was removed and cells were washed twice with PBS containing Ca^{2+} and Mg^{2+} (pH 7.4). Cell were treated with a 4% formaldehyde solution for 20 min to fix them. The nucleus was counterstained with 0.1 μg/mL 4′,6-diamidino-2-phenylindole (DAPI) for 20 min. Finally, the cells were washed again with PBS (pH 7.4) and the cover slips were mounted onto slides. Samples were examined under a LSM700 confocal laser-scanning microscope (Carl Zeiss Microscopy, Jena, Germany).

2.14. Activated Partial Thromboplastin Time Test

An aPTT test was performed as previously described, using a Coatron M1 coagulation analyser (Teco, Neufahrn, Germany) to determine the effect of the formulations on blood coagulation [41]. Briefly, 25 μL of plasma was mixed with 25 μL each of sample and of aPTT reagent followed by addition of an equal volume of pre-warmed 0.025 M calcium chloride solution to activate coagulation. Coagulation was determined spectrophotometrically.

2.15. Haemolysis Assay

To determine the effect of the formulations on blood, human erythrocytes were isolated from fresh blood as described previously [42]. Briefly, following prior consent from the donor, whole blood was collected into EDTA tubes as centrifuged to separate the red blood cell (RBC) pellet. The RBC pellet was washed thrice with PBS buffer (pH 7.4) and diluted to a ratio of 1:50 with PBS. The erythrocytes were incubated together with the formulations in V-bottom microtiter plates placed in an orbital shaker KS4000 IC (IKA Werke) for 1 h at 37 °C. Supernatants were collected from the plates following centrifugation and the absorbance was determined at 540 nm using a plate reader (FLUOstar Optima). PBS (pH 7.4) and 1% Triton X-100 were used as controls.

2.16. Statistical Analysis

All experiments were performed in triplicates and data are presented as mean ± standard deviation, unless otherwise stated. Two-tailed Student's *t*-test was performed to identify statistical significance differences.

3. Results

3.1. Physicochemical Characterisation

Preliminary experiments were performed to determine an ideal stabiliser concentration, homogenisation speed and curcumin content. In emulsion–diffusion–evaporation technique, PVA acts as a stabiliser of the nanodroplets and is one of the widely used emulsifier for polymeric nanoparticle preparations. Both, particle size and size distribution are affected by the concentration of PVA in the aqueous phase during particle preparation [35]. After testing different PVA concentrations, we narrowed down to 2% PVA considering the physicochemical properties of the resultant nanoformulations. The Z-average data from DLS as well as LDA results are summarised in Table 1. CUR-NP and NP showed a narrow size distribution indicating a high reproducibility. No extensive change in particle size was noticeable after curcumin loading. This was also the case for lyophilised samples, thereby showing that lyophilisation has no deleterious effects on the physicochemical characteristics. Furthermore, LDA revealed a slightly negative ζ-potential at pH 7.4, regardless of curcumin loading for both formulations.

Table 1. Hydrodynamic diameter, polydispersity index (PDI) and ζ-potential of curcumin-loaded poly(lactic-*co*-glycolic acid) (PLGA) nanoparticles (CUR-NP) and unloaded PLGA nanoparticles (NP). Lyophilised samples represent samples resuspended after lyophilisation. Hydrodynamic diameters are expressed as a measure of particle size distribution by intensity. All samples were measured in triplicates ($n = 9$, independent formulations) and results are expressed as means ± SD.

Sample	Diameter [nm]	PDI	ζ-potential [mV]
NP	194.7 ± 8.7	0.09 ± 0.05	−5.33 ± 0.88
CUR-NP	203.6 ± 7.8	0.08 ± 0.04	−5.24 ± 0.86
Lyophilised NP	195.0 ± 6.9	0.09 ± 0.05	−5.09 ± 0.73
Lyophilised CUR-NP	201.8 ± 6.0	0.09 ± 0.03	−5.43 ± 0.67

PLGA in an aqueous environment undergoes hydrolysis spanning over several weeks, depending on the ratio of lactic acid to glycolic acid. The PLGA 50:50 used in this study therefore exhibits the fastest degradation [43]. To increase the storage stability the nanoparticles were lyophilised, using PVA as cryoprotectant. Freeze-dried and resuspended nanoparticles showed no extensive change in particle size (CUR-NP +0.3 nm, NP −1.8 nm) or polydispersity index (data not shown).

For the determination of encapsulation efficiency, a direct method of dissolving the purified particles and measuring the actual amount of curcumin available in the nanoparticle formulation was used. Since curcumin is near to insoluble in water, un-encapsulated free curcumin should exist in crystalline form. In this case the crystals would sediment during ultracentrifugation and be present in the pellet. Hence discarding the supernatant would not lead to sufficient separation of nanoparticles from crystalline curcumin. Therefore (in contrast to NP), a first separation step of centrifugation with a relatively low force of 2000× *g* is necessary to sediment the larger crystals. The optimised formulation of CUR-NP showed a relatively high encapsulation efficiency (EE) with an actual loading capacity of 2% (Table 2). The good EE results can be explained by the fact that PLGA and curcumin are both soluble in ethyl acetate, which was employed for the nanoparticle preparation process. Moreover, PLGA's ability to effectively encapsulate curcumin has been demonstrated in previous studies [44–46].

Table 2. Theoretical load, percentage of yield, encapsulation efficiency and loading capacity of curcumin-loaded poly(lactic-*co*-glycolic acid) (PLGA) nanoparticles (CUR-NP) and unloaded PLGA nanoparticles (NP). All samples were measured in triplicates ($n = 9$, independent formulations) and results are expressed as means ± SD.

Sample	Theoretical Load [%] [a]	%Yield	%EE	%LC
NP	-	80.5 ± 2.3	-	-
CUR-NP	2.5	87.5 ± 0.7	80.4 ± 10.6	2.0 ± 0.3

[a] % *w/w* loading of curcumin to polymer; EE, encapsulation efficiency; LC, loading capacity.

3.2. Morphological Characterisation

In all images (Figure 1), round-shaped CUR-NP with a monomodal size distribution are clearly visible. Due to dilution of pure nanoparticle samples, even single CUR-NPs with a smooth surface indicating that the curcumin is completely incorporated could be visualised. The AFM size analysis was in agreement with the DLS measurements. It should however be noted that the differences in size arising from the DLS and AFM measurements is because the hydrodynamic diameter is obtained in aqueous conditions whereas the latter is measured under atmospheric conditions [47]. Furthermore, a PVA corona was noticeable in TEM micrographs (Figure 1C). While free PVA was removed through centrifugation and washing of the redispersed nanoparticles prior to each experiment, it is a known fact, that small quantities of PVA still remain on the surface of PLGA nanoparticles, which cannot be removed even by extensive washing. This strong adsorption of PVA may be caused by hydrophobic bonding of PVA's hydroxyl groups to the acetyl groups of PLGA [48].

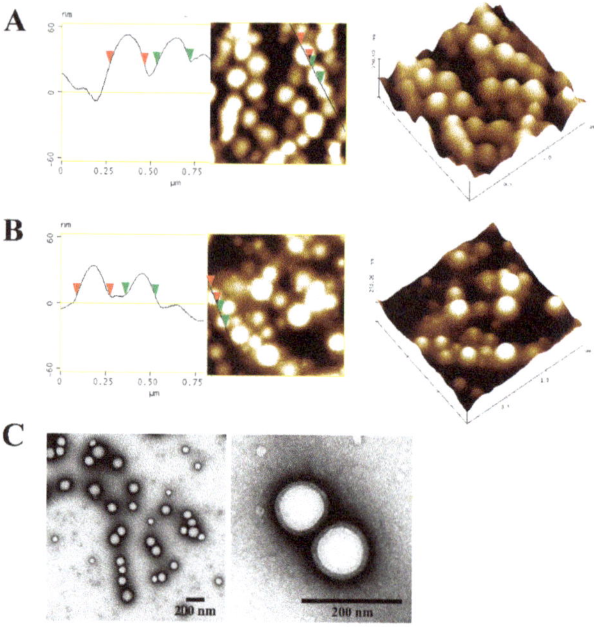

Figure 1. Morphology of representative curcumin-loaded poly(lactic-*co*-glycolic acid) (PLGA) nanoparticles (CUR-NP) shown by atomic force microscopy (AFM) (**A**,**B**) and transmission electron microscopy (TEM) (**C**). (**A**) Freshly prepared nanoparticles, (**B**,**C**) nanoparticles post lyophilisation. All AFM micrographs are presented in height mode. Samples were negatively stained with 2% uranyl acetate prior to TEM measurement. Scale bars in TEM micrographs represent 200 nm.

3.3. *In Vitro Drug Release*

Different mechanisms were reported for the drug release from PLGA nanoparticles: (i) desorption of drug absorbed on the particles surface, (ii) diffusion through the polymer matrix, (iii) erosion of the polymer matrix, and (iv) a combination of erosion and diffusion processes [29,49].

The cumulative drug release of curcumin from CUR-NP at pH 5.5 and 7.4 is shown in Figure 2. Due to curcumin's low water solubility (~0.01 µg/mL at pH 5 and ~0.4 µg/mL at pH 7.3) and low stability at neutral to basic pH, the addition of a solubility-enhancing component was necessary to assure sink conditions and to achieve UV/VIS detectable concentrations [50]. We chose polysorbate 80 for this purpose, since it is reported, that it is capable of increasing the solubility of hydrophobic drugs and protecting them against degradation through the formation of micelles [51].

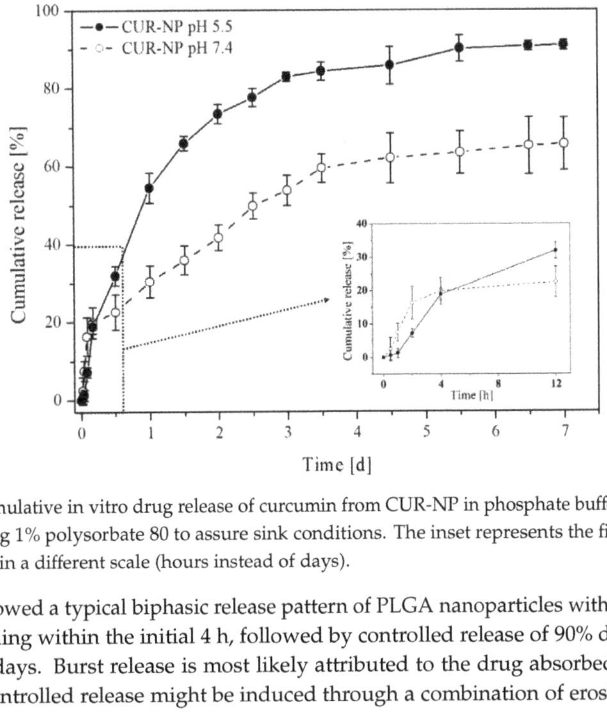

Figure 2. Cumulative in vitro drug release of curcumin from CUR-NP in phosphate buffer (pH 5.5 and 7.4) containing 1% polysorbate 80 to assure sink conditions. The inset represents the first 12 h of the release study in a different scale (hours instead of days).

CUR-NP showed a typical biphasic release pattern of PLGA nanoparticles with a burst release of around 20% loading within the initial 4 h, followed by controlled release of 90% drug loading over the following 7 days. Burst release is most likely attributed to the drug absorbed to the particle's surface, while controlled release might be induced through a combination of erosion and diffusion processes [52,53]. Furthermore, drug release was much faster in acidic medium compared to pH 7.4 which could be caused by a faster degradation rate of PLGA under acidic conditions [33].

Tumour cells often exhibit an extracellular pH of around 6.5, which would lead to a faster release of curcumin in this acidic environment and therefore an increased cytotoxicity compared to healthy tissue [54,55]. Since the pH of endosomes is even lower (pH ~5.5), drug release of internalised nanoparticles would proceed even faster, leading to higher intracellular curcumin concentration and enhanced tumour damage [56]. Both mechanisms could be used for passive tumour-targeting strategy.

3.4. *Serum Stability*

Serum stability assay was performed to evaluate the susceptibility of the formulations. For this purpose, DLS and LDA analyses were performed after incubating the nanoparticles in serum. The experiments were carried at 37 °C in a shaking incubator to simulate physiological conditions. The results show shrinkage in the nanoparticle size after 24 h incubation time in serum (Table 3). This could be related to a release of curcumin bound to the particle surface, which leads to particle erosion [57]. Furthermore, the decrease in particle size could be caused by osmotic forces, as reported previously [58].

The increase in the PDI of the nanoparticles could be attributed to the decrease in homogeneity in the presence of serum. After an initial increase (as compared to DLS measurements from native particles), the ζ-potential decreased to more negative values. The initial increase could be due to the accumulation of plasma proteins on the surface of the NP and the subsequent decrease could be attributed to the drug release (curcumin) onto the surface of the NP [59].

Table 3. Physicochemical changes of poly(lactic-*co*-glycolic acid) (PLGA) nanoparticles containing 0.1 mg/mL curcumin. Particles were incubated for 24 h in serum at a volume ratio of 5:1. The hydrodynamic diameter is expressed as a measure of particle size distribution by intensity. All samples were measured in triplicates ($n = 9$, independent formulations) and data are expressed as means ± SD.

Time [h]	Diameter [nm]	PDI	ζ-potential [mV]
0	183.50 ± 2.11	0.19 ± 0.01	2.19 ± 0.49
1	180.80 ± 3.70	0.19 ± 0.01	−8.30 ± 0.21
4	180.45 ± 3.32	0.22 ± 0.01	−12.65 ± 0.35
24	173.57 ± 2.21	0.22 ± 0.01	−13.30 ± 0.10

3.5. *Irradiation Experiments*

For the photo destructive effect of PDT, SK-OV-3 ovarian carcinoma cells were incubated with different concentrations of CUR-NP und free curcumin for 4 h and were irradiated with different radiation fluence. The efficacy of the irradiation was determined by MTT assay. As shown in Figure 3A, after treatment of cells, curcumin-loaded biodegradable nanoparticles and free curcumin dissolved in DMSO show a high photocytotoxic effect. There is a remarkable difference between the dark (non-irradiated) and irradiated plates. An incubation time of 4 h was found to be ideal among all the formulations, since burst release of around 20% curcumin loading from CUR-NP was found within 4 h of drug release studies (Figure 2). In the initial studies, a radiation fluence of 1.4, 4.3, 8.6 and 13.2 J/cm^2 were tested using CUR-NP and for the subsequent studies, the fluence was narrowed down to 8.6 J/cm^2 for the LED induced PDT (data not shown). Beyond a curcumin concentration of 50 μM, there was no further difference in the effect of PDT. It can also be seen that free curcumin dissolved in DMSO, induced a higher photocytotoxicity effect, which could be attributed to controlled release properties of PLGA nanoparticles, as this might be a rate limiting step for the drug being available for light activation. Nevertheless, since DMSO-dissolved curcumin is not suitable for therapeutic applications and due to its hydrophobic nature, we have used a biodegradable polymeric PLGA nanoformulation to increase the bioavailability of curcumin and to make it suitable for therapeutic applications [60]. The internalisation and subsequent localisation of the photosensitiser within the tumour determine the outcome of the therapy.

3.6. *Subcellular Localisation of Curcumin*

To visualise the effect of PDT on tumour cells, qualitative fluorescence microscopic analysis was performed after irradiation (457 nm; 8.6 J/cm^2) of the cells with 50 μM CUR-NP and free curcumin dissolved in DMSO. The cells were fixed with 4% formaldehyde solution and the cell nucleus was counterstained with 300 nM DAPI. Substantial intracellular localisation of curcumin near the cell's nucleus could be seen in both dark and irradiated samples, as shown in Figure 3B. Upon irradiation, photo-destruction induced by PDT could be clearly observed in the micrographs, which is evident from the nuclear perforation. This might be due to the chromatin condensation and DNA fragmentation caused by the curcumin induced PDT [8]. Comparing free curcumin dissolved in DMSO with curcumin-loaded nanoparticles reveals that CUR-NPs indeed damage the nucleus while only free curcumin alone is not as damaging although the micrographs show a cellular uptake of curcumin by both the samples. This confirms the hypothesis that curcumin particles are internalised by uptake mechanisms such as endocytosis and free curcumin penetrates the cells by diffusion and loses its effect (Section 3.6). It could also be seen that upon irradiation, the fluorescence of free curcumin increased and was more distributed than the CUR-NP. Furthermore, the irradiated blank/untreated cells show no photo-destruction.

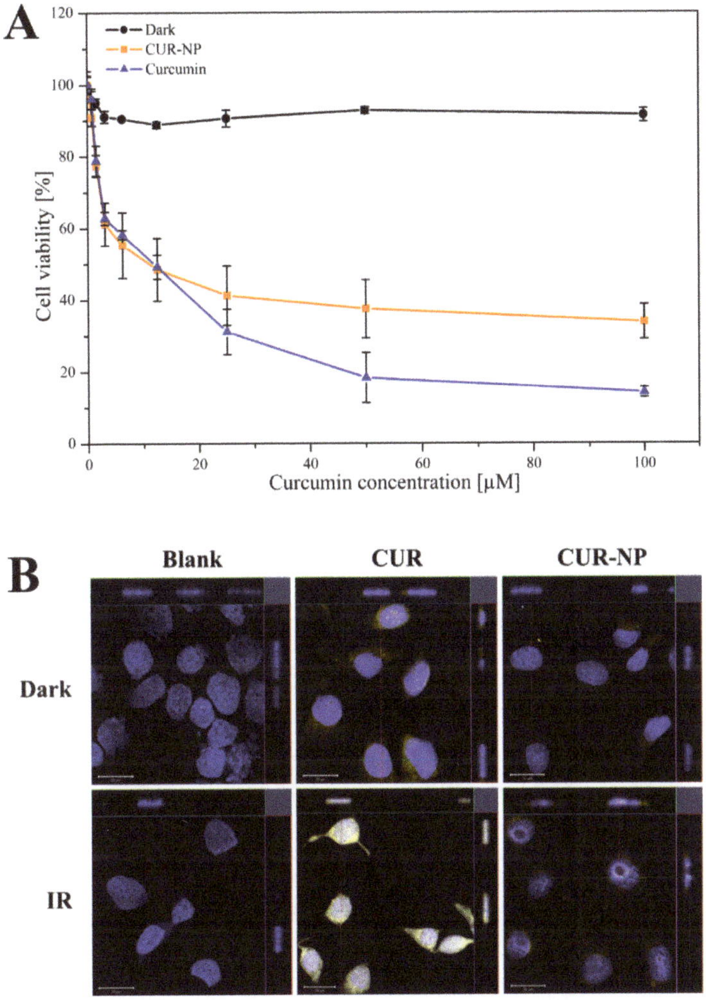

Figure 3. Phototoxic effect of curcumin-loaded poly(lactic-*co*-glycolic acid) PLGA nanoparticles (CUR-NP) and free curcumin dissolved in DMSO (CUR) on SK-OV-3 ovarian carcinoma cells: (**A**) for the MTT-assay, either the nanoformulation or free curcumin was incubated for 4 h at 37 °C and were irradiated at 457 nm with a radiation fluence of 8.6 J/cm^2. Dark was used as negative control and represents cells without irradiation. All samples contain 0.1 mg/mL curcumin and were measured in triplicates (n = 9, independent formulations). Results are expressed as means ± SD. (**B**) CLSM micrographs of SK-OV-3 cells incubated with CUR-NP or free curcumin for 4 h at 37 °C and subsequent irradiation (457 nm, 8.6 J/cm^2). The cell nucleus was counterstained with 0.1 μg/mL DAPI and was fixed with 4% formaldehyde solution. The curcumin concentration in CUR-NP was 50 μM. Nuclear damage is clearly witnessed in the irradiated samples whereas in the dark, the nucleus is intact. Dark was used as negative control and represents cells without irradiation. Scale bars denote 20 μm.

3.7. Cellular Uptake

Different cells use different endocytic pathways for the internalisation of nanoparticles. To study cellular uptake, several different inhibitors of endocytic pathways were utilised. For this purpose, the dynamin-, clathrin- and clathrin-independent endocytosis pathways were considered. Dynasore, a GTPase inhibitor rapidly and reversibly inhibits dynamin activity, which is essential for membrane fission during clathrin-mediated endocytosis (CME). Chlorpromazine also causes a block in CME by inducing the assembly of adaptor proteins and clathrin, leading to the formation of endosomes which then fuse with lysosomes [61]. Chlorpromazine also interferes with the intracellular clathrin processing [62]. Filipin III derived from *Streptomyces filipensis* is a polyene macrolide antibiotic, which inhibits the raft/caveolae endocytosis pathway. It interacts with cholesterol whose presence and state influence endocytic functions [63]. Figure 4 shows a substantial inhibition of CUR-NP by Filipin III and chlorpromazine. The uptake mechanism is dependent on the cell line and hence the data should be regarded cautiously. In our case, chlorpromazine exhibited an 80–90% inhibition of clathrin-dependent endocytosis of CUR-NP and Filipin III a marked 100% inhibition of caveolae-mediated and lipid-raft mediated endocytosis but both could not inhibit free curcumin. Free curcumin dissolved in DMSO does not seem to be internalised by any of these endocytic pathways suggesting a different internalisation mechanism such as diffusion [64]. The inhibition by dynamin dependant endocytosis by Dynasore showed no effect on the CUR-NP.

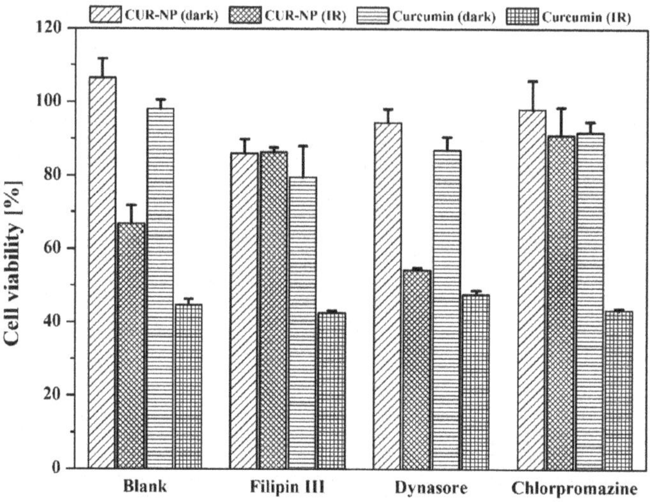

Figure 4. Cellular uptake of curcumin-loaded poly(lactic-*co*-glycolic acid) (PLGA) nanoparticles (CUR-NP) and free curcumin (dissolved in DMSO) in SK-OV-3 cells in presence of three specific inhibitors. After 30 min incubation time with different inhibitors (i.e., 8 μM Filipin III, 80 μM Dynasore and 14 μM chlorpromazine), cells were incubated with 50 μM of either CUR-NP or free curcumin dissolved in DMSO for 4 h at 37 °C and were irradiated at 457 nm with a radiation fluence of 8.6 J/cm^2 (IR). The MTT assay was performed at the end of the experiment to determine the effect of the pathway inhibits on internalisation. Viability of untreated cells was considered as 100%. Dark was used as negative control and represents unirradiated samples. Blank represents cells without any inhibitor. All samples were measured in triplicates (n = 9, independent formulations) and results are expressed as means ± SD.

3.8. Reactive Oxygen Species

ROS generation is the backbone of PDT and is a determining factor in the efficacy of the therapy. As shown in Figure 5, curcumin induced the production of cytotoxic reactive oxygen species after cells were irradiated (50 μM curcumin; 8.6 J/cm^2 radiation fluence). ROS was determined by the measurement of the fluorescence from the lysed cells. Upon incubation with cells, a deacetylated form of DCFDA is oxidised by the resulting ROS to form fluorescent DCF [65]. As expected, curcumin-loaded biodegradable PLGA polymeric nanoparticles exhibited the highest ROS generation followed by free curcumin dissolved in DMSO thereby corresponding to the results of the PDT. The fact that CUR-NP caused the maximum damage to the tumour cells can be explained by the ROS generation, which in this case was intracellular and could be only detected from the cells, which have internalised. Since free curcumin was not internalised by the endocytic pathways, mentioned in Figure 4, the uptake from CUR-NP seems to be higher. As expected, blank (negative control) did not show any ROS generation.

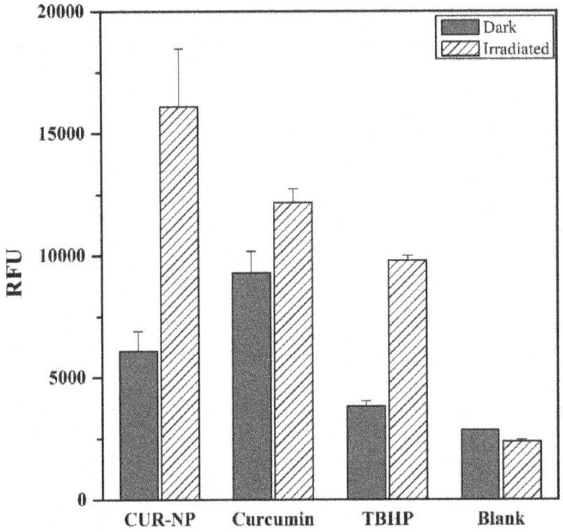

Figure 5. Production of cellular reactive oxygen species (ROS) in response to curcumin-loaded poly(lactic-*co*-glycolic acid) (PLGA) nanoparticles (CUR-NP) or free curcumin (0.1 mg/mL). ROS was measured in SK-OV-3 cells after exposure (for 45 min) to 2′,7′-dichlorofluorescin diacetate (25 μM) and incubation with CUR-NP or free curcumin for 4 h at 37 °C with subsequent irradiation (457 nm, 8.6 J/cm^2). Tert-butyl hydroperoxide (TBHP) was used as positive control and non-irradiated cells served as negative control (Dark), whereas Blank represents untreated cells. Values are represented in relative fluorescence units (RFU). All samples were measured in triplicates (n = 9, independent formulations) and results are expressed as means ± SD.

3.9. *Haemocompatibility*

To study the effect of the nanoformulation on human erythrocytes, haemolysis assay, which determines the amount of haemoglobin released from the erythrocytes upon exposure to NPs (Figure 6A) was performed. To determine the change in coagulation time upon addition of NPs, aPTT was determined (Figure 6B). Upon its release from erythrocytes, haemoglobin reacts with the atmospheric oxygen to form oxyhaemoglobin [66]. The CUR-NP used in the study showed a minimal haemolytic potential. An increase in the aPTT time was also found to be minimal in the case of CUR-NP. The aPTT analysis revealed that the encapsulated curcumin increased the coagulation time by only 4.01 s, whereas upon addition of free curcumin alone, the coagulation time increased to 120.13 s suggesting an interaction of curcumin with the intrinsic proteins or coagulation factors. The normal coagulation time for the plasma was tested to be 32.2 ± 0.1 s and was found to be in the normal range [67]. Coagulation time between 30 and 40 s is considered within acceptable range and coagulation time above 70 s denotes spontaneous and continuous bleeding leaving the patients with the risk of haemorrhage [67–69].

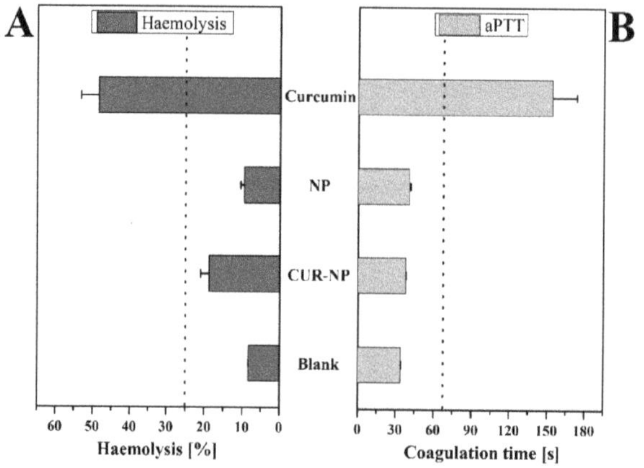

Figure 6. Blood compatibility of curcumin-loaded poly(lactic-*co*-glycolic acid) (PLGA) nanoparticles (CUR-NP), unloaded PLGA nanoparticles (NP) or free curcumin: (**A**) haemolysis assay and (**B**) aPTT test were performed. Where applicable, curcumin was used at a concentration of 0.1 mg/mL. Blank represents erythrocytes in the haemolysis assay and blood plasma in the aPTT test, respectively. Triton X-100 was used as positive control in (**A**) and is considered as 100% haemolysis. All samples were measured in triplicates ($n = 9$, independent formulations) and results are expressed as means ± SD. Dotted lines indicate the tolerance/threshold limits.

4. Conclusions

In this study, curcumin-loaded biodegradable PLGA nanoparticles (CUR-NP) were formulated for photodynamic therapy using a custom manufactured prototype low power LED device. CUR-NP exhibited better physicochemical properties compared to free curcumin, such as improved serum stability. Furthermore, haemocompatibility of curcumin was improved through polymeric encapsulation, minimising the risk of haemorrhages in patients. Successful cellular uptake of CUR-NP was demonstrated with the aid of confocal laser scanning microscopy, which showed photo-destruction and nuclear perforation. Nanoformulation facilitated the use of higher amounts of curcumin, which would otherwise be impossible due to its poor solubility and stability in aqueous media. Curcumin, which elsewise is safe for normal cells, showed cytotoxic effects on tumour cells upon irradiation at a low intensity therefore selectively inhibiting tumour growth. Since production of ROS occurs only upon irradiation of the intracellular photosensitiser, tumour tissues can be targeted with precision. Using LED as an irradiation source offers a number of advantages such as portability, durability and economical compared to lasers. Moreover, LED's have the ability to irradiate a relatively larger area, which could be beneficial for the treatment of large tumours or even skin cancer. The use of visible light of low-energy makes this therapy safer for both the patient and the operator. Finally, the good biocompatibility of this well-documented and highly reproducible formulation process of polymeric nanoparticles are significant advantages of this anti-tumour therapy. Transforming this novel strategy into a safe and effective therapy by combining PDT with other therapies would be our prime research interest in the future.

Author Contributions: Study conceptualization, U.B.; methodology, L.D. and M.R.A.; validation, L.D. and M.R.A.; investigation, L.D. and M.R.A.; data curation, L.D. and M.R.A.; visualisation, S.R.P.; manuscript drafting, L.D. and M.R.A.; manuscript review and editing, U.B., C.E., S.R.P., and M.A.S.; project supervision, J.S.

Funding: This research received no external funding.

Acknowledgments: The authors would like to thank Eva Maria Mohr for her help and technical assistance, Thomas Anders for performing the serum stability studies and Elias Baghdan for his support and motivation. The authors are also grateful to Lumundus GmbH for developing the prototype LED device used in this study and would like to express their gratitude towards Martin Hardt and Sabine Agel from imaging unit of Justus-Liebig-University Gießen.

Conflicts of Interest: The authors declare no conflict of interest.

References

1. Dolmans, D.E.J.G.J.; Fukumura, D.; Jain, R.K. Photodynamic therapy for cancer. *Nat. Rev. Cancer* **2003**, *3*, 380–387. [CrossRef] [PubMed]
2. Ferlay, J.; Shin, H.-R.; Bray, F.; Forman, D.; Mathers, C.; Parkin, D.M. Estimates of worldwide burden of cancer in 2008: GLOBOCAN 2008. *Int. J. Cancer* **2010**, *127*, 2893–2917. [CrossRef] [PubMed]
3. Delaney, T.F.; Sindelar, W.F.; Tochner, Z.; Smith, P.D.; Friauf, W.S.; Thomas, G.; Dachowski, L.; Cole, J.W.; Steinberg, S.M.; Glatstein, E.; et al. Phase I study of debulking surgery and photodynamic therapy for disseminated intraperitoneal tumors. *Int. J. Radiat. Oncol. Biol. Phys.* **1993**, *25*, 445–457. [CrossRef]
4. Pass, H.I.; Delaney, T.F.; Tochner, Z.; Smith, P.E.; Temeck, B.K.; Pogrebniak, H.W.; Kranda, K.C.; Russo, A.; Friauf, W.S.; Cole, J.W.; et al. Intrapleural photodynamic therapy: Results of a phase I trial. *Ann. Surg. Oncol.* **1994**, *1*, 28–37. [CrossRef] [PubMed]
5. Dougherty, T.J.; Gomer, C.J.; Henderson, B.W.; Jori, G.; Kessel, D.; Korbelik, M.; Moan, J.; Peng, Q. Photodynamic Therapy. *JNCI J. Natl. Cancer Inst.* **1998**, *90*, 889–905. [CrossRef] [PubMed]
6. Stewart, F.; Baas, P.; Star, W. What does photodynamic therapy have to offer radiation oncologists (or their cancer patients)? *Radiother. Oncol.* **1998**, *48*, 233–248. [CrossRef]
7. Brown, S.B.; Brown, E.A.; Walker, I. The present and future role of photodynamic therapy in cancer treatment. *Lancet Oncol.* **2004**, *5*, 497–508. [CrossRef]
8. Ahn, J.-C.; Kang, J.-W.; Shin, J.-I.; Chung, P.-S. Combination treatment with photodynamic therapy and curcumin induces mitochondria-dependent apoptosis in AMC-HN3 cells. *Int. J. Oncol.* **2012**, *41*, 2184–2190. [CrossRef]
9. Atchison, J.; Kamila, S.; McEwan, C.; Nesbitt, H.; Davis, J.; Fowley, C.; Callan, B.; McHale, A.P.; Callan, J.F. Modulation of ROS production in photodynamic therapy using a pH controlled photoinduced electron transfer (PET) based sensitiser. *Chem. Commun. (Camb. Engl.)* **2015**, *51*, 16832–16835. [CrossRef] [PubMed]
10. Moan, J.; Berg, K. The Photodegradation of Photodegradation of Porphyrins in Cell can be used to Estimate the Liftime of Singlet Oxygen. *Photochem. Photobiol.* **1991**, *53*, 549–553. [CrossRef]
11. Banerjee, S.; Dixit, A.; Karande, A.A.; Chakravarty, A.R. Remarkable Selectivity and Photo-Cytotoxicity of an Oxidovanadium(IV) Complex of Curcumin in Visible Light. *Eur. J. Inorg. Chem.* **2015**, *2015*, 447–457. [CrossRef]
12. Konopka, K.; Goslinski, T. Photodynamic therapy in dentistry. *J. Dent. Res.* **2007**, *86*, 694–707. [CrossRef] [PubMed]
13. Brancaleon, L.; Moseley, H. Laser and non-laser light sources for photodynamic therapy. *Lasers Med. Sci.* **2002**, *17*, 173–186. [CrossRef] [PubMed]
14. Shishodia, S.; Sethi, G.; Aggarwal, B.B. Curcumin: Getting back to the roots. *Ann. N. Y. Acad. Sci.* **2005**, *1056*, 206–217. [CrossRef] [PubMed]
15. Aggarwal, B.B.; Sundaram, C.; Malani, N.; Ichikawa, H. Curcumin: The Indian solid gold. *Adv. Exp. Med. Biol.* **2007**, *595*, 1–75. [CrossRef] [PubMed]
16. Epstein, J.; Sanderson, I.R.; Macdonald, T.T. Curcumin as a therapeutic agent: The evidence from in vitro, animal and human studies. *Br. J. Nutr.* **2010**, *103*, 1545–1557. [CrossRef] [PubMed]
17. Naksuriya, O.; Okonogi, S.; Schiffelers, R.M.; Hennink, W.E. Curcumin nanoformulations: A review of pharmaceutical properties and preclinical studies and clinical data related to cancer treatment. *Biomaterials* **2014**, *35*, 3365–3383. [CrossRef] [PubMed]
18. Wilken, R.; Veena, M.S.; Wang, M.B.; Srivatsan, E.S. Curcumin: A review of anti-cancer properties and therapeutic activity in head and neck squamous cell carcinoma. *Mol. Cancer* **2011**, *10*, 12. [CrossRef]
19. Kaminaga, Y.; Nagatsu, A.; Akiyama, T.; Sugimoto, N.; Yamazaki, T.; Maitani, T.; Mizukami, H. Production of unnatural glucosides of curcumin with drastically enhanced water solubility by cell suspension cultures of Catharanthus roseus. *FEBS Lett.* **2003**, *555*, 311–316. [CrossRef]
20. Shaikh, J.; Bhosale, R.; Singhal, R. Microencapsulation of black pepper oleoresin. *Food Chem.* **2006**, *94*, 105–110. [CrossRef]
21. Anand, P.; Kunnumakkara, A.B.; Newman, R.A.; Aggarwal, B.B. Bioavailability of curcumin: Problems and promises. *Mol. Pharm.* **2007**, *4*, 807–818. [CrossRef] [PubMed]

22. Agel, M.R.; Baghdan, E.; Pinnapireddy, S.R.; Lehmann, J.; Schäfer, J.; Bakowsky, U. Curcumin loaded nanoparticles as efficient photoactive formulations against gram-positive and gram-negative bacteria. *Colloids Surf. B Biointerfaces* **2019**, *178*, 460–468. [CrossRef] [PubMed]
23. Singh, S.P.; Sharma, M.; Gupta, P.K. Enhancement of phototoxicity of curcumin in human oral cancer cells using silica nanoparticles as delivery vehicle. *Lasers Med. Sci.* **2014**, *29*, 645–652. [CrossRef] [PubMed]
24. Duse, L.; Baghdan, E.; Pinnapireddy, S.R.; Engelhardt, K.H.; Jedelská, J.; Schaefer, J.; Quendt, P.; Bakowsky, U. Preparation and Characterization of Curcumin Loaded Chitosan Nanoparticles for Photodynamic Therapy. *Phys. Status Solidi A* **2018**, *215*, 1700709. [CrossRef]
25. Tsai, W.-H.; Yu, K.-H.; Huang, Y.-C.; Lee, C.-I. EGFR-targeted photodynamic therapy by curcumin-encapsulated chitosan/TPP nanoparticles. *Int. J. Nanomed.* **2018**, *13*, 903–916. [CrossRef]
26. Jiang, S.; Zhu, R.; He, X.; Wang, J.; Wang, M.; Qian, Y.; Wang, S. Enhanced photocytotoxicity of curcumin delivered by solid lipid nanoparticles. *Int. J. Nanomed.* **2017**, *12*, 167–178. [CrossRef] [PubMed]
27. Danhier, F.; Ansorena, E.; Silva, J.M.; Coco, R.; Le Breton, A.; Préat, V. PLGA-based nanoparticles: An overview of biomedical applications. *J. Control. Release* **2012**, *161*, 505–522. [CrossRef] [PubMed]
28. Patel, J.; Amrutiya, J.; Bhatt, P.; Javia, A.; Jain, M.; Misra, A. Targeted delivery of monoclonal antibody conjugated docetaxel loaded PLGA nanoparticles into EGFR overexpressed lung tumour cells. *J. Microencapsul.* **2018**, *35*, 204–217. [CrossRef]
29. Kumari, A.; Yadav, S.K.; Yadav, S.C. Biodegradable polymeric nanoparticles based drug delivery systems. *Colloids Surf. B Biointerfaces* **2010**, *75*, 1–18. [CrossRef]
30. Michlovská, L.; Vojtová, L.; Mravcová, L.; Hermanová, S.; Kučerík, J.; Jančář, J. Functionalization Conditions of PLGA-PEG-PLGA Copolymer with Itaconic Anhydride. *Macromol. Symp.* **2010**, *295*, 119–124. [CrossRef]
31. Hariharan, S.; Bhardwaj, V.; Bala, I.; Sitterberg, J.; Bakowsky, U.; Ravi Kumar, M.N.V. Design of estradiol loaded PLGA nanoparticulate formulations: A potential oral delivery system for hormone therapy. *Pharm. Res.* **2006**, *23*, 184–195. [CrossRef] [PubMed]
32. Anderson, J.M.; Shive, M.S. Biodegradation and biocompatibility of PLA and PLGA microspheres. *Adv. Drug Deliv. Rev.* **1997**, *28*, 5–24. [CrossRef]
33. Jain, R.A. The manufacturing techniques of various drug loaded biodegradable poly (lactide-co-glycolide) (PLGA) devices. *Biomaterials* **2000**, *21*, 2475–2490. [CrossRef]
34. Duse, L.; Pinnapireddy, S.R.; Strehlow, B.; Jedelská, J.; Bakowsky, U. Low level LED photodynamic therapy using curcumin loaded tetraether liposomes. *Eur. J. Pharm. Biopharm.* **2018**, *126*, 233–241. [CrossRef]
35. Ravi Kumar, M.N.V.; Bakowsky, U.; Lehr, C.M. Preparation and characterization of cationic PLGA nanospheres as DNA carriers. *Biomaterials* **2004**, *25*, 1771–1777. [CrossRef]
36. Bhatt, P.; Lalani, R.; Vhora, I.; Patil, S.; Amrutiya, J.; Misra, A.; Mashru, R. Liposomes encapsulating native and cyclodextrin enclosed paclitaxel: Enhanced loading efficiency and its pharmacokinetic evaluation. *Int. J. Pharm.* **2018**, *536*, 95–107. [CrossRef]
37. Schaefer, J.; Schulze, C.; Marxer, E.E.J.; Schaefer, U.F.; Wohlleben, W.; Bakowsky, U.; Lehr, C.-M. Atomic force microscopy and analytical ultracentrifugation for probing nanomaterial protein interactions. *ACS Nano* **2012**, *6*, 4603–4614. [CrossRef]
38. Senior, J.; Gregoriadis, G. Stability of small unilamellar liposomes in serum and clearance from the circulation: The effect of the phospholipid and cholesterol components. *Life Sci.* **1982**, *30*, 2123–2136. [CrossRef]
39. Plenagl, N.; Duse, L.; Seitz, B.S.; Goergen, N.; Pinnapireddy, S.R.; Jedelska, J.; Brüßler, J.; Bakowsky, U. Photodynamic therapy–Hypericin tetraether liposome conjugates and their antitumor and antiangiogenic activity. *Drug Deliv.* **2019**, *26*, 23–33. [CrossRef]
40. Tariq, I.; Pinnapireddy, S.R.; Duse, L.; Ali, M.Y.; Ali, S.; Amin, M.U.; Goergen, N.; Jedelská, J.; Schäfer, J.; Bakowsky, U.; et al. Lipodendriplexes: A promising nanocarrier for enhanced gene delivery with minimal cytotoxicity. *Eur. J. Pharm. Biopharm.* **2019**, *135*, 72–82. [CrossRef]
41. Pinnapireddy, S.R.; Duse, L.; Strehlow, B.; Schäfer, J.; Bakowsky, U. Composite liposome-PEI/nucleic acid lipopolyplexes for safe and efficient gene delivery and gene knockdown. *Colloids Surf. B Biointerfaces* **2017**, *158*, 93–101. [CrossRef] [PubMed]
42. Evans, B.C.; Nelson, C.E.; Yu, S.S.; Beavers, K.R.; Kim, A.J.; Li, H.; Nelson, H.M.; Giorgio, T.D.; Duvall, C.L. Ex vivo red blood cell hemolysis assay for the evaluation of pH-responsive endosomolytic agents for cytosolic delivery of biomacromolecular drugs. *J. Vis. Exp.* **2013**, e50166. [CrossRef] [PubMed]

43. Makadia, H.K.; Siegel, S.J. Poly Lactic-co-Glycolic Acid (PLGA) as Biodegradable Controlled Drug Delivery Carrier. *Polymers (Basel)* **2011**, *3*, 1377–1397. [CrossRef] [PubMed]
44. Yallapu, M.M.; Gupta, B.K.; Jaggi, M.; Chauhan, S.C. Fabrication of curcumin encapsulated PLGA nanoparticles for improved therapeutic effects in metastatic cancer cells. *J. Colloid Interface Sci.* **2010**, *351*, 19–29. [CrossRef]
45. Shaikh, J.; Ankola, D.D.; Beniwal, V.; Singh, D.; Kumar, M.N.V.R. Nanoparticle encapsulation improves oral bioavailability of curcumin by at least 9-fold when compared to curcumin administered with piperine as absorption enhancer. *Eur. J. Pharm. Sci.* **2009**, *37*, 223–230. [CrossRef] [PubMed]
46. Raschpichler, M.; Agel, M.R.; Pinnapireddy, S.R.; Duse, L.; Baghdan, E.; Schäfer, J.; Bakowsky, U. In situ intravenous photodynamic therapy for the systemic eradication of blood stream infections. *Photochem. Photobiol. Sci.* **2019**, *18*, 304–308. [CrossRef] [PubMed]
47. Sitterberg, J.; Ozcetin, A.; Ehrhardt, C.; Bakowsky, U. Utilising atomic force microscopy for the characterisation of nanoscale drug delivery systems. *Eur. J. Pharm. Biopharm.* **2010**, *74*, 2–13. [CrossRef] [PubMed]
48. Murakami, H.; Kobayashi, M.; Takeuchi, H.; Kawashima, Y. Preparation of poly(dl-lactide-co-glycolide) nanoparticles by modified spontaneous emulsification solvent diffusion method. *Int. J. Pharm.* **1999**, *187*, 143–152. [CrossRef]
49. Singh, R.; Lillard, J.W. Nanoparticle-based targeted drug delivery. *Exp. Mol. Pathol.* **2009**, *86*, 215–223. [CrossRef] [PubMed]
50. Tonnesen, H. Studies on curcumin and curcuminoids. XIV. Effect of curcumin on hyaluronic acid degradation in vitro. *Int. J. Pharm.* **1989**, *50*, 91–95. [CrossRef]
51. Shahani, K.; Panyam, J. Highly loaded, sustained-release microparticles of curcumin for chemoprevention. *J. Pharm. Sci.* **2011**, *100*, 2599–2609. [CrossRef] [PubMed]
52. Magenheim, B.; Levy, M.Y.; Benita, S. A new in vitro technique for the evaluation of drug release profile from colloidal carriers—Ultrafiltration technique at low pressure. *Int. J. Pharm.* **1993**, *94*, 115–123. [CrossRef]
53. Baghdan, E.; Raschpichler, M.; Lutfi, W.; Pinnapireddy, S.R.; Pourasghar, M.; Schäfer, J.; Schneider, M.; Bakowsky, U. Nano spray dried antibacterial coatings for dental implants. *Eur. J. Pharm. Biopharm.* **2019**, *139*, 59–67. [CrossRef] [PubMed]
54. Webb, B.A.; Chimenti, M.; Jacobson, M.P.; Barber, D.L. Dysregulated pH: A perfect storm for cancer progression. *Nat. Rev. Cancer* **2011**, *11*, 671–677. [CrossRef] [PubMed]
55. Yang, R.; Shim, W.-S.; Cui, F.-D.; Cheng, G.; Han, X.; Jin, Q.-R.; Kim, D.-D.; Chung, S.-J.; Shim, C.-K. Enhanced electrostatic interaction between chitosan-modified PLGA nanoparticle and tumor. *Int. J. Pharm.* **2009**, *371*, 142–147. [CrossRef] [PubMed]
56. Li, B.; Xu, H.; Li, Z.; Yao, M.; Xie, M.; Shen, H.; Shen, S.; Wang, X.; Jin, Y. Bypassing multidrug resistance in human breast cancer cells with lipid/polymer particle assemblies. *Int. J. Nanomed.* **2012**, *7*, 187–197. [CrossRef]
57. Li, Y.-P.; Pei, Y.-Y.; Zhang, X.-Y.; Gu, Z.-H.; Zhou, Z.-H.; Yuan, W.-F.; Zhou, J.-J.; Zhu, J.-H.; Gao, X.-J. PEGylated PLGA nanoparticles as protein carriers: Synthesis, preparation and biodistribution in rats. *J. Control. Release* **2001**, *71*, 203–211. [CrossRef]
58. Wolfram, J.; Suri, K.; Yang, Y.; Shen, J.; Celia, C.; Fresta, M.; Zhao, Y.; Shen, H.; Ferrari, M. Shrinkage of pegylated and non-pegylated liposomes in serum. *Colloids Surf. B Biointerfaces* **2014**, *114*, 294–300. [CrossRef]
59. Nafee, N.; Schneider, M.; Schaefer, U.F.; Lehr, C.-M. Relevance of the colloidal stability of chitosan/PLGA nanoparticles on their cytotoxicity profile. *Int. J. Pharm.* **2009**, *381*, 130–139. [CrossRef]
60. Hebling, J.; Bianchi, L.; Basso, F.G.; Scheffel, D.L.; Soares, D.G.; Carrilho, M.R.O.; Pashley, D.H.; Tjäderhane, L.; de Souza Costa, C.A. Cytotoxicity of dimethyl sulfoxide (DMSO) in direct contact with odontoblast-like cells. *Dent. Mater.* **2015**, *31*, 399–405. [CrossRef]
61. Schäfer, J.; Höbel, S.; Bakowsky, U.; Aigner, A. Liposome-polyethylenimine complexes for enhanced DNA and siRNA delivery. *Biomaterials* **2010**, *31*, 6892–6900. [CrossRef] [PubMed]
62. Wang, L.H.; Rothberg, K.G.; Anderson, R.G. Mis-assembly of clathrin lattices on endosomes reveals a regulatory switch for coated pit formation. *J. Cell Biol.* **1993**, *123*, 1107–1117. [CrossRef] [PubMed]
63. Orlandi, P.A.; Fishman, P.H. Filipin-dependent Inhibition of Cholera Toxin: Evidence for Toxin Internalization and Activation through Caveolae-like Domains. *J. Cell Biol.* **1998**, *141*, 905–915. [CrossRef] [PubMed]

64. Kunwar, A.; Barik, A.; Mishra, B.; Rathinasamy, K.; Pandey, R.; Priyadarsini, K.I. Quantitative cellular uptake, localization and cytotoxicity of curcumin in normal and tumor cells. *Biochim. Biophys. Acta* **2008**, *1780*, 673–679. [CrossRef] [PubMed]
65. McLennan, H.R.; Degli Esposti, M. The Contribution of Mitochondrial Respiratory Complexes to the Production of Reactive Oxygen Species. *J. Bioenerg. Biomembr.* **2000**, *32*, 153–162. [CrossRef] [PubMed]
66. Dobrovolskaia, M.A.; McNeil, S.E. *Handbook of Immunological Properties of Engineered Nanomaterials*; World Scientific: Singapore, 2016; Volume 1, p. 692.
67. Lenaham, J.G.; Frye, S.; Phillips, G.E. Use of the activated partial thromboplastin time in the control of heparin administration. *Clin. Chem.* **1966**, *12*, 263–268. [PubMed]
68. Greer, J.P. *Wintrobe's Clinical Hematology*; Lippincott Williams & Wilkins: Philadelphia, PA, USA, 2008; p. 1272.
69. Pinnapireddy, S.R.; Duse, L.; Akbari, D.; Bakowsky, U. Photo-Enhanced Delivery of Genetic Material Using Curcumin Loaded Composite Nanocarriers. *Clin. Oncol.* **2017**, *2*, 1323.

pharmaceutics

Article

Factorial Design as a Tool for the Optimization of PLGA Nanoparticles for the Co-Delivery of Temozolomide and O6-Benzylguanine

Maria João Ramalho, Joana A. Loureiro, Manuel A. N. Coelho and Maria Carmo Pereira *

LEPABE—Laboratory for Process Engineering, Environment, Biotechnology and Energy, Faculty of Engineering, University of Porto, Rua Dr. Roberto Frias, 4200-465 Porto, Portugal

* Correspondence: mcsp@fe.up.pt

Received: 22 June 2019; Accepted: 8 August 2019; Published: 10 August 2019

Abstract: Poly(D,L-lactic-*co*-glycolic) (PLGA) nanoparticles (NPs) have been widely studied for several applications due to their advantageous properties, such as biocompatibility and biodegradability. Therefore, these nanocarriers could be a suitable approach for glioblastoma multiforme (GBM) therapy. The treatment of this type of tumours remains a challenge due to intrinsic resistance mechanisms. Thus, new approaches must be envisaged to target GBM tumour cells potentially providing an efficient treatment. Co-delivery of temozolomide (TMZ) and O6-benzylguanine (O6BG), an inhibitor of DNA repair, could provide good therapeutic outcomes. In this work, a fractional factorial design (FFD) was employed to produce an optimal PLGA-based nanoformulation for the co-loading of both molecules, using a reduced number of observations. The developed NPs exhibited optimal physicochemical properties for brain delivery (dimensions below 200 nm and negative zeta potential), high encapsulation efficiencies (EE) for both drugs, and showed a sustained drug release for several days. Therefore, the use of an FFD allowed for the development of a nanoformulation with optimal properties for the co-delivery of TMZ and O6BG to the brain.

Keywords: drug delivery; experimental design; fractional factorial design; O6-methylguanine DNA methyltransferase (MGMT) protein; glioblastoma multiforme

1. Introduction

Nanomedicine has been arousing increasing interest, since it allows the early diagnosis and monitoring of several diseases and also can increase the efficacy of conventional pharmacological treatments by enabling their controlled delivery [1–4]. Several materials have been widely studied for nanoparticles (NPs) development. Among polymeric materials, poly(D,L-lactic-*co*-glycolic) (PLGA) is perhaps the most used, due to being FDA-approved, biocompatible, biodegradable, and having tunable physicochemical properties [5]. Since these polymeric NPs allow the release of drugs in a controlled and sustained manner for long periods, the required drug doses and administration frequency can be minimized, decreasing the toxicity of the encapsulated drug [6]. Additionally, PLGA NPs are up-taken by endocytic mechanism, circumventing the p-glycoprotein-mediated cellular efflux, enabling drug accumulation in the target cells. Also, since this polymer can be easily functionalized with different materials, the design of NPs with diverse targeting moieties can be achieved [7].

Therefore, several efforts have been conducted in seeking the development of PLGA NPs for novel therapies with several therapeutic drugs without reducing their bioavailability or activity. The entrapment of therapeutic drugs in PLGA NPs also allow a reduction in their toxicity, and therefore these NPs have been extensively studied for cancer treatment. Most particularly, several attempts to encapsulate temozolomide (TMZ) in PLGA NPs for brain delivery have been conducted [8]. TMZ, the gold standard treatment for glioblastoma multiforme (GBM), presents several limitations, as it has low bioavailability and high toxicity, reducing its pharmacological activity [9]. Therefore, TMZ encapsulation in a nanocarrier emerges as a suitable approach to increase its chemotherapeutic efficacy, since it will avoid drug elimination after administration and increase its accumulation in the target tissues. GBM is the brain cancer with a high mortality rate, since its classical therapy fails to effectively cure the disease [10]. Thus, it is urgent to find new approaches for its treatment.

In fact, different PLGA delivery nanosystems were developed for TMZ in the last decade. Jain et al., 2014 developed PLGA NPs for TMZ entrapment. Although exhibiting promising results, modifying the NPs' surface with a targeting moiety to improve the specificity of the nanocarriers could be a suitable strategy to enhance its efficiency [11]. Later, Ananta et al., 2016 prepared PLGA NPs that were not able to maintain a favorable sustained release of TMZ [12]. Also, some authors used active targeting strategies to increase the NPs specificity to GBM cells. In fact, Jain et al., 2011 modified the surface of PLGA NPs with transferrin molecules to promote transport across the blood–brain barrier (BBB). The developed nanocarriers enhanced the accumulation and cytotoxic effect of TMZ in mice's brains [13]. Lee et al., 2016 also functionalized PLGA NPs with folate molecules to increase the NPs accumulation in target cells, but the prepared nanosystems did not exhibit good encapsulation efficiency (EE) values [14]. Our group (2018) also developed anti-transferrin receptor monoclonal antibody modified PLGA NPs for the delivery of TMZ. The obtained results showed that PLGA NPs are a promising approach for GBM treatment with TMZ [15].

However, GBM patients usually exhibit low sensitivity to therapy due to intrinsic resistance mechanisms. The O6-methylguanine DNA methyltransferase (MGMT) protein is highlighted as one of the main causes of therapeutic failure of TMZ [16]. Therefore, new strategies to decrease resistance to therapy are necessary. A promising approach is the use of molecules that are able to revert or inhibit these intrinsic resistance mechanisms. Most recently, co-treatment with O6-benzylguanine (O6BG) has been explored to decrease the resistance to TMZ's therapy by binding to the MGMT protein, leading to its inhibition, and consequently hindering the repair of the damaged DNA [17,18]. Nanoencapsulation of O6BG may reduce its toxicity in healthy tissues by targeted delivery. Accumulation of O6BG in healthy tissues is undesirable to avoid inactivation of the MGMT protein in these tissues, and consequently exacerbates the toxicity of alkylating agents, such as TMZ [19].

Also, systemic administration of two free drugs usually leads to infective pharmacological activity and, consequently, treatment failure, due to differences in the biodistribution profile of each drug [20]. Therefore, the entrapment of both drugs in NPs should address this problem. In fact, to increase therapeutic efficiency, the co-loading of TMZ with other molecules in PLGA NPs has also already been studied for glioblastoma therapy. Xu et al., 2016 proposed the co-delivery of TMZ with paclitaxel using PLGA NPs [21]. Until this moment, to the best of our knowledge, the co-encapsulation of TMZ and O6BG in a nanosystem was not reported. Thus, the aim of this work was to prepare a nanocarrier for the co-loading of both drugs.

However, the use of PLGA NPs faces a few limitations because of their poor loading capacity and the typical initial burst release. Additionally, the production of PLGA NPs requires different stages that can present high costs and be difficult to scale-up, such as centrifugation and dialysis. Also, it can be challenging to entrap hydrophilic drugs, since those exhibit a high partition into the aqueous phase during NPs preparation [6]. Therefore, is necessary to optimize the preparation methods. As well, usually the entrapment of two distinct molecules, with different physicochemical properties such as hydrophilicity and molecular weight, can be a challenging task. Though, high encapsulation of both drugs is desirable to increase the nanosystem efficiency and to reduce the amount of administered polymer. Also, it is essential to control the physicochemical properties of the developed NPs, such as size and surface charge, since these parameters control their biological fate, biodistribution, and toxicity, therefore affecting their therapeutic potential [22,23].

Therefore, experimental design could be a suitable approach to optimize the nanoformulation. In fact, in the last years, experimental design has been used for the optimization of drug-loaded NPs [24,25]. This is because a nanoformulation design requires full knowledge of the correlation between the experimental factors and the obtained NPs properties. To obtain an optimized nanoformulation using a conventional screening method (evaluating the effect of one experimental variable at a time) is expensive and time-consuming. Experimental design is therefore a validated and useful tool for the development and optimization of experimental procedures with a lower number of observations while still providing the desired information on the correlation between the experimental and the response variables. The obtained model can then be used for predicting future observations within the original design range [26]. Different types of experimental designs can be used, and in this work, a fractional factorial design (FFD) is proposed. When several experimental variables are being studied, FFD is a rapid and reliable tool, allowing the exploration of a maximum number of variables, while requiring less experimental observations, and still obtaining the desired information. However, since it uses only a partial combination of factors, some information about possible interactions can be lost [27].

Thus, the effect of five experimental variables on the physicochemical properties of the developed PLGA NPs and encapsulation efficiency (EE) of both drugs was studied in this work. The studied experimental factors were the amounts of TMZ, O6BG, surfactant, organic solvent, and polymer.

2. Materials and Methods

2.1. Materials

TMZ (MW 194.15, purity ≥99%) was purchased from Selleck Chemicals (Munich, Germany). O6BG (MW 241.25, purity ≥98%) was obtained from Abcam (Cambridge, UK). PLGA Resomer® RG503H (50:50; MW 24,000–38,000), poly(vinyl alcohol) (PVA) Mowiol® 4-88 (MW 31,000), phosphate buffer saline (PBS) and dichloromethane were acquired from Sigma–Aldrich (St. Louis, MO, USA). Uranyl acetate was provided by electron microscopy sciences (Hatfield, UK).

2.2. Preparation of TMZ+O6BG-Loaded PLGA NPs

For the synthesis of PLGA NPs loading, both TMZ and O6BG, a variation of the single emulsion–solvent evaporation method, were used [28]. Known amounts of PLGA, TMZ, and O6BG were dissolved in a dichloromethane solution. A 2% (*w*/*v*) PVA solution was added drop-by-drop to the prepared organic solution. Then, the solution was vortexed (Genius 3, ika®vortex, Germany) and placed in an ultrasonic bath at an ultrasonic frequency of 45 kHz (Ultrasonic cleaner, VWR™, Kuala Lumpur, Malaysia) to yield an oil-in-water (o-in-w) emulsion.

The emulsion was then poured into a 0.2% (*w*/*v*) PVA solution and maintained in unremitting agitation in a magnetic stirrer (800 rpm, Colorsquid, ika®, Staufen, Germany) until complete organic solvent evaporation (6 h). The suspension was then filtered using a membrane with a pore size of 200 nm, (polyethersulfone membrane syringe filter, VWR, Radnor, PA, USA) and stored at 4 °C overnight to increase the NPs' stability, avoiding their aggregation. After, the samples were centrifuged for 30 min at 14100× *g* (MiniSpin®plus, Eppendorf, Hamburg, Germany), to separate NPs from the non-encapsulated drug. The supernatant containing the non-encapsulated drug was saved for analysis.

2.3. Experimental Design and Data Analysis

High and comparable EE values for each drug are a prerequisite for the co-loading in the same NPs. In preliminary studies, it was verified that the co-encapsulation of TMZ and O6BG reduced the encapsulation of each drug alone. Thus, it was necessary to optimize the entrapment of both drugs. Also, it was important to control the physicochemical properties of the developed NPs, since these parameters influence the therapeutic efficacy and toxicity of the NPs. To obtain an optimized formulation, it is useful to study how the several experimental parameters influence the entire production process.

Therefore, a 2^{5-2} FFD was implemented using the Minitab Statistical Software (Minitab Inc., State college, PA, USA) to determine the effect of different experimental factors on the PLGA NPs features. The studied independent variables were the amount of both used drugs, the quantity of surfactant and organic solvent, and the amount of polymer. A variation of a full factorial design, in which only a subset of the total runs was performed, took place. The chosen factors of interest were varied on two levels (determined in preliminary studies) according to the experimental plan presented in Table 1. Two replicates were conducted for each combination and for center levels, and the order of the experiments was randomly sorted to avoid any bias.

Table 1. Process and formulation parameters of the used FFD.

Parameter	Component	Units	Applied Level		
			Low Level (−1)	Centre Level (0)	High Level (+1)
X_1	m_{TMZ}	mg	0.250	0.625	1
X_2	m_{O6BG}	mg	0.250	0.625	1
X_3	$\%_{PVA}$	% (*w*/*v*)	2	3	4
X_4	V_{DCM}	mL	0.50	0.75	1
X_5	m_{PLGA}	mg	10	15	20

Note: m_{TMZ}—Mass of TMZ; m_{O6BG}—Mass of O6BG; $\%_{PVA}$—percent weight per volume of PVA; V_{DCM}—Volume of dichloromethane; m_{PLGA}—mass of PLGA.

The observed response dependent variables were the NPs size, PdI, zeta potential values, and EE of both TMZ and O6BG. For the experimental design, 18 formulations were prepared, and an outline of the experimental plan and its results is shown in Table 2.

Table 2. Outline of the experimental design and results. The experimental levels (low, centre, and high) are represented by the coded values of −1, 0, and +1, respectively.

Run Order	Coded Independent Variables					Measured Dependent Variables				
	X_1	X_2	X_3	X_4	X_5	Mean Size (nm)	PdI	Zeta Potential (mV)	EE (%)	
									TMZ	O6BG
1	+1	+1	−1	+1	−1	183	0.148	−26.2	55.1	78.5
2	−1	+1	+1	−1	−1	202	0.110	−26.7	38.7	74.1
3	0	0	0	0	0	173	0.115	−23.4	45.1	71.6
4	+1	+1	−1	+1	−1	179	0.165	−24.1	56.7	80.8
5	−1	−1	−1	+1	+1	178	0.132	−24.4	53.5	94.0
6	−1	+1	−1	−1	+1	193	0.158	−22.4	52.3	91.6
7	+1	−1	−1	−1	−1	162	0.158	−27.5	31.8	99.0
8	0	0	0	0	0	178	0.127	− 22.1	44.7	77.3
9	−1	−1	+1	+1	−1	192	0.190	−27.0	42.6	79.7
10	+1	+1	+1	+1	+1	171	0.142	−20.3	27.2	78.5
11	+1	−1	+1	−1	+1	200	0.137	−22.7	40.3	99.8
12	+1	−1	+1	−1	+1	205	0.109	−22.5	46.3	99.6
13	+1	+1	+1	+1	+1	180	0.118	−22.3	28.9	82.6
14	+1	−1	−1	−1	−1	176	0.125	−28.0	36.6	99.5
15	−1	+1	−1	−1	+1	189	0.130	−21.7	50.9	95.5
16	−1	−1	−1	+1	+1	172	0.158	−23.4	54.5	98.0
17	−1	−1	+1	+1	−1	184	0.146	−26.2	46.7	78.1
18	−1	+1	+1	−1	−1	205	0.141	−27.5	36.5	76.9

The applied experimental design accounts for main terms and two-factor interactions terms. The latter refers to two different variables that interact with each other, creating a combined effect on the response that independently would not occur. Therefore, the main effects and the two-factor interactions are accounted for in the used regression model. Thus, regression equations were obtained for each studied dependent variable to quantify the relationship between these and all the experimental independent variables. The experimental data was then fitted to the following polynomial regression Equation (1) [29]:

$$Y = \beta_0 + \sum_{i=1}^{5} \beta_i X_i + \sum \beta_{ij} X_i X_j \tag{1}$$

in which Y is the predicted response; β_0 is the intercept term and the remaining term; $X_{i,j}$ is the studied levels of the independent variables; and $\beta_{j,i}$ is the fitted coefficients for $X_{i,j}$.

The statistical regression models for the different dependent variables were fitted independently. Additionally, the polynomial equations were statistically validated using ANOVA (Analysis of Variance) by statistical significance of coefficients, R^2 values, and normal distribution of the residues. Minitab Statistical Software was also used for the statistical analysis of the data.

2.4. *TMZ+O6BG-Loaded PLGA NPs Physicochemical Characterization*

2.4.1. Dynamic Light Scattering for Size Determination

The mean diameter and size distribution of the prepared NPs were evaluated by dynamic light scattering (DLS). The measurements were performed in a ZetaSizer Nano ZS (Malvern Instruments, Worcestershire, UK). The attained data is given in intensity distribution. The intensity-weighted mean diameter (Z-average) is given. For the optimized formulation, at least three independent measurements were performed, and obtained results are expressed as the mean and standard deviation (SD). Statistical analysis was performed using the *t*-student test, and p-values ≤ 0.05 were considered significant.

2.4.2. Laser Doppler Velocimetry Method for Zeta Potential Determination

The zeta potential values of the prepared NPs were determined by laser doppler velocimetry method. The measurements were also performed in a ZetaSizer Nano ZS (Malvern Instruments, Worcestershire, UK). The analysis was performed using the dielectric constant of water. For the optimized formulation, at least three independent measurements were performed, and obtained results are expressed as the mean and SD. Statistical analysis was performed using the *t*-student test.

2.4.3. Transmission Electron Microscopy for Morphological Analysis

The morphological analysis of the NPs was obtained by transmission electron microscopy (TEM). The NPs were prepared on copper grids (Formvar/Carbon-400 mesh Copper, Agar Scientific, Essex, UK) and negatively stained. For that, 10 µL of samples were stained with 2% (*v*/*v*) uranyl acetate for 45 s, and air-dried. This is a heavy metal salt able to scatter electrons, enhancing the contrast to better visualize the samples [30]. Then, the NPs were visualized using a JEM 1400 electron microscope (Jeol, Tokyo, Japan) at an accelerating voltage of 80 kV.

2.5. *TMZ+O6BG-Loaded PLGA NPs Stability Studies*

The stability of the prepared PLGA NPs was analysed through size and zeta potential variations. PLGA NPs' dispersions in ultrapure water were stored at 4 °C and DLS measurements were performed at different timepoints to evaluate variations in PLGA NPs size and zeta potential values. These measurements were performed weekly, for 6 weeks. For optimized nanoformulation, three independent samples were used.

2.6. *Drug Encapsulation Efficiency of TMZ+O6BG-PLGA NPs*

The TMZ and O6BG EE values of the prepared PLGA NPs were assessed by UV–Vis spectrophotometry, using the following Equation (2):

$$EE = \frac{total\ amount\ of\ drug - amount\ of\ free\ drug}{total\ amount\ of\ drug} \times 100 \qquad (2)$$

Non-encapsulated TMZ and O6BG molecules were separated from the NPs colloidal suspension by centrifugation (30 min, 14100× *g*, MiniSpin®plus, Eppendorf, Germany) and quantified (UV-1700 PharmaSpec UV-Vis spectrophotometer, Shimadzu, Kyoto, Japan) at 240 nm and 329 nm for O6BG and TMZ, respectively. The results were correlated to control samples corresponding to total amount of drug. For the optimized formulation, three independent experiments were conducted, and obtained results are expressed as the mean and SD. Statistical analysis was performed using the *t*-student test.

2.7. *In Vitro Release of TMZ and O6BG from PLGA NPs*

To assess the drug release profile of the developed NPs, in vitro dialysis studies were performed for 20 days at 37 °C. For that, a cellulose dialysis membrane (Float-A-Lyzer G2, CE, 10KDa, SpectrumLabs, Los Angles, CA, USA) was rinsed in ultrapure water for 24 h before the beginning of the experiments and equilibrated with release medium 1 h before the dialysis.

A known amount of TMZ+O6BG-PLGA NPs diluted in 2 mL of release medium was placed into the inner space of the dialysis membrane. The outside space was filled with a known volume of release medium to ensure sink conditions. PBS (pH 7.4, 0.01 M) was used as the release buffer to mimic the physiological salt concentrations and pH of blood plasma. The dialysis membrane was kept in continuous stirring at 200 rpm at 37 °C, simulating the physiological temperature. The amount of drug release at predetermined timepoints (0, 6, and 24 h; and day 3, 6, 8, 9, 11, 13, 16, and 20) was quantified by UV-Vis spectrophotometry (UV-1700 PharmaSpec UV-Vis spectrophotometer, Shimadzu, Kyoto, Japan). A solution of TMZ and O6BG in PBS was used as control. For the optimized formulation, three independent experiments were conducted, and obtained results are expressed as the mean and SD. Statistical analysis was performed using the *t*-student test.

The TMZ and O6BG release curves, representing the percentage of drug released in function of time, were then plotted by the following Equation (3):

$$\%\ drug\ released = \frac{amount\ of\ drug\ released\ at\ time\ t}{amount\ of\ encapsulated\ drug} \times 100 \quad (3)$$

3. Results and Discussion

3.1. *Statistical Analysis of Experimental Data*

The applied experimental design allowed the identification of the experimental factors influencing the physicochemical properties of the NPs, astheir size and the encapsulation efficiency of both drugs on the polymeric matrix. Since zeta potential (≤−20 mV) and PdI values (<0.2) were always inside the desired range, they were not considered for this model.

Previous nanoformulation studies were conducted to identify the five major experimental variables that affect the PLGA NPs properties, such as size and EE of both drugs. The amount of TMZ, O6BG, surfactant, organic solvent, and polymer were chosen and varied in two levels. The levels choice was based on the knowledge acquired in the preliminary experiments. All other parameters, such as type of surfactant and organic solvent, aqueous to organic phase ratio, time of sonication, ultrasonic frequency of sonication, process temperature, and emulsification and evaporation processes, were maintained constantly.

The mathematical model (Equation (1)) was fitted to the data and statistical analysis was performed using ANOVA (Table 3). As Table 3 shows, the model is statistically significant ($p < 0.05$) with insignificant lack of fit ($p > 0.05$) for all the chosen responses. Hence, the linear model was acceptable for response prediction within the range of experimental variables.

Table 3. Results of analysis of variance (ANOVA) for the regression models. Non-significant factors are marked as blue.

Source	Size, Y_1 (nm)				EE of TMZ, Y_2 (%)				EE of O6BG, Y_3 (%)			
	SS	df	*F*-Value	*p*-Value	SS	Df	*F*-Value	*p*-Value	SS	df	*F*-Value	*p*-Value
Model	4192.98	7	17.06	0.0001	1381.55	7	41.63	<0.0001	4311.36	7	35.96	<0.0001
X_1	12.20	1	8.92	0.0199	55.29	1	11.66	0.0244	170.83	1	9.97	0.0586
X_2	117.63	1	14.74	0.0200	41.70	1	8.80	0.0510	190.31	1	11.11	0.0520
X_3	3452.85	1	30.67	<0.0001	194.24	1	40.97	0.0001	1788.94	1	104.45	<0.0001
X_4	49.99	1	1.42	0.1263	46.25	1	9.76	0.0208	134.43	1	7.85	0.0745
X_5	100.42	1	11.41	0.0224	260.82	1	55.02	<0.0001	1843.34	1	107.63	<0.0001
X_2X_3	313.15	1	23.14	0.0057	350.60	1	74.20	<0.0001	173.67	1	10.14	0.0577
X_2X_5	146.75	1	18.42	0.0176	432.64	1	91.26	<0.0001	9.83	1	0.57	0.8480
Residual	351.06	10			47.40	10			171.27	10		
Lack of Fit	46.42	1	1.37	0.2716	2.77	1	0.56	0.4742	29.41	1	1.87	0.2051
Pure Error	304.64	9			44.64	9			141.86	9		
Cor Total	4544.04	17			1428.95	17			4482.63	17		
R^2			0,9227				0,9668				0,9618	

Figure 1 also shows a satisfactory agreement between experimental observations and predicted response, proving that the applied regression model is suitable for the determination of the optimal experimental settings for the preparation of NPs.

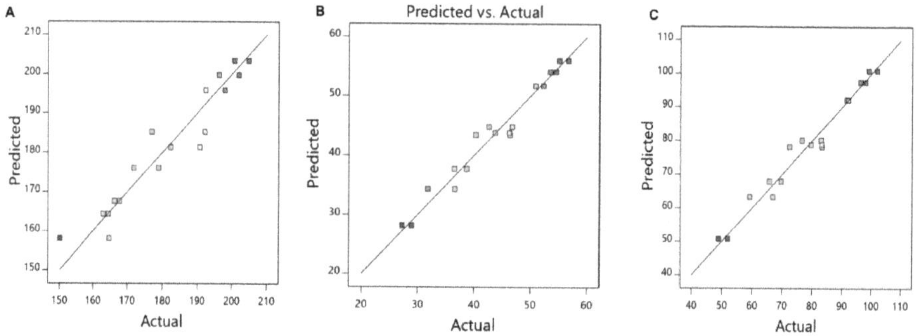

Figure 1. Graphical plots of the experimental values versus the model predicted values for (**A**) mean size, (**B**) EE for TMZ, and (**C**) EE for O6BG. Red dots indicate high values and blue dots low values, respectively.

The statistical analysis of the obtained results also allowed the determination of the regression coefficient (RC) values that describes the direction of the relationship between an experimental variable and the variable response. A negative sign before a factor indicates that the response decreases, whereas a positive sign shows that the response increases with the experimental factor. With the RC values, it was possible to obtain the polynomial regression equation for each studied response, quantifying the relationship between each of the studied experimental variables and the response. The non-significant factors ($p > 0.05$) were excluded from the mathematical model, allowing the reduction in complexity of the attained equations.

Response surface analysis and contour graphs were plotted based on the determined model polynomial function in a three and a two-dimensional model, respectively, illustrating the effect of the chosen significant independent factors on each observed response. The effect of all the independent variables on each dependent variable was studied and the effects of the most significant variables are discussed in detail below.

3.1.1. Effect on NPs' Size

The size of the NPs is a critical property, influencing its half-life, biodistribution, and cellular internalization. The size of the NPs ranged from 162 nm (sample 7) to 205 nm (sample 12). Almost all the studied variables significantly affected the size of the NPs ($p < 0.05$), as shown in Table 3.

The polynomial Equation (4) describes the relationship between the significant studied experimental variables ($p < 0.05$) and the size of the NPs:

$$\begin{aligned} Size = \; & 134.9 - 9.67\, amount\ TMZ\ (mg) + 83.9\, amount\ O6BG\ (mg) \\ & + 15.05\, amount\ PVA\ (\%) + 1.644\, amount\ PLGA\ (mg) \\ & - 13.41\, amount\ O6BG\ (mg) \times amount\ PVA\ (\%) \\ & - 2.541\, amount\ O6BG\ (mg) \times amount\ PLGA\ (mg) \end{aligned} \quad (4)$$

The experimental factor that most significantly affected NPs size was the amount of surfactant, which exhibited a positive effect on the size of the NPs. Thus, an increase in PVA concentration led to an increase in NP size. Although using a high quantity of surfactant can induce the creation and stabilization of smaller NPs due to a reduction of the interfacial tension between the polymer and the external aqueous phase [31], the opposite effect was verified. This may be explained by the increased viscosity of the aqueous phase when increasing the PVA concentration. A higher viscosity decreases the shear stress, originating emulsion droplets with larger sizes. Also, higher amounts of surfactant can promote the coalescence of the NPs, yielding NPs with larger diameters [32]. Also, some studies report that residual PVA remains at the surface, contributing to the size increase [33].

Additionally, the amount of PLGA also exhibited a significantly positive effect on the size of the NPs ($p < 0.05$). Increasing the polymer amount increases the viscosity of the organic phase, decreasing the shear stress, as mentioned above. Also, the augmented viscosity hampers the diffusion of the organic solvent into the aqueous phase, leading to the formation of larger emulsion droplets, originating larger NPs after solvent evaporation [34].

Higher initial loading of O6BG positively affected the NPs size ($p < 0.05$), since it will result in higher drug loading, as will be discussed later. The contrary was observed for TMZ amount ($p < 0.05$), since higher amounts of TMZ will decrease its entrapment, as will also be discussed later.

Response surface analysis and contour graphs were plotted (Figure 2) based on this model polynomial function. Contour and surface response plots allowed visual identification of the optimal levels of each factor, to choose the most suitable values for the development of an optimal formulation. Both response surface and contour plots (Figure 2A,B) showed that the lower the amount of PVA and PLGA, the lower the NPs size, as already predicted by the calculated positive RC values (Equation (4)). Thus, optimal nanoformulation with small dimensions would fall into the low and central levels of both factors.

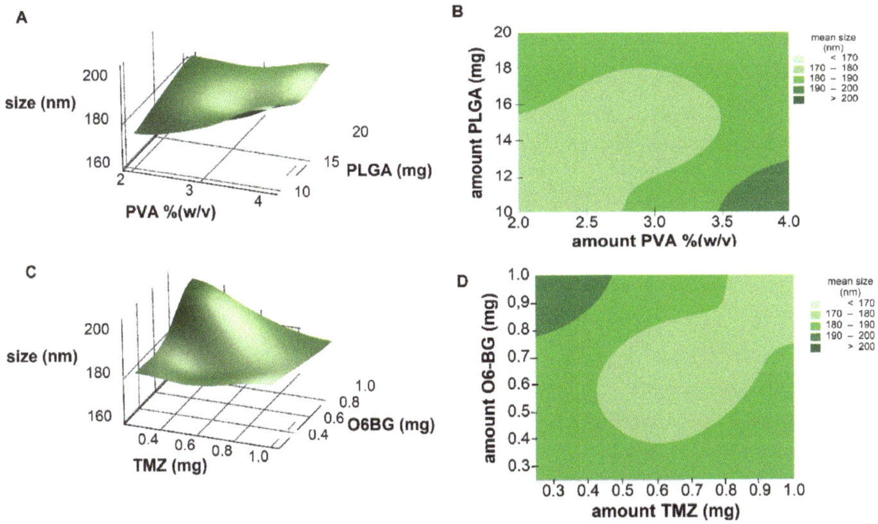

Figure 2. (**A**) Response surface plot and (**B**) contour plot showing the effect of two factors (amount of PVA and PLGA) on the resulting NPs size. (**C**) Response surface plot and (**D**) contour plot showing the effect of two factors (amount of TMZ and O6BG) on the resulting NPs size.

In addition to all the five factors significantly affecting the NPs size, some significant two-factor interactions were observed (Table 3). Interaction plots on Figure 3 show high interaction between two factors. Each point in the interaction plot shows the mean size values at different combinations of factor levels. As the lines are not parallel, with different slopes, the plot indicates that there is an interaction between the two factors [35]. The same is verified in all the attained interaction plots.

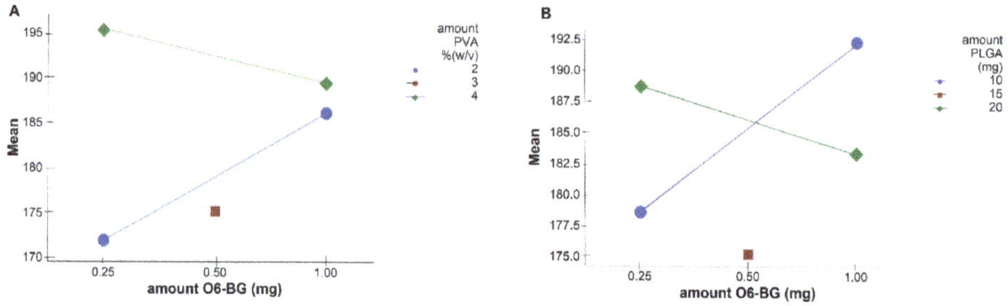

Figure 3. Interaction plots for size data means showing significant two-way interaction terms for the independent variables. Solid blue lines display factor at low level, whereas dashed green lines are the high level of the factors. (**A**) interaction term X_2X_3: amount of O6BG/amount of surfactant, (**B**) X_2X_5: amount of O6BG/amount of PLGA.

The main effects can be confused with the two-factor interactions, which explains the results obtained in Figure 2. Although Equation (4) predicts that the lower the amount of TMZ and the higher the amount of O6BG, the higher the NPs size, optimal values fall into high levels of both TMZ and O6BG. As shown in Figure 3A,B, at low levels of both PLGA and PVA amounts, O6BG positively affects the size of the NPs. On the other hand, at high levels of PLGA and PVA, increasing the O6BG amount decreases the size of the NPs.

3.1.2. Effect on the Encapsulation Efficiency of TMZ

High encapsulation of the drug is desirable to increase the nanosystem efficiency and to reduce the amount of administered polymer. The EE values for TMZ in the prepared NPs ranged from 27.2 (sample 10) to 56.7% (sample 4).

The polynomial Equation (5) describes the relationship between the significant studied experimental variables ($p < 0.05$) and the EE of TMZ:

$$\begin{aligned} EE\%\ TMZ = \ & 11.84 - 8.80\ amount\ TMZ\ (mg) - 3.40\ amount\ PVA\ (\%) \\ & + 7.95\ amount\ DCM\ (mL) + 1.461\ amount\ PLGA\ (mg) \\ & -13.87\ amount\ O6BG(mg) \times amount\ PVA\ (\%) \\ & - 2.153\ amount\ O6BG\ (mg) \times amount\ PLGA\ (mg) \end{aligned} \quad (5)$$

EE values were significantly influenced by almost all the studied independent variables, as shown by the calculated *p*-values on Table 3. All the experimental variables show a positive effect on the EE of TMZ, except PVA concentration and TMZ amount. PLGA had a positive effect on the EE of TMZ. As already mentioned, a higher amount of PLGA polymer results in larger NPs and, consequently, higher encapsulation of the drug. Also, the increased viscosity caused by higher PLGA concentration mentioned above could complicate the diffusion of TMZ molecules into the aqueous phase, enhancing the drug's entrapment into the NPs' polymeric matrix [36]. On the other hand, increased volumes of organic solvent increased drug solubility, therefore increasing drug entrapment.

Also, the encapsulation of TMZ decreased with PVA concentration due to a higher partition of TMZ molecules into the aqueous phase during emulsification, decreasing the EE values. It is reported that drug molecules can diffuse out from the oil nanodroplets and solubilize in PVA micelles at the aqueous phase [37]. Increasing the TMZ amount may cause saturation of the organic phase, leading to the partition of the drug molecules to the aqueous phase, lowering its entrapment in the polymeric matrix.

Response surface analysis and contour graphs (Figure 4) show that the optimal formulation would be prepared with low amounts of PVA and TMZ, and high amounts of PLGA and organic solvent.

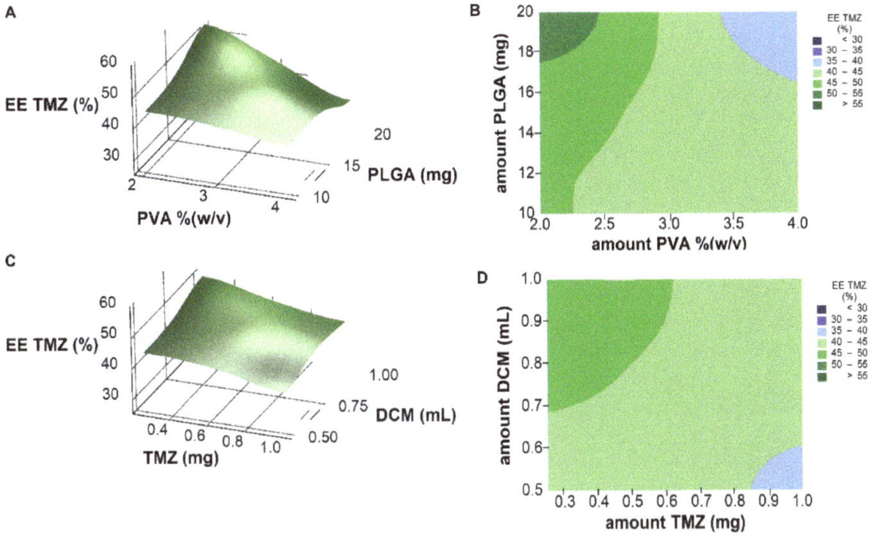

Figure 4. (**A**) Response surface plot and (**B**) contour plot showing the effect of two factors (amount of PVA and PLGA, respectively) on the resulting EE TMZ values. (**C**) Response surface plot and (**D**) contour plot showing the effect of two factors (amount of TMZ and DCM, respectively) on the resulting EE TMZ values.

All five studied variables were part of an extensive interaction system. A significant two-factor interaction of O6BG amount with both PLGA and PVA amounts (Table 3) was verified ($p < 0.05$), as shown in Figure 5.

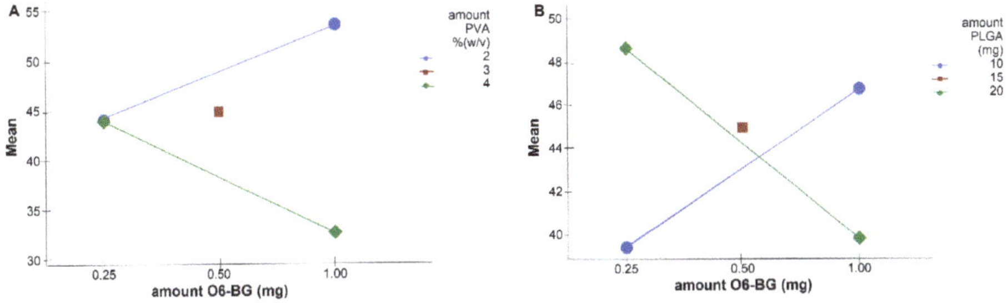

Figure 5. Interaction plots for EE of TMZ data means showing significant two-way interaction terms for the dependent variables. Solid blue lines display factors at a low level, whereas green dashed lines are the high level of the factors. (**A**) interaction term X_2X_3: amount of O6BG/amount of surfactant, (**B**) X_2X_5: amount of O6BG/amount of PLGA.

The described two-factor interaction between the PVA and O6BG amounts shows a significant PVA concentration effect and interaction term (Figure 5A). There is a difference among the means of the two PVA levels, but not a difference in the means among O6BG amount. This proves that although the amount of O6BG did not show a significant direct effect on the EE values, the interaction term exists. The amount of O6BG affected the nanoformulation at both low and high PVA amounts. At low PVA levels, the O6BG amount showed a positive effect in the TMZ encapsulation. The opposite was observed at high PVA amounts.

The same is verified in the two-factor interaction between the PLGA and O6BG amounts (Figure 5B).

3.1.3. Effect on the Encapsulation Efficiency of O6BG

The EE values of O6BG in the prepared NPs ranged from 70.9 (sample 3) to 99.8% (sample 11).

The polynomial Equation (6) describes the relationship between the significant studied experimental variables ($p < 0.05$) and the EE of O6BG:

$$E\%\ O6BG = 104.4 - 4.12\, amount\ PVA\ (\%) + 0.854\, amount\ PLGA\ (mg) \quad (6)$$

PVA and PLGA amounts were the only studied independent variables that significantly influenced the EE values, as shown by the calculated *p*-values on Table 3. While PLGA concentration showed a positive effect on the EE of O6BG, PVA concentration exhibited a negative influence. In fact, increasing PVA concentration hampered the encapsulation of the drug, since it enhanced the partition of O6BG molecules into the aqueous phase during emulsification, decreasing the encapsulation efficiencies values [37]. Also, as already mentioned, higher amounts of PLGA polymer resulted in larger NPs and, consequently, higher encapsulation of the drug. Also, the increased viscosity caused by higher PLGA concentration mentioned above could complicate the diffusion of O6BG molecules into the aqueous phase, enhancing the drug's entrapment into the NPs' polymeric matrix [36].

Response surface analysis and contour graphs (Figure 6) show that predicted optimal nanoformulation with high O6BG molecules entrapment would fall into the high levels of PLGA and low levels of surfactant concentration. No significant two-factor interactions were observed.

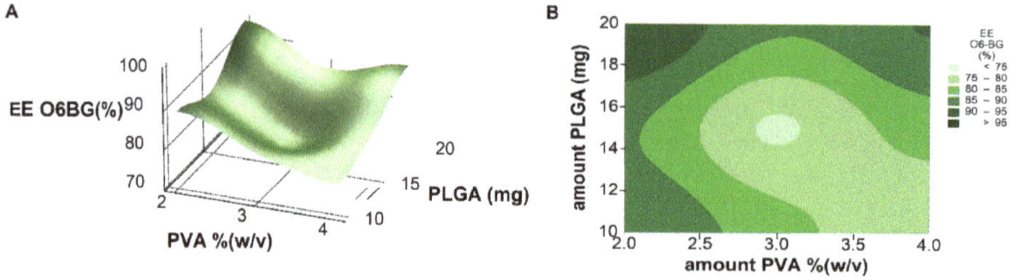

Figure 6. (**A**) Response surface plot and (**B**) contour plot showing the effect of two factors (amount of PLGA and PVA, respectively) on the resulting EE values.

3.2. Nanoformulation Optimization and Physicochemical Characterization

The optimization process was performed by determining the optimal experimental values, which were obtained by solving the three determined polynomial regression equations and grid searching in the response surface graphs or contour plots applying the following criteria: to minimize the particle size (Y_1), and to maximize the EE values for both drugs (Y_2 and Y_3). The optimum levels of the formulation factors are presented in Table 4.

Table 4. Optimal formulation parameters for the PLGA NPs determined by the experimental design.

Parameter	Component	Units	Optimal Value
X_1	m_{TMZ}	mg	1
X_2	m_{O6BG}	mg	1
X_3	$\%_{PVA}$	% (*w/v*)	0.5
X_4	V_{DCM}	mL	1
X_5	m_{PLGA}	mg	15

A formulation checkpoint was prepared according to the predicted model to validate the reliability and the precision of the factorial design, using these optimal experimental conditions. The checkpoint formulation was prepared in triplicates and the properties of the attained NPs were within the range of the predicted values, as shown in Table 5.

Table 5. Validation of the model by comparing the predicted values with the observed experimental values. The experimental data is given as the mean ± SD (n = 3).

	Size (Y_1) (nm)	EE TMZ (Y_2) (%)	EE O6BG (Y_3) (%)
Predicted values	173 (164–180)	68 (63–73)	89 (83–95)
Experimental values	177 ± 4	63 ± 4	90 ± 4
Experimental error	2%	7%	1%

Higher encapsulation verified for O6BG can be explained by its greater affinity for the organic phase, compared with TMZ. While TMZ exhibits a log P value of 0.36, for O6BG, log P is 1.66 (values obtained from Marvin Sketch Calculator software. Chemaxon^{TM},Version 16.4.25, Budapest, Hungary).

The developed NPs exhibit mean dimensions suitable for brain delivery. Although PdI and zeta potential values were not considered for the model, they were also determined for the checkpoint formulations. The developed NPs exhibited zeta potential values of −22 ± 1 mV and PdI values of 0.19 ± 0.01. The high negative zeta potential values are associated with low toxicity [38] and suggest that the NPs are stable [39]. In fact, all the nanoformulations prepared in the 18 runs of the FFD (data not shown) and the checkpoint formulations proved to be stable in storage conditions for at least six weeks, as shown in Figure 7.

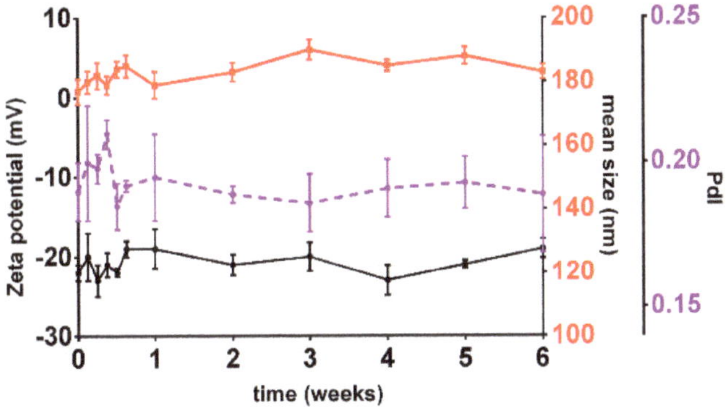

Figure 7. Stability of the prepared checkpoint nanoformulations. Black line: Graphical representation of variations in zeta potential values over 6 weeks (data plotted on the left Y-axis); Red line: Graphical representation of variations in mean sizes over 6 weeks (data plotted on the first right Y-axis); Purple line: Graphical representation of variations in PdI values over 6 weeks (data plotted on the second right Y-axis). Results are given as mean ± SD (n = 3).

The stability of the prepared colloidal suspension NPs is granted by a PVA layer on the NP's surface. During the NPs preparation, the PVA molecules adsorb on the surface of the resultant nanodroplets acting as a mechanic barrier that prevent coalescence and avoid NPs' aggregation [40]. TEM image (Figure 8) shows this stabilizer layer around the NPs' surface. Also, the attained PdI values suggest that the colloidal suspension is monodisperse. Therefore, it can be concluded that the method followed for the preparation of PLGA NPs produced well-stabilized monodisperse O6BG+TMZ-loaded PLGA NPs.

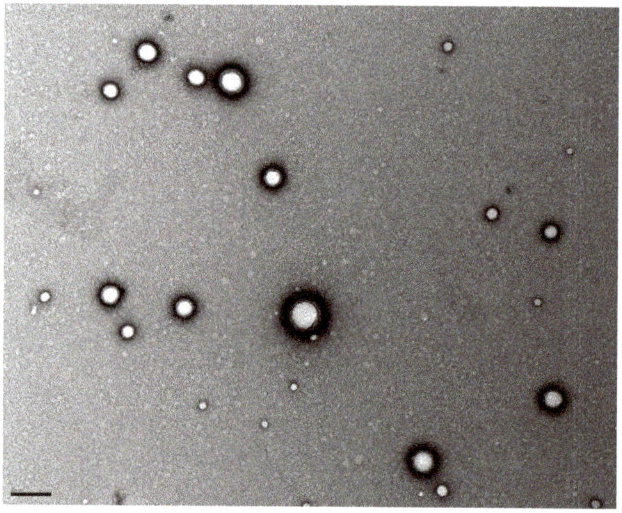

Figure 8. TEM photograph showing the morphology of the developed PLGA NPs. Scale bar is 200 nm.

3.3. Drug Release from the PLGA NPs

For the evaluation of the in vitro release profile of the developed NPs, the dialysis method was used. To simulate the physiological conditions (temperature, salt concentration, and pH), the release studies were conducted at 37 °C in PBS (10 mM, pH 7.4). The release curves were plotted and are shown in Figure 9.

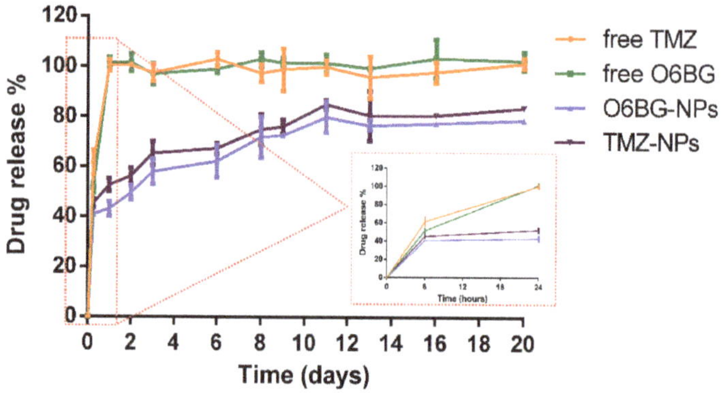

Figure 9. In vitro release of TMZ and O6BG from PLGA NPs in PBS (10 mM, pH 7.4) at 37 °C. Free TMZ and O6BG were used as control. Results are given as mean ± SD (n = 3). An enlargement of the first time points—6 and 24 h—is presented to highlight the burst effect.

PLGA NPs showed an initial burst release for both drugs, due to the presence of the drug molecules adsorbed at the NPs' surface. As Figure 9 depicts, at the first 24 h, 48 ± 3% of the total TMZ and 43 ± 3% of the total O6BG were released. Then, both drugs located at the polymeric matrix of the NPs were released in a slower and controlled manner that was prolonged for 20 days.

A faster release for TMZ was verified in comparison to O6BG. The PLGA NPs exhibited a release of TMZ of 81 ± 1% after 20 days, while for O6BG, only about 79 ± 1% of the entrapped drug was released at day 20 (Figure 9). This may occur due to the higher affinity of TMZ to the aqueous buffer (lower log P values). However, the release profile is similar for the two drugs, due to the erosion of the polymer matrix that allows the release of both molecules.

The obtained release profile proved that the developed PLGA NPs are able to maintain the sustained and controlled release of both TMZ and O6BG, and therefore could be used for the co-delivery of entrapped molecules. Further optimization of the release profile to reduce the initial burst release may improve the efficiency of the nanovehicle.

4. Conclusions

PLGA NPs as drug delivery systems have gained increasing interest in the last years, due to their unique physicochemical properties, such as stability, biocompatibility, and biodegradability. Encapsulating drugs in PLGA NPs allows them to maximize their therapeutic effects and to minimize their side effects. Since the biological fate and toxicity of the NPs depend on their physicochemical properties, the developed nanosystems must meet several criteria to be appropriate for brain delivery.

PLGA NPs for the entrapment of both TMZ and O6BG for further GBM treatment were developed in this work. TMZ is the used drug for GBM chemotherapy, and despite its therapeutics effects being well-known, the drug is not able to effectively cure GBM patients. Intrinsic resistance mechanisms are of one of the major obstacles for GBM treatment, due to the DNA repair by the MGMT protein. To overcome this issue, co-therapy with TMZ and an inhibitor of the MGMT protein, O6BG, was proposed. O6BG is a guanine analogue able to inhibit the activity of the MGMT protein, hampering the DNA repair. Thus, O6BG could decrease the resistance to TMZ's therapy.

Hence, an optimized nanoformulation was developed in this work, and for that, FFD was used. FFD proved to be a suitable approach for the design of co-loaded NPs allowing to optimize the single emulsion solvent evaporation production process with a low replica number in a reliable and precise manner. The regression analysis of the results proved to adequate to identify the main experimental variables affecting the physicochemical properties of the developed NPs. The optimal experimental conditions were chosen, and the optimal nanoformulation was prepared using 1 mg of both drugs, 0.5% (*w*/*v*) of PVA, 1 mL of dichloromethane, and 15 mg of PLGA.

The developed NPs exhibited high encapsulation efficiencies for both drugs and showed a sustained drug release. Thus, it is expected that these nanocarriers will provide an effective brain delivery, allowing the drug to reach the brain at desirable doses, leading to a significant improvement on the GBM treatment.

PLGA NPs provide a novel and potentially efficient approach for the co-administration of TMZ and O6BG, presenting a potential solution for the intrinsic resistance mechanisms to TMZ due to MGMT high expression. Though these NPs could potentially overcome the limitations of the currently available therapies, future in vitro and in vivo tests are necessary to assess the nanosystem efficacy.

Author Contributions: All authors contributed to the conceptualization of the experiments. M.J.R. conducted the experiments, data analyses, and the writing of the manuscript. M.d.C.P contributed to the review and editing of the manuscript.

Funding: This work was financially supported by: project UID/EQU/00511/2019—Laboratory for Process Engineering, Environment, Biotechnology and Energy—LEPABE, funded by national funds through FCT/MCTES (PIDDAC); Project POCI-01-0145-FEDER-006939, funded by FEDER funds through COMPETE2020—Programa Operacional Competitividade e Internacionalização (POCI) and by national funds (PIDDAC) through FCT/MCTES;—Project "LEPABE-2-ECO-INNOVATION"—NORTE-01-0145-FEDER-000005, funded by Norte Portugal Regional Operational Programme (NORTE 2020), under PORTUGAL 2020 Partnership Agreement, through the European Regional Development Fund (ERDF), and FCT doctoral grant- PD/BD/105984/2014.

Acknowledgments: The authors are grateful to Simone Morais for her comments and suggestions during the preparation of this manuscript.

Conflicts of Interest: The authors declare no conflict of interest.

References

1. Hou, W.; Xia, F.; Alfranca, G.; Yan, H.; Zhi, X.; Liu, Y.; Peng, C.; Zhang, C.; de la Fuente, J.M.; Cui, D. Nanoparticles for multi-modality cancer diagnosis: Simple protocol for self-assembly of gold nanoclusters mediated by gadolinium ions. *Biomaterials* **2017**, *120*, 103–114. [CrossRef] [PubMed]
2. Darvishi Cheshmeh Soltani, R.; Jorfi, S.; Safari, M.; Rajaei, M.-S. Enhanced sonocatalysis of textile wastewater using bentonite-supported ZnO nanoparticles: Response surface methodological approach. *J. Environ. Manag.* **2016**, *179*, 47–57. [CrossRef] [PubMed]
3. Dalvand, A.; Nabizadeh, R.; Reza Ganjali, M.; Khoobi, M.; Nazmara, S.; Hossein Mahvi, A. Modeling of Reactive Blue 19 azo dye removal from colored textile wastewater using L-arginine-functionalized Fe3O4 nanoparticles: Optimization, reusability, kinetic and equilibrium studies. *J. Magn. Magn. Mater.* **2016**, *404*, 179–189. [CrossRef]
4. Andrade, S.; Ramalho, M.J.; Loureiro, J.A.; Pereira, M.C. Interaction of Natural Compounds with Biomembrane Models: A Biophysical Approach for the Alzheimer's Disease Therapy. *Colloids Surf. B Biointerfaces* **2019**, *180*, 83–92. [CrossRef] [PubMed]
5. Ramalho, M.J.; Pereira, M.C. Preparation and Characterization of Polymeric Nanoparticles: An Interdisciplinary Experiment. *J. Chem. Educ.* **2016**, *93*, 1446–1451. [CrossRef]
6. Makadia, H.K.; Siegel, S.J. Poly Lactic-co-Glycolic Acid (PLGA) as Biodegradable Controlled Drug Delivery Carrier. *Polymers* **2011**, *3*, 1377–1397. [CrossRef] [PubMed]
7. Danhier, F.; Ansorena, E.; Silva, J.M.; Coco, R.; Le Breton, A.; Preat, V. PLGA-based nanoparticles: An overview of biomedical applications. *J. Control Release* **2012**, *161*, 505–522. [CrossRef]
8. Mujokoro, B.; Adabi, M.; Sadroddiny, E.; Adabi, M.; Khosravani, M. Nano-structures mediated co-delivery of therapeutic agents for glioblastoma treatment: A review. *Mater. Sci. Eng C* **2016**, *69*, 1092–1102. [CrossRef]
9. Ramalho, M.J.; Andrade, S.; Coelho, M.Á.N.; Loureiro, J.A.; Pereira, M.C. Biophysical interaction of temozolomide and its active metabolite with biomembrane models: The relevance of drug-membrane interaction for Glioblastoma Multiforme therapy. *Eur. J. Pharm. Biopharm.* **2019**, *136*, 156–163. [CrossRef]
10. Shergalis, A.; Bankhead, A.; Luesakul, U.; Muangsin, N.; Neamati, N. Current Challenges and Opportunities in Treating Glioblastoma. *Pharmacol. Rev.* **2018**, *70*, 412. [CrossRef]
11. Jain, D.S.; Athawale, R.B.; Bajaj, A.N.; Shrikhande, S.S.; Goel, P.N.; Nikam, Y.; Gude, R.P. Unraveling the cytotoxic potential of Temozolomide loaded into PLGA nanoparticles. *DARU J. Pharm. Sci.* **2014**, *22*, 18. [CrossRef] [PubMed]
12. Ananta, J.S.; Paulmurugan, R.; Massoud, T.F. Temozolomide-loaded PLGA nanoparticles to treat glioblastoma cells: A biophysical and cell culture evaluation. *Neurol. Res.* **2016**, *38*, 51–59. [CrossRef] [PubMed]
13. Jain, A.; Chasoo, G.; Singh, S.K.; Saxena, A.K.; Jain, S.K. Transferrin-appended PEGylated nanoparticles for temozolomide delivery to brain: In vitro characterisation. *J. Microencapsul.* **2011**, *28*, 21–28. [CrossRef] [PubMed]
14. Lee, C.; Ooi, I. Preparation of temozolomide-loaded nanoparticles for glioblastoma multiforme targeting—Ideal versus reality. *Pharmaceuticals* **2016**, *9*, 54. [CrossRef] [PubMed]
15. Ramalho, M.J.; Sevin, E.; Gosselet, F.; Lima, J.; Coelho, M.A.N.; Loureiro, J.A.; Pereira, M.C. Receptor-mediated PLGA nanoparticles for glioblastoma multiforme treatment. *Int. J. Pharm.* **2018**, *545*, 84–92. [CrossRef] [PubMed]
16. Lee, S.Y. Temozolomide resistance in glioblastoma multiforme. *Genes Dis.* **2016**, *3*, 198–210. [CrossRef] [PubMed]
17. Warren, K.E.; Gururangan, S.; Geyer, J.R.; McLendon, R.E.; Poussaint, T.Y.; Wallace, D.; Balis, F.M.; Berg, S.L.; Packer, R.J.; Goldman, S.; et al. A phase II study of O6-benzylguanine and temozolomide in pediatric patients with recurrent or progressive high-grade gliomas and brainstem gliomas: A Pediatric Brain Tumor Consortium study. *J. Neuro Oncol.* **2012**, *106*, 643–649. [CrossRef]
18. Quinn, J.A.; Jiang, S.X.; Reardon, D.A.; Desjardins, A.; Vredenburgh, J.J.; Rich, J.N.; Gururangan, S.; Friedman, A.H.; Bigner, D.D.; Sampson, J.H.; et al. Phase II trial of temozolomide plus o6-benzylguanine in adults with recurrent, temozolomide-resistant malignant glioma. *J. Clin. Oncol.* **2009**, *27*, 1262–1267. [CrossRef]

19. Srivenugopal, S.K.; Rawat, A.; Niture, S.K.; Paranjpe, A.; Velu, C.; Venugopal, S.N.; Rao Madala, H.; Basak, D.; Punganuru, S.R. Posttranslational Regulation of O6-Methylguanine-DNA Methyltransferase (MGMT) and New Opportunities for Treatment of Brain Cancers. *Mini Rev. Med. Chem.* **2016**, *16*, 455–464. [CrossRef]
20. Miao, L.; Guo, S.; Zhang, J.; Kim, W.Y.; Huang, L. Nanoparticles with Precise Ratiometric Co-Loading and Co-Delivery of Gemcitabine Monophosphate and Cisplatin for Treatment of Bladder Cancer. *Adv. Funct. Mater.* **2014**, *24*, 6601–6611. [CrossRef]
21. Xu, Y.; Shen, M.; Li, Y.; Sun, Y.; Teng, Y.; Wang, Y.; Duan, Y. The synergic antitumor effects of paclitaxel and temozolomide co-loaded in mPEG-PLGA nanoparticles on glioblastoma cells. *Oncotarget* **2016**, *7*, 20890–20901. [CrossRef]
22. Hoshyar, N.; Gray, S.; Han, H.; Bao, G. The effect of nanoparticle size on in vivo pharmacokinetics and cellular interaction. *Nanomedicine* **2016**, *11*, 673–692. [CrossRef]
23. Blanco, E.; Shen, H.; Ferrari, M. Principles of nanoparticle design for overcoming biological barriers to drug delivery. *Nat. Biotechnol.* **2015**, *33*, 941. [CrossRef]
24. Wang, F.; Chen, L.; Jiang, S.; He, J.; Zhang, X.; Peng, J.; Xu, Q.; Li, R. Optimization of methazolamide-loaded solid lipid nanoparticles for ophthalmic delivery using Box–Behnken design. *J. Liposome Res.* **2014**, *24*, 171–181. [CrossRef]
25. Mazaheri, H.; Ghaedi, M.; Asfaram, A.; Hajati, S. Performance of CuS nanoparticle loaded on activated carbon in the adsorption of methylene blue and bromophenol blue dyes in binary aqueous solutions: Using ultrasound power and optimization by central composite design. *J. Mol. Liq.* **2016**, *219*, 667–676. [CrossRef]
26. Mäkelä, M. Experimental design and response surface methodology in energy applications: A tutorial review. *Energy Convers. Manag.* **2017**, *151*, 630–640. [CrossRef]
27. Papaneophytou, C.P.; Kontopidis, G. Statistical approaches to maximize recombinant protein expression in Escherichia coli: A general review. *Prot. Expr. Purif.* **2014**, *94*, 22–32. [CrossRef]
28. Ramalho, M.J.; Loureiro, J.A.; Gomes, B.; Frasco, M.F.; Coelho, M.A.; Pereira, M.C. PLGA nanoparticles as a platform for vitamin D-based cancer therapy. *Beilstein J. Nanotechnol.* **2015**, *6*, 1306–1318. [CrossRef]
29. Cun, D.; Jensen, D.K.; Maltesen, M.J.; Bunker, M.; Whiteside, P.; Scurr, D.; Foged, C.; Nielsen, H.M. High loading efficiency and sustained release of siRNA encapsulated in PLGA nanoparticles: Quality by design optimization and characterization. *Eur. J. Pharm. Biopharm.* **2011**, *77*, 26–35. [CrossRef]
30. Franken, L.E.; Boekema, E.J.; Stuart, M.C.A. Transmission Electron Microscopy as a Tool for the Characterization of Soft Materials: Application and Interpretation. *Adv. Sci.* **2017**, *4*, 1600476. [CrossRef]
31. Liu, Y.; Pan, J.; Feng, S.-S. Nanoparticles of lipid monolayer shell and biodegradable polymer core for controlled release of paclitaxel: Effects of surfactants on particles size, characteristics and in vitro performance. *Int. J. Pharm.* **2010**, *395*, 243–250. [CrossRef]
32. Tefas, L.R.; Tomu, Ţ.Ă.I.; Achim, M.; Vlase, L. Development and optimization of quercetin-loaded PLGA nanoparticles by experimental design. *Clujul Med.* **2015**, *88*, 214–223. [CrossRef]
33. Narayanan, K.; Subrahmanyam, V.M.; Venkata Rao, J. A Fractional Factorial Design to Study the Effect of Process Variables on the Preparation of Hyaluronidase Loaded PLGA Nanoparticles. *Enzyme Res.* **2014**, *2014*, 162962. [CrossRef]
34. Dos Santos, K.C.; da Silva, M.F.G.F.; Pereira-Filho, E.R.; Fernandes, J.B.; Polikarpov, I.; Forim, M.R. Polymeric nanoparticles loaded with the 3,5,3′-triiodothyroacetic acid (Triac), a thyroid hormone: Factorial design, characterization, and release kinetics. *Nanotechnol. Sci. Appl.* **2012**, *5*, 37–48. [CrossRef]
35. Calignano, F.; Manfredi, D.; Ambrosio, E.P.; Iuliano, L.; Fino, P. Influence of process parameters on surface roughness of aluminum parts produced by DMLS. *Int. J. Adv. Manuf. Technol.* **2013**, *67*, 2743–2751. [CrossRef]
36. Song, X.; Zhao, Y.; Hou, S.; Xu, F.; Zhao, R.; He, J.; Cai, Z.; Li, Y.; Chen, Q. Dual agents loaded PLGA nanoparticles: Systematic study of particle size and drug entrapment efficiency. *Eur. J. Pharm. Biopharm.* **2008**, *69*, 445–453. [CrossRef]
37. Sharma, N.; Madan, P.; Lin, S. Effect of process and formulation variables on the preparation of parenteral paclitaxel-loaded biodegradable polymeric nanoparticles: A co-surfactant study. *Asian J. Pharm. Sci.* **2016**, *11*, 404–416. [CrossRef]
38. Fröhlich, E. The role of surface charge in cellular uptake and cytotoxicity of medical nanoparticles. *Int. J. Nanomed.* **2012**, *7*, 5577. [CrossRef]

39. Ramalho, M.J.; Loureiro, J.A.; Gomes, B.; Frasco, M.F.; Coelho, M.A.N.; Pereira, M.C. PLGA nanoparticles for calcitriol delivery. In Proceedings of the 2015 IEEE 4th Portuguese Meeting on Bioengineering (ENBENG), Porto, Portugal, 26–28 February 2015; pp. 1–6. [CrossRef]
40. Sahoo, S.K.; Panyam, J.; Prabha, S.; Labhasetwar, V. Residual polyvinyl alcohol associated with poly (D, L-lactide-co-glycolide) nanoparticles affects their physical properties and cellular uptake. *J. Controll. Release* **2002**, *82*, 105–114. [CrossRef]

pharmaceutics

MDPI

Article

Nanoemulsion Structural Design in Co-Encapsulation of Hybrid Multifunctional Agents: Influence of the Smart PLGA Polymers on the Nanosystem-Enhanced Delivery and Electro-Photodynamic Treatment

Urszula Bazylińska [1,*], Julita Kulbacka [2] and Grzegorz Chodaczek [3]

1 Faculty of Chemistry, Wroclaw University of Science and Technology, 50-370 Wroclaw, Poland
2 Department of Molecular and Cellular Biology, Faculty of Pharmacy with Division of Laboratory Diagnostics, Wroclaw Medical University, 50-556 Wroclaw, Poland
3 Łukasiewicz Research Network—PORT Polish Center for Technology Development, 54-066 Wroclaw, Poland
* Correspondence: urszula.bazylinska@pwr.edu.pl; Tel.: +48-71-320-21-83

Received: 27 June 2019; Accepted: 8 August 2019; Published: 11 August 2019

Abstract: In the present study, we examined properties of poly(lactide-*co*-glycolide) (PLGA)-based nanocarriers (NCs) with various functional or "smart" properties, i.e., coated with PLGA, polyethylene glycolated PLGA (PEG-PLGA), or folic acid-functionalized PLGA (FA-PLGA). NCs were obtained by double emulsion (water-in-oil-in-water) evaporation process, which is one of the most suitable approaches in nanoemulsion structural design. Nanoemulsion surface engineering allowed us to co-encapsulate a hydrophobic porphyrin photosensitizing dye—verteporfin (VP) in combination with low-dose cisplatin (CisPt)—a hydrophilic cytostatic drug. The composition was tested as a multifunctional and synergistic hybrid agent for bioimaging and anticancer treatment assisted by electroporation on human ovarian cancer SKOV-3 and control hamster ovarian fibroblastoid CHO-K1 cell lines. The diameter of PLGA NCs with different coatings was on average 200 nm, as shown by dynamic light scattering, transmission electron microscopy, and atomic force microscopy. We analyzed the effect of the nanocarrier charge and the polymeric shield variation on the colloidal stability using microelectrophoretic and turbidimetric methods. The cellular internalization and anticancer activity following the electro-photodynamic treatment (EP-PDT) were assessed with confocal microscopy and flow cytometry. Our data show that functionalized PLGA NCs are biocompatible and enable efficient delivery of the hybrid cargo to cancer cells, followed by enhanced killing of cells when supported by EP-PDT.

Keywords: smart nanocarriers; folic acid; verteporfin; cisplatin; SKOV-3 cells; CHO-K1 cells; electroporation; theranostic cargo; double emulsion approach

1. Introduction

Effective nanocarriers (NCs) for cancer treatment need both passive and active targeting approaches to achieve highly specific drug delivery to cancer cells while avoiding rapid clearance by the mononuclear phagocyte system and cytotoxicity to normal cells [1]. Recently, the field of biomedical applications, including drug encapsulation, has raised much interest, in part due to the advancement of the biomaterials and "smart" polymers, which enable preparation of containers with novel functional properties (e.g., size, charge, interfacial functionalization) by means of nanoemulsion structural design. Nanoemulsion systems (so-called submicron emulsions, parenteral emulsions, or miniemulsions) are referred to in the literature as transparent or translucent (often bluish) isotropic dispersions of water and oil, with nano-domains coexisting in high kinetic equilibrium due to the occurrence of surfactant molecules at the oil/water interface [2,3]. Owing to their small size (usually in the range

20–200 nm), high kinetic stability, and much lower surfactant concentration (typically 3–10%) required for their formation in comparison to microemulsions (generally about 20% and higher, both oil-in-water (o/w) and water-in-oil (w/o), nanoemulsions were found to be very attractive for droplet engineering to obtain efficient nanocarriers with functional properties. The usefulness of these formulations as raw structures for container fabrication based on their surface templating approaches, by means of biocompatible polymers and polyelectrolytes, including very effective solvent evaporation and layer-by-layer techniques, have been proved by many scientific investigations [4–7]. Consequently, the structuration of the nanoemulsion droplets is typically carried out with three general approaches: layering, embedding, and clustering [8]. In the embedding process, the double emulsion method permits encapsulation of both hydrophilic and hydrophobic agents in the double compartment structure [7]. Such a hybrid cargo can be loaded simultaneously or separately in the NC liquid core and protected from the environmental conditions [9]. Additionally, the hybrid molecules may also migrate from the outer to inner phase and form a reservoir that enhances spectroscopic or functional properties of photodynamic therapy (PDT) and leads to synergistic anticancer activity. The biological effectiveness of such multifunctional nanosystems in the combined anticancer therapy may be also intensified by increasing their cellular uptake by the electroporation (EP) approach.

The electroporation method is widely used not only in chemotherapy for drug delivery but also in molecular biology and microbiology for gene transfection. The latest data show combination methods are used after electrochemotherapy, where plasmid electrotransfer is applied for stimulation of the immune response for better recovery [10]. It is also crucial that the electroporation technique significantly shortens the time of exposition to the "therapeutic substance". This approach is definitely safes then chemical methods because it does not introduce any additional components to the reaction environment. The EP approach was shown to efficiently increase both the uptake and the photodynamic activity of some second-generation photosensitive agents from the cyanine (e.g., AlPcS4) and chlorin (e.g., Ce6) families for PDT application [11]. The beneficial impact of EP was also demonstrated in the case of porphyrin-origin dyes (e.g., CoTPPS and $MnTMPyPCl_5$), which were used against drug-resistant breast and colon cancer cells [12]. These and other studies performed with phthalocyanines (AlPc and Pc green) [13] showed the promising effect of EP in overcoming drug resistance. Moreover, electroporation can be used for supported delivery or fusion of nanosystems containing multifunctional cargo, which is not possible with chemical methods. Thus, the EP method was effective not only for drugs and photosensitizers in free form but also for encapsulated ones [14]. Our previous study shows that EP enhances delivery and photodynamic activity of encapsulated cyanine IR-780 combined with flavonoids [15]. Thus, the available data clearly indicate a new trend in the application of multifunctional methods in more specific drug delivery to cancerous cells.

Over the last years, different chemical compounds have been utilized in order to improve the stability of double emulsion. This has included various mixtures of surfactants, two types of emulsifiers, and co-surfactants or copolymers to support the final formulation [16,17]. Nevertheless, as has been recently shown, the formation of the nanoemulsion interface by the solvent evaporation approach using functionalized biopolymers, i.e., polyethylene glycol (PEG)-ylated polyesters, such as poly(D,L-lactic acid) (PDLLA), poly(glycolic acid) (PGA), polycaprolactone (PCL), and poly(lactic-*co*-glycolic acid) (PLGA), enhances the long-term colloidal stability of encapsulated bioactive compounds and extends the functional performance of nanoscopic emulsion formulations, rendering them non-toxic and stable in the bloodstream [7,18]. Furthermore, application of PLGA as one of the most promising Food and Drug Administration (FDA)-approved polymers for human use, for example functionalized by folic acid (FA) molecules with selective affinity to the folate receptor (FR), could be beneficial in treatment of human carcinomas with overexpression of FR, including ovarian, breast, and lung cancer cells [19,20].

Thus, in the present study, we prepared NCs coated with PLGA, PLGA-PEG, or PLGA-FA by means of the double emulsion (water-in-oil-in-water (w/o/w)) evaporation process, which enabled co-encapsulation of a hydrophobic, porphyrin-origin photosensitizing agent—verteporfin (VP) in combination with low-dose cisplatin (CisPt)—a hydrophilic chemotherapeutic drug. Obtained NCs were characterized using a variety of analytical methods and functional assays. Finally, their utility in bioimaging as well as in anticancer applications upon human ovarian cancer SKOV-3 cells and control hamster ovarian fibroblastoid CHO-K1 cells was also analyzed when combined with EP and PDT.

2. Materials and Methods

2.1. Chemicals and Reagents

Poly(lactide-*co*-glycolide), ester endcap L:G 50:50 (PLGA, Mw ~ 25,000–35,000), poly(ethylene glycol) methyl ether-block-poly(lactide-*co*-glycolide) (PEG-PLGA, Mw ~ 5000:20,000), poly(lactide-*co*-glycolide)-folate, L:G 50:50 (FA-PLGA, Mw ~ 25,000–35,000) employed as biocompatible, "stealth" and functionalized "smart" polymers, respectively, for NC stabilization were obtained from PolySciTech® (West Lafayette, IN, USA). Cremophor A25 and didodecyldimethylammonium bromide, di-C_{12}DMAB, applied as a hydrophilic non-ionic and a hydrophobic double chain cationic surfactant, were obtained from BASF Care Creations (Monheim am Rhein, Germany) and Sigma Aldrich (Poznan, Poland), respectively. Verteporfin (VP) and cisplatin (Cis-Pt), utilized as hybrid cargo, were from Sigma-Aldrich (Poznan, Poland). Supplementary chemical compounds were of commercial grade and were used as received. Doubly distilled water was purified using a Milli-Q purification system (Millipore, Bedford, MA, USA).

2.2. Co-Encapsulation of Hybrid Agents in Polymeric Nanocarriers by Double Emulsion (w/o/w) Solvent Evaporation Method

Polymeric NCs stabilized by PLGA, PEG-PLGA, FA-PLGA, and non-ionic and cationic surfactants for co-encapsulation of a therapeutic (CisPt), as well as a diagnostic and therapeutic agent—VP (both in the initial concentration of 130 µM)—were obtained using double emulsion (w/o/w) evaporation method [7,21]. Generally, we emulsified an aqueous internal phase (with CisPt) in dichloromethane (with VP, PLGA at a concentration of 5 mg/mL and di-C_{12}DMAB) at 1:4 ratio using a homogenizer set to 25,000 rpm for 5 min. Next, the primary water-in-oil (w/o) nanoscopic emulsion was poured into 1% Cremophor A25 aqueous solution (a hydrophilic surfactant), stirred in a homogenizer for 10 min (25,000 rpm), and immersed in an ice-water bath to obtain the w/o/w emulsion. Then, we evaporated the organic solvent under reduced pressure in a rotary evaporator (Hei-VAP Value Digital, Heidolph Instruments, Schwabach, Germany) with a rotation speed of 150 rpm for 30 min at 25 °C and polymeric NCs with a PLGA, PEG-PLGA, or FA-PLGA shell. The hybrid cargo were collected the following day.

2.3. Nanocarrier Size, Polydispersity, and Particle Charge

The main physicochemical parameters of NCs, such as hydrodynamic diameter (D_H), polydispersity index (PDI), and particle charge (ζ-potential), were analyzed by means of dynamic light scattering (DLS) and microelectrophoretic methods using Zetasizer Nano Series (Malvern Instruments, Worcestershire, UK) equipped with a He–Ne laser (632.8 nm). DLS measurements were conducted at 25 °C and the detection angle was 173°, as previously described [7,21,22]. Each value was an average of three runs, with at least 10–20 measurements. We applied the DTS (Nano) program for data evaluation.

2.4. Shape and Morphology

The morphology of the obtained NCs was studied by atomic force microscopy (AFM) and transmission electron microscopy (TEM) according to our previous protocols [7,21]. The AFM observations were conducted using a NanoScope Dimension V instrument with an RT ESP tube scanner (Veeco Instruments, Plainview, NY, USA) Samples were analyzed at 0.5 Hz scanning speed using a low-resonance-frequency pyramidal silicon cantilever resonating at 250–331 kHz at a constant force of 20–80 N/m. The resonance amplitude was adjusted manually to the lowest possible amplitude enabling stable imaging within the contamination layer on the surface. We prepared the samples by adsorption of an NC droplet on mica that was freshly cleaved. After 18 h, the excess substrate was removed by rinsing the mica plates in double distilled water for 1 min and drying for 2 h at room temperature. The TEM imaging of NCs was performed with an Field Electron and Ion Company (FEI) Tecnai G2 20 X-TWIN electron microscope (FEI, Brno, Czech Republic) by placing a few drops of diluted NCs on a Cu-Ni grid and leaving the specimens to dry for 20 h at room temperature.

2.5. *Encapsulation Efficiency*

The Ultraviolet - Visible (UV-VIS) absorbance of NCs with encapsulated VP and Cis-Pt was measured with a Metertech SP8001 spectrophotometer with a 1-cm length path thermostated quartz cell in order to evaluate the encapsulation efficiency (*EE*). The hybrid cargo concentration was calculated using calibration curves according to our previous protocol [21,22]. We determined *EE* as follows:

$$EE = \frac{W_{added} - W_{free}}{W_{added}} \times 100\ \%$$

where W_{added} is the amount of VP or CisPt added during the encapsulation procedure, and W_{free} is the amount of free cargo in the supernatant quantified by UV-VIS spectroscopy after separation of NCs by centrifugation process (14,000 rpm for 30 min).

2.6. *Colloidal Stability*

The backscattering (BS) of pulsed near-infrared, IR light (l = 880 nm) was utilized to measure the long-term colloidal stability of NCs (Turbi-ScanLabExpert, Formulaction SA, Toulouse, France) [7]. In general, two synchronous optical sensors (transmission and backscattering detectors) recorded light transmitted through the sample (0° from the incident radiation) and light back-scattered by the sample (1358 from the incident radiation). The scanning of the sample was performed in a cylindrical glass cell at 25 °C by moving along the entire height of the cell. The BS profiles as a function of the sample height were then collected and analyzed using the instrument's software (Turbisoft version 2.0.0.33, Formulaction SA, Toulouse, France). We measured BS for freshly prepared NCs and after 30 days of the sample storage at 25 °C.

2.7. *Cell Lines*

The biological studies were performed on a human ovarian carcinoma cell line resistant to diphtheria toxin, cisplatin, and adriamycin (SKOV-3), and a hamster ovarian fibroblastoid cell line (CHO-K1) used as a model for transport studies in a pulsed electric field due to very low expression of endogenous ionic channels [23]. The SKOV-3 and CHO-K1 cells were purchased from ATCC® (American Type Culture Collection, distr. LGC Standards, Lomianki, Poland), cultured, and prepared according to the conditions described previously by our group [15].

2.8. *Uptake of Encapsulated Hybrid Cargo—Flow Cytometry Analysis*

The ability to internalize free and encapsulated VP/CisPt by CHO-K1 and SKOV-3 cells was analyzed by flow cytometry using fluorescence-activated cell sorter (FACS, Cube-6, SYSMEX EUROPE GmbH, Warsaw, Poland). Cells were harvested on 12-well plates and after obtaining 80% of confluence, appropriate nanosystems were added as follows. Free VP or NCs containing VP/CisPt were added with a final VP concentration equal to 2.0×10^{-6} M. Then, the cells were incubated for 24 h at 37 °C in a humidified atmosphere containing 5% CO_2. In the next step, cells were detached with Trypsin-EDTA (Sigma-Aldrich Merck-Group, Poznan, Poland), washed in PBS, and resuspended in 0.5 mL of PBS. Flow cytometry analysis was performed using a Cube 6 flow cytometer (Sysmex, Warsaw, Poland). The fluorescence of VP was measured with a FL-4-H detector. Data were collected and analyzed by CyView software (Sysmex, Warsaw, Poland).

2.9. Electroporation Protocol

Electropermeabilization of cell membranes alone and with free or co-encapsulated VP and Cis-Pt was performed using Gene Pulser Xcell™ Electroporation System (BioRad Laboratories, Warsaw, Poland). When cells reached 80% of confluency they were trypsinized and centrifuged (5 min, 1000 rpm). Then, cells were counted and resuspended in 200 µL of electroporation buffer (EP buffer of low electrical conductivity of 0.14 S/m) at cell concentration of 3×10^6/mL [15]. Cells were maintained in suspension and pulsed in a cuvette (VWR) with two aluminum plate electrodes (4 mm gap). The following parameters of electroporation were applied: electrical field intensity E(appl) = 500 V/cm, 5 rectangular unipolar pulses of 1 ms duration. The EP conditions were established according to our previous study [14]. The EP experiments were performed using Gene Pulser Xcell™ Electroporation System 165-2660 (BioRad). After the pulse delivery, cells were incubated for 10 min at 37 °C, then gently centrifuged, resuspended in the cell culture medium (DMEM for SKOV-3 cells or HAM's F10 for CHO-K1 cells, Sigma-Aldrich Merck-Group, Poznan, Poland), and further analyzed by confocal microscopy and subjected to photocytotoxicity studies.

2.10. Intracellular Internalization Studies by Confocal Microscopy

The internalization of co-encapsulated VP by cancer and normal cells was studied with confocal microscopy. Briefly, SKOV-3 and CHO-K1 cells were seeded onto glass cover slips in Petri dishes (Sarstedt - distr. Equimed, Wroclaw, Poland) at a density of 1×10^4 cells per cover slip in a CO_2 incubator for 24 h. Next, the cells were treated with NCs at a concentration corresponding to 2.0×10^{-6} M of VP and incubated at 37 °C for 24 h. For the EP-supported uptake, cells were first processed as described in the Section 2.9 and then seeded onto cover slips, thus the time of exposition to nanosystems was only 10 min. After 24 h, all samples were fixed in 4% formaldehyde (Polysciences Inc., Hirschberg an der Bergstrasse, Germany), washed, and placed onto basic glass slides (SuperFrost, Menzel, Braunschweig, Germany) upon mounting in an anti-fade medium (Roth – distr. Linegal Chemicals Sp. z o.o, Warsaw, Poland) with with 4′,6-diamidino-2-phenylindole (DAPI) for nuclei staining. Microscopy was performed on a spinning disk confocal microscope (Cell Observer SD, Zeiss, Oberchochen, Germany). DAPI was visualized with a 405 nm laser and 450/50 emission filter, while VP fluorescence was excited with a 488 laser and collected with a 629/62 nm emission filter.

2.11. Photodynamic Activity Protocol

Photocytotoxicity of NCs in SKOV-3 and CHO-K1 cells was measured after standard photodynamic procedure and PDT combined with the EP protocol described above by cellular mitochondrial activity determined by the (3-(4,5-dimethylthiazol-2-yl)-2,5-diphenyltetrazolium bromide (MTT) colorimetric assay according to the manufacturer's procedure and our previous experiments [12–15]. The cells subject to PDT were irradiated after 24 h for 10 min with light in the range of 630–680 nm; the final energy delivered to the cell monolayer was 10 J/cm^2. The MTT assay was performed after 24 h post irradiation for the NCs loaded with the theranostic cargo at a concentration equivalent to 1.0–5.0 $\times$ 10^{-6} M of encapsulated VP. The measurements were performed on the GloMax® Discover multimode microplate reader (Promega, Madison, WI, USA). The cell viability in each group was expressed as a percentage of the value obtained for control (untreated) cells (average of three experiments).

2.12. Statistical Analysis

The results are presented as means ± standard deviation (SD) values for minimum $n = 3$ repeats. The results were analyzed by two-way ANOVA for multiple comparisons and $\alpha = 0.05$ GraphPad Prism 7.05. The values where $p \leq 0.05$ (marked with *) were considered as statistically significant.

3. Results and Discussion

3.1. Characteristic of "Smart" PLGA Nanocarriers Obtained by Nanoemulsion Structural Design

The combination of biomaterials, "smart" polymers, and drug delivery systems for both therapeutic and diagnostic (theranostic) approaches enables the development of intelligent devices and brings enormous possibilities for biomedical applications [24]. Biodegradable polyesters, such as poly(glycolic acid) (PGA), poly(lactic acid) (PLA), and their copolymers are the favorite synthetic polymers for biomedical and pharmaceutical applications, since they were proved to be useful in the stabilization of different drug delivery systems and have excellent biocompatibility and bioresorbability [25].

With this in mind, we have tested highly biocompatible polymers made of PLGA (the FDA-approved copolymer of PGA and PLA, also functionalized with different moieties), namely PEG and FA. The "smart" co-polymers were used to design NCs co-loaded with VP, playing a dual role of a diagnostic and a therapeutic agent, and CisPt, a supporting chemotherapeutic drug. Engineered NCs with encapsulated hybrid cargo had an improved colloidal stability and therapeutic activity, as well as extended functional performance with unique attributes e.g., non-toxicity, stability in blood circulation, and cancer-targeting ability. Thus, according to the first phase of our general strategy presented in Scheme 1a, PLGA, PEG-PLGA, and FA-PLGA polymers were used for stabilization and structuration of nanoemulsion droplets involving the three-step w/o/w double emulsion evaporation approach, leading to co-encapsulation of the hybrid cargo, i.e., VP and CisPt in the NC's double compartment.

The second step (Scheme 1b) involved EP-supported PDT upon improved internalization of NCs by human ovarian cancer (SKOV-3) and normal ovary fibroblastoid (CHO-K1) cells. The SKOV-3 cells were selected as difficult-to-treat cells, which are extremely resistant to the wide spectrum of cytostatic drugs, but in particular to cisplatin. Consequently, in our study an attempt was made to design special nanosystems to overcome drug resistance phenomena in human ovarian cancer and also very probable secondary resistance to CisPt, and for rapid elimination of the drug from circulation [26]. The PLGA nanocarriers proposed here can significantly diminish this process, causing good bioavailability and the "willingness" of the cell to accept the natural carrier. Furthermore, as has been proved recently, different drug delivery systems with a negatively charged surface are generally less toxic compared to the positively charged ones, but their cellular uptake may be hindered due to the same negative charge present on the surface of target cells. Thus, the electropermeabilization may enhance the transport of theranostic cargo to target (malignant) cells in spite of their negative surface charge [15,27].

Accordingly, three types of PLGA-NCs loaded with theranostic cargo and three control NCs (systems V1-V6, Table 1) were successfully synthesized. The main physicochemical characteristics of the designed NCs are summarized in Table 1. We measured size (hydrodynamic diameter, D_H), polydispersity index (PdI), zeta potential (ζ), and encapsulation efficiency of VP (EE_{VP}) and CisPt (EE_{CisPt}). NCs with different PLGA shells displayed an average size between 187 and 200 nm, PDI of approximately 0.1–0.2, and ζ from −4 mV to −17 mV, proving efficient assembly of NCs with PEG-ylated and FA-functionalized shells [7,28]. Generally, loaded cargo did not significantly change the NC charge (ζ), as observed for control NCs with a FA-PLGA shell, loaded with VP, CisPt, or empty ones. Furthermore, the loaded NCs showed only a slightly larger size and less unimodal size distribution, which was probably caused by the incorporated cargo molecules. The EE was about 95% for VP and 90% for CisPt. The differences in the encapsulation of the hybrid cargo by PLGA, PEG-PLGA, and FA-PLGA shells are presented in Figure 1 as UV-VIS spectra of the co-encapsulated VP and CisPt compared to the control samples with only the cytostatic drug and with only the photosensitizer, as well as empty NCs. In all nanosystems a characteristic peak at 280 nm for CisPt as well as peaks at 340 nm, 415 nm, and 680 nm for VP can be observed, providing evidence of effective encapsulation of both ingredients.

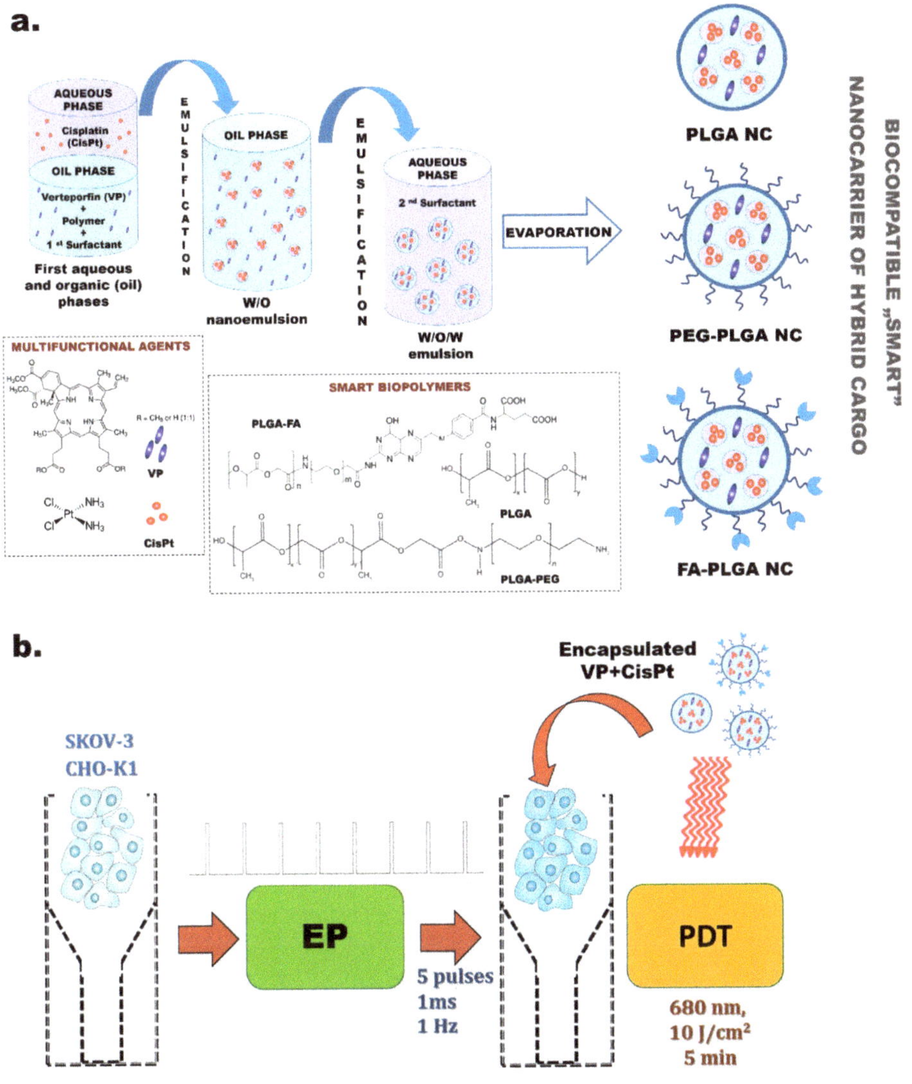

Scheme 1. The general strategy of conducted studies: (**a**) double emulsion (w/o/w) design for co-encapsulation of Verteporfin (VP) and Cisplatin (CisPt) in various Poly(Lactide-*co*-Glycolide) (PLGA) nanocarriers (NCs) and (**b**) electroporation (EP)-supported photodynamic therapy (PDT) using obtained NCs against human ovarian cancer (SKOV-3) and hamster ovarian control (CHO-K1) cells.

Table 1. Physicochemical characteristics of VP and Cis-Pt co-loaded NCs with various PLGA shells.

System	Composition	D_H [nm]	PDI	ζ [mV]	EE_{CisPt}	EE_{VP}
V1	NCs-PLGA + VP + CisPt	193 ± 6	0.16 ± 0.01	−9 ± 1	92 ± 1	97 ± 3
V2	NCs-PLGA-PEG + VP + CisPt	187 ± 5	0.12 ± 0.01	−4 ± 1	88 ± 1	92 ± 1
V3	NCs-PLGA-FA + VP + CisPt	200 ± 7	0.20 ± 0.02	−15 ± 2	90 ± 2	95 ± 3
V4	NCs-PLGA-FA + VP	197 ± 7	0.22 ± 0.02	−16 ± 2	-	96 ± 3
V5	NCs-PLGA-FA + CisPt	194 ± 6	0.25 ± 0.02	−16 ± 2	92 ± 2	-
V6	NCs-PLGA-FA empty	189 ± 5	0.10 ± 0.01	−17 ± 3	-	-

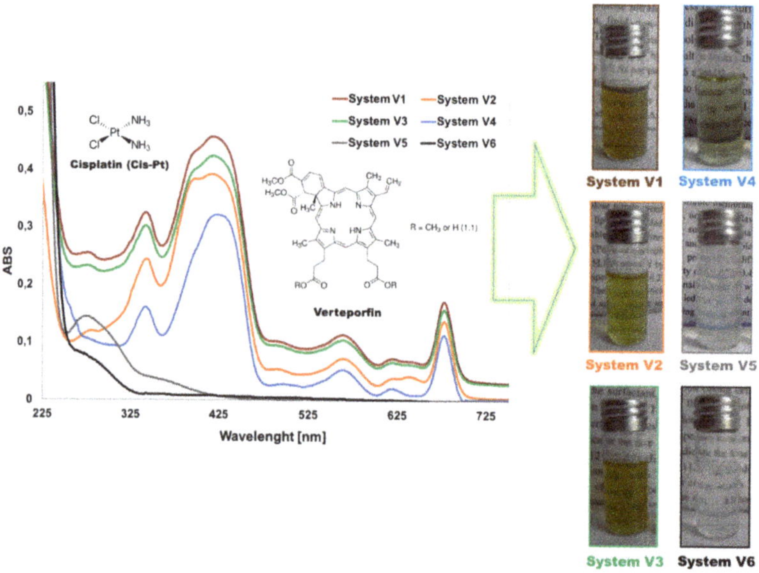

Figure 1. Ultraviolet - Visible (UV-VIS) spectra and images of the VP and CisPt co-loaded nanocarriers with various PLGA shells (samples V1–V3) compared to the control samples loaded only by the cytostatic drug (V5), only by the photosensitizer (V4), and the empty NCs (V6). See descriptions for the systems (V1–V6) in Table 1.

Meanwhile, the correct imaging of NCs is a key parameter in the design of any drug delivery system dedicated to pharmaceutical applications. The obtained PLGA-stabilized NCs were characterized by transmission electron microscopy (TEM) and atomic force microscopy (AFM)—quick, efficient, and relatively non-invasive techniques that can provide evidence on shape and morphology and size distribution of these polymeric nanosystems. The TEM and AFM images of the loaded PLGA-NCs are shown in Figure 2. The TEM imaging demonstrated spherical particles with roughly uniform sizes related to AFM. Furthermore, we observed some differences in morphology as visualized by TEM for nanocarriers prepared with different PLGA shells. In the case of NCs covered by PLGA, the spherical nanoobjects with relatively smoother surface morphology were discovered by both TEM and AFM imaging, while the NCs stabilized by PEG-PLGA and FA-PLGA (Figure 2) had a typical core shell morphology, where the darkest part relates to the denser polymeric/PEG-ylated corona, demonstrating that these shells were successfully formed. The AFM tapping mode scanning presented as 2D and 3D images identified a semi-spherical shape of the NCs, being slightly less regular in the case of NCs stabilized only by PLGA (Figure 2). However, we did not see increased aggregation as it was found by TEM. The NC's size range was smaller than the distribution obtained by the DLS measurement presented in Table 1, as both TEM and AFM were carried out in dry conditions, and the obtained NCs have a tendency to shrink, resulting in losing their primary shape and size [29].

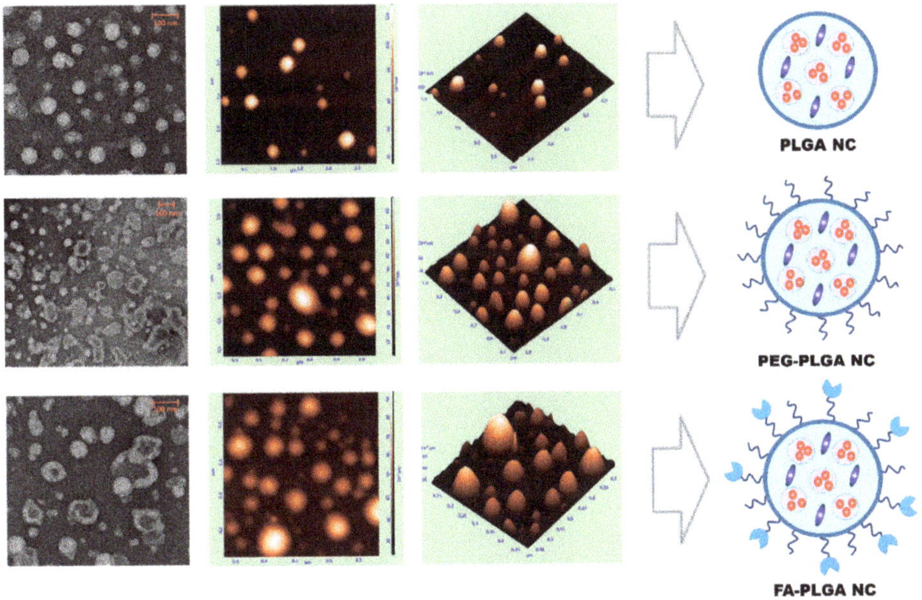

Figure 2. Imaging of VP and CisPt co-loaded PLGA (V1), PEG-PLGA (V2), and FA-PLGA (V3) NCs by means of transmission electron microscopy (TEM) and atomic force microscopy (AFM). See descriptions for the systems (V1–V3) in Table 1.

3.2. Evaluation of Colloidal Stability

The colloidal stability of NCs is one of the most critical factors for any potential biological application, since these nanostructures, when not stabilized electrostatically, are usually metastable due to short-range van der Waals attraction [30]. Consequently, to avoid NC aggregation due to their low colloidal stability, steric or electrostatic repulsion may be applied for stabilization. The literature data indicate the encapsulation of inorganic and organic molecules by surfactants and polymers as the best strategy for the enhancement of any nanostructure colloidal stability and functionality, leading to hybrid core/(polymer-)shell NCs [31]. However, the NC aggregation process is still hard to control, especially in biological environments, which is of essential significance for in vivo applications [6,30].

The detailed estimation of the colloidal stability of different PLGA co-loaded and empty NCs was conducted using turbidimetric method by means of time-dependent BS profiles. The BS levels expressed in % are indicated on the ordinate axis. The investigated formulation level in the measurement vial was expressed in mm and marked on the abscissa axis. By examining the BS profiles (Figure 3), we were able to determine the dynamics of any decomposition processes occurring within the sample. This was achieved through analysis of the distances between the curves in BS profiles of NCs at 0 days (freshly prepared) and after 30 days of their storage at room temperature (Figure 3).

Typically, rapid destabilization phenomena can be recognized by a large distance between the curves, while an overlap of the individual curves indicates the high stability of the analyzed sample and a slow rate of the destabilization process [32]. Based on the graphs shown in Figure 3, we conclude that the studied nanosystems have good colloidal stability, since no macroscopic changes in analyzed samples (aggregation, sedimentation, or creaming processes) were observed at the last day of the performed turbidimetric test.

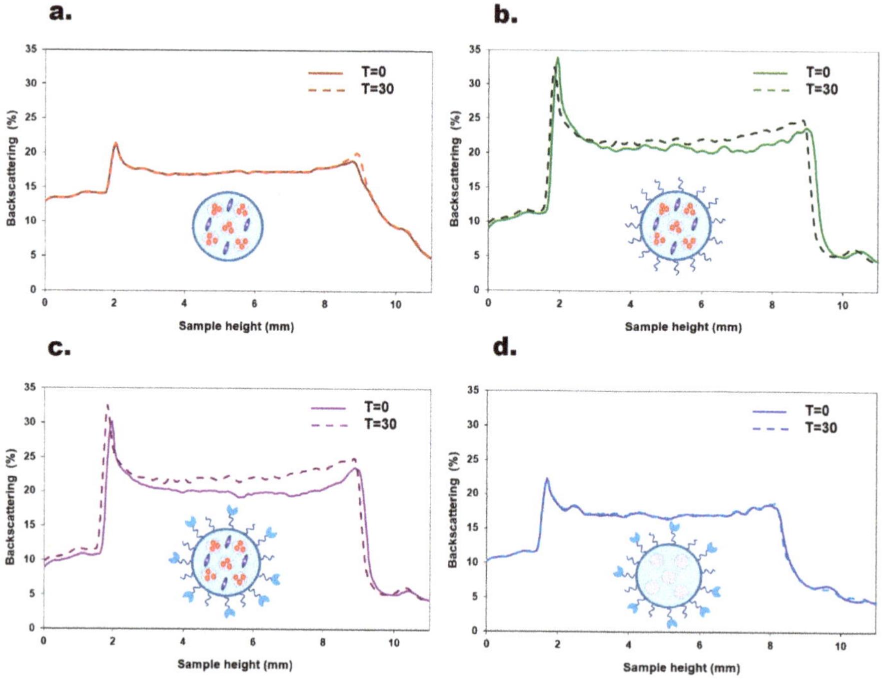

Figure 3. Backscattering profiles of the various PLGA NCs co-loaded with VP and CisPt (**a**–**c**) (V1–V3) and an empty (**d**) (V6) FA-PLGA nanosystem as a function of sample height (mm) analyzed over 30 days ($T = 30$) of the NCs' storage. Solid lines represent measurement at 0 day ($T = 0$, freshly prepared NCs), while dashed lines indicate data obtained after 30 days of storage. See descriptions of the systems in Table 1.

3.3. *Cellular Internalization—Flow Cytometry and Confocal Microscopy Evaluation*

The evaluation of nanosystem uptake by cancer and normal cells is shown in Figure 4. The uptake efficiency of NCs was estimated by flow cytometry after 24 h of exposition. SKOV-3 cells revealed a higher uptake efficiency of VP-loaded NCs than CHO-K1 cells. The strongest fluorescent signal was observed for V3 (VP + CisPt) and V4 (VP) nanosystems, which were functionalized by FA moieties, proving the effective SKOV-3 targeting ability of the "smart" FA-PLGA-coated nanodevices [33]. The improved uptake in the case of nanosystem V3 can be explained by the content of isplaitin, which can slightly sensitize exposed cells and provoke better internalization of VP.

The confocal microscopy was conducted on cells treated with NCs and electroporated. The exposition time to treatment was much shorter and did not exceed 10 min. The acquired microphotographs are presented in Figure 5. Upon EP, the uptake of encapsulated drugs increased, in particular for FA-PLGA-coated NCs (V3 and V4) in both cell lines. Additionally, a stronger fluorescent signal was also found for the V2 (NCs-PLGA-PEG) nanosystem in electroporated ovarian cancer cells. A very low fluorescent signal was detected when cells were not electroporated. Thus, we conclude that EP is favorable when a short time of incubation is required for the therapeutic protocol. Moreover, FA significantly enhanced toxicity against ovarian cancer cells. Our results are in good correspondence with other studies that also indicate functionalization as a promising factor in anticancer protocols [32,34].

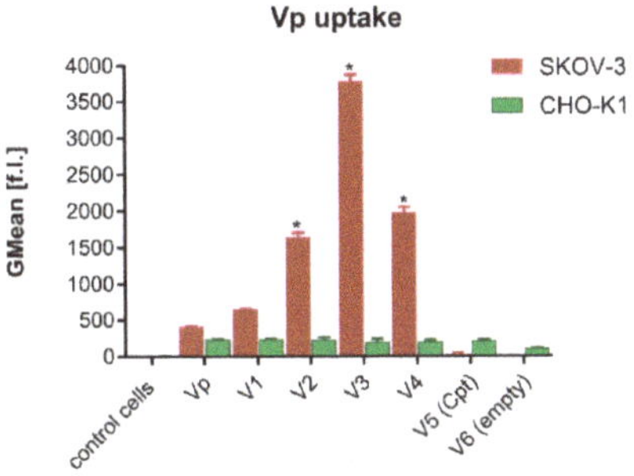

Figure 4. Influence of the nanocarrier surface on the encapsulated cargo uptake by SKOV-3 and CHO-K1 cells evaluated by flow cytometry analysis. The uptake of NCs was expressed as mean fluorescence intensity of VP in the gated cell population after 24 h incubation of cells with the loaded PLGA NCs (V1–V5), control cells (control), empty nanocarriers (V6). and free VP (Vp). See descriptions of the systems in Table 1.

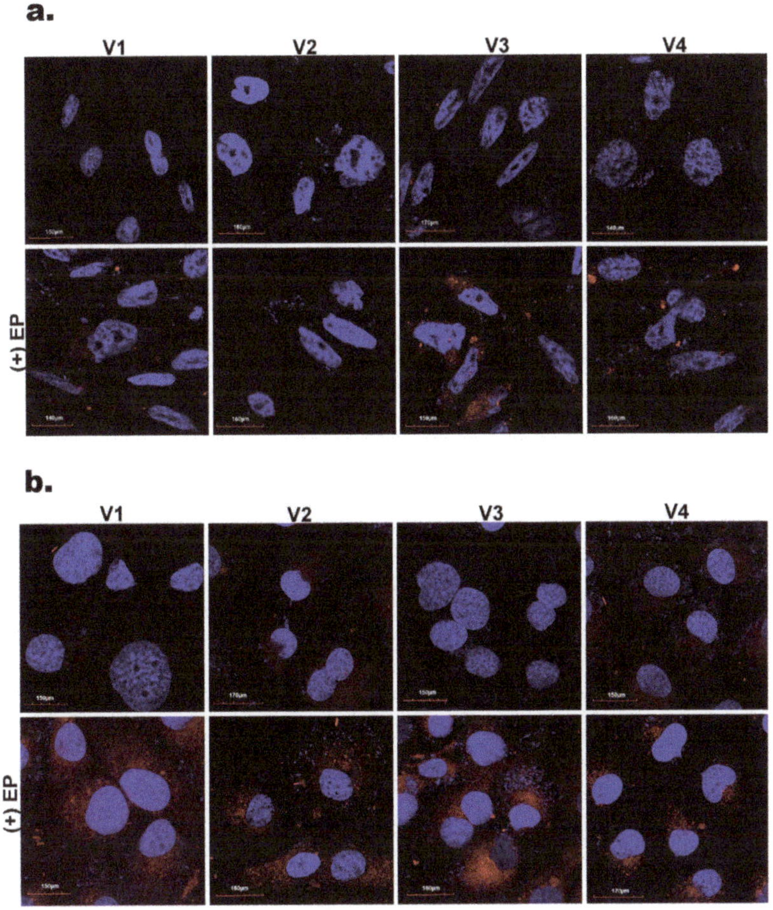

Figure 5. Intracellular distribution of nanosystems after 24 h incubation with NCs (upper panel) or 10 min exposition to NCs after EP (lower panel) in (**a**) CHO-K1 cells and (**b**) SKOV-3 cells.

3.4. Evaluation of PDT and EP-PDT

The results of photocytotoxicity studies for SKOV-3 cells are presented in Figure 6a. The cellular viability was assessed after 24 h of incubation with the NCs followed by irradiation. We found that the photodynamic effect increased proportionally to the concentration of the applied nanosystems. Longer time of incubation induced a significant decrease of cellular viability (60–80% decrease), in particular for the mixed cargo, proving the supportive anticancer effect of CisPt [35]. In Figure 6b,c, both types of cells were first electroporated and then exposed to NCs for 10 min.

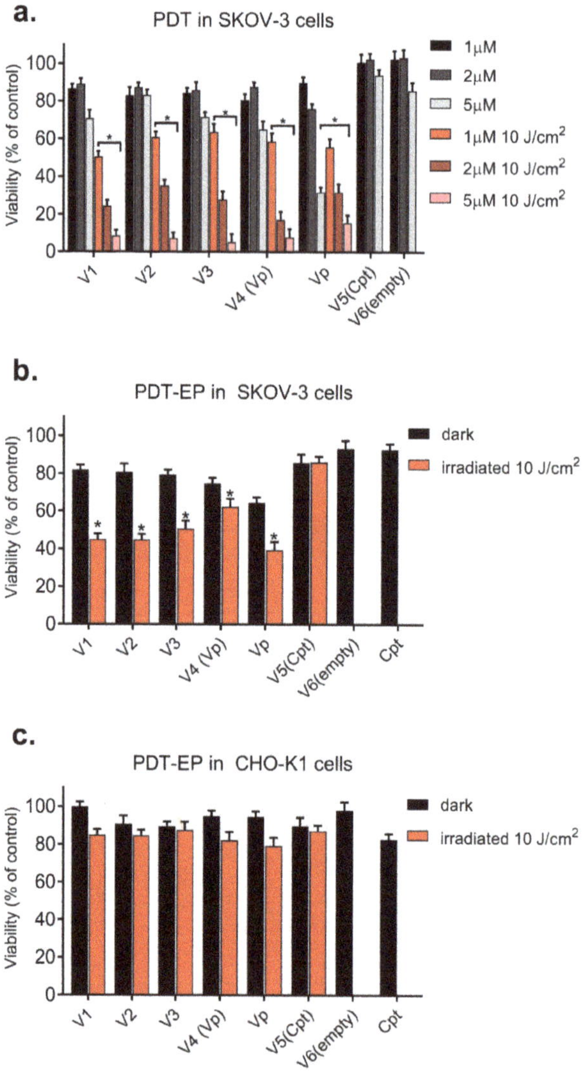

Figure 6. Photodynamic treatment (PDT) in SKOV-3 after 24 h of incubation with NCs (**a**) and photodynamic reaction facilitated by EP after 10 min exposition to nanosystems (C_{VP} = 1 µM) in SKOV-3 (**b**) and CHO-K1 (**c**) cells. The missing red bars for V6 and free Cpt represent non irradiated negative controls. See description of the NCs in Table 1. Data represented as means ± SD for minimum $n = 3$, where * $p \leq 0.05$ was judged as statistically significant.

After irradiation that followed EP, we observed a significant decrease of cellular viability in ovarian cancer cells in a short time, even in the lowest concentration of the applied NCs (C_{VP} = 1 μM). This is in agreement with another study showing that EP enhances photodynamic reaction efficacy in the case of encapsulated drugs [12]. Thus, EP might be a promising and PDT-supporting tool for time-limited anticancer therapies. It is also worth noting that empty NCs (V6) with or without EP showed no toxic effect in SKOV-3 and CHO-K1 cultures, even at higher NC concentration (cell viability above 90%), proving the protective effect of the PLGA shell.

4. Conclusions

In this work, we demonstrate that a rationally designed double emulsion process leads to formulation of long-lasting, biocompatible, and "smart" NCs containing hybrid theranostic cargo with different hydrophobicity. We have designed, engineered, and characterized the effective polymeric nanocontainers (size ~200 nm) for efficient co-encapsulation of Cis-Pt and VP enabling drug delivery and synergistic anticancer activity via standard PDT and EP-enhanced PDT against human ovarian (SKOV-3) cancer cells. Loading hybrid cargo inside the oil core of NCs minimized its interaction with water environment, and thus, no drastic changes in physicochemical properties of the cytostatic drug and photosensitizer were observed. This feature is very favorable from the point of view of any bio-related application. Furthermore, EP with only 5 pulses and short time of loading exposition (10 min) enhanced delivery of encapsulated VP with all types of NCs. The highest photodynamic potential was noticed when VP was co-encapsulated with Cis-Pt and the "smart" NC functionalization with FA led to improved internalization by the SKOV-3 cells. Control CHO-K1 cells were significantly less sensitive to PDT, EP, and EP-PDT. In summary, the presented results reveal that the designed polymeric PLGA shells and colloidal cores in our NCs significantly improve PDT and EP-PDT and warrant future in vivo studies in animal models to fully prove their use for theranostic applications.

Author Contributions: Conceptualization, U.B. and J.K.; methodology, U.B., J.K. and G.C.; software, U.B. and J.K.; validation, U.B., J.K.; formal analysis, U.B., J.K. and G.C.; investigation, U.B., J.K. and G.C.; resources, U.B.; data curation, U.B., J.K. and G.C.; writing—original draft preparation U.B., J.K.; writing—review and editing, U.B., J.K., G.C.; visualization, U.B., J.K.; supervision, U.B.; project administration, U.B.; funding acquisition, U.B.

Funding: This research was funded by the National Science Centre (Poland) within the framework of the SONATA 8 program (No. 2014/15/D/ST4/00808), and by a statutory activity subsidy from the Polish Ministry of Science and Higher Education for the Faculty of Chemistry of Wroclaw University of Science and Technology.

Acknowledgments: We would like to thank Professor Katarzyna Bogunia-Kubik from the Institute of Immunology and Experimental Therapy (Wroclaw, Poland) for facilitating the electroporation tests on the Gene Pulser Xcell™ Electroporation System (BioRad, Warsaw, Poland).

Conflicts of Interest: The authors declare no conflict of interest.

References

1. Alexis, F.; Pridgen, E.; Molnar, L.K.; Farokhzad, O.C. Factors affecting the clearance and biodistribution of polymeric nanoparticles. *Mol. Pharm.* **2008**, *5*, 505–515. [CrossRef] [PubMed]
2. Solans, C.; Izquierdo, P.; Nolla, J.; Azemar, N.; Garcia-Celma, M.J. Nanoemulsions. *Curr. Opin. Colloid Interface Sci.* **2005**, *10*, 102–110. [CrossRef]
3. Bazylińska, U.; Kulbacka, J.; Wilk, K.A. Dicephalic ionic surfactants in fabrication of biocompatiblenanoemulsions: Factors influencing droplet size and stability. *Colloids Surf. A Physicochem. Eng. Asp.* **2014**, *460*, 312–320. [CrossRef]
4. Fornaguera, C.; Feiner-Gracia, N.; Calderó, N.G.; García-Celma, M.J.; Solans, C. PLGA nanoparticles from nano-emulsion templating as imaging agents: Versatile technology to obtain nanoparticles loaded with fluorescent dyes. *Colloids Surf. B Biointerfaces* **2016**, *147*, 201–209. [CrossRef] [PubMed]
5. Fornaguera, C.; Dols-Perez, A.; Calderó, N.G.; García-Celma, M.J.; Camarasa, J.; Solans, C. PLGA nanoparticles prepared by nano-emulsion templating using low-energy methods as efficient nanocarriers for drug delivery across the blood–brain barrier. *J. Control. Release* **2015**, *211*, 134–143. [CrossRef]

6. Bazylińska, U.; Saczko, J. Nanoemulsion-templated polylelectrolyte multifunctional nanocapsules for DNA entrapment and bioimaging. *Colloids Surf. B Biointerfaces* **2016**, *137*, 191–202. [CrossRef]
7. Bazylińska, U. Rationally designed double emulsion process for co-encapsulation of hybrid cargo in stealth nanocarriers. *Colloids Surf. A Physicochem. Eng. Asp.* **2017**, *532*, 476–482. [CrossRef]
8. McClements, J.D. Advances in fabrication of emulsions with enhanced functionality using structural design principles. *Curr. Opin. Colloid Interface Sci.* **2012**, *17*, 235–245. [CrossRef]
9. Ahmed, N.; Michelin-Jamois, M.; Fessi, H.; Elaissari, A. Modified double emulsion process as a new route to prepare submicron biodegradable magnetic/polycaprolactone particles for in vivo theranostics. *Soft Matter* **2012**, *8*, 2554–2564. [CrossRef]
10. Salvadori, C.; Svara, T.; Rocchigiani, G.; Millanta, F.; Pavlin, D.; Cemazar, M.; Lampreht, U.; Sersa, T.G.; Tozon, N.; Poli, A. Effects of electrochemotherapy with cisplatin and peritumoral IL-12 gene electrotransfer on canine mast cell tumors: A histopathologic and immunohistochemical study. *Radiol. Oncol.* **2017**, *51*, 286–294. [CrossRef]
11. Labanauskiene, J.; Gehl, J.; Didziapetriene, J. Evaluation of cytotoxic effect of photodynamic therapy in combination with electroporation in vitro. *Bioelectrochemistry* **2007**, *70*, 78–82. [CrossRef]
12. Kulbacka, J.; Kotulska, M.; Rembiałkowska, N.; Choromańska, A.; Kamińska, I.; Garbiec, A.; Rossowska, J.; Daczewska, M.; Jachimska, B.; Saczko, J. Cellular stress induced by photodynamic reaction with CoTPPS and MnTMPyPCl5 in combination with electroporation in human colon adenocarcinoma cell lines (LoVo and LoVoDX). *Cell Stress Chaperones* **2013**, *18*, 719–731. [CrossRef]
13. Zielichowska, A.; Saczko, J.; Garbiec, A.; Dubińska-Magiera, M.; Rossowska, J.; Surowiak, P.; Choromańska, A.; Daczewska, M.; Kulbacka, J.; Lage, H. The photodynamic effect of far-red range phthalocyanines (AlPc and Pc green) supported by electropermeabilization in human gastric adenocarcinoma cells of sensitive and resistant type. *Biomed. Pharmacother.* **2015**, *69*, 145–152. [CrossRef]
14. Kulbacka, J.; Pucek, A.; Wilk, K.A.; Dubińska-Magiera, M.; Rossowska, J.; Kulbacki, M.; Kotulska, M. The Effect of millisecond pulsed electric fields (msPEF) on intracellular drug transport with negatively charged large nanocarriers made of solid lipid nanoparticles (SLN): In vitro study. *J. Membr. Biol.* **2016**, *249*, 645–661. [CrossRef]
15. Kulbacka, J.; Pucek, A.; Kotulska, M.; Dubińska-Magiera, M.; Rossowska, J.; Rols, M.P.; Wilk, K.A. Electroporation and lipid nanoparticles with cyanine IR-780 and flavonoids as efficient vectors to enhanced drug delivery in colon cancer. *Bioelectrochemistry* **2016**, *110*, 19–31. [CrossRef]
16. Mutaliyeva, B.; Grigoriev, D.; Madybekova, G.; Sharipova, A.; Aidarova, S.; Saparbekova, S.; Miller, R. Microencapsulation of insulin and its release using w/o/w double emulsion method. *Colloids Surf. A Physicochem. Eng. Asp.* **2017**, *521*, 147–152. [CrossRef]
17. Thompson, K.L.; Mable, C.J.; Lane, J.A.; Derry, M.J.; Fielding, L.A.; Armes, S.P. Preparation of pickering double emulsions using block copolymer worms. *Langmuir* **2015**, *31*, 4137–4144. [CrossRef]
18. Chen, S.; Yang, K.; Tuguntaev, R.G.; Mozhi, A.; Zhang, J.; Wang, P.C.; Liang, X.J. Targeting tumor microenvironment with PEG-based amphiphilic nanoparticles to overcome chemoresistance. *Nanomedicine* **2016**, *12*, 269–286. [CrossRef]
19. Kommineni, N.; Mahira, S.; Domb, A.D.; Khan, W. Cabazitaxel-loaded nanocarriers for cancer therapy with reduced side effects. *Pharmaceutics* **2019**, *11*, 141. [CrossRef]
20. Jang, C.; Lee, J.H.; Sahu, A.; Tae, G. The synergistic effect of folate and RGD dual ligand of nanographene oxide on tumor targeting and photothermal therapy in vivo. *Nanoscale* **2015**, *7*, 18584–18594. [CrossRef]
21. Bazylińska, U.; Wawrzyńczyk, D. Encapsulation of TOPO stabilized $NaYF_4$:Er^{3+},Yb^{3+} nanoparticles in biocompatible nanocarriers: Synthesis, optical properties and colloidal stability. *Colloids Surf. A Physicochem. Eng. Asp.* **2017**, *532*, 556–563. [CrossRef]
22. Bazylińska, U.; Zielińska, K.; Saczko, J.; Wilk, K.A. Novel multilayer IR-786-loaded nanocarriers for intracellular delivering: Characterization, imaging, and internalization in human cancer cell lines. *Chem. Lett.* **2012**, *41*, 1354–1356. [CrossRef]
23. Gamper, N.; Stockand, J.D.; Shapiro, M.S. The use of chinese hamster ovary (CHO) cells in the study of ion channels. *J. Pharmacol. Toxicol. Methods* **2005**, *51*, 177–185. [CrossRef]
24. Méndez-Vilas, A.; Solano, A. (Eds.) State of the art of smart polymers: From fundamentals to final applications. In *Polymer Science: Research Advances, Practical Applications and Educational Aspects*, 1th ed.; Formatex Reserach Center: Extremadura, Spain, 2016; pp. 476–487.

25. Azimi, B.; Nourpanah, P.; Rabiee, M.; Arbab, S. Poly (lactic-co-glycolide) fiber: An Overview. *J. Eng. Fibers Fabr.* **2014**, *47*, 47–66.
26. Callaghan, R.; Luk, F.; Bebawy, M. Inhibition of the multidrug resistance P-glycoprotein: Time for a change of strategy? *Drug Metab. Dispos.* **2014**, *42*, 623–631. [CrossRef]
27. Chen, B.; Le, W.; Wang, Y.; Li, Z.; Wang, D.; Ren, L.; Lin, L.; Cui, S.; Hu, J.J.; Hu, Y.; et al. Targeting negative surface charges of cancer cells by multifunctional nanoprobes. *Theranostics* **2016**, *6*, 1887–1898. [CrossRef]
28. Gutiérrez-Valenzuela, C.A.; Guerrero-Germán, P.; Tejeda-Mansir, A.; Esquivel, R.; Guzmán-Z, R.; Lucero-Acuña, A. Folate functionalized PLGA nanoparticles loaded with plasmid pVAX1-NH36: Mathematical analysis of release. *Appl. Sci.* **2016**, *6*, 364. [CrossRef]
29. Lin, P.C.; Lin, S.; Wang, P.C.; Sridhar, R. Techniques for physicochemical characterization of nanomaterials. *Biotechnol. Adv.* **2014**, *32*, 711–726. [CrossRef]
30. Zyuzin, M.V.; Honold, T.; Carregal-Romero, S.; Kantner, K.; Karg, M. Influence of temperature on the colloidal stability of polymer-coated gold nanoparticles in cell culture media. *Small* **2016**, *12*, 1723–1731. [CrossRef]
31. Kowalczuk, A.; Trzcinska, R.; Trzebicka, B.; Müller, A.H.E.; Dworak, A. Loading of polymer nanocarriers: Factors, mechanisms and applications. *Prog. Polym. Sci.* **2014**, *39*, 43–86. [CrossRef]
32. Van der Steen, S.C.; Raavé, R.; Langerak, S.; Van Houdt, L.; Van Duijnhoven, S.M.; Van Lith, S.A.; Massuger, L.F.; Daamen, W.F.; Leenders, W.P.; Van Kuppevelt, T.H. Targeting the extracellular matrix of ovarian cancer using functionalized, drug loaded lyophilisomes. *Eur. J. Pharm. Biopharm.* **2017**, *113*, 229–239. [CrossRef]
33. Siwowska, K.; Schmid, R.M.; Cohrs, S.; Schibli, R.; Müller, C. Folate receptor-positive gynecological cancer cells: In vitro and in vivo characterization. *Pharmaceuticals* **2017**, *10*, 72. [CrossRef]
34. Quarta, A.; Bernareggi, D.; Benigni, F.; Luison, E.; Nano, G.; Nitti, S.; Cesta, M.C.; Di Ciccio, L.; Canevari, S.; Pellegrino, T.; et al. Targeting FR-expressing cells in ovarian cancer with Fab-functionalized nanoparticles: A full study to provide the proof of principle from in vitro to in vivo. *Nanoscale* **2015**, *7*, 2336–2351. [CrossRef]
35. Lange, C.; Bednarski, P.J. Evaluation for synergistic effects by combinations of photodynamic therapy (PDT) with temoporfin (mTHPC) and Pt(II) complexes carboplatin, cisplatin or oxaliplatin in a set of five human cancer cell lines. *Int. J. Mol. Sci.* **2018**, *19*, 3183. [CrossRef]

pharmaceutics

MDPI

Article

Vitamin E-Loaded PLA- and PLGA-Based Core-Shell Nanoparticles: Synthesis, Structure Optimization and Controlled Drug Release

Norbert Varga [1], Árpád Turcsányi [1], Viktória Hornok [1,2] and Edit Csapó [1,3,*]

1 Interdisciplinary Excellence Centre, Department of Physical Chemistry and Materials Science, University of Szeged, Rerrich B. square 1, H-6720 Szeged, Hungary
2 Department of Physical Chemistry and Materials Science, MTA Premium Post Doctorate Research Program, University of Szeged, Rerrich B. Square 1, H-6720 Szeged, Hungary
3 MTA-SZTE Biomimetic Systems Research Group, Department of Medical Chemistry, University of Szeged, Dóm square 8, H-6720 Szeged, Hungary
* Correspondence: juhaszne.csapo.edit@med.u-szeged.hu; Tel.: +36-(62)-544-476

Received: 27 June 2019; Accepted: 16 July 2019; Published: 22 July 2019

Abstract: The (±)-α-Tocopherol (TP) with vitamin E activity has been encapsulated into biocompatible poly(lactic acid) (PLA) and poly(lactide-*co*-glycolide) (PLGA) carriers, which results in the formation of well-defined nanosized (d ~200–220 nm) core-shell structured particles (NPs) with 15–19% of drug loading (DL%). The optimal ratios of the polymer carriers, the TP active drug as well as the applied Pluronic F127 (PLUR) non-ionic stabilizing surfactant, have been determined to obtain NPs with a TP core and a polymer shell with high encapsulation efficiency (EE%) (69%). The size and the structure of the prepared core-shell NPs as well as the interaction of the carriers and the PLUR with the TP molecules have been determined by transmission electron microscopy (TEM), dynamic light scattering (DLS), infrared spectroscopy (FT-IR) and turbidity studies, respectively. Moreover, the dissolution of the TP from the polymer NPs has been investigated by spectrophotometric measurements. It was clearly confirmed that increase in the EE% from ca. 70% (PLA/TP) to ca. 88% (PLGA65/TP) results in the controlled release of the hydrophobic TP molecules (7 h, PLA/TP: 34%; PLGA75/TP: 25%; PLGA65/TP: 18%). By replacing the PLA carrier to PLGA, ca. 15% more active substance can be encapsulated in the core (PLA/TP: 65%; PLGA65/TP: 80%).

Keywords: vitamin E; tocopherol; PLA; PLGA; core-shell nanoparticles; drug delivery; controlled drug release

1. Introduction

The encapsulation of the pharmaceutical ingredients in a macro- or a nanocarrier is a key factor in nanomedicine developments [1–3]. Utilization of drug delivery systems can increase and prolong the efficiency of the active drugs. Due to the good biocompatibility and structural properties, several materials such as proteins and mostly biodegradable polymers, e.g., chitosan, alginate, hyaluronic acid, poly(lactide) or poly(lactic acid) (PLA), polycaprolactone (PCL), poly(trimetilene-carbonate) (PTMC), etc., have been widely used as potential carriers of the nanosized drug delivery systems [4–12]. Many types of PLA copolymers, such as poly(lactide-*co*-glycolide) (PLGA), polylactide-poly(ethylene glycol) (PLA-PEG), poly(caprolactone-ethylene glycol-lactide) (PCELA), etc., are also well-known as drug carriers [13–17]. As a result of copolymerization, the hydrophilicity of the PLA can be remarkably tuned, which opens the possibility of encapsulating active ingredients that have hydrophilic or hydrophobic character into the polymer matrix.

The (±)-α-Tocopherol (TP) is one of the highest biological activity fat-soluble vitamins from the vitamin E family and is composed of eight tocopherols and tocotrienols (alpha (α), beta (β), gamma (γ)

and delta (δ) isomers for both cases) [18]. Several studies focus on the encapsulation of the TP, because vitamin E can prevent and treat many chronic and age-related diseases, e.g., Alzheimer's disease [19]. Furthermore, the TP is described as functioning as an antioxidant, and the higher vitamin E content is associated with a lower risk of several cancer diseases (kidney, lung, bladder, etc.).

Alqahtani et al. synthesized TP-loaded PLGA50 NPs with an average size of ca. 130 nm using Polyvinyl alcohol (PVA) as a stabilizer, but only 4–4.5% of drug loading (in mg drug/100 mg PLGA) was achieved [20]. However, Zigoneanu and coworkers increased the TP loading to 8–16% using the PLGA50 carrier and PVA and sodium dodecyl sulphate (SDS) surfactant stabilizing agents, but it was confirmed that the 86% of the drug dissolved for NPs (d ~220–280 nm) with 8% TP loading, while 36% of the TP drug released from the NPs when the loading was 16% after 1 h [19]. However, Astete et al. synthesized Span80-stabilized TP-containing PLGA50-based nanosized particles in the range of 150–200 nm and the effect of salt concentration on the size and morphology of the NPs has been interpreted, but data for the drug loading and release were not presented [21]. Murugeshu and coworkers also synthesized chitosan/PLGA50-based TP-containing NPs with 8%, 16% and 24% initial loading, but the EE% was only 45–50% and the release studies were carried out in gastrointestinal conditions (pH ~1.50) [22]. Simon et al. successfully encapsulated the TP into PLGA50 using PVA, but only 2.5 mg drug/100 mg NPs can be obtained [23].

In our previous work, the TP as well as the water-soluble derivative of TP (D, α-Tocopherol polyethylene glycol 1000 succinate (TPGS)) and the non-steroidal anti-inflammatory ketoprofen (KP) have been successfully encapsulated into PLA and PLGA75 (lactide:glycolide ratio 75:25) and PLGA65 (lactide:glycolide ratio 65:35) nanocarriers [13]. Instead of the commonly used and above mentioned PLGA50, the PLGA65 and PLGA75 derivatives have been firstly used as carriers for the encapsulation of TP, and the hydrophilicity properties of the carriers and the model drugs on the EE% have been studied in detail. The release measurements of nanocomposites including hydrophobic drugs are very difficult to carry out. In order to facilitate the above-mentioned studies, stabilizing agents such as surfactants have been widely used [5,19]. Non-ionic poloxamer Pluronic F127 was applied previously, which stabilized the drug-containing polymeric NPs by increasing the biocompatibility. The optimal ratios of the carrier, the drug and the stabilizing agent have a dominant effect on the structure and the EE%, as well as the controlled drug release process of the nanosized drug delivery systems, which were not studied previously.

In the present work, the results of the TP-containing nanocomposites, using PLA, PLGA65 and PLGA75 carriers, have been completed and the determinative role of the concentration of the polymer, the TP and the PLUR stabilizing agent on the structure of the drug-containing nanocomposites has been investigated. Moreover, the drug release studies of the prepared TP-containing PLA, PLGA65 and PLGA75 core-shell NPs have also been interpreted.

2. Materials and Methods

2.1. Materials

Polylactide (PLA, Mw = 72,200 ± 15,000 Da) and two poly(lactide-*co*-glycolide) (PLGA) derivatives with a lactide to glycolide ratio at 75:25 (PLGA75, Mw = 69,900 ± 4000 Da) and at 65:35 (PLGA65, Mw = 93,000 ± 1000 Da) were synthetized according to the previously published procedure [13]. Pluronic F127 (PLUR), (±)-α-tocopherol (TP) and sodium phosphate monobasic monohydrate ($NaH_2PO_4 \cdot H_2O$, ≥99%) were obtained from Sigma Aldrich (Budapest, Hungary). Sodium phosphate dibasic anhydrous (Na_2HPO_4, ≥99%) and sodium chloride (NaCl, ≥99%) were purchased from Molar Chemicals (Halásztelek, Hungary). All other reagents and solvents were of analytical grade and used without further purification. The deionized water was obtained by Millipore purification apparatus (18.2 MΩ cm at 25 °C).

2.2. Preparation of TP-Loaded PLA/PLGA NPs

The TP-loaded PLA/PLGA NPs were prepared by nanoprecipitation method (Figure 1). The detailed experimental conditions were presented previously [13]. Briefly, the PLA/PLGA and TP with increasing concentrations (see Table 1) were dissolved in 1.5 mL of acetone, which was dropped slowly (10 μL/5 s) into the aqueous solution of the PLUR stabilizer (15 mL) under room temperature using magnetic stirring with 1000 rpm. In the interest of the evaporation of acetone, the prepared samples were further stirred (350 rpm) for two days. The dispersion was centrifugated with 40 mL MQ water at 12,000 rpm (t = 15 min, T = 25 °C). After the supernatant was removed, the NPs were redispersed in MQ water and the washing methods were repeated two times. The obtained NPs samples were freeze-dried by liquid nitrogen and lyophilized (by Christ Alpha 1-2 LDplus apparatus).

Table 1. The concentration of the components, the average particle diameter, the polydispersity index (PI), the encapsulation efficiency (EE%) and the drug loading (DL%) of the TP-loaded PLA NPs.

	Acetone Phase		Aqueous Phase				
Sample	c_{PLA} (mg·mL^{-1})	c_{TP} (mg·mL^{-1})	c_{PLUR} (mg·mL^{-1})	d_{DLS} ± SD [1] (nm)	PI ± SD	EE%	DL%
PLA concentration dependence	1.25	2.5	0.1	120 ± 33	0.120 ± 0.043	–	–
	2.5	2.5	0.1	156 + 28	0.039 ± 0.012	–	–
	5.0	2.5	0.1	179 ± 35	0.082 ± 0.049	–	–
	10.0	2.5	0.1	201 ± 38	0.095 ± 0.026	69.11	14.73
TP concentration dependence	10.0	0	0.1	188 ± 37	0.092 ± 0.059	–	–
	10.0	0.5	0.1	189 ± 34	0.048 ± 0.028	91.28	4.36
	10.0	1.0	0.1	192 ± 30	0.062 ± 0.032	75.61	7.02
	10.0	2.5	0.1	201 ± 38	0.095 ± 0.026	69.11	14.73
	10.0	5.0	0.1	252 ± 53	0.073 ± 0.033	66.15	24.85
PLUR concentration dependence	10.0	2.5	0	179 ± 40	0.315 ± 0.040	72.19	15.29
	10.0	2.5	0.05	178 ± 21	0.304 ± 0.095	98.34	19.73
	10.0	2.5	0.1	201 ± 38	0.095 ± 0.026	69.11	14.73
	10.0	2.5	0.5	206 ± 36	0.089 ± 0.065	57.94	12.65
	10.0	2.5	1.0	212 ± 36	0.066 ± 0.029	40.75	9.24

[1] The experimental error of the peak maximum is below 2.5%.

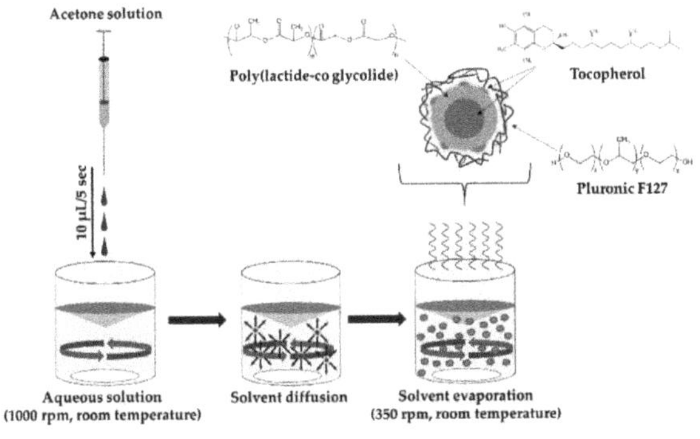

Figure 1. Schematic representation of the preparation of TP-loaded PLGA NPs stabilized by PLUR using nanoprecipitation technique.

2.3. Characterization Methods

The particle size was determined by dynamic light scattering (DLS) with Horiba Sz-100 (HORIBA Jobin Yvon, Longjumeau, France) equipped with a diode pumped frequency doubled (532 nm, 10 mW) laser. The measurements were carried out at 25 ± 0.1 °C with 90° of detection angle in every case.

The transmission electron microscopy (TEM) images were obtained by Jeol JEM-1400plus equipment (JEOL Ltd., Tokyo, Japan) at 120 keV accelerating voltage. The Fourier transform infrared (FT-IR) spectra of the PLA/PLGA-TP NPs were registered by a Jasco FT/IR-4700 with ATR PRO ONE Single-reflection accessory (ABL&JASCO, Budapest, Hungary). The experiments were performed at room temperature from 600 cm^{-1} to 3600 cm^{-1}. The resolution of the spectra was 2 cm^{-1}, which was determined by 128 interferograms.

The particles were dissolved in 1,4-dioxane to determine the EE% and DL% of the composites. The absorbance spectra of the prepared solutions were measured by a Shimadzu UV-1800 UV-Vis double beam spectrophotometer. The spectra were registered in the range of 200–500 nm using 1 cm quartz cuvette at room temperature. The characteristic absorbance band of the TP appeared at 294 nm. The concentration of the encapsulated drug was determined from the calibration curve (Figure S1). The data of the EE% in *w*/*w*% were calculated by the total drug mass used in the synthesis of the NPs (Equation (1)), while DL% was calculated by the mass of the NPs (Equation (2)).

$$EE\% = \frac{encapsulated\ mass\ of\ drug}{total\ mass\ of\ drug\ in\ synthesis} \times 100 \quad (1)$$

$$DL\% = \frac{encapsulated\ mass\ of\ drug}{total\ mass\ of\ the\ nanoparticles} \times 100 \quad (2)$$

2.4. Determination of the Solubility Properties of TP Drug

In order to determine the solubility of TP, 100 μL of acetone solutions of TP (5 mg·mL^{-1}) was added dropwise to 10 mL of PLUR solutions (0.1–1.2 mg·mL^{-1}), and the turbidity was followed with a Precision Bench Turbidity Meter LP2000 (Hanna Ins. Service Kft., Szeged, Hungary). The experiments were performed in a pure aqueous medium and in a phosphate (PBS) buffer (pH = 7.4, 0.9% NaCl) solution at 25 °C and 37 °C.

2.5. Critical Micelle Concentration (cmc) Studies

The critical micelle concentration (cmc) of the applied PLUR was determined by inverse titration method in Krüss K100MK2 type surface tension equipment. The computer-controlled apparatus was supplied with a thermostat and an automatic burette. The Wilhelmy-plate method was applied. A volume of 50 mL of a 1.6 mM surfactant solution was titrated with MQ water or PBS in an aliquot of 10 mL in 40 steps. Each experimental point was the average of at least 5 measurements.

2.6. In Vitro Release Study

The in vitro studies of the different TP-loaded PLA/PLGA NPs were carried out by a UV-Vis spectrophotometer (500–200 nm). The release experiments were performed at 37 °C and a PBS buffer (pH = 7.4, NaCl 0.9%) containing 1 mg·mL^{-1} of PLUR was used, which facilitated the easier feasibility of the release studies [5]. The TP-loaded samples were placed into the cellulose membrane (Sigma Aldrich) with 5 mL of PLUR/PBS medium inserted into 35 mL of a dissolution phase. During the measurements, 3 mL of the release media were taken at specified intervals to measure the released concentration of TP at 268 nm.

The release curves of the TP can be fitted by different kinetic models, such as the first order, Korsmeyer–Peppas, Peppas–Sahlin and Weibull models [24–27]. The measured points were fitted with a nonlinear regression by the QtiPlot 0.9.8.9 svn 2288 program. During the calculation session, the program finds the best fitting function for the measured points. The results consist of the fitted parameters, their standard deviation and the goodness of fitting (root mean squared error).

Depending of the release chemical condition (such as temperature, buffer solution, ionic strength, etc.), the shape of the polymers and the solubility of the drugs, the dissolution curves can be described with different kinetic models. For our calculation, the following kinetic models were used.

First order equation is a frequently used kinetic model. This formula is well applicable for drugs where the dissolution is continuously changing over time and depends only on the concentration.

$$c_t = ce^{-kt} \quad (3)$$

where c_t is the concentration of the solid drug in the matrix of the carried system at t time, c_0 is the initial concentration of the drug and k is the first order release constant.

The Korsmeyer–Peppas kinetic formula is a semi-empirical power law equation where the shape of the polymer matrix (such as film, cylinder or sphere) can be taken into account in the release curve.

$$\frac{c_t}{c_0} = k_m t^n \quad (4)$$

where c_t is the concentration of the dissolved drug at t time, c_0 is the initial concentration of drug, k_m is the kinetic constant and n is the diffusion dissolution index (for sphere shaped particles $n = 0.42$ for the diffusion-controlled mechanism, $n = 1$ for the Case II relaxation controlled mechanism and $0.42 \leq n \leq 1$ for both of them).

The Weibull equation is a general empirical formula which can be used for all release profiles.

$$c_t = 1 - exp\left(\frac{-(t - T_i)^b}{a}\right) \quad (5)$$

where c_t is the concentration of the released component in t time, T_i is the lag time between the initial of measurement and the release of drug (in most cases $T_i = 0$), a is the time scale of the process and b is the shape parameter (shape of the release curve is exponential if $b = 1$, parabola if $b < 1$ or sigmoid if $b > 1$).

The kinetic formula reported by Peppas and Sahlin specifies the diffusion and the relaxation contribution in the drug dissolution process.

$$\frac{c_t}{c_0} = k_1 t^m + k_2 t^{2m} \quad (6)$$

where c_t is the concentration of the dissolved drug in the t time, c_0 is the initial concentration of the drug and k_1, k_2 and m are constants: k_1 is the Fick diffusion contribution, k_2 is the Case II relaxation contribution and m is the diffusion exponent (sphere shaped: $m = 0.43$, Fick diffusion mechanism; $m = 0.85$, Case II relaxation transport mechanism; $0.43 \leq m \leq 0.85$, anomalous transport mechanism).

3. Results

3.1. Effect of the Component Concentrations on the Core-Shell Structure

In order to determine the role of the component quantities of the nanocomposites on the size and the structure, as well as on the EE% the concentration, only one building block (polymer carrier, drug or stabilizing agent) has been modified during the synthesis while the other parameters have been kept constant. The average particle diameters of the NPs were measured by DLS, and the results are summarized in Table 1. In case of PLA concentration dependence, regardless of the amount of the TP ($c = 2.5$ mg·mL^{-1}), the diameter of the NPs permanently increases from 120 nm ($c_{PLA} = 1.25$ mg·mL^{-1} in aceton phase) to ca. 200 nm ($c_{PLA} = 10$ mg·mL^{-1} in the acetone phase). The morphology and the structure of the NPs have been investigated by TEM images as well (Figure 2). The images clearly represent that the structure of the TP/PLUR NPs in the absence of a polymer shows less amorphous structures and TP crystal-like objects are observed. Moreover, we established that an increase in the polymer concentration results in the formation of a well-defined core-shell structure at $c_{PLA} = 10.0$ mg·mL^{-1}. At lower polymer concentrations, this structure is not formed, and because of the low polymer concentration, the purification (centrifugation) of the NPs is impracticable; thus, the determination of the EE% was not possible. For TP that is concentration-dependent, the diameters show a slightly increasing tendency to 2.5 mg·mL^{-1}, but for a 5 mg·mL^{-1} amount of TP, a higher size is obtained (d_{DLS} = 252 nm) (Table 1). The TEM images also confirm this observation (Figure 2). The well-defined core-shell structure is formed at 2.5 mg·mL^{-1} of TP quantity. With a further increase in the TP amount, the core, including the drug, shows a rather crystallized structure instead of the previously confirmed amorphous form. The EE% and the drug loading were determined for all composites. We obtained that the value of the EE% decreased from 91.28% (c_{TP} = 0.5 mg·mL^{-1}) to 66.15% (c_{TP} = 5 mg·mL^{-1}), while the DL% increased from 4.36% (c_{TP} = 0.5 mg·mL^{-1}) to 24.85% (c_{TP} = 5 mg·mL^{-1}). Considering the expected size of the NPs for optimal nanosized drug delivery systems (ca. 200 nm) as well as the crystallization of the core, the 2.5 mg·mL^{-1} amount of TP will be used for further studies at 10.0 mg·mL^{-1} of PLA polymer concentration. Besides optimizing the amount of carrier and active drug, the ratio of the stabilizing PLUR surfactant was also studied. It was clearly confirmed that in the absence of PLUR, the TP molecules were not capsulated into the polymer core, only the binding of the TP drugs onto the surface of the polymer shell is observed (Figure 2). Furthermore, it was established that the smallest particle diameter (d = 178 nm), as well as the highest DL% (ca. 20%), is obtained at 0.05 mg·mL^{-1} of PLUR concentration. Increase in the PLUR amount resulted in a higher particle size (212 nm) as well as a lower DL% values (9%). Most probably, the increase in the PLUR concentration facilitated the solubility of the hydrophobic TP, thus hindering the encapsulation process. In order to confirm the above-mentioned phenomena, turbidity measurements were carried out (Figure S2). According to the TEM images (Figure 2) and the turbidity studies (Figure S2), we can conclude that a low quantity (0.1 mg·mL^{-1}) of PLUR surfactant is advantageous for the formation of PLA-based core-shell NPs, but the presence of a higher amount of PLUR (>0.1 mg·mL^{-1}) results in the decrease of the EE%. Moreover, the presence of a higher amount of stabilizer (≥1.0 mg·mL^{-1}) causes aggregation (Figure 2).

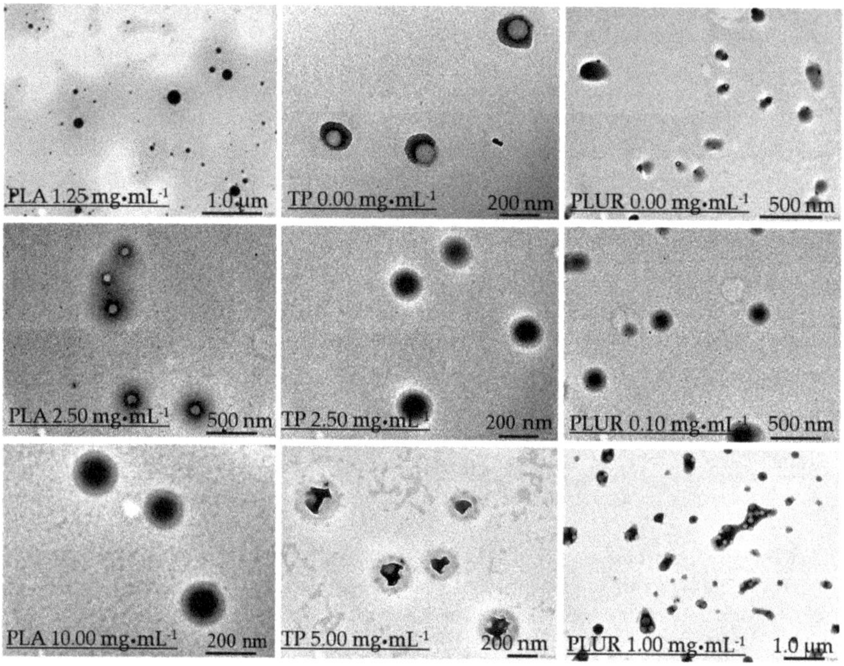

Figure 2. Representative TEM images of the TP-containing PLA NPs using different component concentrations (PLA: 1.25–10.0 mg·mL^{-1}; TP: 0–5.0 mg·mL^{-1} and PLUR: 0–1.0 mg·mL^{-1}).

Besides the determination of the optimal ratio of the composite building blocks, the TP was encapsulated in PLGA75 and PLGA65 copolymers that have increasing hydrophilicity. During the NPs synthesis, the previously optimized concentrations of the polymer carrier (10.0 mg·mL^{-1}), the PLUR (0.1 mg·mL^{-1}) and the TP (2.5 mg·mL^{-1}) were used. Based on DLS studies, we established that the particle size increases from 203 nm to 226 nm with a decrease of the lactide part in the polymer (Figure 3A). Using the optimized component quantities, the core-shell structure was confirmed for PLGA75 and PLGA65-based TP-containing NPs (Figure 3B). The EE% and DL% has been determined for the PLGA75 and PLGA65-based system as well, and we observed that the replacement of the PLA carrier to PLGA75 and PLGA65 polymers resulted in the increase in the EE% to 75.72% and 87.69%, respectively. In addition, the drug loading also increased from 14.73% (PLA) to 15.92% (PLGA75) and to 17.98% (PLGA65) with decreasing lactide content. The higher EE% and DL% can be explained by the fact that the precipitation of the more hydrophilic PLGA carriers is slower than that for PLA and TP, which helps the formation of the well-defined core-shell structure [13].

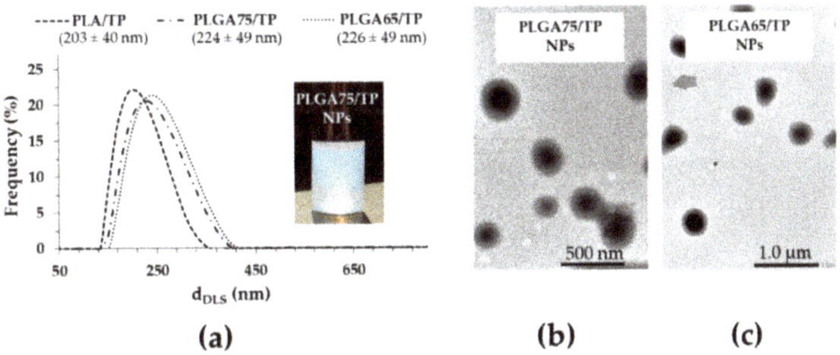

Figure 3. The particle size distribution of the TP-loaded PLA and PLGA NPs (**a**), and representative TEM images of the PLGA75–(**b**) and PLGA65–(**c**) based composites.

3.2. Structural Characterization of the TP-Loaded PLA and PLGA Core-Shell NPs

To determine the interaction between the PLA and TP, the composites have been examined by infrared spectroscopy measurements. Figure 4 displays the spectra of the PLA-based NPs in the absence (Figure 4A) and in the presence of TP at different concentrations (Figure 4B,C). The TP-sensitive bands appear in the range of 3050–2800 cm^{-1} and 1150–1000 cm^{-1}. It is obvious that the intensity of all the determinative bands systematically increase by increasing TP content. At 2994, 2944 and 2968 cm^{-1}, the asymmetric and symmetric CH stretching vibrations of the –CH_2 and –CH_3 groups of the drug appear. Due to the increasing TP concentration, these bands become more intense, indicating the presence of TP molecules in the polymer NPs. The carbonyl group of the PLA appears at 1750 cm^{-1} [28,29]. The TP does not contain a C=O group, thus this band could originate only from the polymer. In the fingerprint region ($\nu \leq 1500$ cm^{-1}), the deformation and bending vibration of the –CH_3 and –CH_2 groups appear at 1453, 1381 and 1267 cm^{-1}, while the band at 1181 cm^{-1} attributes to the stretching mode of the C–O–C (ester). At around 1086 cm^{-1}, further bands of the C–O–C stretching vibration can be observed, which have shoulders (symmetric and asymmetric stretching vibrations). Because of the Ar–O–C group in the TP, the intensity of this C–O–C symmetric stretching vibration at around 1050 cm^{-1} is increased. The band at 865 cm^{-1} and 751 cm^{-1} is characteristic of the polymer carrier, and no shift is observed. The IR measurements performed for PLGA75 and PLGA65 resulted in similar spectra. A strong irreversible interaction between the PLA (or PLGA75, PLGA65) and the TP cannot be discovered (Figures S3 and S4), which facilitates the spontaneous release of the TP active drug from the polymer NPs, but the presence of the TP in the different composites were definitely confirmed by IR.

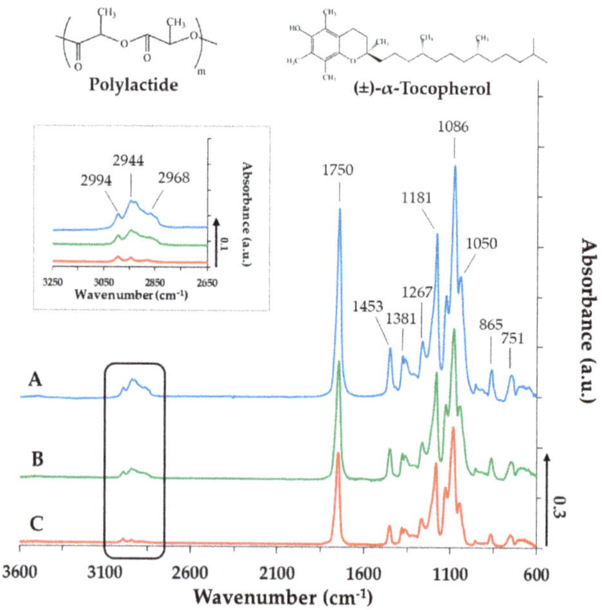

Figure 4. IR spectra of the TP-loaded PLA NPs (A: c_{TP} = 0 mg·mL^{-1}; B: c_{TP} = 2.5 mg·mL^{-1}; C: c_{TP} = 5 mg·mL^{-1}, c_{PLA} = 10 mg·mL^{-1} and c_{PLUR} = 0.1 mg·mL^{-1}).

Turbidimetric measurements were performed to investigate the interaction between the TP and the PLUR stabilizing agent in a pure aqueous solution and in PBS at 25 °C and 37 °C (pH = 7.4, 0.9% NaCl); the results are presented in Figure 5. In MQ water (Figure 5A), the turbidity of TP is systematically decreased till ca. 0.9 mg·mL^{-1} of PLUR concentration. Over time, more and more TP can dissolve in this medium. In contrast, in the PBS solution at 25 °C and at 37 °C, the titration curves exhibit steeper decreasing intensity, which may be due to the reduced critical micellization concentration (cmc) of PLUR in the presence of salt. In the case of the PBS solution, the turbidity remains constant from 0.7 mg·mL^{-1} at 25 °C, while an increase in the temperature to 37 °C 0.6 mg·mL^{-1} value is observed. It is important to mention that the solubility of the TP scarcely depends on time at 37 °C, which allows the possible use of the PLUR stabilizing agent for in vitro drug release measurements.

It is well known that the surfactant affects the solubility of the TP drug in the absence of a polymer carrier. Namely, above cmc, due to the micellization ability of the PLUR, the solubilization of the drug dominates forming TP-loaded individual micelles, while below cmc, only the solubility increases. Accordingly, the cmc of the PLUR was measured by surface tension measurements (Figure 6). In the literature, very different values can be found; the obtained range of cmc is 2–7 mg·mL^{-1} [30,31]. Similar values were measured, but we determined that at 25 °C, the obtained cmc is decreased from 4.92 mg·mL^{-1} to 1.99 mg·mL^{-1} in the presence of the phosphate buffer. If the temperature rises from 25 °C to 37 °C, these values are further decreased. We can conclude that the concentration of the PLUR in the composites is significantly lower than the cmc (4.92 mg·mL^{-1}) value, which excludes the presence of TP-loaded individual micelles.

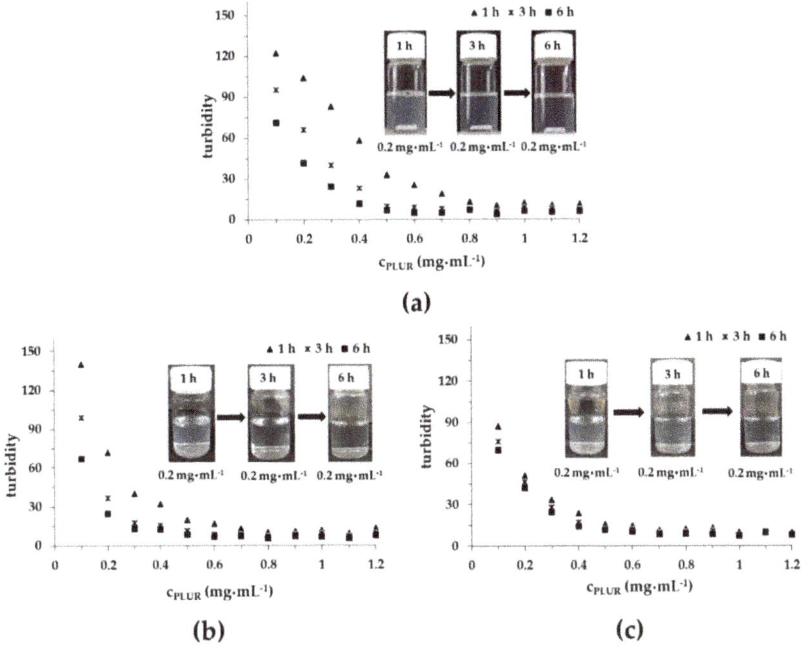

Figure 5. The turbidity of the TP in the PLUR solution at 25 °C in an aqueous medium (**a**), in a PBS buffer at 25 °C (**b**) and at 37 °C (**c**) (pH = 7.4, 0.9 *w*/*w*% NaCl) (c_{TP} = 0.05 mg·mL^{-1}).

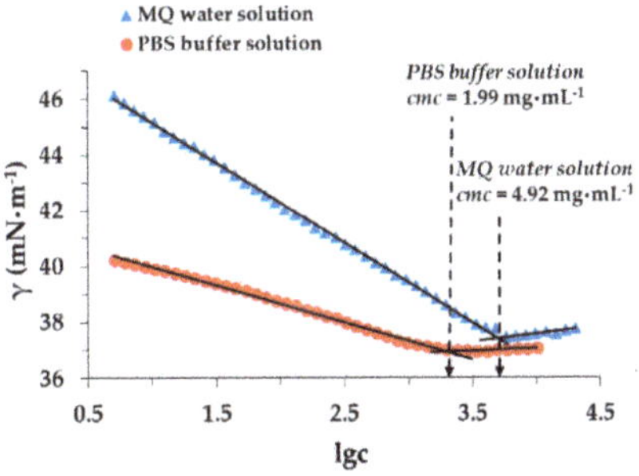

Figure 6. Determination of PLUR cmc at 25 °C in MQ water and in a PBS solution.

3.3. *In Vitro Drug Release Experiments*

The determination of the exact TP amount released from the different composites was carried out by the UV-Vis spectrophotometric method. The spectra of the TP and the calibration curve are presented in Figure 7. The absorption bands of the TP appear in the UV range.

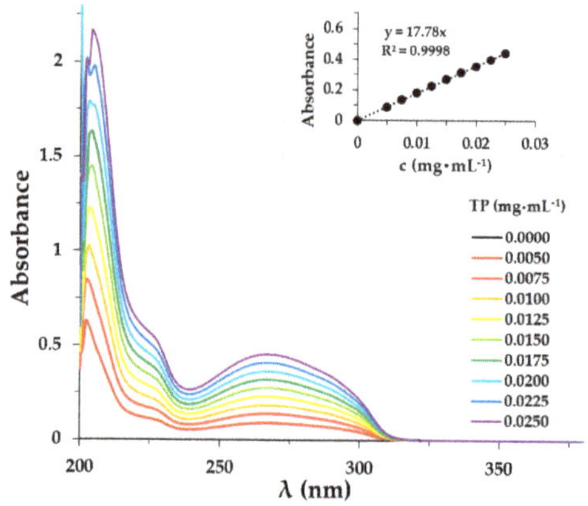

Figure 7. UV spectra of the TP in the PBS solution (c_{PLUR} = 1.0 mg·mL^{-1}, 0.9 *w*/*w*% NaCl).

After the characterization of the TP-loaded PLA and PLGA NPs, the mechanism of the drug release was investigated. The drug dissolution profiles and the fitting of these curves by different kinetic models are demonstrated in Figure 8 (Figure S5). The suitability of several models like Korsmeyer–Peppas, Peppas–Sahlin, the first-order and the Weibull models were investigated. The release curves clearly show that a high amount of TP is retained in the polymers after 7 h. It was also observed that in the first half an hour, the dissolution of the drug occurs relatively quickly, but after that, measurable slow dissolution is observed. Moreover, we found that the active substance is released slowly with a decrease of the lactide part (PLA, 35.0%, PLGA75, 28.3%, PLGA65, 19.8%) in 7 h. The slowest dissolution occurred in the carrier-free TP (15%). Because of the higher hydrophobicity, a higher amount of the non-encapsulated drug can be attached to the surface of the particles, which confirms the above-mentioned dissolution order [13]. Thanks to the application of the nanosized drug carrier systems (polymer NPs), the TP molecules can bind to the enhanced specific surface area of the NPs, which facilitates the dissolution of more TP, in contrast to the bulk TP.

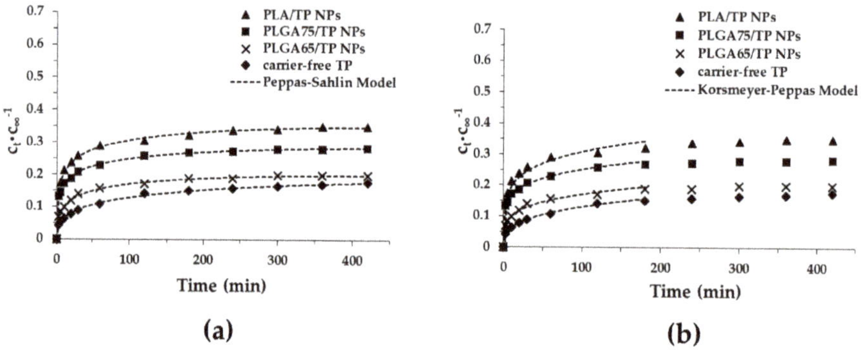

Figure 8. Release profiles and different kinetic models-predicted (Peppas-Sahlin model (**a**), Korsmeyer-Peppas model (**b**)) for the release curves of TP from PLA and PLGA NPs in the PBS buffer (pH = 7.4, 0.9 *w*/*w*% NaCl).

Taking into account the coefficient of the determination (R^2), the Peppas–Sahlin model was the best kinetic formula for our systems (Table 2). The values of the Case II relaxation contribution (k_2) were negative in all cases; therefore, this model does not provide complete information about the dissolution of the TP, but the low diffusion exponent (m) is referred to for the Fickian release.

Table 2. Determined parameters of the TP release by fitting several kinetic equations.

Peppas–Sahlin Formulation	k_1 (min^{-m})	k_2 (min^{-2m})	m	R^2
PLA/TP NPs	0.1371	−0.01346	0.260	0.9974
PLGA75/TP NPs	0.1165	−0.01189	0.248	0.9986
PLGA65/TP NPs	0.0572	−0.00412	0.317	0.9974
TP	0.0304	−0.00124	0.366	0.9978
Korsmeyer–Peppas Formulation	k_m (min^{-n})	n	R^2	
PLA/TP NPs	0.1339	0.182	0.9898	
PLGA75/TP NPs	0.1120	0.175	0.9982	
PLGA65/TP NPs	0.0592	0.223	0.9904	
TP	0.0310	0.311	0.9977	
Weibull Formulation	a	b	R^2	
PLA/TP NPs	6.52	0.178	0.9896	
PLGA75/TP NPs	7.85	0.166	0.9819	
PLGA65/TP NPs	15.45	0.220	0.9898	
TP	28.19	0.285	0.9945	
First Order Formulation	k (min^{-1})	R^2		
PLA/TP NPs	0.0094	0.9792		
PLGA75/TP NPs	0.0113	0.9532		
PLGA65/TP NPs	0.0130	0.9514		
TP	0.0079	0.9856		

The second-best kinetic model was the Korsmeyer–Peppas model where the diffusion dissolution index (n) gives the information about the diffusion and the erosion of the matrix. The values of the n are increased from n = 0.182 (PLA) to n = 0.223 (PLGA65), which is referred to for the diffusion-controlled quasi Fickian drug release. Furthermore, it is important to note that the value of the diffusion dissolution index is lower than 0.42. This is caused by the high polydispersity of the particles (because they are lyophilized) and the very low degradation of the polymers [32]. Presumably, the Case II relaxation contribution from the Peppas–Sahlin model will be low by these effects.

During the slow degradation of the PLA/PLGA carrier, the drug diffuses with difficulty from the core of the particles. The measured and calculated results clearly stated that the released drug originates from the surface region of the particles. Based on this, we could calculate the approximate quantity of the TP in the core of the particles after 7 h: 65.0% (PLA), 71.7% (PLGA75) and 80.2% (PLGA65) from the encapsulated mass of the TP; therefore, a significant amount of the active ingredient can be encapsulated inside the particles.

4. Conclusions

In the work presented here, the successful encapsulation of the hydrophobic α-Tocopherol, one of the determinative natural forms of vitamin E, was carried out using the PLA, PLGA75 and PLGA65 biocompatible polymer carriers by increasing hydrophilicity. To the best of our knowledge, we first proved the formation of well-defined nanosized TP-core PLA/PLGA-shell structured nanocomposites. Optimization of the experimental conditions, such as optimal concentration of drug, polymer carrier and stabilizing PLUR non-ionic surfactant, resulted in the formation of core-shell NPs within the diameter range of ~200–220 nm. For the PLA-based system 14.7% of drug loading was achieved, where most of the TP molecules are encapsulated in the core (65%), while the remaining part of the active ingredient is located on the surface of the polymer shell. By replacing the hydrophobic PLA to PLGA copolymers, both the drug loading (PLGA75: 16%, PLGA65: 18%) as well the EE% can be increased (PLA/TP: 69.1%; PLGA75/TP: 75.7%; PLGA65/TP: 87.7%). Furthermore, the PLGA-based composites contain 71.7% and 80.2% of the encapsulated TP in the core. Considering the slower precipitation ability of the PLGA copolymers in contrast to the PLA and the active ingredient, the higher encapsulation efficiency can be explored. Besides the preparation and the characterization of the composites, the drug release was also studied. The dissolution curves clearly show that depending on the polymer, more than 65–80% of the TP is contained in the composites after 7 h. In the first half an hour, the dissolution of the drug occurs relatively quickly, but after that, measurable slow dissolution was observed. Moreover, we found that the active substance is released slowly with decreasing lactide part (PLA, 35.0%; PLGA75, 28.3%; PLGA65, 19.8%) in 7 h. Thanks to the biocompatibility, cost-effectivity and tuneable hydrophilic properties, the PLA/PLGA polymers are potential candidates for controlled drug release and for the encapsulation of hydrophobic TP or similar sized and structured molecules in core-shell nanosized particles.

Supplementary Materials: The following are available online at http://www.mdpi.com/1999-4923/11/7/357/s1. Figure S1: UV spectra of TP in 1,4-dioxane at different concentrations, Figure S2: The turbidity of TP (a) and PLA (b) in PLUR solution at 25°C in aqueous medium (c_{TP} = 0.25 mg·mL^{-1}, c_{PLA} = 0.25 mg·mL^{-1}), Figure S3: IR spectra of the TP- free (a) and TP- loaded (b) PLGA75 NPs (c_{TP} = 2.5 mg·mL^{-1}, c_{PLA} = 10 mg·mL^{-1}, c_{PLUR} = 0.1 mg·mL^{-1}), Figure S4: IR spectra of the TP- free (a) and TP- loaded (b) PLGA65 NPs (c_{TP} = 2.5 mg·mL^{-1}, c_{PLA} = 10 mg·mL^{-1}, c_{PLUR} = 0.1 mg·mL^{-1}), Figure S5: Release profiles and different kinetic models-predicted (First Order model (a), Weibull model (b)) release curves of TP from PLA and PLGA NPs in PBS solution (pH = 7.4, 0.9 *w*/*w*% NaCl).

Author Contributions: Conceptualization, E.C.; methodology, N.V.; investigation, N.V. and Á.T. and V.H.; data curation, Á.T.; writing—original draft preparation, E.C.; writing—review and editing, V.H.; supervision, E.C.

Funding: University of Szeged Open Access Fund.

Acknowledgments: The research was supported by the National Research, Development and Innovation Office-NKFIH through the project GINOP-2.3.2-15-2016-00060. V.H. acknowledges the Premium Post Doctorate Research Program of the Hungarian Academy of Sciences for the financial support. This paper was supported by the János Bolyai Research Scholarship of the Hungarian Academy of Sciences (E. Csapó). The Ministry of Human Capacities, Hungary grant TUDFO/47138-1/2019-ITM is acknowledged.

Conflicts of Interest: The authors declare no conflict of interest.

References

1. Kumari, A.; Yadav, S.K.; Yadav, S.C. Biodegradable polymeric nanoparticles based drug delivery systems. *Colloids Surf. B Biointerfaces* **2010**, *75*, 1–18. [CrossRef] [PubMed]
2. Nair, L.S.; Laurencin, C.T. Biodegradable polymers as biomaterials. *Prog. Polym. Sci.* **2007**, *32*, 762–798. [CrossRef]
3. Rancan, F.; Papakostas, D.; Hadam, S.; Hackbarth, S.; Delair, T.; Primard, C.; Verrier, B.; Sterry, W.; Blume-Peytavi, U.; Vogt, A. Investigation of Polylactic Acid (PLA) Nanoparticles as Drug Delivery Systems for Local Dermatotherapy. *Pharm. Res.* **2009**, *26*, 2027–2036. [CrossRef] [PubMed]
4. Chanphai, P.; Vesper, A.R.; Bariyanga, J.; Bérubé, G.; Tajmir-Riahi, H.A. Review on the delivery of steroids by carrier proteins. *J. Photochem. Photobiol. B Biol.* **2016**, *161*, 184–191. [CrossRef] [PubMed]
5. Janovák, L.; Turcsányi, Á.; Bozó, É.; Deák, Á.; Mérai, L.; Sebők, D.; Juhász, Á.; Csapó, E.; Abdelghafour, M.M.; Farkas, E.; et al. Preparation of novel tissue acidosis-responsive chitosan drug nanoparticles: Characterization and in vitro release properties of Ca2+ channel blocker nimodipine drug molecules. *Eur. J. Pharm. Sci.* **2018**, *123*, 79–88. [CrossRef] [PubMed]
6. Sarmento, B.; Ferreira, D.; Veiga, F.; Ribeiro, A. Characterization of insulin-loaded alginate nanoparticles produced by ionotropic pre-gelation through DSC and FTIR studies. *Carbohydr. Polym.* **2006**, *66*, 1–7. [CrossRef]
7. Csapó, E.; Szokolai, H.; Juhász, Á.; Varga, N.; Janovák, L.; Dékány, I. Cross-linked and hydrophobized hyaluronic acid-based controlled drug release systems. *Carbohydr. Polym.* **2018**, *195*, 99–106. [CrossRef] [PubMed]
8. Lassalle, V.; Ferreira, M.L. PLA Nano- and Microparticles for Drug Delivery: An Overview of the Methods of Preparation. *Macromol. Biosci.* **2007**, *7*, 767–783. [CrossRef]
9. Jia, W.J.; Gu, Y.C.; Gou, M.L.; Dai, M.; Li, X.Y.; Kan, B.; Yang, J.L.; Song, Q.F.; Wei, Y.Q.; Qian, Z.Y. Preparation of Biodegradable Polycaprolactone/Poly(ethylene glycol)/Polycaprolactone (PCEC) Nanoparticles. *Drug Deliv.* **2008**, *15*, 409–416. [CrossRef]

10. Zhang, Y.; Zhuo, R. Synthesis and drug release behavior of poly(trimethylene carbonate)–poly(ethylene glycol)–poly(trimethylene carbonate) nanoparticles. *Biomaterials* **2005**, *26*, 2089–2094. [CrossRef]
11. Leach, J.B.; Schmidt, C.E. Characterization of protein release from photocrosslinkable hyaluronic acid-polyethylene glycol hydrogel tissue engineering scaffolds. *Biomaterials* **2005**, *26*, 125–135. [CrossRef] [PubMed]
12. Jiang, X.; Xin, H.; Ren, Q.; Gu, J.; Zhu, L.; Du, F.; Feng, C.; Xie, Y.; Sha, X.; Fang, X. Nanoparticles of 2-deoxy-d-glucose functionalized poly(ethylene glycol)-co-poly(trimethylene carbonate) for dual-targeted drug delivery in glioma treatment. *Biomaterials* **2014**, *35*, 518–529. [CrossRef] [PubMed]
13. Varga, N.; Hornok, V.; Janovák, L.; Dékány, I.; Csapó, E. The effect of synthesis conditions and tunable hydrophilicity on the drug encapsulation capability of PLA and PLGA nanoparticles. *Colloids Surf. B Biointerfaces* **2019**, *176*, 212–218. [CrossRef] [PubMed]
14. Kiss, É.; Gyulai, G.; Pénzes, C.B.; Idei, M.; Horváti, K.; Bacsa, B.; Bősze, S. Tuneable surface modification of PLGA nanoparticles carrying new antitubercular drug candidate. *Colloids Surf. A Physicochem. Eng. Asp.* **2014**, *458*, 178–186. [CrossRef]
15. Duse, L.; Agel, M.R.; Pinnapireddy, S.R.; Schäfer, J.; Selo, M.A.; Ehrhardt, C.; Bakowsky, U. Photodynamic Therapy of Ovarian Carcinoma Cells with Curcumin-Loaded Biodegradable Polymeric Nanoparticles. *Pharmaceutics* **2019**, *11*, 282. [CrossRef] [PubMed]
16. Molina, I.; Li, S.; Martinez, M.B.; Vert, M. Protein release from physically crosslinked hydrogels of the PLA/PEO/PLA triblock copolymer-type. *Biomaterials* **2001**, *22*, 363–369. [CrossRef]
17. Hu, Y.; Jiang, X.; Ding, Y.; Zhang, L.; Yang, C.; Zhang, J.; Chen, J.; Yang, Y. Preparation and drug release behaviors of nimodipine-loaded poly(caprolactone)–poly(ethylene oxide)–polylactide amphiphilic copolymer nanoparticles. *Biomaterials* **2003**, *24*, 2395–2404. [CrossRef]
18. Duhem, N.; Danhier, F.; Préat, V. Vitamin E-based nanomedicines for anti-cancer drug delivery. *J. Control. Release* **2014**, *182*, 33–44. [CrossRef]
19. Zigoneanu, I.G.; Astete, C.E.; Sabliov, C.M. Nanoparticles with entrapped α-tocopherol: Synthesis, characterization, and controlled release. *Nanotechnology* **2008**, *19*, 105606. [CrossRef]
20. Alqahtani, S.; Simon, L.; Astete, C.E.; Alayoubi, A.; Sylvester, P.W.; Nazzal, S.; Shen, Y.; Xu, Z.; Kaddoumi, A.; Sabliov, C.M. Cellular uptake, antioxidant and antiproliferative activity of entrapped α-tocopherol and γ-tocotrienol in poly(lactic-co-glycolic) acid (PLGA) and chitosan covered PLGA nanoparticles (PLGA-Chi). *J. Colloid Interface Sci.* **2015**, *445*, 243–251. [CrossRef]
21. Astete, C.E.; Dolliver, D.; Whaley, M.; Khachatryan, L.; Sabliov, C.M. Antioxidant Poly(lactic-*co*-glycolic) Acid Nanoparticles Made with α-Tocopherol–Ascorbic Acid Surfactant. *ACS Nano* **2011**, *5*, 9313–9325. [CrossRef] [PubMed]
22. Murugeshu, A.; Astete, C.; Leonardi, C.; Morgan, T.; Sabliov, C.M. Chitosan/PLGA particles for controlled release of α-tocopherol in the GI tract via oral administration. *Nanomedicine* **2011**, *6*, 1513–1528. [CrossRef] [PubMed]
23. Simon, L.C.; Stout, R.W.; Sabliov, C. Bioavailability of Orally Delivered Alpha-Tocopherol by Poly(Lactic-Co-Glycolic) Acid (PLGA) Nanoparticles and Chitosan Covered PLGA Nanoparticles in F344 Rats. *Nanobiomedicine* **2016**, *3*, 8. [CrossRef] [PubMed]
24. Benkő, M.; Varga, N.; Sebők, D.; Bohus, G.; Juhász, Á.; Dékány, I. Bovine serum albumin-sodium alkyl sulfates bioconjugates as drug delivery systems. *Colloids Surf. B Biointerfaces* **2015**, *130*, 126–132. [CrossRef] [PubMed]
25. Dash, S.; Murthy, P.N.; Nath, L.; Chowdhury, P. Kinetic modeling on drug release from controlled drug delivery systems. *Acta Pol. Pharm.* **2010**, *67*, 217–223.
26. Costa, P.; Sousa Lobo, J.M. Modeling and comparison of dissolution profiles. *Eur. J. Pharm. Sci.* **2001**, *13*, 123–133. [CrossRef]
27. Peppas, N.A.; Sahlin, J.J. A simple equation for the description of solute release. III. Coupling of diffusion and relaxation. *Int. J. Pharm.* **1989**, *57*, 169–172. [CrossRef]
28. Silva, M.E.S.R.; Freitas, R.F.S.; Sousa, R.G. Synthesis, Characterization, and Study of PLGA Copolymer in Vitro Degradation. *J. Biomater. Nanobiotechnol.* **2015**, *6*, 8–19. [CrossRef]
29. Che Man, Y.B.; Ammawath, W.; Mirghani, M.E.S. Determining α-tocopherol in refined bleached and deodorized palm olein by Fourier transform infrared spectroscopy. *Food Chem.* **2005**, *90*, 323–327. [CrossRef]

30. Gyulai, G.; Magyar, A.; Rohonczy, J.; Orosz, J.; Yamasaki, M.; Bosze, S.; Kiss, E. Preparation and characterization of cationic Pluronic for surface modification and functionalization of polymeric drug delivery nanoparticles. *Express Polym. Lett.* **2016**, *10*, 216–226. [CrossRef]
31. Bouchemal, K.; Agnely, F.; Koffi, A.; Ponchel, G. A concise analysis of the effect of temperature and propanediol-1, 2 on Pluronic F127 micellization using isothermal titration microcalorimetry. *J. Colloid Interface Sci.* **2009**, *338*, 169–176. [CrossRef] [PubMed]
32. Ritger, P.L.; Peppas, N.A. A simple equation for description of solute release I. Fickian and non-fickian release from non-swellable devices in the form of slabs, spheres, cylinders or discs. *J. Control. Release* **1987**, *5*, 23–36. [CrossRef]

pharmaceutics MDPI

Article

Monitoring the Fate of Orally Administered PLGA Nanoformulation for Local Delivery of Therapeutic Drugs

Lucia Morelli [1], Sara Gimondi [2], Marta Sevieri [2], Lucia Salvioni [1], Maria Guizzetti [1], Barbara Colzani [1], Luca Palugan [3], Anastasia Foppoli [3], Laura Talamini [2], Lavinia Morosi [2], Massimo Zucchetti [2], Martina Bruna Violatto [2], Luca Russo [2], Mario Salmona [2], Davide Prosperi [1,4], Miriam Colombo [1,*,†] and Paolo Bigini [2,*,†]

1 NanoBioLab, Dipartimento di Biotecnologie e Bioscienze, Università di Milano-Bicocca, Piazza della Scienza 2, 20126 Milano, Italy; luciaagnese.morelli@libero.it (L.M.); lucia.salvioni@unimib.it (L.S.); maria.guizzetti@gmail.com (M.G.); colzanib@gmail.com (B.C.); davide.prosperi@unimib.it (D.P.)
2 IRCCS Mario Negri Institute for Pharmacological Research; Via Mario Negri 2, 20157 Milano, Italy; sara.gimondi@gmail.com (S.G.); m.sevieri@campus.unimib.it (M.S.); laura.talamini@marionegri.it (L.T.); lavinia.morosi@marionegri.it (L.M.); massimo.zucchetti@marionegri.it (M.Z.); martina.violatto@marionegri.it (M.B.V.); luca.russo@marionegri.it (L.R.); mario.salmona@marionegri.it (M.S.)
3 Dipartimento di Scienze Farmaceutiche, Università di Milano, Via Giuseppe Colombo 71, 20133 Milano, Italy; luca.palugan@unimi.it (L.P.); anastasia.foppoli@unimi.it (A.F.)
4 Nanomedicine Laboratory, ICS Maugeri S.p.A. SB, Via S. Maugeri 10, 27100 Pavia, Italy
* Correspondence: miriam.colombo@unimib.it (M.C.); paolo.bigini@marionegri.it (P.B.); Tel.: +39-0264483388 (M.C.); +39-0239014221 (P.B.)
† These authors contributed equally to this work.

Received: 31 October 2019; Accepted: 4 December 2019; Published: 6 December 2019

Abstract: One of the goals of the pharmaceutical sciences is the amelioration of targeted drug delivery. In this context, nanocarrier-dependent transportation represents an ideal method for confronting a broad range of human disorders. In this study, we investigated the possibility of improving the selective release of the anti-cancer drug paclitaxel (PTX) in the gastro-intestinal tract by encapsulating it into the biodegradable nanoparticles made by FDA-approved poly(lactic-*co*-glycolic acid) (PLGA) and coated with polyethylene glycol to improve their stability (PLGA-PEG-NPs). Our study was performed by combining the synthesis and characterization of the nanodrug with in vivo studies of pharmacokinetics after oral administration in mice. Moreover, fluorescent PLGA-nanoparticles (NPs), were tested both in vitro and in vivo to observe their fate and biodistribution. Our study demonstrated that PLGA-NPs: (1) are stable in the gastric tract; (2) can easily penetrate inside carcinoma colon 2 ($CaCo_2$) cells; (3) reduce the PTX absorption from the gastrointestinal tract, further limiting systemic exposure; (4) enable PTX local targeting. At present, the oral administration of biodegradable nanocarriers is limited because of stomach degradation and the sink effect played by the duodenum. Our findings, however, exhibit promising evidence towards our overcoming these limitations for a more specific and safer strategy against gastrointestinal disorders.

Keywords: PLGA-NPs; nanomedicine; gastrointestinal tract; paclitaxel; in vivo imaging

1. Introduction

The generation of therapeutic methods that can improve the organ specificity of many different types of therapeutic agents is one of the major challenges of 21st-century pharmacology [1]. The availability of transport systems that can successfully improve the targeted release of administered substances would be, indeed, crucial in mitigating or even virtually eliminating the toxic side effects of already existing therapeutics. These improved transport systems can be used to limit the off-target spread and, consequently, reduce the potential side effects deriving from an unnecessary tropism [2]. Several strategies have been employed in the last years to improve the accuracy of drug targeting. Quite recently, our group and others have demonstrated that the systemic administration of anti-inflammatory and immunomodulatory agents, such as corticosteroids, can have a safer and more effective pharmacological profile when linked to a nanometric vector [3,4]. The potential prospects of nanocarriers for pharmacological purposes is extremely versatile complex. It is widely known that even the slightest modification of basic physicochemical strictures, such as the dimension, shape, and surface of the material itself, can indeed lead to a myriad of different results—causing a very vast range of reactions amongst the cells and organs of the host [5]. One of the primary and certainly most prominent, therapeutic strategies is the incorporation of the anti-neoplastic drug doxorubicin into liposomes for the treatment of tumors [6]. The first generation of nanoparticles (NPs) used for these purposes are mainly liposomes, polymeric micelles, and conjugates.

Since the turn of the century, the encapsulation of anti-neoplastic cytotoxic molecules inside NPs has been occurring with greater frequency. In 2004, an original formulation for the creation of nanocarriers to transport the anti-cancer drug paclitaxel (PTX) was developed [7,8]. In that case, different from other carriers, the biodegradable copolymer poly(lactic-*co*-glycolic acid) (PLGA) was utilized. In a more recent study, furthermore, it was shown that the systemic administration of PTX inside poly(lactic-*co*-glycolic acid)-polyethylene glycol-nanoparticles (PLGA-PEG-NPs), when functionalized with ANTI-EGFR, dramatically slows disease progression in triple-negative breast cancer-bearing mice, thus confirming the high efficacy, safety, and tenability of this material [9].

Nanoparticles utilizing FDA-approved PLGA, exhibit some important advantages. In particular, their biocompatibility causes low systemic toxicity in comparison to other polymers [10] and is highly versatile, allowing them to encapsulate various drugs, including small molecules and therapeutic biologics [11–13]. The importance of PLGA nanocarriers in biomedicine is further supported by evidence of its potential benefits in assisting drug transportation across biological barriers [14,15]. Most of their applications concern intravenous administration [16], but, in the last decade, many studies have been published regarding their potential efficacy in oral administration of drugs, even if the absorption efficiency needs to be more deeply investigated and optimized [17].

Although the majority of studies dealing with biodistribution, disease targeting, and therapeutic efficacy of NPs-based drug delivery systems has been conducted using systemic administration of nanoformulation, more recently the exploration of alternative routes, including topical [18,19], intranasal [20], intratracheal [21], and oral administration [22], are gaining increasing recognition.

Building on these advances, this work aimed to characterize the biodistribution of PTX alone or encapsulated in PEGylated PLGA-NPs after oral administration in healthy mice. This approach stems from the hope that, in so doing, the passage of the drug from the gastrointestinal (GI) tract to the bloodstream might be reduced, although not completely eliminating its release in the gut. While the active gastroduodenal absorption of PTX on oral ingestion is quite low, it is not possible to totally exclude systemic toxicity of both circulating cells and filter organs in the case of high doses or repeated cycles of chemotherapy. PLGA-NPs could, therefore, be useful to both protect the drug against gastric degradation and, at the same time, reduce the uptake by mucosae villi or Peyer's patches and the consequent release into the blood, while PEG grafting is expected to improve both the NPs stability and mucus-permeating properties in the GI tract [23,24]. In this study, PLGA-PEG-NPs were orally administered either loaded with PTX or labeled with a fluorophore (Rhodamine B), which allowed for the longitudinal tracking of NPs in both gastrointestinal and peripheral organs. This study could be of interest to understand the potentials of local delivery of anti-cancer agents in GI tumors by the oral administration of NPs, minimizing nonspecific absorption.

2. Materials and Methods

2.1. *Nanoparticles Synthesis*

2.1.1. Materials

PLGA (Resomer® RG 504 H, 50:50 lactide:glycolide, acid terminal, MW 38,000–54,000 g/mol), paclitaxel (PTX, MW 853.91 g/mol), rhodamine B (RhB, MW 479.01 g/mol); polyvinlyl alcohol (PVA, MW 9000–10,000 g/mol, 80% hydrolyzed); *N,N*′-dicyclohexylcarbodiimide (DCC, MW 206.33 g/mol, 99%); *N*-hydroxysuccinimide (NHS, MW 115.09 g/mol, 98%); dimethylamino pyridine (DMAP, MW 122.17 g/mol, 99%); ninhydrin (MW 178.14 g/mol); ethanolamine (MW 61.08 g/mol, ≥98%); dichloromethane (DCM, ≥99.8%); dimethyl sulfoxide (DMSO, ≥99.9%); acetonitrile anhydrous (ACN, 99.8%); methanol (MeOH, ≥99.9%) were purchased from Sigma–Aldrich Co (St Louis, MO, USA). Chloroform (≥99.8%) was purchased from Honeywell Riedel-de-Haën™ (Charlotte, NC, USA)). Poly (ethylene glycol) diamine (NH_2-PEG-NH_2, 6000 Da) was obtained from Rapp Polymer GmbH, (Tuebingen, Germany).

2.1.2. PLGA-PEG-RhB Synthesis

For activation of the carboxylic groups of PLGA, the polymer was dissolved in DCM (15 mL, 40 mg/mL) and the reaction was performed in a round flask for 4 h (250 rpm, RT, under inert gas) after the addition of DCC (30 mg) and NHS (15 mg). Then, the solution was diluted with diethyl ether, and the polymer was collected after precipitation (20 min, 6200× *g*, 4 °C) and evaporation of the residual organic phase. Next, the conjugation with NH_2-PEG-NH_2 (molar ratio PEG/PLGA = 2.7) was performed in 10 mL of $CHCl_3$ overnight (250 rpm, RT, under inert gas). To get rid of the unconjugated PEG, methanol (30 mL) was added, and the polymer was collected after precipitation twice [25]. The residual organic phase was then evaporated under reduced pressure, and the weight yield calculated (68.99 ± 14.94%). The degree of labeling was determined using the colorimetric ninhydrin assay (primary amine detection): Notably, 15 mg of the product (PLGA-PEG) was solubilized in DMSO (400 μL) then diluted with 100 μL of ninhydrin solution (3.5 mg/mL). Ethanolamine (18.75 mg/mL) and PLGA solutions were used as positive and negative controls, respectively. The samples were then incubated for 1 h at 65 °C (400 rpm), and the colored conjugate was detected by UV-Vis spectroscopy (Spectrophotometer Fluormax-Horiba Scientific, Rome, Italy) (λ = 600 nm), and the yield calculated using a calibration curve of ethanolamine standards. The obtained degree of labeling was 71.90 ± 12.70%.

Finally, RhB (molar ratio RhB/PLGA = 6.25) was added to the PLGA-PEG solution in $CHCl_3$ (15 mL) with an excess of DCC and DMAP. The reaction was carried out overnight (250 rpm, RT, under inert gas). At the end of the reaction, the solution was evaporated, and the product was solubilized with DCM (10 mL) and purified thrice by precipitation (20 min, 4 °C, 6200× *g*) after diethyl ether addition (20 mL). The reaction yield was 38.16 ± 13.50%. RhB linked was detected by fluorescence spectrometry ($\lambda_{exitation}$ 555 nm; $\lambda_{emission}$ 574 nm) and the degree of labeling calculated by the comparison with a calibration curve (1.43 ± 0.46 μg/mg).

2.1.3. Synthesis of PLGA-PEG-RhB-NPs and PTX-PLGA-PEG-RhB-NPs

NPs were synthesized according to the single emulsion method. PLGA-PEG-RhB solution (1 mL, 25 mg/mL in DCM) was emulsified with 8 mL of PVA 2% solution by sonicating twice (Sonifier Sound, Branson Ultrasonics, Shanghai, China; 30 s and 38% intensity) in an ice bath. The product was transferred immediately into a solution of PVA 2% (16 mL), allowing the organic solvent to evaporate (4 h, RT, 750 rpm). For PTX-PLGA-PEG-RhB-NPs synthesis, 5 mg of PTX was dissolved along with the polymer. After curing, NPs were collected by centrifugation at 19,500× *g*, 20 min, 4 °C (Heraeus Fresco 21; Thermo Fisher Scientific, Göteborg, Sweden), washed thrice with double distilled water and freeze-dried through an Alpha 1–2 LD freeze drier (Christ, Memmingen, Germany) at

0.500 mbar, −53 °C, 12 h, without the addition of any cryoprotectant. The process yield was calculated after the freeze-drying process, as the ratio between collected NPs and starting raw materials.

The amount of PTX loaded in the NPs was determined by HPLC analysis with UV detection (Waters Associates, Milford, MA, USA, model 2487 Variable Wavelength Detector, Wavelength: 230 nm). Briefly, 0.1 mL of the NPs solution was spiked with 5 µg of IS and extracted with 0.5 mL of CH_3CN. After vortex for 10 s, samples were centrifuged at 13,000 rpm for 10 min. The organic phase was separated and dried under nitrogen, and the residues were dissolved with 250 µL of the mobile phase. Fifty microliters of the reconstituted samples was injected into the HPLC system. The apparatus was equipped with a Symmetry C18 column (5 µm, 4.6 × 150 mm), the mobile phase was composed of 50% ammonium acetate buffer (0.01 M, pH 5), 40% acetonitrile, and 10% methanol with a flux rate of 1.3 mL/min and 30 min run time. NPs solution without drug was used to prepare the calibration curve by the addition of PTX in the range 10–100 µg/mL.

2.2. *Nanoparticles Characterization*

2.2.1. Dynamic Light Scattering (DLS) and Zeta Potential Measurements

All NPs were characterized in terms of size, size distribution, and Zeta potential through dynamic light scattering (DLS, Zetasizer Nano ZS; Malvern Instruments, Cambridge, UK) in ultrapure water at 1 mg/mL, at 25 °C. The scattered light from the NPs in suspension was used to calculate NPs' hydrodynamic diameter considering medium viscosity. NPs size distribution was described by the polydispersity index (PDI), where PDI ≤ 0.2 corresponds to the monodisperse NPs population.

2.2.2. Scanning Transmission Electron Microscopy (STEM) Analysis

PLGA-PEG-RhB-NPs were observed by a Zeiss SEM-FEG Gemini 500, operating at 30 kV in scanning transmission electron microscopy (STEM) mode (Zeiss, Germany). The NPs suspension was deposited onto a formvar-coated 200-mesh copper grid (Ted Pella, CA, USA), negatively stained with 1% uranyl acetate and allowed to dry before examination.

2.2.3. In Vitro Release Studies

PTX-PLGA-PEG-RhB-NPs equivalent to 8.75 µg of PTX were suspended in 1 mL of Tween/PBS and incubated at 37 °C with constant agitation. At predetermined time points (0.5, 1, 2, 4, 6, and 24 h), the suspension was centrifuged at 19,500× *g* for 15 min to separate NPs pellets and supernatants. PTX in the collected supernatants was analyzed with HPLC equipped with a UV detector (1260 Infinity II Series, Agilent Technologies, Palo Alto, CA, USA) and an Atlantis C18 column (25 cm × 4.6 mm, particle size 5 µm) (Supelco, St. Louis, MO, USA). The mobile phase was a mixture of acetonitrile and water (50:50) run in the isocratic mode at a flow rate of 1 mL/min. PTX was detected at 227 nm. PTX quantitation was performed using a calibration curve in a range of 1.25–40 µg/mL, and the results were expressed as a percentage of cumulative release (mean ± SD; $n = 4$).

2.2.4. Nanoparticles Stability in Different Buffers

The hydrodynamic dimension of the nanocarrier was analyzed in different conditions: Artificial saliva (pH 7.6; KH_2PO_4 1.9 mM, $NaHCO_3$ 17 mM, Na_2HPO_4 1.8 mM, NaCl 8.5 mM, KSCN 3.4 mM, Urea 2.2 mM, alpha-amylase 0.007 mM), intestinal fluid (pH 6; KH_2PO_4 49.96 mM, NaOH 0.0154 M, pancreatin 1 g/L), and gastric juice (pH 1.2; NaCl 300 mM, pepsin 0.9 mM, HCl 840 mM) that simulate the media in GI tract. The NPs (15 mg/mL) were incubated for 0, 1, 2, 4, 6, and 24 h and analyzed by DLS. Upon establishing the size stability, the samples were further analyzed with fluorescence spectroscopy ($\lambda_{exitation}$ 555 nm; $\lambda_{emission}$ 574 nm) to monitor the stability of RhB conjugation. The release of RhB after 24 h was then calculated after PLGA-PEG-RhB-NPs removal by centrifugation (mean ± SD; $n = 4$).

2.3. Cells

Carcinoma colon cells ($CaCo_2$) are a cell line derived from human colorectal adenocarcinoma and were purchased from ATCC® HTB-37™. $CaCo_2$ cells were grown as a single layer in adhesion in DMEM-High Glucose (Dulbecco's Modified Eagle Medium High Glucose-Biowest, Nuaillé, France) with the addition of 10% of fetal bovine serum (FBS) and 1% L-glutamine (200 mM), 100 U/mL penicillin, 0.1 mg per mL streptomycin. Cells were maintained at 37 °C and in a 5% CO_2 humidified atmosphere. The ability of PLGA-PEG-RhB-NPs and free RhB to be internalized by $CaCo_2$ cells was evaluated. For each condition, the experiment was conducted in triplicate. Cells were seeded at a density of 40,000 cells/well on 13 mm diameter slides in 24-well plates. Forty-eight hours after sowing, $CaCo_2$ cells were incubated with PLGA-PEG-RhB-NPs (100 μg/mL) and with RhB free (0.06 μg/mL) for 1, 4, and 24 h, while the wells destined for control did not receive any treatment. After incubation, the cells were washed with phosphate-buffer saline (PBS) three times and fixed with a 4% paraformaldehyde solution dissolved in PBS (0.1 M, pH 7.4) for 40 min and the vital nuclear dye Hoechst 33258 (2 μg/mL) was added to each well for 45 min. The slides thus obtained were assembled and analyzed with the Olympus BX51 epifluorescence microscope (Olympus, Tokyo, Japan). All acquisition parameters, including laser settings, were kept constant during all scans. To evaluate a possible cytotoxic effect of PLGA-PEG-RhB-NPs and free RhB, the metabolic activity of $CaCo_2$ cells was evaluated by RealTime-Glo™ MT Cell Viability Assay (Promega, Madison, WI, USA). Cells were seeded at 16,000 cells/well in 96 opaque-walled tissue culture plates and maintained at 37 °C. The RealTime-Glo reagents were added at the same time as the test compounds, according to the manufacturer's protocol. At selected time points (4 and 24 h), the cell viability was monitored by a plate-reading luminometer (GloMax® Discover Microplate Reader, Promega, Madison, WI, USA). For each condition, 6 replicates were prepared. The viability was expressed as a percentage compared to non-treated cells.

2.4. Animals

All procedures involving animals and their health were conducted so as to minimize the number of mice used and their collateral suffering, in accordance with institutional guidelines, national laws (DL n. 24, 4 March 2014; Authorization No. 19/2008 A) and international laws and agreements (EEC Council Directive 2010/63, 6 August 2013; NIH Guide to the Care and Use of Laboratory Animals, US National Research Council, 2008). The research project was first reviewed by the Internal Ethics Committee of the Mario Negri Institute and was subsequently approved by the ministry and designated with the before mentioned code (42/2016-PR).

All animals were housed in SPF (specific pathogen-free) conditions. The housing rooms had a temperature of 22 ± 1 °C, relative humidity values ± of 50 ± 10% and a 12 h light/dark cycle. Furthermore, the animals were kept in cages with free access to water and food.

2.5. Treatments

As regards pharmacokinetics, six weeks old CD1 mice were treated by oral gavage with 20 mg/kg of Cremophor PTX (n = 16) or PTX-PLGA-PEG-RhB-NPs (n = 16), three animals were treated with saline solution as control. Blood was collected in heparinized tubes, 30 min, 1, 4, and 24 h after treatment (4 animals for each group) and centrifuged to obtain plasma. After the mice were sacrificed, the stomach, duodenum, colon, and liver were collected and stored at −20 °C until analysis.

For the biodistribution studies, we recruited 5 animals for each experimental group. To reduce background fluorescence due to the food, mice were fed an AIN-76A diet without alfalfa (Mucedola s.r.l., Settimo Milanese, Italy) for two weeks before the analyzes. The dose of the different formulations was standardized on the quantity of RhB present and fixed at 0.6 mg/kg mouse based on previous experimental studies. Vehicle treated mice received the same volume of saline solution. Mice were sacrificed at 1, 4, or 24 h to follow the fate of RhB. The sacrifice was performed by cervical dislocation, and the stomach, intestine, liver, and blood were collected to perform ex vivo analyses. Plasma for fluorometric analysis was obtained from the blood collected. Finally, a piece of liver and GI tracts of three animals for each experimental group were frozen at −80 °C for cryostatic sections and histological analysis.

2.6. Molecular Imaging and Histology

The in vivo biodistribution of the different formulations was monitored over time using an optical fluorescence imaging system (IVIS Lumina XRMS, PerkinElmer, Waltham, MA, USA). Ex vivo scans of organs from mice sacrificed at 1, 4, and 24 h after the treatment were performed by the same instrument. The following acquisition parameters were used: Excitation filter: 580 nm, emission filter: 620 nm, exposure time: Auto, binning factor: Medium, f/Stop: 2, Field of View: D (for the gastro line—intestinal), C (for peripheral organs). Very importantly, the Living Image Software 4.3.1 (Perkin Elmer, Waltham, MA, USA) conjugated with the spectral unmixing system was used to separate the RhB signal from tissue autofluorescence, image processing, and fluorescence signal quantification analysis.

Longitudinal sections of 20 μm of thickness were prepared and then, after adhesion in glass slides, were incubated with a PBS solution of Hoechst 33258 (2 μg/mL, Sigma–Aldrich) for 45 min and, after three washes in PBS, observed at the Microscopy Virtual Slide (Olympus, Tokyo, Japan), to obtain rapid organ scans of the whole section with high anatomical resolution.

2.7. NPs Characterization from Homogenates and Biological Fluids

The fluorescence of PLGA-PEG-RhB-NPs was analyzed in tissues explanted from previously treated animals. The analyzed tissues were liver, stomach, intestine, and plasma. Each tissue sample was weighed, homogenized in PBS 1X according to a 1:4 weight ratio, and centrifuged at 1200 rpm for 10 min at 4 °C. For all samples, the analysis was performed using the Infinite® M200 multimode plate reader exciting at the wavelength of 500 nm and recording the signal at a range of emission wavelength from 550 to 560 nm. For the stability studies in solutions mimicking gastrointestinal fluids, the NPs were incubated in stock solutions as previously described [26].

2.8. Pharmacokinetics

The total concentration of PTX in the different biological matrices was determined by HPLC-UV, as previously described [27]. For the determination of PTX in organs, tissues were previously homogenized in 0.2 M CH_3COONH_4 pH 4.5. Each study sample (0.3 mL for plasma and 0.5 mL for homogenate tissues) was assayed together with five points of a standard calibration curve prepared in the corresponding control biological matrix obtained from untreated mice at concentrations ranging from 0.05 to 5 μg/sample. The limits of quantification (LOQ) were 0.16 μg/mL and 0.6 μg/g for plasma and organs, respectively.

2.9. Statistical Analysis

All data were expressed as mean ± SD, Student's *t*-test and *p* values were done using the GraphPad Prism version 6.00 for Windows (Graph-Pad Software, San Diego, CA, USA).

3. Results and Discussion

3.1. NPs Synthesis and Characterization

The protocol for the polymer modification was set up, and PLGA-PEG-RhB was employed to synthesize the NPs to be used as carriers for PTX. The covalently bound RhB chromophore allowed the tracking of the PLGA fate in vivo. The single emulsion method allowed us to obtain both unloaded and PTX-loaded NPs, having controlled size and homogenous size distributions (Table 1). As measured by DLS analysis, the encapsulation of PTX did not significantly change the dimensional features of NPs. Moreover, a negative Z-potential was observed as opposed to PLGA-PEG-NPs, where the free amines of PEG determined the surface properties. However, a small difference in net negative charge values between PLGA-PEG-RhB-NPs and PTX-PLGA-PEG-RhB-NPs was attributable to the intramolecular reorganization of portions of the polymer chains due to hydrophobic interaction with PTX. PTX loaded NPs were further characterized by STEM, which showed pseudospherical shaped nanoparticles with the size around 200 nm (Figure S1). PTX encapsulation efficiency detected by HPLC analysis was 10.87 ± 1.13%, while the calculated loading efficiency was 4.77 ± 0.15%, in line with previous studies [28]. The drug release performance was evaluated, and the test was conducted in PBS containing 0.2 *v*/*v* % Tween 80, considering both its poor solubility and the analysis detection limits [29]. As previously reported with similar experimental settings, a fast PTX dissolution was observed (80% within 4 h) (Figure S2) [30].

Table 1. NPs characterization and process yield. Data represent mean ± SD (n = 3).

Formulations	Size (nm ± SD)	PDI	Z-Potential (mV ± SD)	Process Yield (%, *w*/*w* ± SD)
PLGA-PEG-NPs	190.3 ± 12.7	0.078 ± 0.032	+17.9 ± 5.9	53.3 ± 7.7
PLGA-PEG-RhB-NPs	205.6 ± 18.6	0.168 ± 0.030	−18.2 ± 1.5	50.1 ± 13.4
PTX-PLGA-PEG-RhB-NPs	201.6 ± 26.2	0.205 ± 0.032	−13.3 ± 1.7	39.3 ± 7.3

Stability in Mimicking Biological Fluids

The first study was carried out to verify the potential role of PEG on the fate of PLGA-NPs in solutions mimicking saliva, gastric, and proximal intestinal fluid, respectively (Figure 1A,B). Longitudinal measurement of the NPs size by DLS was possible because, diversely from the serum, they are very poor in macromolecules, and the interference on the recording is almost zero. Figure 1A shows that the presence of the PEG confers to PLGA-NPs a long-lasting stability for at least 24 h after incubation. In contrast, the lack of PEG rapidly induced aggregation in NPs incubated in gastric fluid and, to a lesser extent in intestinal fluid. Since we aimed to conserve considerable stability up until the large intestine, we decided to carry on our studies using pegylated PLGA-NPs exclusively.

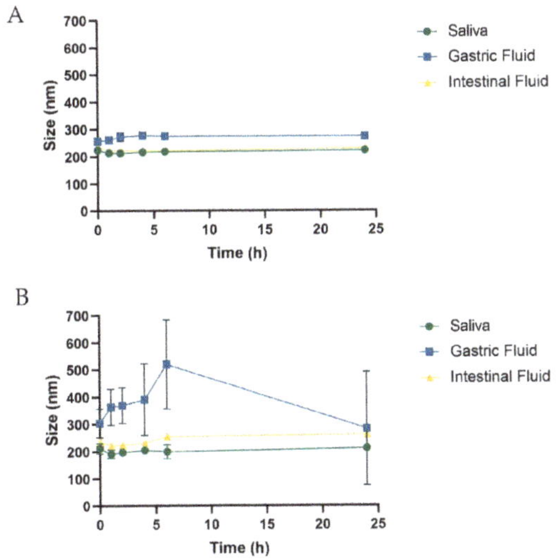

Figure 1. Graphics showing the diameter, measured by dynamic light scattering (DLS), after incubation of (**A**) pegylated or (**B**) not pegylated poly(lactic-*co*-glycolic acid)-nanoparticles (PLGA-NPs) in artificial fluids mimicking the saliva (green lines), the gastric juice (blue lines), or proximal intestinal fluid (yellow line). For all conditions, the measurements were carried out at 1, 2, 4, 6, 24 h from the starting point.

Then, the stability of the RhB conjugation was performed by measuring the dye release in the same conditions (Figure 2A). After 24 h of incubation, the released RhB was below 15% in all cases, suggesting the system reliability for in vitro and in vivo NPs tracking. Additionally, as reported in Figure 2B, the fluorescence intensity of PLGA-PEG-RhB-NPs was not greatly affected by the incubation media showing that, once conjugated, the dye abolishes its pH-dependent emission properties [31].

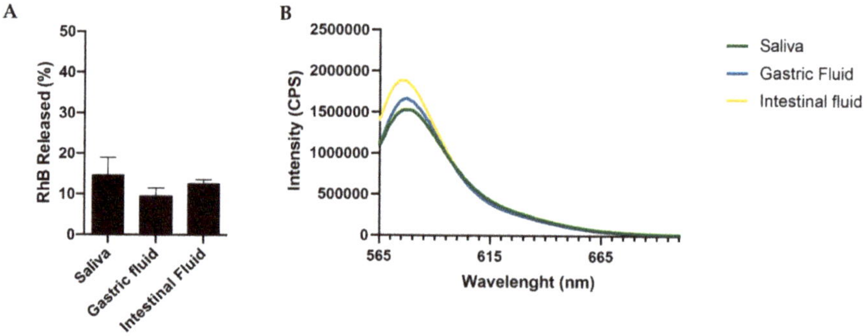

Figure 2. Graphics showing (**A**) the amount of RhB released from PLGA-PEG-RhB-NPs after 24 h incubation in artificial fluids mimicking the saliva, the gastric juice, or proximal intestinal fluid (mean ± SD; $n = 5$); (**B**) emission spectra of poly(lactic-*co*-glycolic acid)-polyethylene glycol-RhB-nanoparticles (PLGA-PEG-RhB-NPs) incubated in different media.

3.2. Pharmacokinetics

To understand the possible influence of the nanoformulate on the biodistribution of drugs, we performed a pharmacokinetic study of the well-known anti-cancer agent, paclitaxel, administered orally as a free drug (PTX) or loaded into NPs (PTX-PLGA-PEG-RhB-NPs). PTX release Figure 3 shows the levels of PTX measured in the stomach (A), in the duodenum (B) and in the colon (C) of mice sacrificed 30 min, 1 and 4 h after treatment with PTX free (blue bars) or with PTX-PLGA-PEG-RhB-NPs (red bars). The drug concentration in the samples collected at 24 h after treatment resulted under the limit of detection. Overall, the drug concentration measured in all the gastrointestinal tissues was comparable between the two formulations and higher in the stomach and duodenum compared to the colon. Despite the comparable tissue drug level, interestingly, the nano-formulation reduced the systemic absorption of the drug. The measurement of the plasmatic levels, in fact, showed a clear reduction of systemic absorption of PTX when administered as PTX-PLGA-PEG-RhB-NPs, leading to a consequent reduction of liver accumulation, as shown in Figure 3D,E. Similarly to previous results achieved by our group [32,33], no evidence of acute toxicity was observed in mice receiving the single administration of PTX-PLGA-PEG-RhB-NPs.

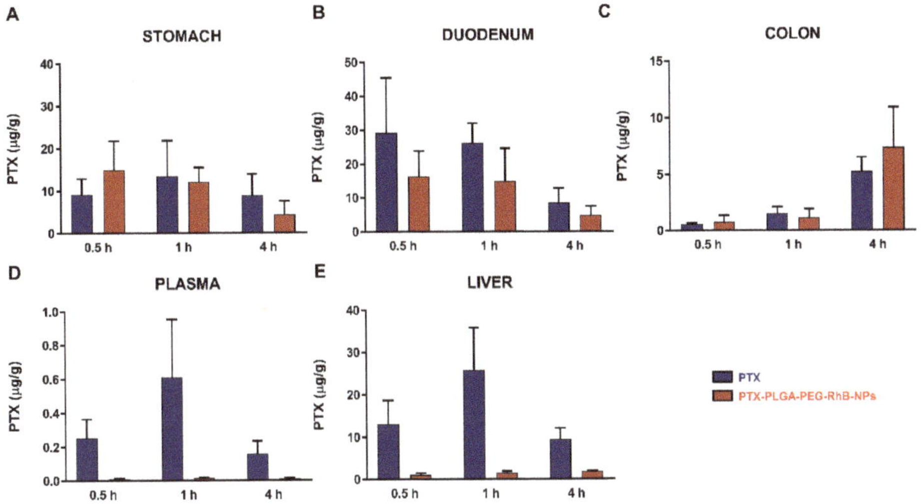

Figure 3. Comparison of the paclitaxel (PTX) distribution in (**A**) stomach, (**B**) duodenum, (**C**) colon, (**D**) plasma, and (**E**) liver of mice after administration of a single treatment of PTX or PTX-PLGA-PEG-RhB-NPs (20 mg/kg p.o.). The bars are the mean value ± SD (n = 4).

3.3. NPs Biodistribution and Nanosafety

To better understand the mechanisms that govern the different behavior of the free and NPs-encapsulated drug, we decided to monitor the NPs transit at the digestive tract level, the interactions with the gastric and intestinal structures, the possible transition to the bloodstream and penetration into filter organs by marking PLGA with a fluorescent molecule, RhB. Since the monitoring of biodegradable NPs through an indirect method, such as fluorescence analysis, presents the risk of artifacts caused by the dye release or degradation [34], all studies were carried out by comparing animals that received PLGA-PEG-RhB-NPs with those animals treated with the same dose of free RhB. As the in vitro studies suggested, the stability and reliability of this indirect approach (see Figure 2), enable us to obtain further evidence from the biodistribution study that explains the different behavior between an orally administered small molecule and our nanocarrier.

Figure 4A shows the distribution of the signal associated with the GI tract in animals sacrificed 1, 4, and 24 h after treatment with the vehicle (left) or RhB (upper blue panel), and PLGA-PEG-RhB-NPs (lower red panel), respectively. The power of the excitation laser was set in vehicle-treated mice to avoid any possible overlapping between RhB and tissue autofluorescence. In each panel, it is possible to see an upper region shaped like a sack, which represents the stomach. The small intestine is the snake-like shape in the middle. The enlargement in the last part of the intestine includes cecum, colon, and rectum and is called "large intestine". The signal associated with RhB is clearly detectable in both groups receiving the same amount of dye. As expected, in both groups, it is also possible to see a progressive decay of signal and a shift from the distal part of the GI tract. However, in mice treated with NPs, the presence of the dye was more persistent, in particular in the distal part of the small intestine and in the large intestine. The quantification of the signal (Figure 4B) confirmed this observational study: In particular, the levels of the signal in the intestine at the 4th hour after the treatment was markedly higher in animals receiving the nanoformulation.

To better understand the interaction between PLGA-PEG-RhB-NPs and the GI tract structures, histological evaluation was carried out, also exploiting the presence of RhB to track them along the anatomic path and their development in time. Even in this case, the same doses of RhB free were injected into a further group of mice to exclude any possible misinterpretation of the results due to the possible loss of NPs stability and the release of the dye.

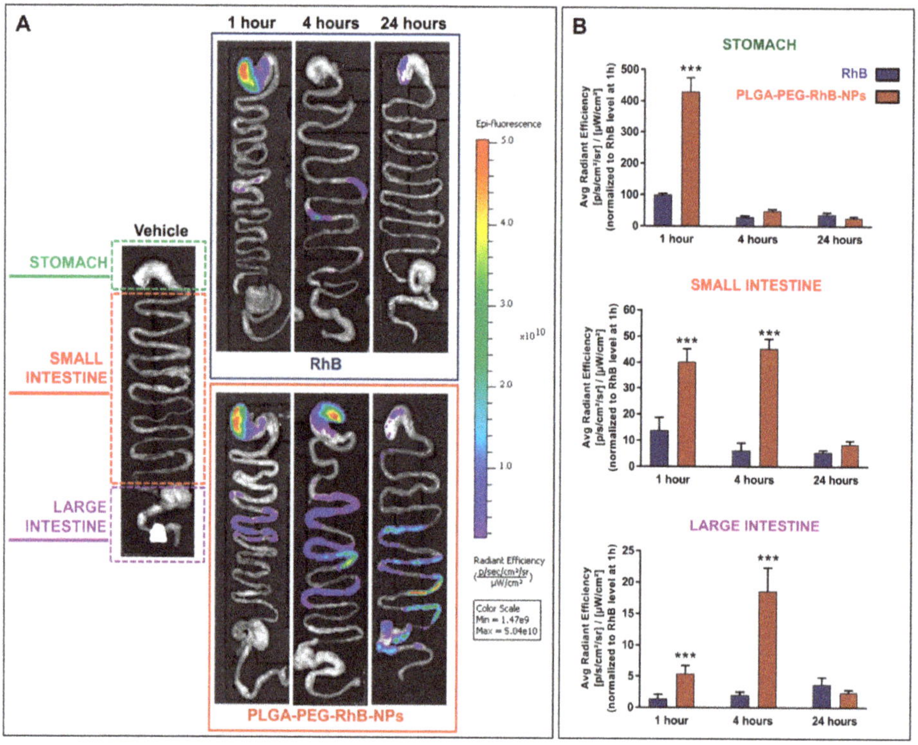

Figure 4. Signal distribution in the gastrointestinal (GI) tract of mice treated with RhB or PLGA-PEG-RhB-NPs. (**A**) Representative ex vivo scanning of the excised organs after washing with saline solutions to remove debris of feces. Animals were sacrificed at 1, 4, and 24 h after the treatment with the same dose of RhB. In the left column, a vehicle-treated mouse was shown to demonstrate the lack of the auto fluorescent component in this analysis. The interval of fluorescence signal intensity associated with the scale of colors is reported on the right. Five animals for each experimental group were used. (**B**) The quantification of the signal associated with the treatment was performed, dividing each sample into three tracts, as shown in the panel. The bars are the value of signal normalized to 100, considering the mean value measured in the stomach of mice treated with RhB alone and sacrificed 1 h after the oral administration. The bars are the mean value ± SD ($n = 5$). The Student's t-test was used to compare the levels between the two groups for each time point. *** $p < 0.0001$.

Figure 5A, upper panels, shows representative images taken from gastric sections of mice sacrificed 1, 4, and 24 h after PLGA-PEG-RhB-NPs administration. Although a progressive reduction of the signal can be seen, it is important to underline that the anatomical localization of the signal remains almost exclusively confined outside the gastric cells that are characterized by the intense blue staining due to the presence of the nuclear dye Hoechst 33258. Opposite, the RhB alone, Figure 5A lower panels, deeply penetrated inside the gastric parenchyma as clearly evidenced by the purple staining due to the merge between the red and the blue signal. This is more pronounced at the first hour and, interestingly, it does not involve the whole structure of the stomach, but it is almost confined to the superficial region of the gastric mucosa (left part of the picture at 1 h). At the 4th hour, the signal is lower but more penetrated and homogeneously spread in the parenchyma, whereas after one day, as already demonstrated by ex vivo quantification (Figure 4B), the fluorescent intensity strongly decreased. The different pattern of staining shown at the 4th and 24th hour strongly suggests that, in spite of the gastric activity, PLGA-PEG-RhB-NPs remain stable enough to avoid the release of the free dye. This is in line with the results reported in Figure 1A by DLS in solutions mimicking gastric juice.

The stability of the nanoparticles inside the stomach is essential to transport any encapsulated drug to the intestine. In Figure 4B, the signal measurement along the whole small intestine was evaluated, whereas histological analysis was focused on the more proximal part of the intestine, the duodenal region. The duodenum is one of the most critical portions of the GI for the absorption of metabolites and drugs. The active uptake by mucosae villi and Peyer's patches allows the absorption of many substances into the bloodstream and their consequent systemic distribution. Even if this process is required to provide energy and nutrients and to distribute therapeutic agents orally administered, it can be a hurdle for a localized gut delivery. In Figure 3, we have reported that the encapsulation in PLGA-PEG-RhB-NPs dramatically reduces the PTX absorption. This suggests that these kinds of NPs are able to pass through this first part of the intestine, maintaining their stability. Representative images from coronal sections of the duodenum from mice sacrificed 1 h after the treatment (lower panel on the left and higher magnification right in Figure 5B) shows that NPs are in the lumen and inside the intervilli space but are not absorbed by mucosa. A deeper interaction with villi can be seen at the 4th hour after treatment. However, even in this case, the red and blue signals are close but do not overlap. An overlapping was indeed clearly seen in distinction from mice treated with the same amount of RhB free at least up to the first 4 h after ingestion. A higher magnified picture furthermore confirms the restricted interaction between the red signal and the peripheral region of villi 1 h after treatment. As reported in the measurement of the intestinal levels by ex vivo scanning (Figure 4B), an almost complete disappearance of the red signal can be seen 24 h after the treatment in both experimental groups.

By histology, we found that RhB-free treated mice showed deep red staining in villi, whereas the red signal remained in the lumen of the intestinal tube in animals receiving RhB with NPs. Since enterocytes in villi are tightly connected to the vessels, it is, therefore, possible to hypothesize that RhB can easily penetrate the circulatory tree. To confirm that NPs can dramatically reduce the passage from the small intestine to the bloodstream, we compared the RhB levels both in plasma and in the liver of mice treated with PLGA-PEG-RhB-NPs or RhB free. Figure 6A, where plasmatic levels of RhB-related signals were normalized to the value measured in mice treated with RhB-free during the first hour of analysis, clearly reveals that the nanoformulation (red bars) leads to an almost complete abolishment of hematic absorption of RhB. This striking difference between the two groups supports the hypothesis that these NPs are stable and almost completely eradicate the absorption of themselves and of the relevant cargo by gastric and intestinal mucosae. The fast and quite elevated half-life of RhB in the blood led to an expected accumulation of the dye in the main filter organ, the liver, Figure 6B. Similar to the results obtained from the blood, the animals treated with RhB exclusively showed a well detectable red signal in liver sections (Figure 6C).

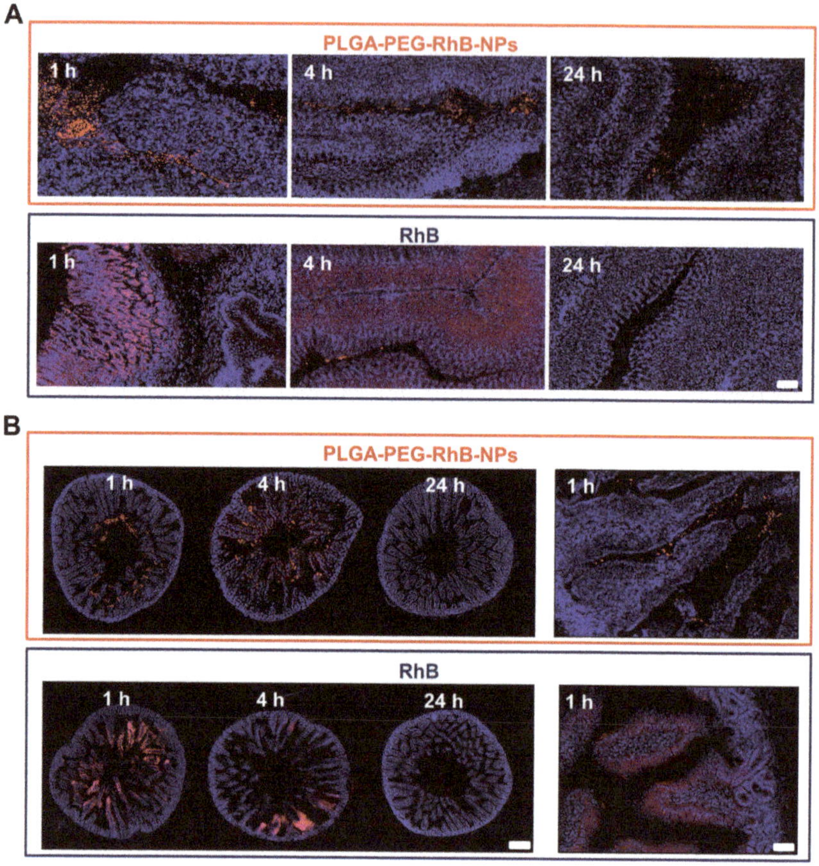

Figure 5. Representative images showing the distribution of RhB in (**A**) stomach and (**B**) duodenum of mice treated with either PLGA-PEG-RhB-NPs or RhB alone. (**A**) In the upper panels, the localization of signal (red) in gastric sections (blue) is shown 1, 4, and 24 h after the oral administration of PLGA-PEG-RhB-NPs. The same procedure has been used to track the presence of the dye in mice receiving the same amount of RhB (lower panels). Scale bar 100 μm. (**B**) Representative images of the duodenum are shown of PLGA-PEG-RhB-NPsand RhB-treated mice in upper and lower panels, respectively. Scale bar 200 μm. The thicker and more intense blue staining in the periphery of the sections represents the basal layer of the duodenum where the exchanges of tissue/blood occur. The red signal is more concentrated to the center, likely corresponding to the lumen close to the apical side of the villi. A higher magnified picture from a mouse sacrificed 1 h after the ingestion of NPs confirms the weak interaction between NPs and villi. In the duodenum of RhB-treated mice, a deep overlapping between the villi and the RhB was observed at both 1 and 24 h after treatment. A higher magnified image confirms the penetration of the dye into the external side of the villi. Scale bar 50 μm.

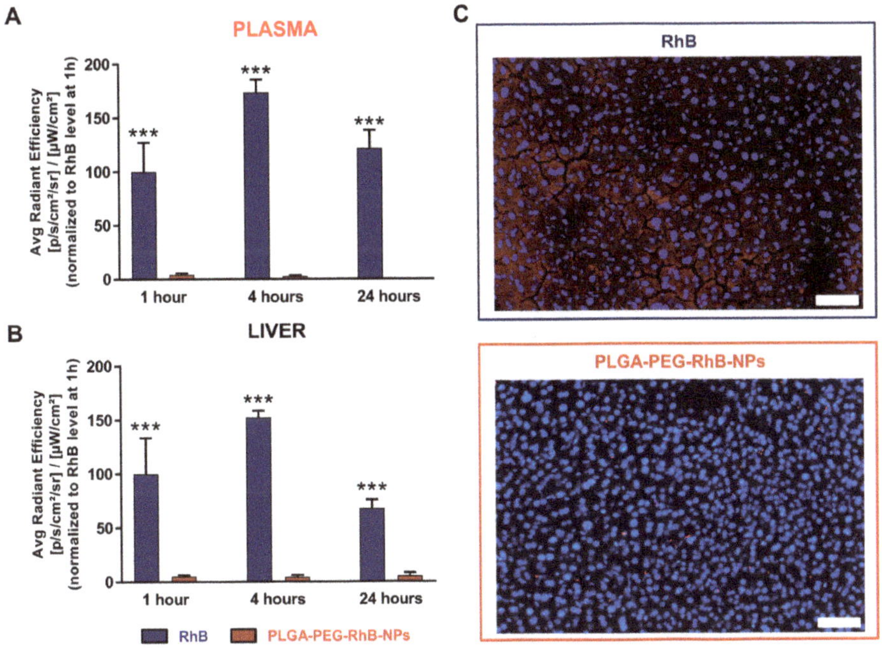

Figure 6. Signal measurement in (**A**) the plasma and (**B**) the liver of mice orally treated with RhB (blue bars) or PLGA-PEG-RhB-NPs (red bars). Animals were sacrificed at 1, 4, and 24 h after the treatment with the same dose of RhB. Five animals for each experimental group were used. In both graphics, quantification of the signal was normalized to the RhB level at 1 h and expressed as 100. The bars are the mean value ± SD ($n = 5$). The Student's t-test was used to compare the levels between the two groups for each time point. *** $p < 0.0001$. (**C**) Representative images of the signal related to the dye in a section of liver from mice treated with PLGA-PEG-RhB-NPs (upper panel) or RhB alone (lower panel), both sacrificed 1 h after the treatment. Scale bar 50 μm.

The last part of the study was carried out to investigate if these NPs were able to penetrate into $CaCo_2$ cells and where they localize inside the cells. This experiment was aimed at exploring the possible application of our results to future local treatment of colorectal cancer. Figure 7A shows the progressive process of internalization of NPs (orange spots). The quantification of the occupied area of NPs inside the cell cytoplasm is reported in Figure 7B. Progressive penetration of NPs occurs and, at the 4th hour after incubation, they already occupied the 2% to 3% of the whole cell area. Although relatively low, this percentage in terms of a potential release of a therapeutic cargo cannot be considered negligible. The red arrow in Figure 7A and the higher magnified picture in Figure 7D confirm that NPs are deeply penetrated (orange spots) and the deep internalization of NPs inside the cell cytoplasm starting from the 4th hour of incubation. Moreover, the cell viability assay confirmed the safety of the materials we selected for this study. Indeed, neither RhB nor PLGA-PEG-RhB-NPs alone modify the healthiness of the cells for the whole duration of the treatment (Figure 7C).

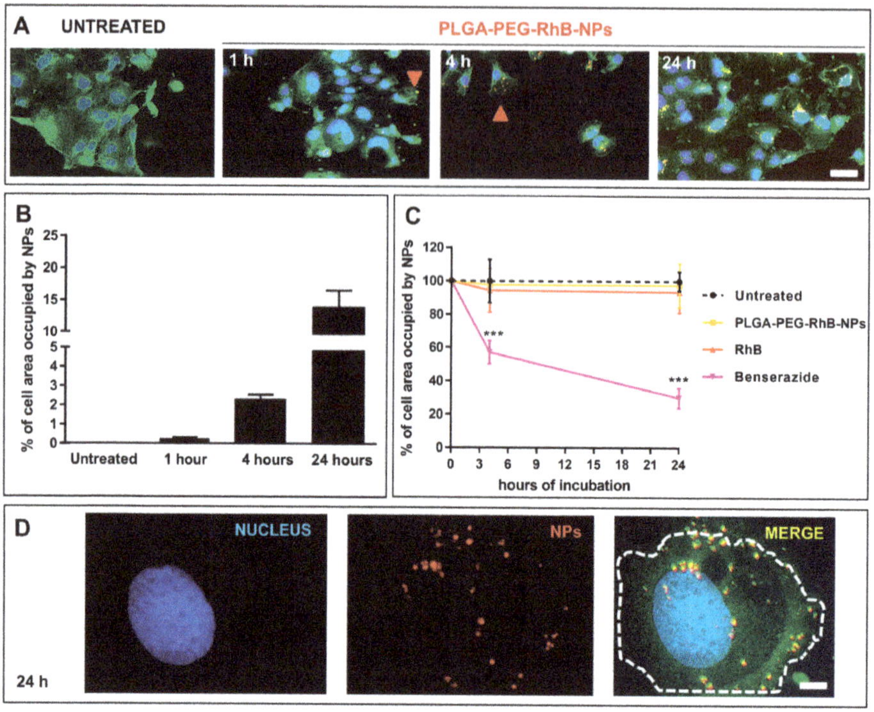

Figure 7. Longitudinal study to evaluate PLGA-PEG-RhB-NPs internalization in $CaCo_2$ cells. (**A**) Low magnified pictures showing the progressive penetration of NPs in cells. Representative images have been selected from non-treated cells (NT, left panel). For each image, the nuclei were stained with Hoechst 33258 (blue), whereas high excitation with the laser at the wavelength of 488 nm will allow unveiling the border of the cells by exploiting their auto fluorescent profile (green). Starting from the 1st hour, it is possible to see orange spots obtained by the merge between the red signal referred to the RhB and the green background. These spots became more evident 4 and 24 h after incubation. Scale bar 70 μm. (**B**) Quantification of the percentage of the area (pixels) occupied by NPs for every single cell at the different time-points. Data are expressed as mean value ± SD, $n = 10$. (**C**) Quantification of the viability of the cells of $CaCo_2$ after exposure to RhB (orange line), PLGA-PEG-RhB-NPs (yellow line), and Benserazide as inner control (purple line) measured by RealTime-Glo™ MT Cell Viability Assay (Promega kit) 4 and 24 h after incubation The values obtained from non-treated cells were normalized to 100 for each time point. Values are expressed as mean ± SD ($n = 6$). The Student's *t*-test was used to compare the levels among the two groups for each time point. *** $p < 0.0001$ compared to NT. (**D**) Higher magnification pictures showing the same field of view achieved 24 h after PLGA-PEG-RhB-NPs incubation. In the left panel, the cell nucleus, in the middle panel, the red spots associated with NPs, and in the right panel, the merge between the three channels. The dotted line is the border of the cell. Scale bar 15 μm.

4. Conclusions

The current study sought to evaluate the effect of the nanocarrier on the transport of a drug and its behavior within the gastrointestinal tract and absorption to the bloodstream. It is important to note, however, that although RhB was originally used as a tracer to visualize NPs. The results obtained by comparing the biodistribution in mice of the free fluorophore and linked to the NPs is of further relevance for future developments. It is, in fact, interesting how PLGA-PEG-NPs manage to preserve the gastroduodenal absorption by using chemically different molecules along with different loading approaches to the nanocarrier. The low systemic exposure and, at the same time, equivalent drug concentration at the intestinal level, with even a trend to increase in the colon 4 h after the treatment, could have a significant positive outcome on the safety of a wide range of drugs targeting inflammatory and neoplastic diseases.

Supplementary Materials: The following are available online at http://www.mdpi.com/1999-4923/11/12/658/s1, Figure S1: STEM image of PTX-PLGA-PEG-RhB-NPs, Figure S2: Release kinetics of PTX loaded NPs in PBS containing 0.2 *v*/*v*% Tween 80. PTX-PLGA-PEG-RhB-NPs equivalent to 8.75 μg of PTX were suspended in 1 mL of release medium and incubated at 37 °C with constant agitation. At predetermined time points (0.5, 1, 2, 4, 6 and 24 h), the NPs were separated by centrifugation and the supernatant analyzed by HPLC. The results are the mean value ± SD ($n = 4$).

Author Contributions: L.M. (Lucia Morelli), L.S., M.G., B.C. performed nanoparticles synthesis, functionalization, and characterization; S.G., M.S. (Marta Sevieri), L.T., L.R., M.B.V. performed in vivo experiments; L.M. (Lavinia Morosi) performed pharmacokinetic analysis; L.P., A.F. did the data curation; D.P., M.S. (Mario Salmona) gave critical comments on the manuscript; D.P., M.C., P.B., M.Z. designed the research and provided funding; P.B., M.C. wrote the manuscript; all authors revised the manuscript critically.

Funding: This research was funded by Academic Funding Unimib 2018 to M.C.; the Italian Ministry of University and Research (MIUR) through grant "Dipartimenti di Eccellenza-2017" to University of Milano Bicocca, Department of Biotechnology and Biosciences, Direzione Generale Ricerca, Innovazione, Università, Export e Internazionalizzazione of the Regione Lombardia.

Acknowledgments: We thank Tiziano Catalani for the STEM image.

Conflicts of Interest: The authors declare no conflict of interest.

References

1. Peer, D.; Karp, J.M.; Hong, S.; Farokhzad, O.C.; Margalit, R.; Langer, R. Nanocarriers as an emerging platform for cancer therapy. *Nat. Nanotechnol.* **2007**, *2*, 751–760. [CrossRef] [PubMed]
2. Heath, J.R. Nanotechnologies for biomedical science and translational medicine. *Proc. Natl. Acad. Sci. USA* **2015**, *112*, 14436–14443. [CrossRef] [PubMed]
3. Violatto, M.B.; Casarin, E.; Talamini, L.; Russo, L.; Baldan, S.; Tondello, C.; Messmer, M.; Hintermann, E.; Rossi, A.; Passoni, A.; et al. Dexamethasone Conjugation to Biodegradable Avidin-Nucleic-Acid-Nano-Assemblies Promotes Selective Liver Targeting and Improves Therapeutic Efficacy in an Autoimmune Hepatitis Murine Model. *ACS Nano* **2019**, *13*, 4410–4423. [CrossRef] [PubMed]
4. Bartneck, M.; Scheyda, M.K.; Warzecha, K.T.; Rizzo, Y.L.; Hittatiya, K.; Luedde, T.; Storm, G.; Trautwein, C.; Lammers, T.; Tacke, F. Fluorescent cell-traceable dexamethasone-loaded liposomes for the treatment of inflammatory liver diseases. *Biomaterials* **2015**, *37*, 367–382. [CrossRef] [PubMed]
5. Hoshyar, N.; Gray, S.; Han, H.; Bao, G. The effect of nanoparticle size on in vivo pharmacokinetics and cellular interaction. *Nanomedicine* **2016**, *11*, 673–692. [CrossRef]
6. Uziely, B.; Jeffers, S.; Isacson, R.; Kutsch, K.; Wei-Tsao, D.; Yehoshua, Z.; Libson, E.; Muggia, F.M.; Gabizon, A. Liposomal doxorubicin: Antitumor activity and unique toxicities during two complementary phase I studies. *J. Clin. Oncol.* **1995**, *13*, 1777–1785. [CrossRef]
7. Feng, S.; Mu, L.; Win, K.Y. Nanoparticles of Biodegradable Polymers for Clinical Administration of Paclitaxel. *Curr. Med. Chem.* **2004**, *11*, 413–424. [CrossRef]
8. Bhatt, P.; Lalani, R.; Vhora, I.; Patil, S.; Amrutiya, J.; Misra, A.; Mashru, R. Liposomes encapsulating native and cyclodextrin enclosed paclitaxel: Enhanced loading efficiency and its pharmacokinetic evaluation. *Int. J. Pharm* **2018**, *536*, 95–107. [CrossRef]
9. Venugopal, V.; Krishnan, S.; Palanimuthu, V.R.; Sankarankutty, S.; Kalaimani, J.K.; Karupiah, S.; Kit, N.S.; Hock, T.T. Anti-EGFR anchored paclitaxel loaded PLGA nanoparticles for the treatment of triple negative breast cancer.In-vitro and in-vivo anticancer activities. *PLoS ONE* **2018**, *13*, e0206109. [CrossRef]
10. Kamaly, N.; Yameen, B.; Wu, J.; Farokhzad, O.C. Degradable controlled-release polymers and polymeric nanoparticles: Mechanisms of controlling drug release. *Chem. Rev.* **2016**, *116*, 2602–2663. [CrossRef]
11. Colzani, B.; Pandolfi, L.; Hoti, A.; Iovene, P.A.; Natalello, A.; Avvakumova, S.; Colombo, M.; Prosperi, D. Investigation of antitumor activities of trastuzumab delivered by PLGA nanoparticles. *Int. J. Nanomed.* **2018**, *13*, 957–973. [CrossRef] [PubMed]

12. Park, J.; Fong, P.M.; Lu, J.; Russell, K.S.; Booth, C.J.; Saltzman, W.M.; Fahmy, T.M. PEGylated PLGA nanoparticles for the improved delivery of doxorubicin. *Nanomedicine* **2009**, *5*, 410–418. [CrossRef] [PubMed]
13. Ramezani, M.; Ebrahimian, M.; Hashemi, M. Current Strategies in the Modification of PLGA-based Gene Delivery System. *Curr. Med. Chem.* **2017**, *24*, 728–739. [CrossRef] [PubMed]
14. Li, J.; Sabliov, C.M. PLA/PLGA nanoparticles for delivery of drugs across the blood-brain barrier. *Nanotechnol. Rev.* **2013**, *2*, 241–257. [CrossRef]
15. Gamboa, J.M.; Leong, K.W. In vitro and in vivo models for the study of oral delivery of nanoparticles. *Adv. Drug Deliv. Rev.* **2013**, *65*, 800–810. [CrossRef]
16. Danhier, F.; Ansorena, E.; Silva, J.M.; Coco, R.; Le Breton, A.; Préat, V. PLGA-based nanoparticles: An overview of biomedical applications. *J. Control. Release* **2012**, *161*, 505–522. [CrossRef]
17. Wagner, A.M.; Gran, M.P.; Peppas, N.A. Designing the new generation of intelligent biocompatible carriers for protein and peptide delivery. *Acta Pharm. Sin. B* **2018**, *8*, 147–164. [CrossRef]
18. Santini, B.; Zanoni, I.; Marzi, R.; Cigni, C.; Bedoni, M.; Gramatica, F.; Palugan, L.; Corsi, F.; Granucci, F.; Colombo, M. Cream formulation impact on topical administration of engineered colloidal nanoparticles. *PLoS ONE* **2015**, *10*, e0126366. [CrossRef]
19. Musazzi, U.M.; Santini, B.; Selmin, F.; Marini, V.; Corsi, F.; Allevi, R.; Prosperi, D.; Cilurzo, F.; Colombo, M.; Minghetti, P. Impact of semi-solid formulations on skin penetration of iron oxide nanoparticles. *J. Nanobiotechnol.* **2017**, *15*, 14. [CrossRef]
20. Cunha, S.; Amaral, M.H.; Lobo, J.M.S.; Silva, A.C. Lipid Nanoparticles for Nasal/Intranasal Drug Delivery. *Crit Rev. Ther. Drug Carr. Syst.* **2017**, *34*, 257–282. [CrossRef]
21. Codullo, V.; Cova, E.; Pandolfi, L.; Breda, S.; Morosini, M.; Frangipane, V.; Malatesta, M.; Calderan, L.; Cagnone, M.; Pacini, C.; et al. Imatinib-loaded gold nanoparticles inhibit proliferation of fibroblasts and macrophages from systemic sclerosis patients and ameliorate experimental bleomycin-induced lung fibrosis. *J. Control. Release* **2019**, *310*, 198–208. [CrossRef] [PubMed]
22. Salvioni, L.; Fiandra, L.; Del Curto, M.D.; Mazzucchelli, S.; Allevi, R.; Truffi, M.; Sorrentino, L.; Santini, B.; Cerea, M.; Palugan, L.; et al. Oral delivery of insulin via polyethylene imine-based nanoparticles for colonic release allows glycemic control in diabetic rats. *Pharmacol. Res.* **2016**, *110*, 122–130. [CrossRef] [PubMed]
23. Suk, J.S.; Xu, Q.; Kim, N.; Hanes, J.; Ensign, L.M. PEGylation as a strategy for improving nanoparticle-based drug and gene delivery. *Adv. Drug Deliv. Rev.* **2016**, *99*, 28–51. [CrossRef] [PubMed]
24. Inchaurraga, L.; Martín-Arbella, N.; Zabaleta, V.; Quincoces, G.; Peñuelas, I.; Irache, J.M. In vivo study of the mucus-permeating properties of PEG-coated nanoparticles following oral administration. *Eur. J. Pharm. Biopharm.* **2015**, *97 Pt A*, 280–289. [CrossRef]
25. Aggarwal, S.; Gupta, S.; Pabla, D.; Murty, R.S.R. Gemcitabine-loaded PLGA-PEG immunonanoparticles for targeted chemotherapy of pancreatic cancer. *Cancer Nanotechnol.* **2013**, *4*, 145–157. [CrossRef]
26. Lazzari, S.; Moscatelli, D.; Codari, F.; Salmona, M.; Morbidelli, M.; Diomede, L. Colloidal stability of polymeric nanoparticles in biological fluids. *J. Nanopart. Res.* **2012**, *14*, 920. [CrossRef]
27. Fruscio, R.; Lissoni, A.A.; Frapolli, R.; Corso, S.; Mangioni, C.; D'Incalci, M.; Zucchetti, M. Clindamycin-paclitaxel pharmacokinetic interaction in ovarian cancer patients. *Cancer Chemother. Pharmacol.* **2006**, *58*, 319–325. [CrossRef]
28. Enlow, E.M.; Luft, J.C.; Napier, M.E.; DeSimone, J.M. Potent engineered PLGA nanoparticles by virtue of exceptionally high chemotherapeutic loadings. *Nano Lett.* **2011**, *11*, 808–813. [CrossRef]
29. Abouelmagd, S.A.; Sun, B.; Chang, A.C.; Ku, Y.J.; Yeo, Y. Release Kinetics Study of Poorly Water-Soluble Drugs from Nanoparticles: Are We Doing It Right? *Mol. Pharm.* **2015**, *12*, 997–1003. [CrossRef]
30. Gullotti, E.; Yeo, Y. Beyond the imaging: Limitations of cellular uptake study in the evaluation of nanoparticles. *J. Control. Release* **2012**, *164*, 170–176. [CrossRef]
31. Zhang, W.; Shi, K.; Shi, J.; He, X. Use of the fluorescence of rhodamine B for the pH sensing of a glycine solution. *Proc. SPIE* **2016**, 10155. [CrossRef]
32. Capasso Palmiero, U.; Morosi, L.; Lupi, M.; Ponzo, M.; Frapolli, R.; Zucchetti, M.; Ubezio, P.; Morbidelli, M.; D'Incalci, M.; Bello, E.; et al. Self-Assembling PCL-Based Nanoparticles as PTX Solubility Enhancer Excipients. *Macromol. Biosci.* **2018**, *18*, 1800164. [CrossRef]

33. Colombo, C.; Morosi, L.; Bello, E.; Ferrari, R.; Licandro, S.A.; Lupi, M.; Ubezio, P.; Morbidelli, M.; Zucchetti, M.; D'Incalci, M.; et al. PEGylated Nanoparticles Obtained through Emulsion Polymerization as Paclitaxel Carriers. *Mol. Pharm.* **2016**, *13*, 40–46. [CrossRef]
34. Tenuta, T.; Monopoli, M.; Kim, J.; Salvati, A.; Dawson, K.; Sandin, P.; Lynch, I. Elution of Labile Fluorescent Dye from Nanoparticles during Biological Use. *PLoS ONE* **2011**, *6*, e25556. [CrossRef]

pharmaceutics

MDPI

Article

Risperidone-Loaded PLGA–Lipid Particles with Improved Release Kinetics: Manufacturing and Detailed Characterization by Electron Microscopy and Nano-CT

Christopher Janich [1], Andrea Friedmann [2], Juliana Martins de Souza e Silva [3], Cristine Santos de Oliveira [3], Ligia E. de Souza [1], Dan Rujescu [4], Christian Hildebrandt [5], Moritz Beck-Broichsitter [5], Christian E. H. Schmelzer [2] and Karsten Mäder [1,*]

1 Institute of Pharmacy, Faculty of Biosciences, Martin Luther University Halle-Wittenberg, 06120 Halle (Saale), Germany; christopher.janich@pharmazie.uni-halle.de (C.J.); ligia.souza@pharmazie.uni-halle.de (L.E.d.S.)
2 Department of Biological and Macromolecular Materials, Fraunhofer Institute for Microstructure of Materials and Systems IMWS, 06120 Halle (Saale), Germany; andrea.friedmann@imws.fraunhofer.de (A.F.); christian.schmelzer@imws.fraunhofer.de (C.E.H.S.)
3 Institute of Physics, Martin Luther University Halle-Wittenberg, 06120 Halle (Saale), Germany; juliana.martins@physik.uni-halle.de (J.M.d.S.eS.); cristine.santos-de-oliveira@physik.uni-halle.de (C.S.d.O.)
4 Department of Psychiatry, Psychotherapy and Psychosomatics, Martin Luther University Halle-Wittenberg, 06120 Halle (Saale), Germany; dan.rujescu@uk-halle.de
5 MilliporeSigma a Business of Merck KGaA, 64293 Darmstadt, Germany; christian.hildebrandt@merckgroup.com (C.H.); moritz.beck-broichsitter@merckgroup.com (M.B.-B.)
* Correspondence: karsten.maeder@pharmazie.uni-halle.de; Tel.: +49-345-55-25167

Received: 31 October 2019; Accepted: 2 December 2019; Published: 9 December 2019

Abstract: For parenteral controlled drug release, the desired zero order release profile with no lag time is often difficult to achieve. To overcome the undesired lag time of the current commercial risperidone controlled release formulation, we developed PLGA–lipid microcapsules (MCs) and PLGA–lipid microgels (MGs). The lipid phase was composed of middle chain triglycerides (MCT) or isopropylmyristate (IPM). Hydroxystearic acid was used as an oleogelator. The three-dimensional inner structure of Risperidone-loaded MCs and MGs was assessed by using the invasive method of electron microscopy with focused ion beam cutting (FIB-SEM) and the noninvasive method of high-resolution nanoscale X-ray computed tomography (nano-CT). FIB-SEM and nano-CT measurements revealed the presence of highly dispersed spherical structures around two micrometres in size. Drug release kinetics did strongly depend on the used lipid phase and the presence or absence of hydroxystearic acid. We achieved a nearly zero order release without a lag time over 60 days with the MC-MCT formulation. In conclusion, the developed lipid-PLGA microparticles are attractive alternatives to pure PLGA-based particles. The advantages include improved release profiles, which can be easily tuned by the lipid composition.

Keywords: controlled release; PLGA; risperidone; microparticles; microcapsules; oleogels; electron microscopy; three-dimensional X-ray imaging; nano-CT; biodegradable polymers; hydroxy-stearic acid

1. Introduction

The use of parenteral controlled release drug delivery systems (CR-DDS) is a highly attractive and rational way to improve drug therapy. Potential benefits include decrease of administration frequency, increase of bioavailability and less side effects. Polylactic acid (PLA) and poly(lactic-*co*-glycolic acid

(PLGA) products have existed for several decades and are, by far, the most widely used polymers for CR-DDS [1,2]. The major clinical use is the peptide-based treatment of GnrH-dependent diseases such as breast and prostate cancer. Other important and increasingly used applications include the local delivery of drugs to the eye [3], the brain [4], or the inner ear [5,6]. An important application of CR-DDS is also the controlled release of antipsychotic compounds to treat schizophrenia. Long-acting parenteral controlled release of antipsychotic drugs overcomes the poor compliance of patients with oral administration [7,8]. Risperidone is an important drug in the treatment of schizophrenia. Several oral and one parenteral controlled release formulations are available and a PLGA based CR-DDS formulation is marketed (Risperdal® Consta®). According to the treatment schedule [9], patients initially have to take tablets, although the CR-DDS has already been injected. This situation is not a favorable scenario of pharmacotherapy, because schizophrenia patients do frequently not comply and double medication should be avoided. The release mechanisms of Risperdal® Consta® have been studied in detail by the group of Burgess and good in vitro and in vivo correlations have been observed by means of scaling factors [10,11]. Initially, a lag phase with almost no release is observed, followed by a faster release. The group clearly showed that the release onset is related to the water penetration and polymer degradation-induced drop of the polymer glass transition temperature T_g to 37 °C.

Several publications describe alternative PLGA formulations for the controlled release of risperidone. However, commonly, the presence of lag times, sigmoidal release curves, and rather short release times (<1 month) have been observed [10–15]. Souza et al. tested different formulations with single or repeated doses in vivo [16]. For single dose injections, peak maxima were observed after around one week. Unfortunately, no in vitro release data were published in the paper [16]. Therefore, it is clearly still desirable to develop biodegradable DDS with better risperidone release kinetics. The CR-DDS should provide a constant release over the 1–2 months without a lag time and the need for initial oral comedication. It should also be cost effective. Unfortunately, the selection possibility for biodegradable materials for parenteral controlled release is rather limited. In addition to PLA and PLGA, there are few other polymers and lipids. Liquid lipids are used clinically, but they provide only release control for a short period of few days. Solid lipids are also attractive as parenteral depot systems [17], but their use is often linked with problems of polymorphic transitions and limited enzymatic degradability due to the crystalline structure of the lipid. The use of polymer–lipid hybrid structures—compared to pure polymer or pure lipid-based DDS—offers additional opportunities. For example, increased drug loads may be achieved by the lipid component and the release rate might be controlled by the polymer shell. The incorporation of liquid oil into a microparticle might compromise the mechanical stability of the particle and could potentially lead to oily leakage and loss of release control. The incorporation of solid lipids will overcome this problem, but this advantage is at the cost of drug solubility in the crystalline lipid matrix and the problem of polymorphic transitions. It is desirable to combine the advantages of a liquid lipid phase with high mechanical stability. Therefore, we decided to investigate oleogels as possible alternatives. A suitable oleogelator for parenteral applications is 12-hydroxystearic acid (HS). Hydroxystearic acid induces gel formation in lipids due to its precipitation as a HS-fiber network [18,19]. Preclinical studies on injectable oleogels demonstrate the good biocompatibility and degradability of HS-oleogels in vivo [20,21].

We selected PLGA Expansorb® 75-7E from Merck KGaA (Darmstadt, Germany), which has the following properties: lactide/glycolide (L/G) ratio of 75/25, a molecular weight (MW) 80–115 kDa, and the end groups are esterified. The physical chemical properties of Expansorb®75-7E were similar to the PLGA in the market product Risperdal Consta® which contains a lactide/glycolide (L/G) ratio of 78/22, a molecular weight (MW) ~111 kDa and the end groups are esterified [22]. As the oil phase, we incorporated isopropyl myristate (IPM) or middle chain triglycerides (MCT) to form a microcapsule (MC) structure. Oily components with saturated fatty acids were used to avoid the danger of oxidation problems connected with unsaturated fatty acids.

The properties and the release kinetics of the micro-DDS will depend on the general structure and the physicochemical state of the drug and the excipients. Possible structures of (drug-free) micro-DDS are shown in Figure 1.

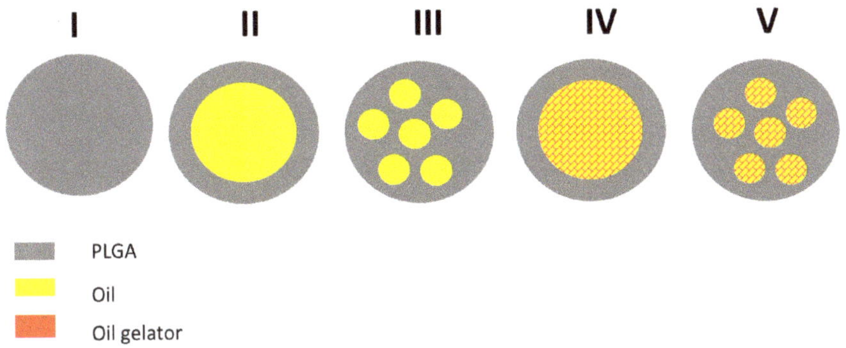

Figure 1. Possible structures of poly(lactic-*co*-glycolic acid (PLGA) and PLGA/lipid microparticles: I: PLGA microparticle; II: microcapsule with core–shell structure; III: microcapsule with multiple oily droplets; IV: microgel with core–shell structure; microgel with multiple gelled lipid domains.

PLGA microparticles are expected to form a homogeneous polymer matrix (structure I). Microcapsules composed of PLGA and oil could show structures with a single oily core and a PLGA shell (structure II) or multiple oily droplets encapsulated in PLGA (structure III). The addition of the oil gelator will increase the viscosity, but again, the oily phase could exist as a single core (structure IV) or as multiple cores (structure V). The drug could be dissolved in the polymer and the oil and/or exist as an own solid (crystalline or amorphous) phase. A rational development of the DDS requires an appropriate characterization. For particle sizing of micro-DDS, static light scattering is the standard method. We decided to use X-ray diffraction for the detection of crystalline structures in our DDS. In addition, we wanted to explore the microstructure inside the DDS. For this purpose, we used two complementary imaging techniques, which were focus ion beam scanning electron microscopy (FIB-SEM) and Zernike phase-contrast X-ray nano-computed tomography (nano-CT). By means of focused ion-beam preparation, we could cut the particles in situ and we were able to visualize the morphology of the microstructure at the location of the cut. In addition, we performed nano-CT experiments. Due to the high penetration depth of X-rays, nano-CT provided high-resolution three-dimensional images of the DDS and we were able to visualize their internal structures at any virtual slice without the need for cutting or slicing them.

In summary, the main goals of this work were:

(i) development of risperidone loaded PLGA–lipid microcapsules (MCs) and PLGA–lipid microgels (MGs) with an optimized release profile (no lag time, constant release rate)
(ii) characterization of the three-dimensional inner structure of risperidone-loaded MCs and MGs by using focused ion beam (FIB) preparation and three-dimensional X-ray imaging (nano-CT).

2. Materials and Methods

2.1. Materials

Expansorb DLG 75-7E, batch number C100010962, (Merck, Darmstadt, Germany) PVA 4-88 (Merck, Darmstadt, Germany) Risperidone 95+% (Activate Scientific, Shanghai, China), Pioneer MCT (middle chain triglycerides) batch number: 1132229 (Hansen & Rosenthal, Hamburg, Germany); isopropylmyristate Ph-Eur. batch number: 1407015-01, (Euro OTC Pharma GmbH, Bönen, Germany); dichloromethane 99.9% (Carl-Roth, Karlsruhe, Germany); hydroxystearic acid was a gift from Alberdingk Boley GmbH, (Krefeld, Germany). Risperdal Consta® 50 mg batch number: GCSK002,

(Janssen-Cilag GmbH, Neuss, Germany); 2.5 mL GASTIGHT® syringe (Hamilton Germany GmbH, Planegg, Martinsried, Germany), syringe pump (Pump 11 Elite, Harvard Apparatus, Holliston, MA, USA).

2.2. *Methods*

2.2.1. Preparation of PLGA Microcapsules (MC) and PLGA Microgels (MG)

Risperidone-loaded microcapsules were prepared using an oil-in-water (o/w) emulsion solvent extraction/evaporation technique.

For this purpose, 100 mg PLGA 75-7E, 100 mg of the oil phase (MCT or IPM), and 50 mg of Risperidone were weighted into a 4 mL glass vial and dissolved into the final volume of 2 mL dichloromethane by vortexing. The organic solutions were drawn up into a 2.5 mL GASTIGHT® syringe (Hamilton Germany GmbH, Planegg, Martinsried, Germany) and injected with a flow rate 5.2 µL/min by a syringe pump (Pump 11 Elite, Harvard Apparatus, Holliston, MA, USA) directly into 300 mL of aqueous PVA 4-88 (Merck, Darmstadt, Germany) solution 0.25% (*w/v*) stirred with a Rotating Paddle, USP at 240 rpm at room temperature. The aqueous phase contains double distilled and filtrated water with 0.25% polyvinyl alcohol PVA 4-88 (Merck, Darmstadt, Germany). When particle formation was completed, the suspensions were transferred into a 500 mL round bottom flask and the remaining DCM were removed under vacuum 40 mbar for 20 min. Thereafter, sedimented microcapsules were transferred by a glass pipette into a 15 mL tube and the supernatants were removed carefully. Then, the microcapsules were washed thrice with sterile filtrated double distilled water, filled up to the final volume of 1 mL and lyophilized.

2.2.2. Lyophilization

For freeze-drying, the samples were frozen rapidly with liquid nitrogen at −196 °C and lyophilised on a Christ Alpha 2–4 freeze dryer (Martin Christ Gefriertrocknungsanlagen GmbH, Osterode am Harz, Germany) in combination with a Vacuubrand RC 6 vacuum pump (Vacuubrand GmbH, Wertheim, Germany) for 24 h. The chamber was evacuated to 0.05 mbar, corresponding to −48 °C on the sublimation curve of ice. The prepared microcapsules were stored at −20 °C until further use.

2.2.3. Static Light Scattering Measurements

Routine measurements of the size distribution were carried out by laser diffractometry on a Mastersizer 2000 (Malvern Instruments, Malvern, United Kingdom) combined with a Hydro 2000 S wet dispersion unit. The samples were measured at a laser obscuration of 5%, corresponding to 11%–15% obscuration of the blue laser, in purified water. A series of five runs was evaluated by the Mastersizer 2000 software version 5.60, using the Mie theory, assuming spherical particles with a refractive index of 1.44, absorption of 0.001, and a refractive index of 1.33 for the dispersant as optical properties.

2.2.4. Powder XRD

XRD analysis was performed on a Bruker D8-Advance diffractometer, equipped with a one-dimensional silicon strip detector (LynxEye) operating with Cu Kα radiation. Diffraction was measured from $2\theta = 5°$ to 70° with a step size of 0.01° and a counting time of 1 s.

2.2.5. High-Performance Liquid Chromatography (HPLC)

A HPLC Agilent 1100 Series (Agilent, Santa Clara, CA, USA) was used. Separations were carried out using a BDS-Hypersil® C18 5 µm 250 × 4.6 column (Thermo Fisher Scientific, Waltham, MA, USA) and a mobile phase of 75% (*v/v*) methanol/25% (*v/v*) water, 1% (*v/v*) tetramethylammoniumhydroxid pentahydrate solution 25% *w/v*. Runs were carried out at 0.5 mL/min over 16 min and the absorption at 280 nm was recorded. The retention time of risperidone was 7.5 min. Injection volumes of 20 µL were used for determination of drug loading.

2.2.6. Drug Loading and Encapsulation Efficiency (EE)

Risperidone-loaded microcapsules were weighed and diluted in methanol to get a final concentration of 0.5mg/mL. All the samples were sonicated for 10 min to accelerate the dissolving of MC. Thereafter all samples were diluted 0.05 mg/mL and filtered (Minisart SRP 4, 0.45 μm syringe filter, Sartorius Stedim Biotech GmbH, Göttingen, Deutschland). The concentration of risperidone was determined via HPLC. Drug loading and EE were calculated as following equations from literature: Drug loading = (weight of drug entrapped/weight of MC used) × 100%; EE = (experimental drug loading/theoretically drug) × 100% [23].

2.2.7. In Vitro Drug Release Assay

The in vitro release of risperidone from MCs was carried out in a heated bath shaker (Memmert GmbH + Co. KG, Schwabach, Germany) at 37 °C and constant shaking at 50 rpm. A total of 6 mg of formulation I-IV and 3 mg of Risperdal Consta® were suspended in 1 mL of 10 mmol phosphate buffer saline (PBS) at pH 7.4 and transferred into a Spectra Por Float-A_Lyzer ® G2 1000kD. The float-A Lyzer was placed into a 100 mL measuring cylinder and filled up with PBS to the final volume of 60 mL so that the float-A Lyzer is submerged in PBS. To maintain sink conditions, the buffer solution was drawn off half (half change) when the risperidone concentration reaches 7%–8% of the solubility in PBS. Drug quantity was determined by UV-1800 spectrophotometer (Shimadzu, Kyoto, Japan). All measurements were performed in triplicate and the measurement points were represented as the mean ± SD.

2.2.8. Scanning Electron Microscopy and Focused Ion Beam Preparation

High resolution investigations of the microparticle morphology were done by scanning electron microscopy (SEM) using a Quanta 3D FEG (field emission gun) instrument from FEI Company, (Hillsboro, OH, USA). The microparticles were carefully placed on SEM sample holders using carbon tape and subsequently coated with a thin layer of platinum using a magnetron sputtering system (HVD, Dresden, Germany) to achieve a conducting surface. The SEM images were obtained under high vacuum conditions using an acceleration voltage of 5 keV, a working distance of 6–10 mm and electron beam currents of 12 pA determined with an Everhart–Thornley detector.

The microscope is designed as a dual beam (electron and ion beam) which gives the possibility to perform FIB preparations in the same device. Here, the gallium ion beam and the electron beam operate independently of each other. The point of coincidence of the two beams is located at a working distance of 10 mm. The angle between both beams is 52°. To allow vertical cutting with the FIB, the sample was tilted by this angle. Thus observation of the cross-sections with the electron beam was also done at an angle of 52°. After screening the samples by conventional SEM, the particle of interest was selected for cross-sectional preparation (target preparation). As the first step in cross-section preparation at predefined areas of interest coarse material was ablated with gallium ions accelerated at 30 kV with ion currents of 15 nA. After this coarse milling step lower beam currents of Afterwards, 1–0.5 nA were used to polish the cross sections. The samples were not cooled during FIB milling. The patterning conditions used conform to the standard patterning conditions for silicon materials. The SEM observation can be done during ion milling or after subsequent milling steps.

2.2.9. Nano-CT

X-ray imaging experiments were performed at the Institute of Physics, Martin Luther University Halle-Wittenberg, in a Carl Zeiss Xradia 810 Ultra (Cr source, 5.4 keV, for instrument details see [24]). Phase-contrast imaging mode was used with a phase-ring positioned near the back focal plane of the zone plate. In each experiment, a total of 901 projections with a field-of-view (FOV) of 65 μm were obtained over 180°, with an exposure time of 15 s per projection, a detector binning of 2, and voxel size of 128 nm. Image reconstruction was performed by filtered back-projection algorithm using the software integrated into the Xradia 810 Ultra. For each sample, two or three imaging experiments were performed, moving the sample vertically to enable the visualization of one entire microsphere in one single direction (from now on named Z-direction). For each sample's datasets, the brightness and contrast were adjusted in ImageJ® and the datasets were then stitched together using the Merge module of the Thermo Fisher Scientific Avizo® software (version 9.2, Thermo Fisher Scientific, Hillsboro, OR, USA). Median denoising and nonlocal mean filter were applied to the datasets. The 3D renderings presented here were created using either the commercial software arivis Vision4D® (version 2.12.6, arivis AG, Rostock, Germany) or Avizo®. For the dispersed internal spherical structures (ISS) evaluation, a volume located in the center of the 3D image was extracted consisting of 325 × 325 × Z voxels. Z varies among the samples, as it is almost equal to the number of pixels of the microsphere diameter, but is smaller than it to prevent voxels that do not consist of sample to be added to the calculation. The results obtained here represent the ratio of ISS compared to the total extracted volume of the sample (PLGA+ISS). For binarization of the datasets into PLGA and ISS, automatic thresholding with IsoData criterion was used, followed by island removal for the removal of small clusters of voxels (<15). A surface was generated from the segmented dataset in order to smooth sharp voxel edges and the specific surface area (SSA) values obtained represent the ratio of the constructed surface area (SA) to the volume (V) for each sample.

3. Results and Discussion

3.1. *Preparation of MCs and MGs*

For the preparation of MCs and MGs, the conventional oil-in-water (o/w) emulsion solvent evaporation method was used [2,25]. All components were dissolved in dichloromethane. The components of the four different formulations are listed in the Table 1. In general, the composition followed a 2:2:1 mass ratio of PLGA/Lipid/Risperidone. Therefore, the theoretical drug load was 20% [m/m]. High encapsulation efficiencies (>92%) were achieved, as seen in Table 1.

Table 1. Composition and encapsulation efficiency of Risperidone-loaded PLGA microcapsules (MC) and microgels (MG).

Sample	PLGA (mg)	Risperidone [mg]	MCT (mg)	IPM (mg)	HS (mg)	% Encapsul. Efficiency [+/−% SD]
MC-IPM	100	50	0	100	0	92.62 [0.94]
MG-IPM	100	50	0	90	10	100.82 [5.50]
MG-MCT	100	50	90	0	10	96.05 [3.98]
MC-MCT	100	50	100	0	0	92.79 [3.23]

MC = microcapsule, MG = microgel, PLGA = poly(lactide-*co*-glycolide), MCT = middle chain triglycerides, IPM = isopropylmyristate, HS = hydroxystearic acid.

To get reproducible particle sizes, it was important to keep all size-influencing parameters constant, e.g., the o-phase volume and solvent, the concentration and type of polymer, the volume of the continuous phase and the type and concentration of stabilizer, the temperature, and the stirring speed [26]. After lyophilization the MCs did not agglomerate and were resuspendable in water.

3.2. *Characterization*

3.2.1. Particle Size Measurements

The particle size and morphology of MC are key factors which potentially affect EE, drug release rate, and the biodistribution [27,28]. The optimal size range for microparticles with controlled release characteristics is considered to be 10–200 μm. Particles in the lower micron range will show a faster drug release, and they are also more easily phagocytosed by immune cells [29]. Sizes larger than

200 µm are difficult to inject and will show more inhomogeneous degradation [30]. The particle size distribution is shown in Figure 2.

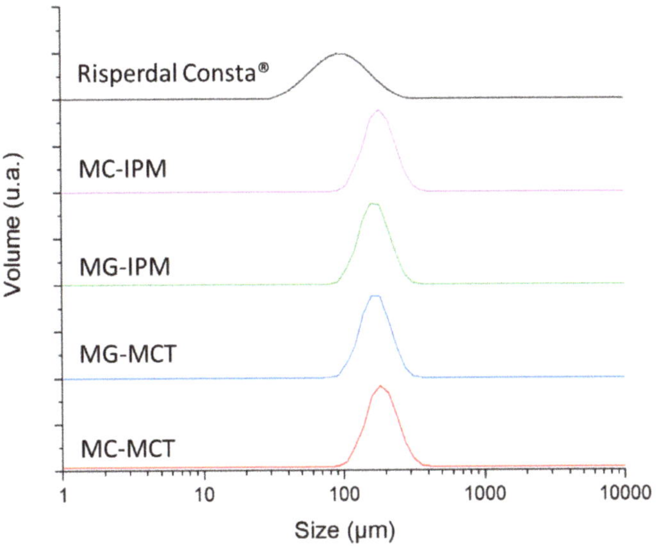

Figure 2. Volume-averaged particle sizes of microcapsules, microgels, and the reference product.

The particle size of the microcapsule formulations MC-IPM and MC-MCT are very similar ($D_{(0.5)}$ = 172.8 µm and 169.5 µm). The particle sizes for the corresponding microgels MG-IPM and MG-MCT were smaller compared to the microcapsules, but again very similar between IPM and MCT ($D_{(0.5)}$ = MG-IPM 157.2 µm; MG-MCT 155.9 µm), as seen Figure 2 and Table 2. It can be concluded, therefore, that the incorporation of hydroxystearic acid leads to slightly smaller particles, probably due to the impact on the surface tension during particle formation. The determined $D_{(0.5)}$ of the market product is 89.1 µm, which is in line with a size range between 25 and 150 µm reported in the literature [31]. The particle size distribution of the commercial product is considerably broader compared to our formulations (Span Risperdal Consta® 1.165 vs. span values between 0.612 and 0.637), as seen in Table 2.

Table 2. Particle size distributions determined by static light scattering.

Sample	D(0.1) (µm)	D(0.5) (µm)	D(0.9) (µm)	Mean $D_{(4,3)}$ (µm)	Span	Uniformity
Risperdal Consta®	50.7	89.1	154.5	96.9	1.165	0.36
MC-IPM	126.1	172.8	236.3	178.1	0.637	0.202
MG-IPM	116.4	157.2	212.6	161.2	0.612	0.193
MG-MCT	115.3	155.9	211.8	160.2	0.620	0.196
MC-MCT	123.7	169.5	230.8	174.4	0.631	0.197

3.2.2. Scanning Electron Microscopy and Focused Ion Beam Investigations

SEM images of Risperdal Consta® are shown in Figure 3. The market product Risperdal Consta® shows a high heterogeneity of particle size, shape, and surface structures. Surprisingly, structures showing partial engulfment, as seen in Figure 3d, and collapsed structures are visible, as seen in Figure 3c. Larger magnification of the surface indicated the presence of anisotropic material, which are most likely drug crystals, as seen in Figure 3e,f.

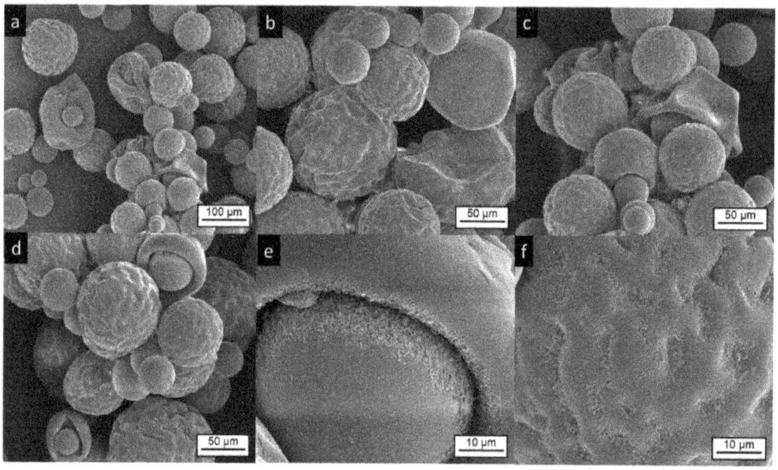

Figure 3. Electron microscopic pictures of Risperdal Consta®. The pictures at lower magnifications (**a–d**) show the presence of different particle sizes and shapes. Most particles show wrinkled surfaces; some have a smooth surface. High magnification shows the presence of small anisotropic material on the surface (**e**,**f**).

The microcapsules and microgels do not show the irregular structures of the commercial product. Most of MCT microcapsules are spherical, some show an elongated shape, which probably results from the coalescence of two particles, as seen in Figure 4a, white stars. The surface of the MC-MCT microcapsules shows a pattern similar to golf balls, as seen in Figure 4b,c. Very rarely, small anisotropic structures are found on the particle surface, as seen in Figure 4c, top middle, which most likely represent risperidone crystals. Cutting of the particles by focused ion beam (FIB) provides insights into the internal structure, as seen in Figure 4b,d. A multicore structure with dimensions in the lower micron range becomes visible. Within the internal structures, anisotropic shapes are visible which most likely present precipitated drug crystals, as seen in Figure 4d.

The electron microscopic pictures of the corresponding microgel MG-MCT shows spherical particles with a structured surface that appears smoother compared to the MC-MCT surface (Figure 5). In contrast to the MC-MCT microcapsules, we could not detect elongated shapes, as indicated by white stars in Figure 4a, which emerge from the coalescence of two particles. The pictures obtained by FIB cutting of the particles show internal multicore structures with dimensions in the lower micron range, as seen in Figure 5b,d.

IPM-loaded microcapsules (MC-IPM) have a spherical shape, as seen in Figure 6a. The surface appears similar to the MC-MCT capsules. Very rarely, small crystals can be detected on the particle surface, as seen in Figure 6c. After FIB cutting, multicore structures in the lower micron range become visible, as seen in Figure 6b,d.

The electron microscopy investigation on MG-IPM microgels shows the presence of spherical particles, as seen in Figure 7a. Higher magnifications of the particle surface show surface dips ("golf ball structure") similar to the MG-MCT microgels. In addition, elongated "hair-like" structures are visible, as seen in Figure 7c,d, which are not detectable in the other samples. Most likely, they are mainly composed of hydroxystearic acid, which is known to induce gel formation by elongated structures. Again, FIB cutting was used to explore the interior of the particles. Multicore microdomains became visible, as seen in Figure 7b, similar to the MG-MCT microgel particles with dimensions in the lower micron range.

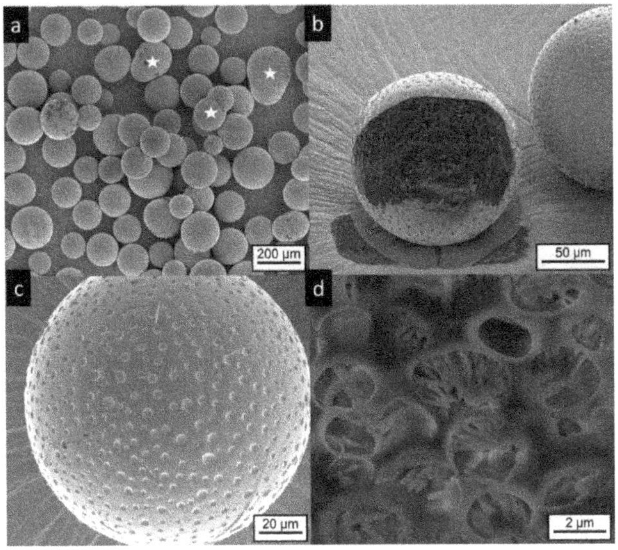

Figure 4. Electron microscopic pictures of MC-MCT microcapsules, (for composition details see Table 1). Intact particles are presented at lower (**a**,**c**) higher magnification. Focused ion beam (FIB) cut particles are shown at lower magnification in (**b**) and after deeper FIB milling and higher magnification in (**d**). The white stars in (**a**) indicate anisotropic particles, which were most likely formed by coalescence of two particles.

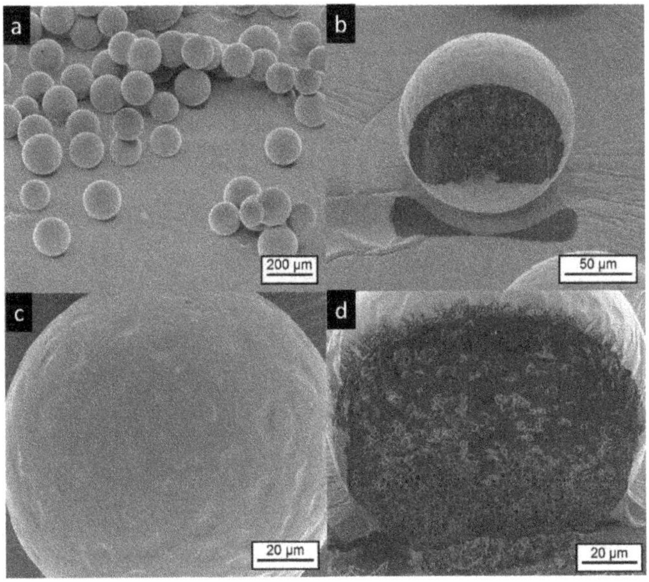

Figure 5. Electron microscopic pictures of MG-MCT microgels (for composition see Table 1). Intact particles are presented at lower (**a**,**c**) higher magnification. FIB cut particles are shown at lower magnification in (**b**) and at higher magnification at (**d**).

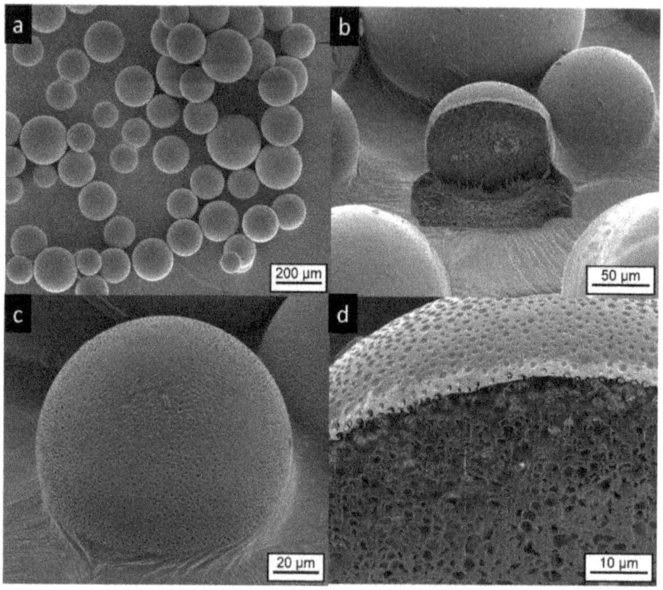

Figure 6. Electron microscopic pictures of MC-IPM microcapsules (composition details are given in Table 1). Intact particles are presented at lower (**a**,**c**) higher magnification. FIB cut particles are shown at lower magnification in (**b**) and at higher magnification at (**d**).

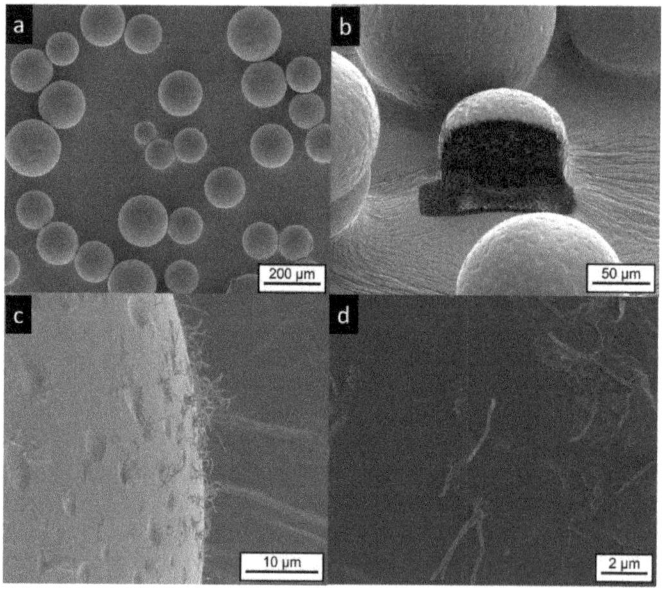

Figure 7. Electron microscopic pictures of MG-IPM microgels (sample composition is given in Table 1). Intact particles are presented at lower (**a**,**b**) and higher (**c**,**d**) magnifications. A FIB cut particle is shown in (**b**). Elongated, "worm like" structures (most likely precipitated hydroxyl-stearic acid) are visible in (**c**,**d**).

The crystallinity of the formulations was evaluated by X-ray diffraction. The diffractograms of PLGA, the pure drug risperidone, the commercial product, and the developed microcapsules and microgels are shown in Figure 8. It is known that the drug can exist in different polymorphs [32–34]. As expected, the polymer PLGA does not show sharp peaks, but only broad lines, indicating an amorphous state. The pure drug shows multiple sharp peaks. The X-ray diffractograms of the commercial product Risperdal Consta® and the developed formulations show a similar pattern, but they are different from the pure drug. From the measurements, it can be concluded that risperidone exists in all formulations at least partially in the crystalline state. The crystalline state in the formulation is different from the crystal structure of the parent drug, but very similar to the crystal form of risperidone, which has been observed after micronization with supercritical fluids [33].

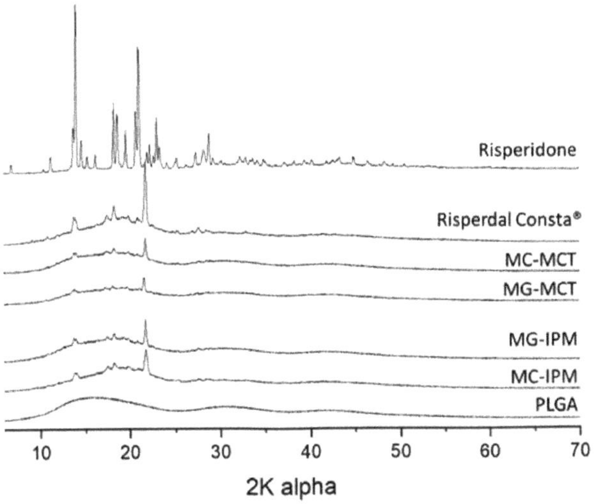

Figure 8. Wide angle X-ray diffractograms of microcapsules, microgels, Risperdal Consta®, and the pure drug risperidone.

4. X-Ray Nano-Imaging (Nano-CT)

X-ray imaging comprises a class of nondestructive techniques that are used to visualize the internal structure of different materials. In conventional X-ray imaging, differences in the X-ray absorption cross-section of different materials results in the image contrast [35,36]. Therefore, if the different parts composing one sample show only small differences in the X-ray attenuation, a poor image contrast will be obtained in conventional absorption-based X-ray imaging. This is specially the case for samples composed of elements with low electron density, making the characterization of the structures and visualization of the sample features a very hard task. Alternatively, phase-contrast imaging methods enable the visualization of features of weakly absorbing objects [35,37,38]. X-ray phase-contrast imaging is based on a different physical principle, and the contrast of the image is generated by the detection of the changes in the phase of the wavefront [36,39]. In this work, we imaged the different microspheres using Zernike phase-contrast imaging in a benchtop nano-CT scanner (Xradia Ultra 810). Phase-contrast nano-CT was used to characterize the inner structure of the commercial product Risperdal Consta® (Figure 9) and the developed microcapsules and microgels (Figure 10). Videos Supplementary Videos S1–S5 show the inner 3D structure of the microparticles. The diameter and volume distributions of the internal structures are presented in Supplementary Figures S1 and S2.

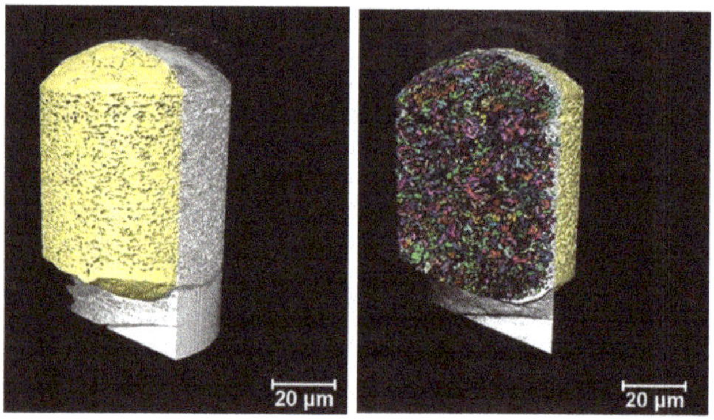

Figure 9. Volumetric representation the inner part of one reconstructed microsphere of Risperdal Consta® imaged with nano-CT. The left picture shows the X-ray signal in gray and the microsphere PLGA matrix pseudocolored in yellow. On the right, the small colored structures represent the dispersed internal spherical structures (ISS), which can be attributed to the dispersed drug particles. The volume of the ISS is about 13% of the total volume of the particle.

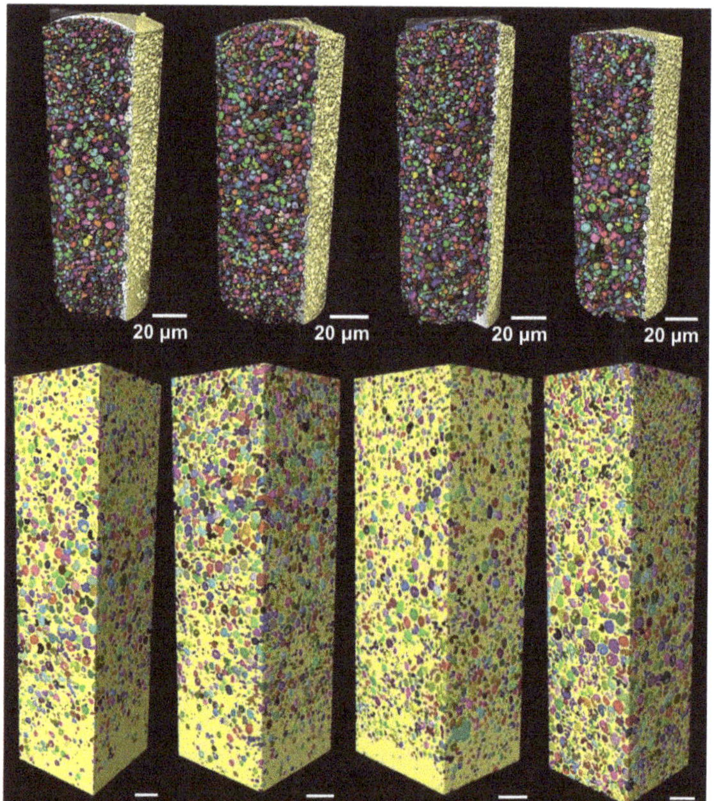

Figure 10. Volumetric representation of the inner part of the microspheres imaged with nano-CT. On the top from left to right, the MC-IPM, MG-IPM, MG-MCT, and MC-MCT microspheres pseudocolored in yellow are shown, with the dispersed internal spherical structures (ISS) represented in different colors. On the bottom, the volume used for the estimations done with nano-CT are shown for all samples.

All microgels and microcapsules show internal spherical structures (ISS) in the submicrometer range, with sphericity values close to unit. However, the volume percentage of the ISS in the microcapsules is between 35% and 46% and is, therefore, much higher than the volume percentage of the ISS phase in the Risperdal Consta® particles. The ISS phase can be attributed to the dispersed lipid phase for the microgels and microcapsules. In contrast, the ISS consists of risperidone only for the Risperdal Consta® system. A quantitative treatment of the 3D images indicates that the ISS have diameter around 2 μm for lipid-based systems and around 800 nm for the Risperdal Consta® product, as seen in Table 3. The ISS diameter and volume distributions are shown in Figures S1 and S2.

Table 3. Volume, diameter and sphericity of the internal dispersed phase (ISS), assessed by nano-CT.

Sample	Volume ISS	ISS Diameter	ISS Sphericity
Risperdal Consta®	13%	0.8 ± 0.2 μm	0.95 ± 0.02
MG-MCT	35%	2.0 ± 0.9 μm	0.8 ± 0.2
MC-MCT	46%	2.2 ± 0.8 μm	0.7 ± 0.1
MC-IPM	39%	1.9 ± 1.3 μm	0.8 ± 0.2
MG-IPM	43%	2.3 ± 1.0 μm	0.8 ± 0.3

Compared to other imaging techniques, Zernike phase-contrast nano-CT has the advantage of producing high-resolution three-dimensional images of low-absorbing samples, which enables the estimation of the values shown here. As with most high-resolution imaging techniques, nano-CT has the disadvantage of producing images limited to a small field of view, thus, to obtain the images of one single microparticle, it was necessary to image the sample two or three times and stitch the datasets. One of the benefits of phase-contrast nano-CT for imaging low-absorbing samples is that it does not require any special sample preparation procedure or induce any sample damage within the imaged area, in contrast with FIB-SEM. Overall, the results obtained by FIB-SEM correlate well with those obtained by nano-CT, and the same structures were observed. Both methods indicate the presence of a high percentage of internal structures (ISS) with sizes around two micrometers. Therefore, the microcapsules and microgels represent structures III and V of Figure 1, respectively.

5. Drug Release

The in vitro release of risperidone from the commercial product in the developed formulations is shown in Figure 11. For all formulations, a low percentage of the drug (<10%) is initially released. Most likely, drug molecules on the surface (which were observed by electron microscopy) are released very quickly. After the initial release, there is almost no release detectable for two weeks for Risperdal Consta®. Thereafter, the release accelerates. The developed formulations show different release profiles. The fastest release was observed for the MG-IPM microgel. The slowest release rate observed for the lipid formulations was for the MC-IPM microcapsules system. Therefore, it can be concluded that the oleogelator hydroxystearic acid had a strong impact on the release kinetics. The impact of the oleogelator hydroxystearic acid is also visible for the MCT systems, but to a smaller extent. Again, the microparticles with the incorporated oleogel (MG-MCT) release faster compared to the HS-free microcapsules (MC-MCT). The desired release profile was obtained with the MC-MCT microparticles: an almost linear release profile over two months with no lag time has been achieved.

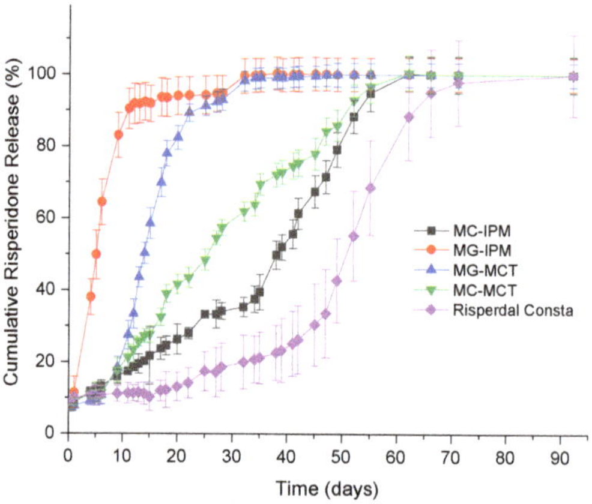

Figure 11. In vitro release profiles of Risperidone microcapsules (MC), microgels (MG), and Risperdal Consta®.

6. Conclusions

The results of our study show that lipid–PLGA hybrid systems are interesting alternatives to pure PLGA microparticles. The saturated liquid lipids MCT and IPM were incorporated in the form of microcapsules and microgels. Hydroxystearic acid (HS) was used as an oleogelator. The particle structure was investigated by FIB-SEM and nano-CT imaging. Both methods support the existence of a highly dispersed internal phase with spherical particle sizes of around 2 micrometers. The noninvasive nano-CT imaging enabled the production of 3D representations of the microparticles' structures in high-resolution images. After image processing and segmentation, it was possible to estimate the amount, size distribution, and sphericity of the internal structures. The presence of hydroxystearic acid accelerated the drug release. The drug release kinetics can be tuned by the lipid composition. An almost ideal release profile for risperidone was achieved with PLGA–MCT microcapsules (MC-MCT). In contrast to Risperdal Consta®, no lag time and an almost linear release over 60 days was observed. Therefore, the developed systems are suitable candidates to overcome the need for double medication (injected microparticles and tablets), which is currently needed due to the release lag time of Risperdal Consta®.

Supplementary Materials: The following are available online at http://www.mdpi.com/1999-4923/11/12/665/s1, Figure S1: Diameter distribution of the internal spherical structures (ISS) determined by nano-CT. Figure S2: Volume distribution of the internal structures (ISS) determined by nano-CT. Video S1: Video of the internal structure obtained by nano-CT for Risperdal Consta. Video S2: Video of the internal structure obtained by nano-CT for MC-IPM. Video S3: Video of the internal structure obtained by nano-CT for MG-IPM. Video S4: Video of the internal structure obtained by nano-CT for MG-MCT. Video S5: Video of the internal structure obtained by nano-CT for MC-MCT.

Author Contributions: Conceptualization, K.M., C.J., L.E.d.S., D.R., C.H., M.B.-B.; Particle manufacturing and drug release studies: C.J., L.E.d.S., K.M. Nano-CT measurements and data evaluation: J.M.d.S.eS. and C.S.d.O. F.I.B.-SEM: A.F., C.S.d.O. writing—original draft preparation, C.J., K.M.; writing—review and editing, C.J., A.F., J.M.d.S.eS., C.S.d.O., D.R., C.H., M.B.-B., C.H., C.E.H.S., supervision: K.M.

Funding: JMSS thanks DFG (WE 4051/21-1). This study was supported by funds from MilliporeSigma a Business of Merck KGaA, Darmstadt, Germany. CEHS acknowledges funding from the Fraunhofer Internal Program under Grant No. Attract 069-608203. The funding sponsors had no role in the design, execution, interpretation, or writing of the study.

Acknowledgments: The authors gratefully acknowledge Toni Buttler (Institute of Chemistry, Martin Luther University Halle-Wittenberg, Germany) for carrying out the powder-XRD measurements and Carl Zeiss for the loan of the phase-ring.

Conflicts of Interest: C.H. and M.B.-B. are employees of MilliporeSigma a Business of Merck KGaA, Darmstadt, Germany.

References

1. Jain, A.; Kunduru, K.R.; Basu, A.; Mizrahi, B.; Domb, A.J.; Khan, W. Injectable formulations of poly(lactic acid) and its copolymers in clinical use. *Adv. Drug Deliv. Rev.* **2016**, *107*, 213–227. [CrossRef] [PubMed]
2. Makadia, H.K.; Siegel, S.J. Poly Lactic-*co*-Glycolic Acid (PLGA) as biodegradable controlled drug delivery carrier. *Polymers* **2011**, *3*, 1377–1397. [CrossRef] [PubMed]
3. Bhat, P.V.; Goldstein, D.A. Dexamethasone intravitreal implant (allergan) for the treatment of noninfectious uveitis. *Expert Opin. Orphan Drugs* **2014**, *2*, 301–310. [CrossRef]
4. Hänggi, D.; Etminan, N.; Macdonald, R.L.L.; Steiger, H.J.H.J.; Mayer, S.A.S.A.; Aldrich, F.; Diringer, M.N.M.N.; Hoh, B.L.B.L.; Mocco, J.; Strange, P.; et al. NEWTON: Nimodipine microparticles to enhance recovery while reducing toxicity after subarachnoid hemorrhage. *Neurocrit. Care* **2015**, *23*, 274–284. [CrossRef]
5. Plontke, S.K.; Glien, A.; Rahne, T.; Mäder, K.; Salt, A.N. Controlled release dexamethasone implants in the round window niche for salvage treatment of idiopathic sudden sensorineural hearing loss. *Otol. Neurotol.* **2014**, *35*, 1168–1171. [CrossRef]
6. Mäder, K.; Lehner, E.; Liebau, A.; Plontke, S.K.S.K. Controlled drug release to the inner ear: Concepts, materials, mechanisms, and performance. *Hear. Res.* **2018**, *368*, 49–66. [CrossRef]
7. Citrome, L. Long-acting injectable antipsychotics update: Lengthening the dosing interval and expanding the diagnostic indications. *Expert Rev. Neurother.* **2017**, *17*, 1029–1043. [CrossRef]
8. Chou, Y.H.; Chu, P.C.; Wu, S.W.; Lee, J.C.; Lee, Y.H.; Sun, I.W.; Chang, C.L.; Huang, C.L.; Liu, I.C.; Tsai, C.F.; et al. A systemic review and experts' consensus for long-acting injectable antipsychot. *Clin. Psychopharmacol. Neurosci.* **2015**, *13*, 121–128. [CrossRef]
9. Available online: https://www.accessdata.fda.gov/drugsatfda_docs/label/2007/021346s015lbl.pdf (accessed on 4 December 2019).
10. Rawat, A.; Bhardwaj, U.; Burgess, D.J. Comparison of in vitro-in vivo release of Risperdal® Consta® microspheres. *Int. J. Pharm.* **2012**, *434*, 115–121. [CrossRef]
11. Shen, J.; Choi, S.; Qu, W.; Wang, Y.; Burgess, D.J. In vitro-in vivo correlation of parenteral risperidone polymeric microspheres. *J. Control. Release* **2015**, *218*, 2–12. [CrossRef]
12. Andhariya, J.V.; Shen, J.; Wang, Y.; Choi, S.; Burgess, D.J. Effect of minor manufacturing changes on stability of compositionally equivalent PLGA microspheres. *Int. J. Pharm.* **2019**, *566*, 532–540. [CrossRef] [PubMed]
13. Wu, Z.; Zhao, M.; Zhang, W.; Yang, Z.; Xu, S.; Shang, Q. Influence of drying processes on the structures, morphology and in vitro release profiles of risperidone-loaded PLGA microspheres. *J. Microencapsul.* **2019**, *36*, 21–31. [CrossRef] [PubMed]
14. Turek, A.; Borecka, A.; Janeczek, H.; Sobota, M.; Kasperczyk, J. Formulation of delivery systems with risperidone based on biodegradable terpolymers. *Int. J. Pharm.* **2018**, *548*, 159–172. [CrossRef] [PubMed]
15. Zhao, J.; Wang, L.; Fan, C.; Yu, K.; Liu, X.; Zhao, X.; Wang, D.; Liu, W.; Su, Z.; Sun, F.; et al. Development of near zero-order release PLGA-based microspheres of a novel antipsychotic. *Int. J. Pharm.* **2017**, *516*, 32–38. [CrossRef] [PubMed]
16. D'Souza, S.; Faraj, J.; Deluca, P.; D'Souza, S.; Faraj, J.; Deluca, P. Microsphere delivery of Risperidone as an alternative to combination therapy. *Eur. J. Pharm. Biopharm.* **2013**, *85*, 631–639. [CrossRef] [PubMed]
17. Vollrath, M.; Engert, J.; Winter, G. Long-term release and stability of pharmaceutical proteins delivered from solid lipid implants. *Eur. J. Pharm. Biopharm.* **2017**, *117*, 244–255. [CrossRef] [PubMed]
18. Fameau, A.-L.; Houinsou-Houssou, B.; Novales, B.; Navailles, L.; Nallet, F.; Douliez, J.-P. 12-Hydroxystearic acid lipid tubes under various experimental conditions. *J. Colloid Interface Sci.* **2010**, *341*, 38–47. [CrossRef]
19. Co, E.; Marangoni, A.G. The formation of a 12-hydroxystearic acid/vegetable oil organogel under shear and thermal fields. *J. Am. Oil Chem. Soc.* **2013**, *90*, 529–544. [CrossRef]
20. Mäder, K.; Windorf, M.; Kutza, J. Injizierbare Depotformulierungen zur Kontrollierten Wirkstofffreisetzung. WO2015062571, 29 October 2014.
21. Lampp, L.; Rogozhnikova, O.Y.; Trukhin, D.V.; Tormyshev, V.M.; Bowman, M.K.; Devasahayam, N.; Krishna, M.C.; Mäder, K.; Imming, P. A radical containing injectable in-situ-oleogel and emulgel for prolonged in-vivo oxygen measurements with CW EPR. *Free Radic. Biol. Med.* **2019**, *130*, 120–127. [CrossRef]
22. Garner, J.; Skidmore, S.; Park, H.; Park, K.; Choi, S.; Wang, Y. A protocol for assay of poly(lactide-*co*-glycolide) in clinical products. *Int. J. Pharm.* **2015**, *495*, 87–92. [CrossRef]

23. Hu, Z.; Liu, Y.; Yuan, W.; Wu, F.; Su, J.; Jin, T. Effect of bases with different solubility on the release behavior of risperidone loaded PLGA microspheres. *Colloids Surf. B Biointerfaces* **2011**, *86*, 206–211. [CrossRef] [PubMed]
24. Available online: https://www.zeiss.com/content/dam/Microscopy/us/download/pdf/technical-notes/x-ray-microscopy/en_44_013_027_tech-note_ultra-nanotomography.pdf (accessed on 4 October 2019).
25. O'Donnell, P.B.; McGinity, J.W. Preparation of microspheres by the solvent evaporation technique. *Adv. Drug Deliv. Rev.* **1997**, *28*, 25–42. [CrossRef]
26. Wischke, C.; Schwendeman, S.P. Principles of encapsulating hydrophobic drugs in PLA/PLGA microparticles. *Int. J. Pharm.* **2008**, *364*, 298–327. [CrossRef] [PubMed]
27. Han, F.Y.; Thurecht, K.J.; Whittaker, A.K.; Smith, M.T. Bioerodable PLGA-based microparticles for producing sustained-release drug formulations and strategies for improving drug loading. *Front. Pharmacol.* **2016**, *7*, 185. [CrossRef] [PubMed]
28. Barrow, W.W. Microsphere technology for chemotherapy of mycobacterial infections. *Curr. Pharm. Des.* **2004**, *10*, 3275–3284. [CrossRef] [PubMed]
29. Dawes, G.J.S.; Fratila-Apachitei, L.E.; Mulia, K.; Apachitei, I.; Witkamp, G.J.; Duszczyk, J. Size effect of PLGA spheres on drug loading efficiency and release profiles. *J. Mater. Sci. Mater. Med.* **2009**, *20*, 1089–1094. [CrossRef]
30. Anderson, J.M.; Shive, M.S. Biodegradation and biocompatibility of PLA and PLGA microspheres. *Adv. Drug Deliv. Rev.* **2012**, *64*, 72–82. [CrossRef]
31. D'Souza, S.; Faraj, J.A.; Giovagnoli, S.; DeLuca, P.P. Development of risperidone PLGA microspheres. *J. Drug Deliv.* **2014**, 620464. [CrossRef]
32. Krochmal, B.; Diller, D.; Dolitzky, B.-Z.; Aronhime, J. Preparation of Risperidone. WO2002012200A9, 8 August 2001.
33. Bagratashvili, V.N.; Bogorodskiy, S.E.; Egorov, A.M.; Krotova, L.I.; Mironov, A.V.; Parenago, O.O.; Pokrovskiy, O.I.; Ustinovich, K.B.; Chizhov, P.S.; Prokopchuk, D.I.; et al. Polymorphism of risperidone in supercritical fluid processes of micronization and encapsulation into aliphatic polyesters. *Russ. J. Phys. Chem. B* **2017**, *11*, 1163–1172. [CrossRef]
34. Karabas, I.; Orkoula, M.G.; Kontoyannis, C.G. Analysis and stability of polymorphs in tablets: The case of Risperidone. *Talanta* **2007**, *71*, 1382–1386. [CrossRef]
35. Pfeiffer, F.; Weitkamp, T.; Bunk, O.; David, C. Phase retrieval and differential phase-contrast imaging with low-brilliance X-ray sources. *Nat. Phys.* **2006**, *2*, 258–261. [CrossRef]
36. Kaulich, B.; Thibault, P.; Gianoncelli, A.; Kiskinova, M. Transmission and emission x-ray microscopy: Operation modes, contrast mechanisms and applications. *J. Phys. Condens. Matter* **2011**, *23*, 083002. [CrossRef] [PubMed]
37. Davis, T.J.; Gao, D.; Gureyev, T.E.; Stevenson, A.W.; Wilkins, S.W. Phase-contrast imaging of weakly absorbing materials using hard X-rays. *Nature* **1995**, *373*, 595–598. [CrossRef]
38. Núñez, J.A.; Goring, A.; Hesse, E.; Thurner, P.J.; Schneider, P.; Clarkin, C.E. Simultaneous visualisation of calcified bone microstructure and intracortical vasculature using synchrotron X-ray phase contrast-enhanced tomography. *Sci. Rep.* **2017**, *7*, 13289. [CrossRef] [PubMed]
39. Olivo, A.; Castelli, E. X-ray phase contrast imaging: From synchrotrons to conventional sources. *Riv. Nuovo Cim.* **2014**, *37*, 467–508.

pharmaceutics

MDPI

Article

Dry Tablet Formulation of PLGA Nanoparticles with a Preocular Applicator for Topical Drug Delivery to the Eye

Woo Mi Ryu [1,†], Se-Na Kim [2,†], Chang Hee Min [1] and Young Bin Choy [1,2,3,*]

1 Interdisciplinary Program in Bioengineering, College of Engineering, Seoul National University, Seoul 08826, Korea; wmryu@snu.ac.kr (W.M.R.); ddf5209@snu.ac.kr (C.H.M.)
2 Institute of Medical & Biological Engineering, Medical Research Center, Seoul National University, Seoul 03080, Korea; ksn777@snu.ac.kr
3 Department of Biomedical Engineering, Seoul National University College of Medicine, Seoul 03080, Korea
* Correspondence: ybchoy@snu.ac.kr; Tel.: +82-2-740-8592
† These authors contributed equally to this work.

Received: 25 October 2019; Accepted: 2 December 2019; Published: 4 December 2019

Abstract: To enhance ocular drug bioavailability, a rapidly dissolving dry tablet containing alginate and drug-loaded poly(lactic-*co*-glycolic acid) (PLGA) nanoparticles was proposed. For hygienic and easy administration of an accurate drug-dose with this tablet, the use of a preocular applicator was suggested. Herein, a dry tablet was prepared by embedding dexamethasone-loaded PLGA nanoparticles in alginate, which was deposited on the tip of the applicator. The nanoparticles were loaded with 85.45 µg/mg drug and exhibited sustained drug release for 10 h. To evaluate in vivo efficacy, dexamethasone concentration in the aqueous humor was measured after topical administration of the dry tablet, with the applicator, to rabbit eyes and was compared to that achieved with Maxidex®, a commercially-available dexamethasone eye drops. When applied with the preocular applicator, the dry tablet containing alginate could be fully detached and delivered to the eye surface. In fact, it showed up to 2 h of nanoparticle retention on the preocular surface due to tear viscosity enhancement, causing an estimated 2.6-fold increase in ocular drug bioavailability compared to Maxidex®. Therefore, the preocular applicator combined with a dry alginate tablet containing PLGA nanoparticles can be a promising system for aseptically delivering an accurate dose of ophthalmic drug with enhanced bioavailability.

Keywords: PLGA; nanoparticles; alginate; ophthalmic drug delivery; dexamethasone

1. Introduction

Topical drug administration is an easy route for ocular drug delivery. However, an ophthalmic drug, prepared as a solution and suspension, is known to clear very rapidly from the eye surface (<3 min), leading to a short preocular residence time, ultimately limiting drug bioavailability [1,2]. Less than 5% of a topically-applied drug can reach the interior target tissue of the eye [3,4]. Micro- or nano-particles have thus been suggested as ophthalmic drug carriers to achieve prolonged preocular residence time as well as sustained drug release [5]. Nevertheless, when formulated in the aqueous suspension, an additional fluid added to the eye would accelerate tear clearance, which would also expedite the drainage of the drug-loaded particles [6]. Therefore, an alternative strategy has been suggested and it includes a dry tablet formulation, where the drug-loaded particles are embedded in a tablet medium of a rapidly-dissolving polymer [7]. When applied to the eye surface, the tablet medium would dissolve in the tear fluid to free the drug-loaded particles on the preocular surface. The dissolution of the tablet medium would thus increase tear viscosity to delay tear clearance, which in turn would improve the preocular retention of drug-loaded particles. However, applying a dry tablet to the sensitive eye may not be convenient for patients. In addition, this administration route might not be hygienic, as the tablet would be retrieved with the bare fingers [8,9].

Herein, a combined system of a preocular applicator and a rapidly-dissolving dry tablet containing drug-loaded nanoparticles was proposed. The applicator was designed to contain two compartments (i.e., a handle plus a tip to hold the dry tablet). Thus, for administration, a dry tablet on the tip could approach and touch the eye surface while the handle is held by one's fingers. The tablet would then be separated from the tip and its medium would be dissolved in the tear fluid to free the drug-loaded nanoparticles. This would provide an easy way to topically administer the dry tablet and importantly, allow tablets to remain aseptic during administration.

To test the system and strategy, a preocular applicator using the biocompatible polydimethylsiloxane (PDMS) was prepared in this study. The shape and dimension of the applicator herein was designed to be similar to that of the commercially-available, one-time use applicator of artificial tear fluid based on patients' familiarity [10]. The commercially-available applicator is basically a container of liquid; however, a dry tablet formulation was loaded at the tip of the applicator proposed in this work. For the formulation, poly(lactic-*co*-glycolic acid) (PLGA) nanoparticles were prepared to be loaded with dexamethasone, a widely used corticosteroid drug for the treatment of eye inflammation [11,12], to achieve sustained drug release. The nanoparticles were then suspended in a solution containing a mixture of water-soluble polymers, polyvinyl alcohol (PVA), and alginate, which was then freeze-dried on top of the applicator tip to produce a rapidly-dissolving dry tablet. PLGA has been widely used for sustained drug delivery due to its high biocompatibility and degradability to non-toxic by-products [13,14]. PVA has already been approved for clinical use in various ophthalmic applications [15,16]. In the present study, alginate was also employed in the tablet medium as a viscosity enhancer of tear fluids. Alginate is a highly-biocompatible polymer [17] that can form a gel in the presence of the multivalent cation, Ca^{2+}, which is abundant in the tear fluid [18,19]. Due to this unique property, when dissolved in the tear fluid, alginate can further increase tear viscosity, thereby synergistically improving the preocular retention of the drug-loaded nanoparticles.

Dexamethasone-loaded PLGA nanoparticles (DX/NP) were prepared by the solid-in-oil-in-water (S/O/W) emulsion method and characterized by scanning electron microscopy (SEM) and dynamic light scattering (DLS), which were employed to assess nanoparticle size and morphology. The cytotoxicity of DX/NP was examined using human corneal epithelial cells (HCECs). To examine the effect of alginate on tear viscosity enhancement, two distinct tablet formulations containing DX/NP were prepared with media containing both PVA and alginate (DX/NP AL_TAB) and PVA alone (i.e., without alginate) (DX/NP TAB), respectively. For in vivo experiments, these tablet formulations were applied to rabbit eyes, and drug concentration in the aqueous humor was assessed and compared to the commercially-available ophthalmic dexamethasone medication, Maxidex®, (i.e., an aqueous suspension of dexamethasone itself).

2. Materials and Methods

2.1. Materials

PLGA (lactic acid:glycolic acid = 50:50; i.v. = 48,000) was purchased from Evonik Industry (Essen, Germany). Dexamethasone and Nile red were purchased from Tokyo Chemical Industry (Tokyo, Japan). PVA (87–89% hydrolyzed), phosphate buffered saline (PBS) tablets, tween 80, trifluoroacetic acid (TFA; >99%), and calcium chloride dihydrate (>99%) were obtained from Merck (St. Louis, MO, USA). Dichloromethane (DCM; >99.5%), *N*,*N*-dimethylformamide (DMF; >99.5%), and acetone (>99.5%) were supplied by DaeJung (Siheung-si, Korea). Acetonitrile (ACN; >99.9%) was purchased from J.T. Bakers (Phillisburg, NJ, USA). PDMS (Sylgard 184) was obtained from Sewang Hitech Silicone (Bucheon-si, Korea). Alcaine® (0.5% proparacaine hydrochloride ophthalmic solution) and Maxidex®

(0.1% dexamethasone ophthalmic suspension) were purchased from Alcon-Couvreur (Fort Worth, TX, USA).

2.2. Preparation of Drug-Loaded Nanoparticles

DX/NP was prepared via S/O/W emulsification [20]. Briefly, 300 mg PLGA and 100 mg dexamethasone were dissolved in 7 mL DCM. The resulting solution was then added to 8 mL of 1% *w/v* PVA solution, which was emulsified with a sonicator (Sonic Dismembrator Model 500, Fisher Scientific, Illkirch-Graffenstaden, France) at 160 W for 10 min. The emulsion was transferred into 100 mL of 1% *w/v* PVA solution at room temperature and stirred at 400 rpm under vacuum (−10 psi) for 60 min for solvent evaporation. The final suspension was then centrifuged at 200 rpm for 10 min, where the precipitates were eliminated and only the suspension at the top was collected. The collected suspension was then washed with deionized (DI) water three times via centrifugation at 13,000 rpm for 10 min. Thereafter, the suspension was lyophilized for 20 h to obtain dry DX/NP. PLGA nanoparticles loaded with Nile red (NR/NP) instead of dexamethasone were also prepared to evaluate its in vivo preocular retention properties. As such, 5 mg Nile Red and 500 mg PLGA were dissolved in 7 mL DCM for emulsification.

2.3. Preparation of Tablet-Loaded Preocular Applicators

A preocular applicator with a dry tablet on the tip was prepared, as depicted in Figure 1. To prepare a preocular applicator, two different molds were first made to fabricate a handle and tip (Figure 1a). In the molds, a mixture of a PDMS base and its curing agent (10:1, *v/v*) was poured and cured slightly at 80 °C for 2 h. Then, each of the constituent pieces was assembled and cured further at 80 °C for another 2 h to allow for their bonding. In this work, two different tablets, with media containing both PVA and alginate, and PVA alone, were produced to yield DX/NP AL_TAB and DX/NP TAB, respectively. To prepare the DX/NP AL_TAB, 20 mg/mL DX/NP were suspended in a solution containing 0.1% *w/v* PVA and 2% *w/v* alginate. Above this alginate concentration, the solution became too viscous to properly distribute the DX/NP. To prepare the DX/NP TAB, 20 mg/mL DX/NP were suspended in a solution of 0.1% *w/v* PVA only. Twenty µL of the resulting suspension was then poured in the reservoir that was made by tightly fitting a cylindrical connector with open ends to a tip of the applicator (Figure 1b). The whole piece was then rapidly frozen using liquid nitrogen, which was then lyophilized for 5 h at 0.01 mbar with the collector temperature set at −80 °C (FreeZone 6 Dryer system, Labconco, Kansas City, MO, USA) [21–23]. After that, a connector was carefully removed to produce the preocular applicator loaded with the tablet formulation. To evaluate the in vivo preocular retention properties, NR/NP was embedded instead of DX/NP to produce tablets of NR/NP AL_TAB and NR/NP TAB, prepared under the same condition employed for DX/NP AL_TAB and DX/NP TAB, respectively.

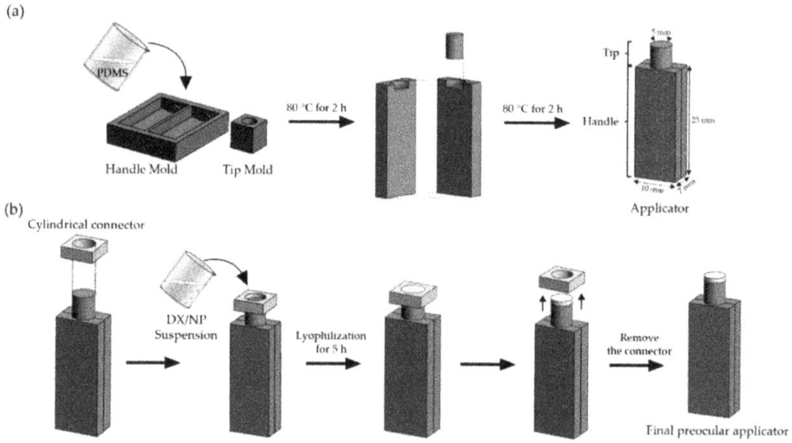

Figure 1. Schematic illustration of the fabrication of the DX/NP tablet-loaded applicator. (**a**) Fabrication of the PDMS applicator. (**b**) Tablet preparation on the tip of the applicator.

2.4. Characterization

The morphology of DX/NP was examined by SEM (JSM-7800F Prime, JEOL, Tokyo, Japan). Its size distribution was determined using DLS (ELS-2000ZS, Otsuka Electronics, Osaka, Japan) and a zetasizer (Nano ZS, Malvern, UK) with a particle suspension prepared in DI water [24]. To measure the drug loading amount, the solution of dexamethasone that was fully extracted from the DX/NP and tablet was respectively prepared. Briefly, 4 mg of DX/NP was completely dissolved in 4 mL DMF. The tablet at the applicator tip was fully immersed in 1 mL DMF, sonicated for 2 h, and centrifuged at 13,000 rpm. The supernatant was then diluted with ACN in a 1:1 ratio. Drug concentration in each resulting solution was measured using high-performance liquid chromatography (HPLC; Agilent 1260 series, Agilent Technologies, Santa Clara, CA, USA) with a Diamonsil C_{18} column (5 μm, 150 × 4.6 mm). Column temperature and absorbance wavelength were set at 37 °C and 240 nm, respectively. Injection volume and flow rate were 20 μL and 1.5 mL/min, respectively. The mobile phase was comprised of 0.1% TFA and ACN mixed in a 65:35 ratio. Given with the loading amount of the DX/NP, the encapsulation efficiency (EE) was calculated by the equation.

$$\text{EE}\ (\%) = \frac{\text{Amount of drug loaded in nanoparticles}}{\text{Initial amount of drug}} \times 100 \tag{1}$$

The thicknesses of the tablets herein were measured using a caliper (ABSOLUTE Digimatic Caliper, Mitutoyo, Kanagawa, Japan). To assess the particle distribution in the tablet, the NR/NP TAB and NR/NP AL_TAB were imaged with a fluorescence microscope (Leica DMI4000 B, Leica Microsystems, Wetzlar, Germany). To examine the effect of the tablet medium, DX/NP AL_TAB and DX/NP TAB were each immersed in 1 mL of the medium mimicking tear fluid, i.e., PBS containing Ca^{2+} (10 mM, pH 7.4, [Ca^{2+}] = 39.4 μg/mL) [25], for 10 min at 37 °C. The viscosity of the resulting solution was then measured using a rheometer (Advanced Rheometric Expansion System, Rheometric Scientific, New Castle, DE, USA), where gap separation, temperature, and shear rate were set at 0.8 mm, 25 °C, and 100 s^{-1}, respectively.

2.5. In Vitro Drug Release Study

To examine the in vitro drug release profiles, 400 μg DX/NP, two different tablets containing the DX/NP (i.e., DX/NP TAB and DX/NP AL_TAB) and Maxidex®, all of which contained the same amount of about 35 μg DX were each placed in a dialysis membrane bag (SnakeSkin™ Dialysis Tubing, 10 kDa,

Thermo Scientific, USA). The bag was then immersed in 5 mL pH 7.4 PBS containing 39.4 µg/mL Ca^{2+} and 0.5% *w/v* Tween 80 to meet the sink condition of DX [26]. While being incubated at 37 °C, at scheduled times, 1 mL of the release medium was collected and the same volume of fresh PBS was added back. The amount of released dexamethasone was measured using HPLC as described above.

2.6. Cytotoxicity Evaluation

The in vitro cytotoxicity of DX/NP was evaluated using HCECs (PCS-700-010, ATCC, Manassas, VA, USA). HCECs were grown in a corneal epithelial cell basal medium (PCS-700-030, ATCC, USA) with supplements (PCS-700-040, ATCC) at 37 °C in a humidified environment with 5% CO_2. Prior to the assay, HCECs were seeded in a 96-well plate at 1×10^5 cells/well and grown for 24 h. Subsequently, 100 µL of the DX/NP suspension, which was prepared in the cell growth medium at concentrations of 5, 10, 25, 50, 100, 250, 500, and 1000 µg/mL, was added to each well and incubated at 37 °C for 24 h. The medium was then completely removed and replaced with 100 µL of fresh medium. Thereafter, 10 µL of an EZ-Cytox solution was added to each well and incubated at 37 °C for 2 h under dark conditions. Cell viability was measured using a microplate reader, with absorbance and reference wavelengths of 450 nm and 600 nm, respectively (VersaMax ELISA Microplate Reader; Molecular Devices, San Jose, CA, USA).

2.7. Animal Experiments

In vivo experiments were conducted with the healthy eyes of male New Zealand White rabbits (weight 2.1–2.5 kg). Rabbits were granted free access to food and water and were housed in a controlled environment: temperature; 21 ± 1 °C, humidity; 55 ± 1%, and light/dark cycle; 12 h/12 h. The in vivo experimental protocols were approved by the Institutional Animal Care and Use Committee at Seoul National University Hospital (IACUC No. 19–0133, date: 30 July 2019).

First, the in vivo preocular retention properties of the nanoparticles were evaluated after topical administration to the eye. The preocular retention properties were assessed as reported in our previous studies, with slight modifications [27,28]. Briefly, the NR/NP AL_TAB or NR/NP TAB on the applicator tip was applied on the lower cul-de-sac of the left eye of rabbits. At scheduled times, the rabbit eye was anesthetized by topical administration of a drop of Alcaine® and the entire preocular surface was thoroughly wiped with a surgical sponge (PVA spear; Sidapharm, Thessaloniki, Greece) to collect the NR/NP. The sponge was then fully immersed in 5 mL DMF and sonicated for 2 h to fully dissolve the NR/NP. The amount of Nile red in the sample was measured using HPLC-mass spectroscopy (LC-MS) with a Polaris 5 C_{18}-A (2.7 µm pore size, 4.6 × 150 mm) and the following conditions: column temperature, 30 °C; absorbance wavelength, 243 nm; injection volume, 20 µL; and flow rate, 0.45 mL/min. The mobile phase was prepared by mixing 0.1% formic acid and ACN in the ratio, 55:45. For statistics, four animals (i.e., one left eye for each rabbit) were assigned per time point for each formulation. With the NR/NP AL_TAB and NR/NP TAB, the in vivo profile of tablet disintegration was also assessed. For this, the eye was imaged using a digital camera (Galaxy S10, Samsung, Seoul, Korea) at 0 and 30 s after topical administration of the tablet to rabbit eyes.

To assess the in vivo ocular drug bioavailability, each of the three formulations (i.e., the DX/NP AL_TAB and DX/NP TAB in the applicator, and 35 µL Maxidex®) with the same dose of dexamethasone (ca. 35 µg dexamethasone) was administered onto the lower cul-de-sac of rabbit eyes. At scheduled times, the rabbit was anesthetized with a subcutaneous injection of a cocktail containing 20 mg/kg ketamine and 10 mg/kg xylazine. Thereafter, approximately 100 µL of aqueous humor (AH) was collected using a 31 G needle (BD Ultra-Fine II, Becton Dickinson and Company, Franklin Lakes, NJ, USA). Drug concentration in AH was analyzed by LC-MS as described above. For statistics, three animals (i.e., the left eye of each rabbit) were assigned per time point for each formulation.

2.8. *Statistical Analysis*

Statistical analysis was performed using the amount of particles remaining on the preocular surface and drug concentration in AH by the Mann–Whitney U-test. A p-value < 0.05 was considered to indicate statistical significance (SPSS version 22, IBM, Armonk, NY, USA).

3. Results

3.1. *Characterization of the Formulations*

DX/NP was prepared using the emulsion method. As a result, the drug-loaded nanoparticles were found to display a spherical shape, as shown in Figure 2a. Particle diameter was 336.92 ± 5.56 nm (Figure 2b) while drug-loading amount was 85.45 ± 5.44 µg/mg (i.e., EE = c.a. 25.6%). The size and shape of NR/NP were similar to those of DX/NP (Figure S1 in the Supplementary Information). As shown in Figure 2c, the zeta potential of the blank PLGA nanoparticles was measured to be −3.9 mV, which was shifted to −27.2 mV with the DX/NP due to a negative charge of the encapsulated drug, DX [29]. Figure 3a shows the tablets prepared in the present experiment (i.e., the DX/NP TAB and DX/NP AL_TAB). They displayed a cylindrical shape and were well attached at the tip applicator. As shown in Figure 3b, the tablets loaded with the NR/NP exhibited a evenly-distributed fluorescent signal throughout the medium, implying a homogenous distribution of the particles in the tablet. As the same amount of DX/NP was embedded during tablet preparation, the drug loading amount per tablet was similar between DX/NP TAB and DX/NP AL_TAB (34.11 ± 0.48 and 34.89 ± 0.28 µg, respectively). For the same reason, the thicknesses of the DX/NP TAB and DX/NP AL_TAB were also similar, which were measured to be 1.02 ± 0.03 and 1.02 ± 0.06 mm, respectively. When tablets were dissolved in Ca^{2+}-containing PBS, the solution with DX/NP AL_TAB (i.e., the tablet containing 400 µg alginate) had a higher viscosity of 0.93 Pa s than that with DX/NP TAB (i.e., the tablet without alginate) of 0.01 Pa s. However, both tablets herein were observed to be disintegrated in 30 s by dissolution in tear fluid when topically administered to rabbit eyes (Figure 3c).

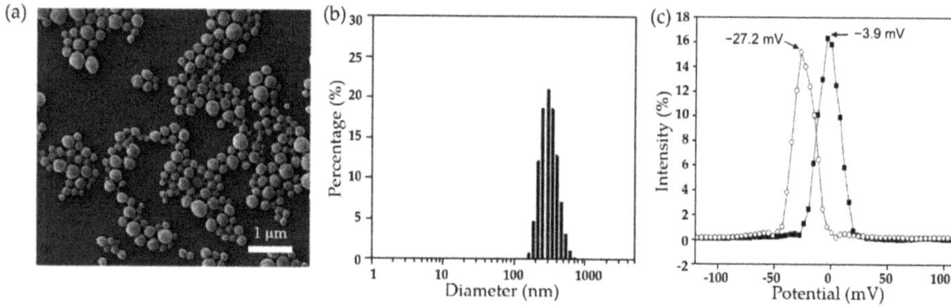

Figure 2. Characterization of DX/NP. (**a**) Representative scanning electron micrographs of DX/NP. (**b**) Particle size distribution of DX/NP measured by the dynamic light scattering (DLS) method; Polydispersity index = 0.056. (**c**) Zeta potentials of the blank PLGA nanoparticles (■) and DX/NP (○).

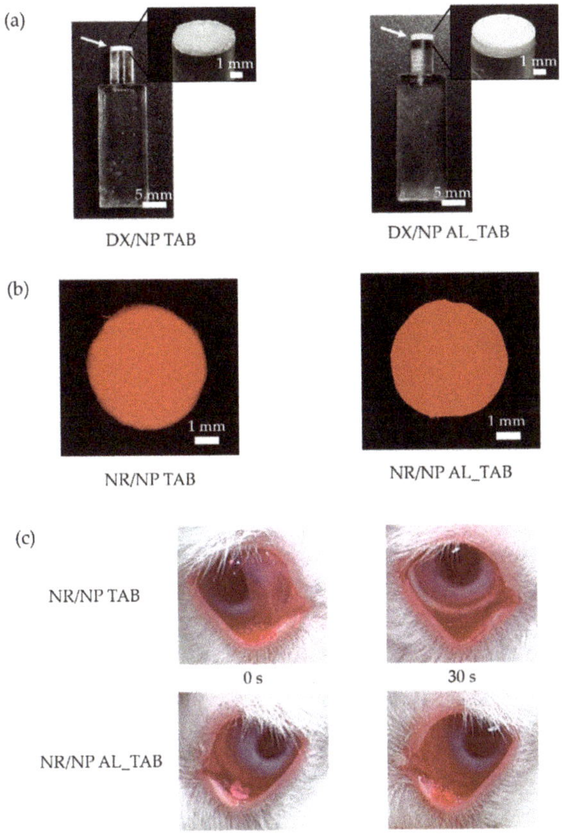

Figure 3. Characterizations of tablets. (**a**) Representative optical images of the applicator with the DX/NP TAB and DX/NP AL_TAB. White arrows indicate the dry tablets prepared on the applicator tip. (**b**) Representative fluorescence images of the NR/NP TAB and NR/NP AL_TAB. (**c**) Disintegration profiles of the NR/NP TAB and NR/NP AL_TAB administered topically to rabbit eyes.

3.2. *In Vitro Evaluation Results*

Figure 4 shows the in vitro drug release profiles of DX/NP and the tablet formulations, which displayed sustained drug release for up to 10 h. The release profiles did not differ among the formulations as the tablet medium dissolved rapidly to achieve almost instantaneous freeing of DX/NP. Drug release was slightly more suppressed with DX/NP AL_TAB, which could be due to a slight increase in the viscosity of the release medium in a dialysis membrane bag owing to the interaction between algnate and Ca^{2+}. On the other hand, Maxidex®, which is basically a suspension of DX, exhibited almost complete dissolution of the drug in 2 h. When tested with HCECs, the DX/NP exhibited a cell viability greater than 90% at all testing concentrations (Figure 5), suggesting that the DX/NP was non-cytotoxic.

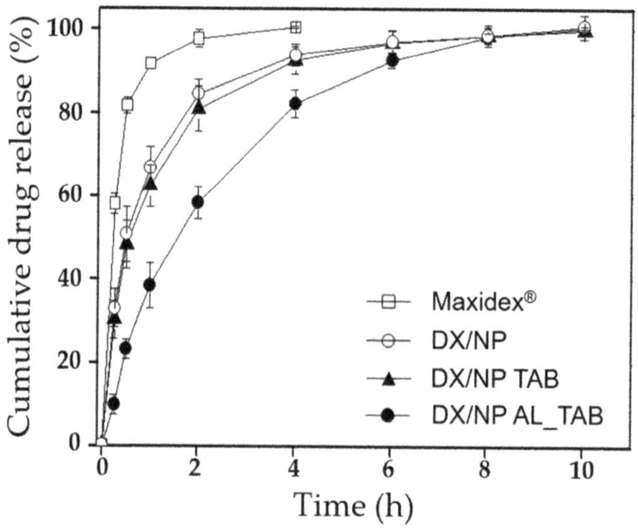

Figure 4. In vitro drug release profiles of Maxidex® (□), DX/NP (○), DX/NP TAB (▲), and DX/NP AL_TAB (●).

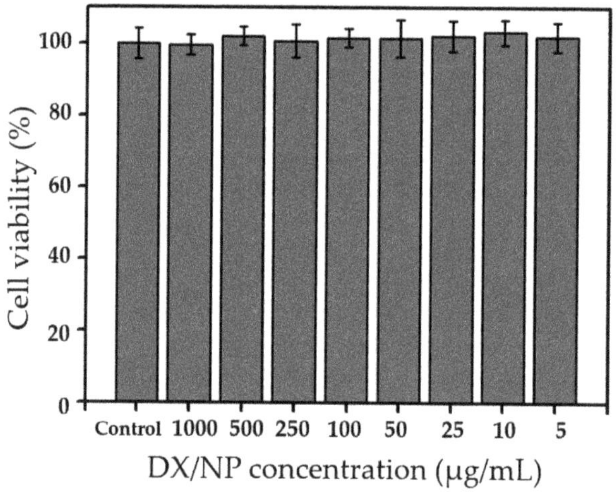

Figure 5. In vitro cytotoxicity of DX/NP on human primary corneal epithelial cells.

3.3. *In Vivo Experimental Results*

To examine the effect of alginate on preocular retention of the nanoparticles, two different tablets embedded with the Nile red-loaded nanoparticles (i.e., NR/NP TAB and NR/NP AL_TAB) were employed for in vivo evaluation. As shown in Figure 6, by incorporating alginate in the tablet, the amount of nanoparticles remaining at the preocular surface increased. For the tablet without alginate, the average percent of remaining particles was relatively low (i.e., 4.8 and 0.2% at 15 min and 1 h, respectively). However, owing to the presence of alginate in the tablet, 33 and 12% of particles remained at 15 min and 1 h, respectively, and could still be detected until 2 h. As conventional eye drops are known to completely disappear in 5 min [1], such findings suggest a greater improvement in drug retention at the eye surface. For the tablet with alginate, there was a high variability in particle retention observed at 1 h. Right after the tablet medium dissolved in the tear fluid to interact with Ca^{2+}

ions, the tear viscosity increased to allow for an evident retention of the NP at the preocular surface. However, as the time progressed, the tear fluid with alginate would be continuously diluted with a newly-generated fresh tear, which would vary greatly depending on the subject. After 2 h, almost all particles would be cleared from the preocular surface.

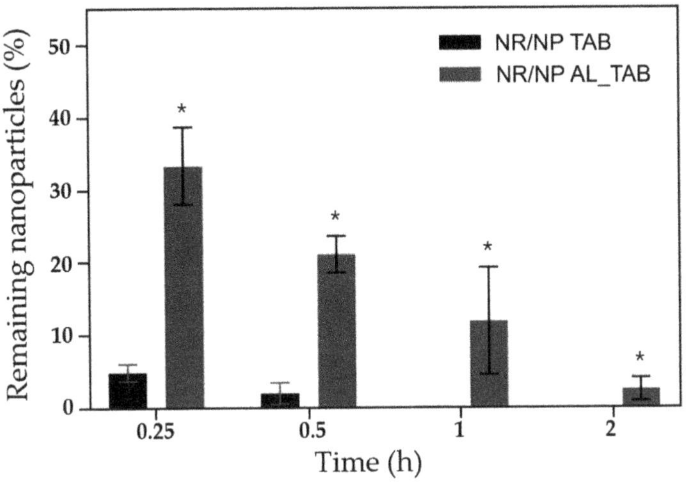

Figure 6. In vivo preocular retention profiles of NR/NP TAB (■) and NR/NP AL_TAB (■) on rabbit eyes. * indicated statistical significance between the formulations at each time point ($p < 0.05$).

The pharmacokinetic profiles of the tablet formulations were assessed and compared with that of the commercially-available dexamethasone eye drops, Maxidex®. Figure 7 and Table 1 display the drug concentration profiles in the AH and their pharmacokinetic parameters, respectively. At 30 min, Maxidex® had a C_{max} of 285.93 ng/mL; however, this value rapidly decreased to an undetectable level at 6 h after the administration. Interestingly, albeit formulated in a dry tablet, the DX/NP TAB without alginate exhibited a much lower drug bioavailability. In addition, its area under the drug concentration-time curve (AUC) and Cmax were less than a half of those with Maxidex®. Such finding could be due to a lower amount of drug exposure at the eye surface, where the DX/NP disappeared relatively rapidly and only a small portion of drug within the particles was actually released during the early period post-administration. On the other hand, 100% drug would be exposed right after a bolus administration with Maxidex®.

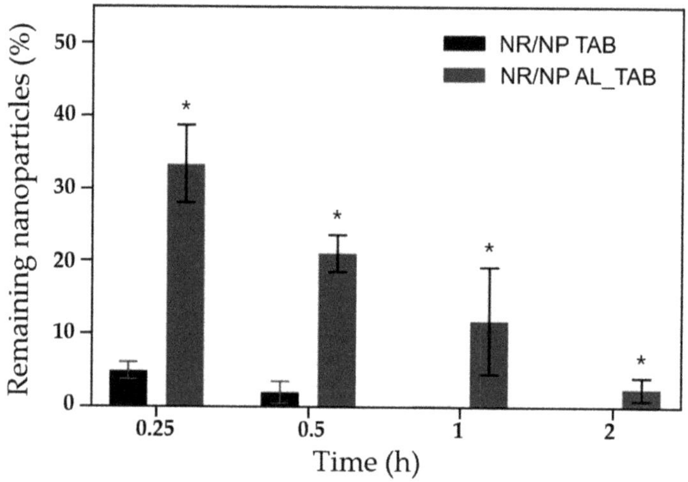

Figure 7. Dexamethasone concentration in the AH of rabbit eyes after the administration of Maxidex® (○), DX/NP TAB (□), and DX/NP AL_TAB (▲). * DX/NP AL_TAB was significantly different from Maxidex®.

Table 1. Pharmacokinetic parameters of dexamethasone in the AH.

Formulations	T_{max} (h)	C_{max} (ng·mL^{-1})	AUC [a] (ng·h·mL^{-1})
Maxidex®	0.5	285.93	388.51
DX/NP TAB	1	62.18	165.93
DX/NP AL_TAB	2	370.33	981.23

[a] calculated using the trapezoidal rule.

The highest drug bioavailability was observed with the DX/NP AL_TAB. In fact, its AUC was 2.5-fold greater than that of Maxidex®. By incorporating alginate in the tablet, it appeared to increase the viscosity of the tear fluid, enabling a larger amount of DX/NP to be retained in the preocular space for a longer period. During this period, the DX/NP would release and expose a significant amount of dexamethasone in a sustained manner. As a result, the drug concentration in tear would gradually increase before being absorbed into the AH to shift the time to reach C_{max} (T_{max}) to a later point (i.e., 2 h) and further increase the value of C_{max} itself, compared to that of Maxidex®. Due to prolonged retention and sustained drug release, drug clearance from the AH was delayed, demonstrating that drug concentration in the AH was statistically significantly higher than that of Maxidex® at 4–10 h ($p < 0.05$). All tablets tested in this experiment were almost instantaneously fully detached and released from the applicator tip to the preocular surface when they were in contact with the lower cul-de-sac of rabbit eyes (Figure S2 in Supplementary Information).

4. Discussion

Topical drug administration is considered to be an easy route for ophthalmic drug delivery. However, for conventional formulations of drops and suspensions, the additional liquid accelerates tear clearance to further lower drug bioavailability in the eye [30,31]. To overcome this, a dry tablet loaded with microparticles had been proposed as topical drug-delivery formulations [7]. The tablet was prepared by compressing the medium of mannitol together with drug-loaded microparticles composed of PLGA and a mucoadhesion promoter, PEG. Thus, dissolution of mannitol increased the tear viscosity to allow for a prolonged time of interaction between PEG in particles and mucin in the preocular surface. However, the tablet with a high density appeared to dissolve in tear fluid in minutes, during which a relatively large tablet could cause eye irritation. Therefore, the formulation was upgraded to be able to dissolve rapidly in tear fluid and for this, the dry tablet was prepared by freeze-drying [7]. In this way, the tablet medium could possess a porous structure and thus, after administration, it dissolved almost instantaneously in tear fluid to free the drug-loaded particles in the preocular surface. With this strategy, the particles still maintained a preocular retention property, as well as showing sustained drug release, hence improved drug bioavailability in the eye.

Herein, a rapidly-dissolving dry tablet formulation embedded with drug-loaded PLGA nanoparticles (i.e., the DX/NP) was proposed again for topical drug delivery to the eye. Owing to the presence of alginate in the tablet medium, tear viscosity increased because of the interaction between alginate and Ca^{2+}. This hampered tear clearance, leading to a higher preocular retention of DX/NP (Figure 6). Therefore, in addition to residing longer in the preocular space, DX/NP could continuously release the drug to tear fluids, providing more time for drug diffusion into the eye and eventually improving ocular drug bioavailability (Figure 7).

In the present study, the tablets were prepared by freeze-drying the DX/NP suspension; this allowed the tablet medium of the water-soluble polymers to be highly porous for rapid dissolution of the tablet in the tear fluid. Considering commercialization, the moisture content in the tablet may need to be minimized further to provide with an appropriate shelf-life of the product. Owing to the nano-sizes of DX/NP, the sensitive eye surface would not be irritated [32]. Therefore, with the alginate content employed for the tablet herein, any sign of eye irritation or discomfort was observed during in vivo experiments. By using the applicator, the tablet could be delivered without being touched, thereby achieving a hygienic application. PDMS was employed as the constituent material for the applicator because of its biocompatibility, should the applicator touch the eye surface. More importantly, PDMS is inherently hydrophobic and it is known to possess a low surface release energy [33]. Therefore, the dry tablet composed of hydrophilic polymers could be separated almost instantaneously without loss when wet with tear fluids (Figure S2), thereby enabling the delivery of an accurate drug dose to the eye using the tablet formulation.

5. Conclusions

Herein, a formulation consisting of a dry tablet containing alginate and drug-loaded PLGA nanoparticles is derived to improve the bioavailability of drugs that are topically delivered to the eye. Owing to the ease and hygienic administration of the developed tablet formulation, the use of a preocular applicator is suggested. Based on the hydrophobicity and low surface release energy of the constituent material, PDMS, the applicator could almost instantaneously cause full separation of the dry tablet composed of hydrophilic polymers when applied to the eye surface. After topical delivery, the tablet medium dissolved rapidly to free the drug-loaded nanoparticles in the tear fluid, where alginate could react with Ca^{2+} to increase its viscosity. Such findings suggest that the drug-loaded PLGA nanoparticles can be better retained in the preocular space and the drug can be subsequently released in a sustained manner, ultimately enhancing ocular drug bioavailability. Therefore, it is concluded that the combination of a dry tablet formulation with an alginate medium, drug-loaded PLGA nanoparticles, and a preocular applicator is a promising strategy to achieve patient-friendly, topical drug delivery to the eye, with enhanced drug availability.

Supplementary Materials: The following are available online at http://www.mdpi.com/1999-4923/11/12/651/s1; Figure S1: Characterizations of NR/NP; Figure S2: Images of rabbit eyes.

Author Contributions: Conceptualization, Y.B.C.; Methodology, Y.B.C., W.M.R., and S.-N.K.; Validation, W.M.R. and S.-N.K.; Animal experiments, W.M.R, S.-N.K., and C.H.M.; Formal analysis, W.M.R. and S.-N.K.; Investigation, W.M.R. and S.-N.K.; Writing—original draft preparation, W.M.R. and S.-N.K.; Writing—review and editing, W.M.R., S.-N.K., and Y.B.C.; Visualization, W.M.R. and S.-N.K.; Supervision, Y.B.C.; Project administration, Y.B.C.; Funding acquisition, Y.B.C.

Funding: This research was supported by grants from the Korea Health Technology R&D Project through the Korea Health Industry Development Institute (KHIDI) funded by the Ministry of Health & Welfare, Republic of Korea (grant no. HI15C1744), the Technology Innovation Program (10060067, Technical development of nanosystem-based technology for drug therapy preventing side effects after cataract surgery) funded by the Ministry of Trade, Industry, and Energy (MI, Korea), and the Bio & Medical Technology Development Program of the National Research Foundation (NRF) funded by the Ministry of Science, ICT & Future Planning (2015M3A9E2030129).

Conflicts of Interest: The authors declare no conflict of interest.

References

1. Lee, V.H.; Robinson, J.R. Topical ocular drug delivery: Recent developments and future challenges. *J. Ocul. Pharmacol.* **1986**, *2*, 67–108. [CrossRef]
2. Lang, J.C. Ocular drug delivery conventional ocular formulations. *Adv. Drug Deliv. Rev.* **1995**, *16*, 39–43. [CrossRef]
3. Sigurdsson, H.H.; Konráðsdóttir, F.; Loftsson, T.; Stefansson, E. Topical and systemic absorption in delivery of dexamethasone to the anterior and posterior segments of the eye. *Acta Ophthalmol. Scand.* **2007**, *85*, 598–602. [CrossRef]
4. Järvinen, K.; Järvinen, T.; Urtti, A. Ocular absorption following topical delivery. *Adv. Drug Deliv. Rev.* **1995**, *16*, 3–19. [CrossRef]
5. Tsai, C.H.; Wang, P.Y.; Lin, I.; Huang, H.; Liu, G.S.; Tseng, C.L. Ocular Drug Delivery: Role of Degradable Polymeric Nanocarriers for Ophthalmic Application. *Int. J. Mol. Sci.* **2018**, *19*, 2830. [CrossRef]
6. Zhu, H.; Chauhan, A. A mathematical model for ocular tear and solute balance. *Curr. Eye Res.* **2005**, *30*, 841–854. [CrossRef]
7. Choy, Y.B.; Patel, S.R.; Park, J.H.; McCarey, B.E.; Edelhauser, H.F.; Prausnitz, M.R. Mucoadhesive Microparticles in a Rapidly Dissolving Tablet for Sustained Drug Delivery to the Eye. *Investig. Ophthalmol. Vis. Sci.* **2011**, *52*, 2627–2633. [CrossRef]
8. Dart, J.K.G.; Stapleton, F.; Minassian, D. Contact lenses and other risk factors in microbial keratitis. *Lancet* **1991**, *338*, 650–653. [CrossRef]
9. Radford, C.F.; EWoodward, G.; Stapleton, F. Contact lens hygiene compliance in a universitypopulation. *J. Br. Contact Lens Assoc.* **1993**, *16*, 105–111. [CrossRef]
10. Refresh. REFRESH®Classic. Available online: https://www.refreshbrand.com/Products/refresh-classic (accessed on 15 October 2019).
11. Graham, R.O.; Peyman, G.A. Intravitreal injection of dexamethasone: Treatment of experimentally induced endophthalmitis. *Arch. Ophthalmol.* **1974**, *92*, 149–154. [CrossRef]
12. Tan, D.T.; Chee, S.P.; Lim, L.; Lim, A.S. Randomized clinical trial of a new dexamethasone delivery system (Surodex) for treatment of post-cataract surgery inflammation. *Ophthalmology* **1999**, *106*, 223–231. [CrossRef]
13. Makadia, H.K.; Siegel, S.J. Poly Lactic-*co*-Glycolic Acid (PLGA) as Biodegradable Controlled Drug Delivery Carrier. *Polymers* **2011**, *3*, 1377–1397. [CrossRef] [PubMed]
14. Jain, R.A. The manufacturing techniques of various drug loaded biodegradable poly(lactide-*co*-glycolide) (PLGA) devices. *Biomaterials* **2000**, *21*, 2475–2490. [CrossRef]
15. Lee, S.S.; Hughes, P.; Ross, A.D.; Robinson, M.R. Biodegradable implants for sustained drug release in the eye. *Pharm. Res.* **2010**, *27*, 2043–2053. [CrossRef]
16. Davis, J.; Gilger, B.; Robinson, M. Novel approaches to ocular drug delivery. *Curr. Opin. Mol. Ther.* **2004**, *6*, 195–205.
17. Draget, K.I.; Skjåk-Bræk, G.; Smidsrød, O. Alginate based new materials. *Int. J. Biol. Macromol.* **1997**, *21*, 47–55. [CrossRef]
18. Braccini, I.; Pérez, S. Molecular Basis of Ca^{2+}-Induced Gelation in Alginates and Pectins: The Egg-Box Model Revisited. *Biomacromolecules* **2001**, *2*, 1089–1096. [CrossRef]
19. Fu, S.; Thacker, A.; Sperger, D.M.; Boni, R.L.; Buckner, I.S.; Velankar, S.; Munson, E.J.; Block, L.H. Relevance of rheological properties of sodium alginate in solution to calcium alginate gel properties. *Aaps Pharmscitech* **2011**, *12*, 453–460. [CrossRef]
20. Villanueva, J.R.; Bravo-Osuna, I.; Herrero-Vanrell, R.; Martínez, I.T.M.; Navarro, M.G. Optimising the controlled release of dexamethasone from a new generation of PLGA-based microspheres intended for intravitreal administration. *Eur. J. Pharm. Sci.* **2016**, *92*, 287–297. [CrossRef]
21. Labconco. FreeZone 6 Liter -84C Console Freeze Dryer. Available online: https://www.labconco.com/product/freezone-6-liter-84c-console-freeze-dryer-11/6303 (accessed on 26 November 2019).

22. Loan, T.; Easton, C.; Alissandratos, A. DNA amplification with in situ nucleoside to dNTP synthesis, using a single recombinant cell lysate of *E. coli*. *Sci. Rep.* **2019**, *9*, 1–8. [CrossRef]
23. Ojha, T.; Pathak, V.; Drude, N.; Weiler, M.; Rommel, D.; Rütten, S.; Lammers, T. Shelf-Life Evaluation and Lyophilization of PBCA-Based Polymeric Microbubbles. *Pharmaceutics* **2019**, *11*, 433. [CrossRef]
24. Sun, C.; Wang, X.; Zheng, Z.; Chen, D.; Wang, X.; Shi, F.; Wu, H. A single dose of dexamethasone encapsulated in polyethylene glycol-coated polylactic acid nanoparticles attenuates cisplatin-induced hearing loss following round window membrane administration. *Int. J. Nanomed.* **2015**, *10*, 3567.
25. Bright, A.M.; Tighe, B.J. The composition and interfacial properties of tears, tear substitutes and tear models. *J. Br. Contact Lens Assoc.* **1993**, *16*, 57–66. [CrossRef]
26. Kim, J.; Peng, C.-C.; Chauhan, A. Extended release of dexamethasone from silicone-hydrogel contact lenses containing vitamin E. *J. Control. Release* **2010**, *148*, 110–116. [CrossRef]
27. Park, C.G.; Kim, M.J.; Park, M.; Choi, S.Y.; Lee, S.H.; Lee, J.E.; Choy, Y.B. Nanostructured mucoadhesive microparticles for enhanced preocular retention. *Acta Biomater.* **2014**, *10*, 77–86. [CrossRef]
28. Kim, S.N.; Park, C.G.; Huh, B.K.; Lee, S.H.; Min, C.H.; Lee, Y.Y.; Choy, Y.B. Metal-organic frameworks, NH2-MIL-88(Fe), as carriers for ophthalmic delivery of brimonidine. *Acta Biomater.* **2018**, *79*, 344–353. [CrossRef]
29. Araújo, J.; Gonzalez, E.; Egea, M.A.; Garcia, M.L.; Souto, E.B. Nanomedicines for ocular NSAIDs: Safety on drug delivery. *Nanomed. Nanotechnol. Biol. Med.* **2009**, *5*, 394–401. [CrossRef]
30. Doane, M.G. Blinking and the Mechanics of the Lacrimal Drainage System. *Ophthalmology* **1981**, *88*, 844–851. [CrossRef]
31. Garaszczuk, I.K.; Montes Mico, R.; Iskander, D.R.; Expósito, A.C. The tear turnover and tear clearance tests—A review. *Expert Rev. Med. Devices* **2018**, *15*, 219–229. [CrossRef]
32. Choy, Y.B.; Park, J.-H.; Prausnitz, M.R. Mucoadhesive microparticles engineered for ophthalmic drug delivery. *J. Phys. Chem. Solids* **2008**, *69*, 1533–1536. [CrossRef]
33. Jin, X.; Strueben, J.; Heepe, L.; Kovalev, A.; Mishra, Y.K.; Adelung, R.; Staubitz, A. Joining the Un-Joinable: Adhesion Between Low Surface Energy Polymers Using Tetrapodal ZnO Linkers. *Adv. Mater.* **2012**, *24*, 5676–5680. [CrossRef]

pharmaceutics

MDPI

Article

A Combinatorial Cell and Drug Delivery Strategy for Huntington's Disease Using Pharmacologically Active Microcarriers and RNAi Neuronally-Committed Mesenchymal Stromal Cells

Emilie M. André [1], Gaëtan J. Delcroix [2,3], Saikrishna Kandalam [1], Laurence Sindji [1] and Claudia N. Montero-Menei [1,*]

[1] CRCINA, UMR 1232, INSERM, Université de Nantes, Université d'Angers, F-49933 Angers, France; emilie.andre01@gmail.com (E.M.A.); kandalamsai@gmail.com (S.K.); laurence.sindji@univ-angers.fr (L.S.)

[2] Geriatric Research, Education, and Clinical Center, and Research Service, Bruce W. Carter Department of Veterans Affairs Medical Center, Miami, FL 33125, USA; gdelcroix@nova.edu

[3] College of Allopathic Medicine, Nova Southeastern University, Fort Lauderdale, FL 33125, USA

* Correspondence: claudia.montero-menei@univ-angers.fr; Tel.: +33 2 44688536

Received: 18 June 2019; Accepted: 2 October 2019; Published: 12 October 2019

Abstract: For Huntington's disease (HD) cell-based therapy, the transplanted cells are required to be committed to a neuronal cell lineage, survive and maintain this phenotype to ensure their safe transplantation in the brain. We first investigated the role of RE-1 silencing transcription factor (REST) inhibition using siRNA in the GABAergic differentiation of marrow-isolated adult multilineage inducible (MIAMI) cells, a subpopulation of MSCs. We further combined these cells to laminin-coated poly(lactic-*co*-glycolic acid) PLGA pharmacologically active microcarriers (PAMs) delivering BDNF in a controlled fashion to stimulate the survival and maintain the differentiation of the cells. The PAMs/cells complexes were then transplanted in an ex vivo model of HD. Using Sonic Hedgehog (SHH) and siREST, we obtained GABAergic progenitors/neuronal-like cells, which were able to secrete HGF, SDF1 VEGFa and BDNF, of importance for HD. GABA-like progenitors adhered to PAMs increased their mRNA expression of NGF/VEGFa as well as their secretion of PIGF-1, which can enhance reparative angiogenesis. In our ex vivo model of HD, they were successfully transplanted while attached to PAMs and were able to survive and maintain this GABAergic neuronal phenotype. Together, our results may pave the way for future research that could improve the success of cell-based therapy for HDs.

Keywords: tissue engineering; Huntington's disease; siRNA; nanoparticles; microcarriers; mesenchymal stromal cells

1. Introduction

Huntington's disease (HD) is a genetic neurodegenerative disorder caused by the abnormal repetition of CAG nucleotides in the Huntingtin (HTT) gene. This leads to a pathological expansion of polyglutamine (polyQ) and aggregation of the mutated HTT protein in the brain, more specifically in the striatum [1,2]. HD is characterized by a progressive degeneration of striatal GABAergic medium spiny projection neurons, followed by a progressive degeneration extending throughout the brain [3]. Clinically, this results in involuntary movements, cognitive impairment and psychiatric manifestations, culminating in death around 15–20 years after the onset of motor symptoms [4,5]. Currently, there is no proven medical therapy to alleviate the onset or progression of HD [6].

Mesenchymal stromal cells (MSCs), have emerged for clinical transplantation studies due to their easy availability, their immune-modulatory properties [7] and their capacity to release neurotrophic factors and to create a neuroprotective microenvironment [8]. Clinical trials using MSCs in the central nervous system are now underway for many neurological disorders and have shown the feasibility of this approach [9]. Pre-clinical studies with HD models have shown improvement in behavior and a reduced lesion volume after MSC implantation. These beneficial effects could be explained by the secretion of neurotrophic factors including brain-derived neurotrophic factor (BDNF), ciliary neutrotrophic factor, nerve growth factor, insulin-like growth factor 1 and epidermal growth factor (EGF) [10,11]. The trans-differentiation of MSCs into a neural/neuronal lineage, even if possible in vitro, is far less efficient in vivo and their functional maturity remains too scarce [12]. The rationale for their transplantation does not lie on their capacity to replace the damaged neuronal cells, but on their ability to neuroprotect and repair by their paracrine effects on the surrounding environment [8]. Even though they are not used in a cell replacement strategy demanding electrophysiolocally functional connections, if the cells are to be safely used in clinical trials, it is important that they blend into the microenvironment of the brain and present the same neuronal phenotype. Therefore, MSCs are required to be committed to a neuronal cell lineage and maintain this phenotype to ensure their safe transplantation in the brain. In this regard, REST/NRSF a repressor transcription factor functioning as a master negative regulator of neurogenesis by binding to a specific DNA domain named RE1 motif is an interesting target to inhibit and thus induce a neuronal specification [13,14]. It was thus recently shown that the silencing of REST, obtained by a recombinant lentivirus carrying a small interfering RNA (siRNA) for REST, induced a neural/neuronal differentiation of MSCs [15,16].

Nanoparticles have been developed to efficiently and safely deliver siRNA both in vitro and in vivo and to avoid insertional gene mutations and viral toxicity issues. We recently designed lipid nanocapsules (LNC) able to encapsulate siRNAs complexed to lipids, thus protecting the siRNA from degradation. LNCs consisting of a lipid liquid core of triglycerides and a rigid shell of lecithin and polyethylene glycol (PEG) can be formulated by a simple and easily industrialized solvent-free process based on the phase inversion of an emulsion [17,18]. They have a high stability and can destabilize lysosome's membranes by a proton sponge effect [19]. We also demonstrated that LNCs associated with siREST in MSCs were able to induce their neuronal commitment with a better efficiency than the commercial reagent Oligofectamine® [9]. However, MSCs are a heterogeneous population presenting different differentiation properties. A homogeneous subpopulation of MSCs, termed marrow-isolated adult multilineage inducible (MIAMI) cells, which present a unique genetic profile expressing several pluripotency markers (Oct4, Sox2, Nanog, SSEA4), share many proteins with embryonic stem cells, secrete more tissue repair factors than MSCs, protect the neurovasculature and can be induced to differentiate into cells from all three germ layers [20–22], emerge as a good alternative for cell therapy studies. MIAMI cells can be specified into the neuronal lineage with epidermal growth factor (EGF) and fibroblast growth factor-2 (FGF-2) [23] and induced to an immature neuronal dopaminergic phenotype presenting appropriate electrophysiological properties, regardless of the donor age, with a three-step protocol in vitro requiring neurotrophins [24,25]. In addition, laminin (LM) was shown to enhance the neuronal differentiation of these cells [8]. However, further efforts are needed to maintain a differentiated phenotype and increase engraftment/survival after transplantation in the brain.

Pharmacologically active microcarriers (PAMs) are biodegradable and biocompatible poly(lactic-*co*-glycolic acid) PLGA-based microspheres covered with extracellular matrix molecules (ECM) such as fibronectin or laminin providing an adequate three-dimensional (3D) biomimetic surface for the transplanted cells and thus enhancing their survival [22,24]. Moreover, PAMs can also release in a controlled manner an encapsulated growth factor allowing a better cell engraftment by stimulating the transplanted cells and/or the microenvironment [22,26]. In this context, PAMs transporting human stem cells and delivering different growth factors have been shown to be beneficial in several animal models of neurological disorders, cartilage and cardiac pathologies [27–30]. In this regard, BDNF, which is involved in neuronal GABAergic differentiation and neuronal survival [30–32] may maintain the differentiated phenotype and increase the survival of the transported MIAMI cells. Moreover, in the case of HD, several studies demonstrated that the expression of BDNF is reduced in the patient's brains [33]. Promising results show that BDNF supplementation increases the survival of enkephalin-immunoreactive striatal neurons, reduces striatal interneuronal loss and improves motor function in HD animal models [31,34,35]. PAMs would have the benefit to prevent BDNF degradation while providing for its controlled release over time, hence maximizing the possible benefits of BDNF for neuroprotection in HD.

In this study, we aim to first improve the neuronal differentiation of a subpopulation of MSCs, the MIAMI cells, as well as to study their behavior after combining them with PAMs releasing BDNF in vitro and on organotypic brain cultures. We first evaluated the impact of siREST LNC transfection on the neuronal commitment of EGF/FGF-2 pre-treated MIAMI cells. We set up a differentiation protocol based mainly on sonic hedgehog (SHH) treatment followed by BDNF as previously described [36,37], to evaluate the GABAergic differentiation potential of these cells in vitro; GABAergic cells being the predominant cell type present in the striatum. We thus also studied the effects of the REST nanocapsule-silencing on MIAMI cells induction towards the GABAergic phenotype. The secretome and neuronal marker expression of MIAMI cells adhered onto LM-coated PAMs releasing BDNF was also studied in order to assess their potential therapeutic effect. We finally evaluated their behavior in an ex vivo organotypic model of HD. Therefore, the novelty of our paper lies in the combinatorial strategy using siREST silencing nanoparticles and cell transporting biodegradable polymeric PAMs releasing BDNF to achieve neuronally-committed MSCs releasing growth factors for a safe neuroprotective/neurorepair strategy for HD.

2. Materials and Methods

2.1. siRNA-LNCs

LNCs were formulated, as previously described [38] by mixing 20% *w/w* Labrafac® WL 1349 (caprylic-capric acid triglycerides, Gatefossé S.A. Saint-Priest, France), 1.5% *w/w* Lipoid S75-3® (Lecithin, Ludwigshafen, Germany), 17% *w/w* Kolliphor® HS 15 (Polyethylene glycol-15-Hydroxystearate PEGHS BASF, Ludwigshafen, Germany), 1.8% *w/w* NaCl (Prolabo, Fontenay-sous-Bois, France) and 59.8% *w/w* water (obtained from a Milli-Q system, Millipore, Paris, France) together under magnetic stirring. Briefly, three temperature cycles between 60 and 95 °C were performed to obtain phase inversions of the emulsion. A subsequent rapid cooling and dilution with ice cold water (1:1.4) at the last phase inversion temperature led to blank LNC formation. For liposome preparation, a cationic lipid DOTAP (1,2-dioleyl-3-trimethylammoniumpropane) (Avanti® Polar Lipids Inc., Alabaster, AL, USA), solubilized in chloroform, was mixed at a 1/1 molar ratio with the neutral lipid DOPE (1,2-dioleyl-sn-glycero-3-phosphoethanolamine) (Avanti® Polar Lipids Inc.) to obtain a final concentration of 30 mM of cationic lipid. After chloroform vacuum evaporation, the lipid film was rehydrated and liposomes sonicated. A simple equivolume mix of liposomes and siRNA resulted in lipoplexes characterized by a charge ratio of 5 between the positive charge of lipids and the negative charge of nucleic acids. To obtain siRNA-LNCs, the water introduced at the last phase inversion temperature was replaced by lipoplexes, i.e., REST siRNA: (sense sequence: 5′-CAG-AGU-UCA-CAG-UGC-UAA-GAA -3′; Eurogentec, Seraing, Belgium) and control (scrambled) siRNA (sense sequence: 5′-UCUACGAGGCACGAGACUU-3′; Eurogentec) complexed with cationic liposomes in a defined charge ratio as described above. To avoid the possible denaturation of siRNA the addition of lipoplexes was performed at 40 °C.

2.2. Fluorescent siRNA-LNCs-DiD

To formulate fluorescent siRNA-LNCs, a solution of DiD (1,1′-dioctadecyl-3,3,3′,3′-tetramethylindodicarbocyanine perchlorate; em. = 644 nm; exc. = 665 nm) (Invitrogen, Cergy-Pontoise, France) solubilized in acetone at 25 mg/mL was prepared.

For in vitro experiments, the DiD concentration was fixed at 200 µg/mL of LNC suspension or corresponding to 1.36 mg of DiD per grams of Labrafac®. The adequate volume of DiD solubilized in acetone was incorporated in Labrafac® and acetone was evaporated at room temperature. The formulation process was unchanged, and formulation was stored at 4 °C, protected from light. For siRNA fluorescent LNCs, a fluorescent Alexa 488 siRNA (Eurogentec) was used.

2.3. BDNF-Releasing, Laminin (LM)-Coated PAMs

Synthesis and characterizations of PLGA-P188-PLGA polymer were performed using Synbio3 platform supported by GIS IBISA and ITMO Cancer. BDNF-releasing PAMs were prepared as previously described using a solid/oil/water emulsion solvent extraction-evaporation method [30]. Briefly, BDNF and human serum albumin were first nanoprecipitated separately and nanoprecipitated proteins were dispersed in the organic phase containing the polymer at a protein loading of 1 μg of protein and 5 μg of human serum albumin/mg of PAMs. The suspension was emulsified in a poly(vinyl alcohol) aqueous phase and after solvent extraction in an aqueous phase, the microspheres were filtered and freeze-dried. Blank microspheres, without protein, were prepared following a similar process. To obtain LM-covered PAMS (LM-PAMs), PLGA-P188-PLGA microspheres were coated with LM and poly-D-Lysine (PDL) as previously described [29]. Briefly, the coating solutions prepared in Dulbecco's Phosphate-Buffered Saline (DPBS) were mixed under rotation with the microspheres at a final concentration of the coating molecules of 16 μg/mL of LM and 24 μg/mL of PDL (corresponding to a 40:60 ratio of LM:PDL). In vitro BDNF release from PAMs was performed as previously described by incubation of 5mg PAMs in citrate buffer and dosage by ELISA of collected fractions of the supernatant over time [30].

2.4. LNC and PAM Characterization

The size and Zeta potential of LNCs ($N = 3$) were measured by using the Dynamic Light Scattering (DLS) method using a Malvern Zetasizer® apparatus (Nano Series ZS, Malvern Instruments S.A., Worcestershire, UK) after dilution at a ratio of 1:200 with deionized water. PAM's size was measured with a Multisizer® coulter counter (Beckman Coulter, Roissy France), zeta potential was measured by DLS [30]. The laminin surface was characterized by confocal microscopy (Leica TCS SP8, France) after LM immunostaining as previously described [30]. Lyophilized PAMs were incubated for 30 min at room temperature (RT) under 15 rpm stirring in DPBS containing 4% bovine serum albumin (BSA), 0.2% Tween 20 (DPBS BT). After washing, anti-LM mouse monoclonal antibody (Sigma-Aldrich, St-Louis, MO, USA, 100 μg/mL in DPBS) was added for 1.5 h under rotation at 37 °C. After washing, biotinylated anti-mouse IgG antibody (2.5 μg/mL in DPBS BT) was added for 1 h, at RT, washed and incubated with streptavidin–fluoroprobe 547 (1:1000 in DPBS) at RT, for 40 min. ($N = 3$, $n = 3$)

2.5. MIAMI E/F Cells

MIAMI cells were isolated from human bone marrow (Lonza, donor #3515) and expanded on fibronectin (Sigma-Aldrich) coated flasks at 120 cells/cm^2 in low oxygen tension (3% of O_2 and 5% of CO_2) in Dulbecco's Modified Eagle Medium-low glucose (DMEM, Gibco, Life Technologies, Paisley, UK), supplemented with 3% of Foetal Bovine serum (FBS), 100 μM of ascorbic acid and a mixture of lipids, as previously described [20]. Using the same density of cells and culture condition, a 10 days treatment with an addition of 20 ng/mL of EGF and 20 ng/mL of FGF-2 (both from R&D systems, Lille, France) and 5 μg/mL of Heparin (Sigma-Aldrich) was used to enhance neuronal specification [23] and these cells were named MIAMI E/F cells. Every three days, half of the culture medium was replaced.

2.6. MIAMI E/F Cell Transfection

MIAMI E/F cells were seeded at 2000 to 3000 cells per cm^2 in wells coated with LM (2 μg/cm^2, Sigma-Aldrich). Experiments were performed in Opti-MEM® media (Life technologies, France). SiRNA-LNCs were incubated with cells at 37 °C in a humidified atmosphere with 3% O_2 and 5% CO_2 for 4 h before serum addition. Cells were harvested after appropriate time and assayed for mRNA expression levels by RT-qPCR or protein expression level by immunofluorescence.

2.7. LNC Cell Time Retention

MIAMI cells (2000 to 3000 cells per cm^2) were seeded on glass slides coated with LM (2 µg/cm^2, Sigma-Aldrich). SiRNA fluorescent LNCs were incubated with cells at 37 °C in a humidified atmosphere with 3% O_2 and 5% CO_2 for 4 h in Opti-MEM® media before DPBS washing and paraformaldehyde fixation (4% of paraformaldehyde during 15 min at 4 °C) or FBS addition. At Day 0 and before fixation, 100nM LysoTracker Red (Molecular Probes, Eugene, OR, USA.) was added to the media. After washing, cells were visualized from Day 0 to Day 6 post-transfection using a fluorescence confocal multispectral imaging, FCSI (Leica TCS SP8).

2.8. MIAMI E/F Cell Neuronal Differentiation

MIAMI E/F transfected with siRNA-LNC cells were seeded (2000–3000 cells per cm^2) on 175 cm^2 cell culture flasks for Step 1 and on glass slides on 6 well plates for Step 2 coated with LM (2 µg/cm^2, Sigma-Aldrich, St-Louis, MO, USA) and different conditions tested to obtain the best two-step GABAergic differentiation protocol (Figure 1). The first step was performed with DMEM/F12 (Glutamax, Gibco, Life Technologies, Paisley, UK) supplemented with 5% of N2 (1X) (both from Gibco, Life Technologies), and 200 ng/mL of sonic hedgehog (SHH, Peprotech, Rocky Hill, USA) for 14 days and 35 mL of media was used per flask. These cells are named MIAMI-SHH. Second step: Neurobasal media (Neurobasal, Gibco, Life Technologies) supplemented with or without 1 mM of Valproic acid (Sigma-Aldrich) and with or without 30 ng/mL of BDNF (Peprotech) for 14 days and 3 mL of media was used per well. Length and surface area were quantified using MetaVue software®. 6 pictures from each condition (24 in total) were taken with a 10× objective and used to determine total area and length. Only cells responding to the treatment (with neurite-like structures) were analyzed in this experiment.

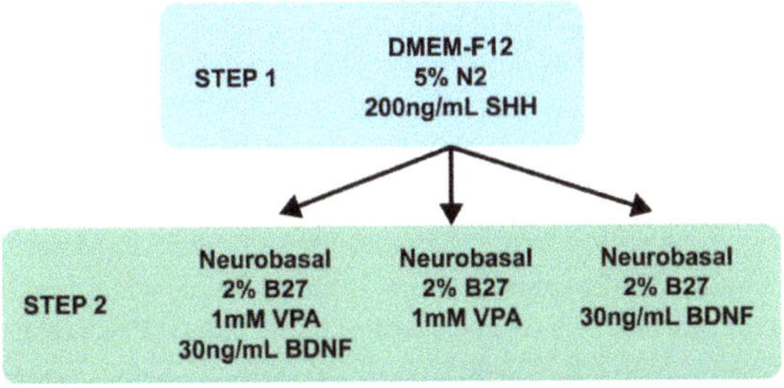

Figure 1. Media tested for the GABAergic differentiation of marrow-isolated adult multilineage inducible (MIAMI) cells.

2.9. Formation of PAMs-Cell Constructs

Cell adhesion studies were performed based on previous published protocols [22,24]. At the end of Step 1, MIAMI-SHH-siREST cells were detached and resuspended in DMEM-F12 supplemented with either 3% FBS (Lonza, Verviers, Belgium), 5% of N2 (1X) (Gibco, Life Technologies, Paisley, UK) or 2% of B27 (1X) (Gibco, Life Technologies). Lyophilized PAMs (0.5 mg) were resuspended in coated microcentrifuge tubes (Sigmacote, Sigma) containing DMEM-F12 (Gibco, Life Technologies), and mixed with 0.5 mL of cell suspension (2.5 × 10^5 cells/0.50 mg PAMs). The mixture was then gently flushed and plated in 1.9 cm^2 Costar ultra low adherence plate (#3473, Corning, Avon, France). Plates were incubated at 37 °C during 4 h for MIAMI E/F, to allow cell attachment on PAM surface (covered with FN). PAMs/cell aggregates were pelleted by centrifugation at 200 g for 2 min. Cell adhesion to PAM surface was assessed by microscopic observation and cells adhered to PAMs were quantified using the Cyquant cell proliferation assay (CyQuant Cell proliferation Assay kit, Invitrogen). Complexes were further studied using light and confocal microscopy and scanning electron microscopy. Samples were prepared for scanning electron microscopy analysis as previously described [24].

2.10. Preparation of Brain Organotypic Slices

Organotypic cultures were prepared as previously described by our team [39]. Briefly, Albinos wild-type Sprague–Dawley rats from the SCAHU (Service Commun d'Animalerie Hospitalo-Universitaire, N°49007002, Angers University, France) were used. Postnatal 9–11 (P9–11) days old Sprague–Dawley pups were used to generate the ex vivo model of HD. Animals were rapidly euthanized after anesthesia, by intraperitoneal injection of 80 mg/kg of ketamine (Clorketam 1000, Vetoquirol, Lure, France) and 10 mg/kg of xylazine (Rompum 2%, Bayer Health Care, Kiel, Germany). Brains were removed and rapidly dissected before being glued onto the chuck of a vibratome cooled with a bath of Gey's balanced salt solution supplemented with 6.5 mg/L of glucose and antibiotics. 400 μm thick slices were cut using a vibratome (Motorized Advance Vibroslice MA752, Campdem instruments) in different configurations to obtain a progressive degeneration of the GABAergic medium spiny neurons (MSNs). Each hemisphere was mechanically separated to culture 8 organotypic brain slices total (4 from each hemispheres). Obtained slices were next transferred to 30 mm diameter semi-porous membrane inserts (Millicell-CM, Millipore, Guyancourt, France) within a 6-well plate, containing Neurobasal medium (Gibco, Life Technologies) supplemented with 6.5 mg/L of glucose, 1 mM of L-glutamine, 1x B27 supplements (Gibco, Life Technologies) and antibiotics. Slices were incubated at 37 °C and 5% CO_2 up to 30 days and half of the medium was removed every 2–3 days. Each slice was cultured on a single membrane to increase their survival over time.

2.11. Injection of the Cells-PAMs Constructs into Organotypic Slides

Three days after organotypic slice preparation, the cells-PAMs were injected into the striatum using a 22-gauge Hamilton needle (Hamilton, Bonaduz, Switzerland) connected to a micromanipulator. The total injection volume consisted of 2 μL of culture medium containing approximately 75,000 cells alone or adhering to 0.1 mg of PAMs. The injections were done at an infusion rate of 0.5 μL/minute. The needle was left in place for 5 min before removal to avoid the cells being expelled from the organotypic slices.

2.12. Reverse Transcription and Real Time Quantitative PCR

Experiments were performed following the guidelines of the PACeM core facility ("Plate-forme d'Analyse Cellulaire et Moléculaire", Angers, France). Total RNA of cells was extracted, purified using RNeasy Microkit (Qiagen, Courtaboeuf, France), treated with DNase (10 U DNase I/μg total RNA) and the concentration determined with a ND-2000 NanoDrop (Thermo Fisher Scientific, Wilmington, DE, USA). RNA integrity was verified on Experion RNA StdSens chip (Bio-Rad). First strand complementary DNA (cDNA) synthesis was performed with a SuperScriptTM II Reverse Transcriptase (Invitrogen), in combination with random hexamers, according to the manufacturer's instructions. cDNAs were purified (Qiaquick PCR purification kit, Qiagen) and 3 ng of cDNA mixed with MaximaTM SYBR Green qPCR Master Mix (Fermentas) and primer mix (sense and antisense at 0.3 μM, Table 1 (Eurofins MWG Operon, Ebersberg, Germany) in a final volume of 10 μL. Amplification was carried out on a LightCycler 480 (Roche): denaturation step at 95 °C for 10 min and 40 cycles of 95 °C for 15 s, 60 °C for 30 s. Specificity of the primers was controlled. The GeNormTM freeware (http://medgen.ugent.be/-jvdesomp/genorm/) was used to choose GAPDH and ACTB, as the most stable housekeeping genes. The relative transcript quantity (Q) was determined by the delta Cq method: Q = E(Cq min in all the samples tested − Cq of the sample), where E is related to the primer efficiency (E = 2 if the primer efficiency = 100%). Relative quantities (Q) were normalized using the multiple normalization method (Vandesompele et al., 2002). Q normalized = Q/(geometric mean of the stable housekeeping genes Q). The 2(−ΔΔCt) method was retained, using a housekeeping gene and gene of interest (Livak and Schmittgen, 2001) tested on control sample and treated sample.

Table 1. Primer sequences used for RT-qPCR.

Gene	Full name	NM number	Sequences
ACTB	Actin	NM_001101.3	F: CCAGATCATGTTTGAGACCT R: GGCATACCCCTCGTAGAT
BDNF	Brain-derived neurotrophic factor	NM_001143816	F: CAAACATCCGAGGACAAGG R: TACTGAGCATCACCCTGG
DARPP32	Dopamine- and cAMP-regulated phosphoprotein	NM_181505	F: GAGAGCCTCAGGAGAGGG R: CTCATTCAAATTGCTGATAGACTGC
Dlx2	Distal-less homeobox 2	NM_004405	F: GACCTTGAGCCTGAAATTCG R: ACCTGAGTCTGGGTGAGG
GAD67	Glutamic Acid Decarboxylase 67	NM_000817	F: GGTGGCTCCAAAAAATCAAAGC R: CAATGTCAGACTGGGTAGCG
GAPDH	glyceraldehyde-3-phosphate dehydrogenase	NM_001289745.1	F: CAAAAGGGTCATCATCTCTGC R: AGTTGTCATGGATGACCTTGG
GDNF	Glial cell line-derived neurotrophic factor	NM_011675.2	Qiagen, ref #QT00001589
Pax6	Paired box 6	NM_000280	F: TTTCAGCACCAGTGTCTACC R: TAGGTATCATAACTCCGCCC
NGF	Nerve growth factor	NM_002506	Qiagen, ref #QT00043330
REST	RE1-silencing transcription factor	NM_001193508.1	F: ACTCATACAGGAGAACGCC R: GTGAACCTGTCTTGCATGG
VEGFA	Vascular endothelial growth factor A	NM_001204384	F: CAGCGCAGCTACTGCCATCCA R: CAGTGGGCACACACTCCAGGC
ACTB	Actin	NM_001101.3	F: CCAGATCATGTTTGAGACCT R: GGCATACCCCTCGTAGAT

2.13. Immunocytofluorescence

After treatment, cells were fixed with 4% paraformaldehyde (PFA, Sigma, St Louis, MO, USA) in DPBS (Lonza, Verviers, Belgium) pH 7.4 during 15 min at 4 °C, washed and non-specific sites were blocked with DPBS, Triton 0.1% (DPBS-T, Triton X-100, Sigma), bovine serum albumin 4% (BSA, Fraction V, PAA Lab, Austria), normal goat serum 10% (NGS, Sigma) during 45 min at RT. A mouse anti-human β3-tubulin (2 μg/mL, clone SDL.3D10, Sigma), a mouse anti-human neurofilament medium (NFM, 1:50, clone NN18, Sigma-Aldrich), a monoclonal rabbit anti-human dopamine- and cAMP-regulated neuronal phosphoprotein (DARPP32, 0,6 μg/mL, clone EP721Y, Abcam, Paris, France), a mouse anti-glutamic acid decarboxylase-67 antibody (GAD67, 2 μg/mL, clone 1G10.2, Millipore SA), and a rabbit anti-GABA transporter 1 (GAT1 500 ng/mL Millipore SA) were used to characterize cell differentiation. Cells were incubated overnight with the primary antibody diluted into DPBS-T, BSA 4% at 4 °C. After washes, slices were incubated with the biotinylated mouse (2.5 μg/mL, Vector Laboratories, Burlingame, CA, USA) or rabbit secondary antibody (7.5 μg/mL, Vector Laboratories) for 1 hour at RT. Then slices were washed and incubated with Streptavidin Fluoroprobes 488 or 547H (Interchim, Montluçon, France) diluted 1:1000 or 1:500 respectively in DPBS for 1 h before mounting and observation using a fluorescence microscope.

2.14. Immunofluorescence

Immunofluorescence of brain slices was performed as previously described [39]. Immunofluorescence was performed using antibodies against human mitochondria (hMito) (10 ng/mL, mitochondrial cytochrome C oxidase subunit II, Abcam), rabbit anti-human dopamine- and cAMP-regulated neuronal phosphoprotein DARPP32 (DARPP32, 0.6 μg/mL, clone EP721Y, Abcam) and a mouse anti-GAD67 (5 μg/mL, clone 1G10.2, Millipore SA). Isotypic controls were performed for each antibody. Free-floating slices were incubated in 1% DPBS-T (Sigma-Aldrich). After pre-blocking for 4 h with 4% BSA (fraction V, PAA Laboratories, Piscataway, NJ, USA), 10% NGS (Sigma-Aldrich) in DPBS-T, slices were incubated for 48 h at 4 °C with monoclonal antibodies against DARPP32, GAD67 or hMito with diluted in 4% BSA DPBS-T. After washing, the sections were incubated for 2 h slices with the biotinylated mouse (2.5 μg/mL, Vector Laboratorie,) or rabbit secondary antibody (7.5 μg/mL, Vector Laboratories) at room temperature. Then, slices were washed and incubated for 2 h with Streptavidin Fluoroprobes 488 or 547H (Interchim) diluted 1:500 or 1:1000 respectively in DPBS. Finally, the sections were washed mounted using fluorescent mounting medium (Dako, Carpinteria) and observed in a confocal microscopy.

2.15. MIAMI Cell Secretome Analysis

A Luminex® Multiplex secretome assay was used to quantify cytokines and growth factors secreted by MIAMI cells under different conditions. 72 h before the experiment, media was completely removed and replaced by media without serum for MIAMI E/F or appropriate media for the other conditions. The media was collected and frozen at 72 h. The secretome of MIAMI cells was compared to the secretome of E/F pre-treated MIAMI cells cultured for 7–10 days. The secretome of E/F MIAMI cells was then analyzed after exposure to different conditions: after commitment into GABA-like progenitors (transfected by the si-REST and exposure during 14 days to SHH, and adhered on PAMS during 72H with Neurobasal media and 1 mM of Valproic acid. Eight human growth factors (Brain-derived neurotrophic factor (BDNF), beta polypeptide-Nerve growth factor (b-NGF), stem cell factor (SCF), Leukemia inhibitory factor (LIF), Hepatocyte growth factor (HGF), Placenta growth factor-1 (PlGF-1), Stromal cell-derived factor 1 alpha (SDF-1alpha), Vascular endothelial growth factor-A (VEGFa)) were quantified using 2 Luminex® Multiplex assay panels: ProcartaPlex® Human Chemokine Panel I 9 Plex (#EPX090-12187-901, ThermoFisher), ProcartaPlex® Human Chemokine Panel 11Plex (#EPX110-12170-901, ThermoFisher). Samples were centrifuged at 4 °C for 10 min at 10,000 g and prepared as per the manufacturer's recommendations using a Bio-Plex Pro wash station (Bio-Rad, Hercules, CA). No sample dilution was performed. Quantification of growth factors was performed on a Magpix apparatus (Bio-Rad) and analyzed with the Bio-Plex Manager Version 3.0 software (Bio-Rad). Appropriate media were used as control and to determine the background. Background subtraction was performed with appropriate media depending on the sample, $N = 2$, $n = 2$. N represents one independent experiment done at one moment using 375,000 MIAMI cells adhered to 0.5mg of PAMs for each condition tested, n is the number of samples for each condition.

2.16. Data Analysis

Data are presented as the mean value of three independent experiments +/− standard deviation (SD) unless otherwise stated. Significant differences between samples were determined using an ANOVA test, followed by Dunnett's multiple comparison tests, unless otherwise stated. Luminex data was analyzed with Kruskal–Wallis test followed by pair-wise comparison. Threshold *p*-value was set to 0.05 and significant differences was depicted with a "*".

3. Results

3.1. MIAMI E/F Cell Transfection

LNCs carrying REST siRNA of around 85 nm size and a positive surface charge of +7 mV were used for the incorporation of siREST into MIAMI E/F cells. To better understand the increase of neuronal markers in MSCs with siREST-LNCs [38], fluorescence confocal multispectral imaging was used to characterize LNC uptake and siRNA delivery on MIAMI E/F cells (Figure 2A). Triple-labeled LNCs, siRNA and lysosomes were generated, with respectively DiD, Alexa-488 and lysotracker to follow the distribution and localization of siRNA-LNCs over time. Four hours after the contact between LNCs and MIAMI E/F cells, referred to as Day 0, a heterogeneous distribution of LNCs within MIAMI E/F cells was observed (Figure 2A). LNCs (DiD-positive) co-localized with lysosome staining from Day 2 to Day 6 but the same co-localization of the siRNA (Alexa 488) was not observed confirming the results previously published by our team [40]. The number of fluorescent cells for the LNCs and siRNA decreased progressively from Day 0 until Day 9 (Figure 2B). Twelve percent of cells were still DiD-positive at Day 6 (Figure 2B). REST mRNA was quantified by RT-qPCR (Figure 2C) and down-regulated to 70.7% with the siREST at Day 5 when compared to the control cells (siCTRL), receiving a scrambled siRNA.

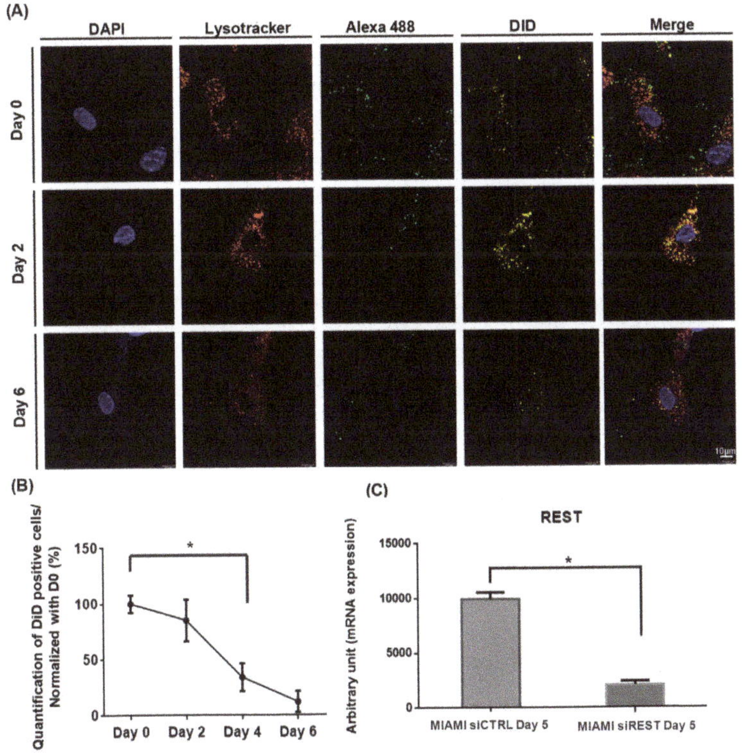

Figure 2. Cellular uptake and cell time retention of siRNA-LNC in MIAMI E/F cells. (**A**) After 4 h incubation, confocal microscopy was performed on MIAMI cells with siRNA-LNCs. Cells were fixed on glass slide and nucleus and lysosome staining were performed with DAPI (blue) and lysotracker (Red). Double fluorescent probes were used to follow siRNA and LNCs: lipophilic DiD (yellow) and Alexa488 siRNA (green). Scale bar represents 10µm. Analysis confirmed the internalization of siRNA-LNCs and its presence until Day 6. (**B**) DiD-positive cells representing the LNC-positive cells were counted using imageJ. 6 images per conditions in 10× objective were selected. (**C**) The expression of REST, measured by RT-qPCR, was decreased significatively (Fold decrease: 4.80 ± 1.80) at Day 5 after transfection compared to the control. $N = 3$. * Significantly different means at $p < 0.05$.

3.2. *MIAMI E/F Cells Commitment into GABA-Like Progenitors*

To ensure that the transplanted cells have a similar phenotype to the neurons in the striatum and thus warrant the safety of their transplantation in that area, the MIAMI E/F cells have been differentiated into GABA-like neurons. Cell density during the first steps is important and the confluence has to be maintained under 30% to induce cell differentiation. The MIAMI E/F cells were transfected with siREST or with a negative control, scrambled siRNA, followed by sonic hedgehog (SHH) treatment (Figure 3A). Some of the MIAMI E/F cells exhibited small neurite-like structures. Fourteen days after transfection, most of the MIAMI-SHH-siREST cells exhibited very long neurite-like structures. The majority of the MIAMI-SHH-siREST cells presented a total length of around 600 µm. In comparison MIAMI-SHH siCTRL cells showed a flat morphology (Figure 3B). REST expression, detected by RT-qPCR, was diminished over time during the treatment when compared to the MIAMI E/F cells. As expected at this time-point no difference in REST expression was observed between the cells transfected with siREST and siCtrl. Pax 6 and Dlx2, transcription factors detected in the ventral telencephalon, were expressed by MIAMI E/F cells and decreased slightly 14 days after the treatment (Figure 3C), The

GABAergic marker, GAD67, was slightly increased following the treatment, at the mRNA and protein level (Figure 3C) and not detected in MIAMI E/F cells, as for DARPP32. No major differences were detected between the siREST and siCtrl conditions at this long-term time point, although there was a tendency in the MIAMI-SHH-siREST cells to express more GAD67 at the mRNA and protein level (Figure 3C,D). The early and late neuronal markers, β3-tubulin and NFM, respectively were expressed by the MIAMI-SHH-siREST cells and were only very slightly detected in MIAMI-SHH-siCtrl as shown by the immunofluorescence staining (Figure 3D).

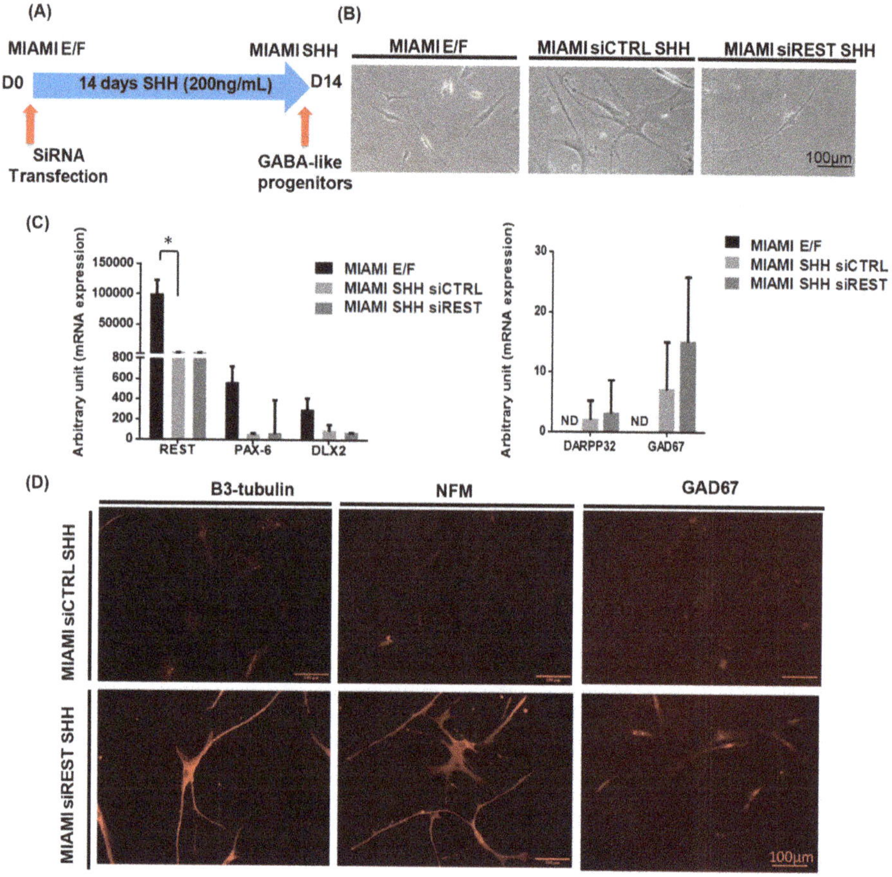

Figure 3. MIAMI E/F cell commitment into GABA-like progenitors. (**A**) For GABA-like progenitors commitment, a simple protocol has been designed using siREST and SHH. Transfection of MIAMI E/F was performed with 250 ng/mL of siCtrl and siREST-LNCs. (**B**) After the culture period of 14 days, MIAMI-SHH-siREST emitted long neuritis exhibiting a neuron-like morphology compared to the flat morphology of MIAMI-SHH-siCtrl. (**C**) The expression of neuronal commitment genes (REST, PAX-6, and Dlx2) and of genes of GABAergic-like neurons (DARPP32 and GAD67) was quantified at 14 days by RT-qPCR in MIAMI E/F, MIAMI-SHH-siREST and MIAMI-SHH-siCTRL cells ($N = 3$). (**D**) After immunofluorescence staining, the neuronal commitment of MIAMI-SHH-siREST cells was confirmed by the expression of β3-tubulin and NFM and a very slight expression of GAD67, all of which are not detected in MIAMI-SHH-siCTRL cells. ND: not detected. * Significantly different means at $P < 0.05$.

3.3. *MIAMI E/F Cell Commitment into GABAergic-Like Neurons*

The GABA-like progenitors were exposed to Valproic acid (VPA) from Day 14 for seven days and BDNF from Day 21 for seven more days or during 15 days to VPA and BDNF in the same time, without SHH (Figure 4A) in order to obtain GABAergic-like neurons. At this step, no morphological change was observed between cells transfected with siCtrl or siREST, but the cells differentiated using VPA and BDNF in the same time exhibited longer neurite-like structure than the cells receiving the protocol using VPA and then BDNF (data not shown). In the same way, the expression of GABA markers detected by RT-qPCR (Figure 4B) is higher with the protocol using VPA together with BDNF compared to the sequential protocol (VPA then BDNF). At the end of the differentiation period, a low expression of Dlx2 and REST was detected by RT-qPCR (data not shown). Moreover, the inhibition of REST had no effect on the expression of GABAergic neuron markers with a tendency to be diminished for GAD67. The majority of the GABA-like neurons were positive β3-tubulin and some for DARPP32, GAD 67 and GAT1 after immunofluorescence, a feature of medium spiny neurons (Figure 4C).

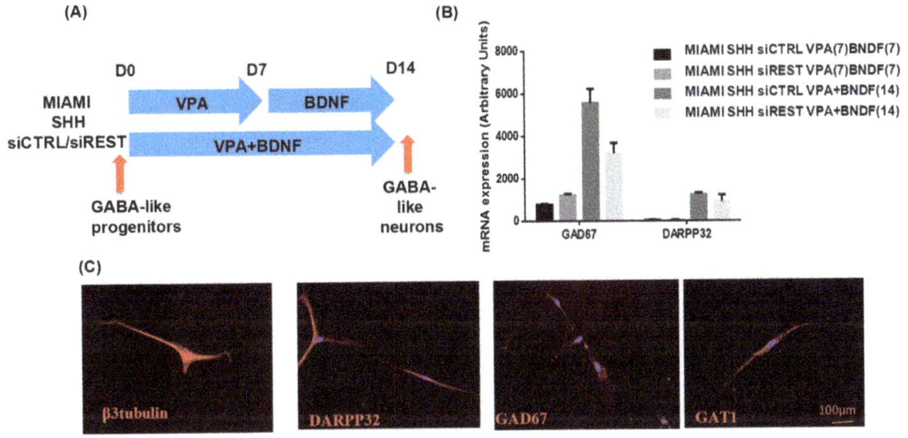

Figure 4. Differentiation and Characterization of GABA-like neurons. (**A**) Schematic procedure of GABAergic differentiation. For both protocols, the concentration of VPA was 10 mM and 30 ng/mL of BDNF). (**B**) The characterization of the differentiation was realized by RT-qPCR. The expression of genes DARPP32 and GAD67 was quantified at the end of the differentiation ($N = 2$, $n = 1$). (**C**) In vitro immunofluorescence against β3-tubulin, DARPP32, GAD67, GAT1 on GABA-like neurons was performed.

3.4. *Adherence of MIAMI E/F SHH-siREST on PAMs*

The particle size of PAMs measured using a Multisizer Coulter Counter was around 30 µm, as also observed using SEM (Figure 5A). LM completely covered the whole surface of PAMs in small patches (Figure 5B) and a continuous release of BDNF was observed over time. Up to 3000 ng/mL of BDNF was progressively released during 40 days from 5 mg PAMs (Figure 5C). During the commitment and the differentiation, no serum and no antibiotics were used. In order to respect these conditions, different media for cell adherence were tested (Figure 5D). Unfortunately, a high proportion of cells were observed floating in the media or adhered to the plastic, particularly with N2 and also with B27 media (Figure 5D). On the contrary, with the 3% serum, the cells adhered nicely onto the PAMs (Figure 5D). Different times of adherence were tested, and it was observed that the adherence of the GABA-like progenitors increased over time; it was better after 6 h and 72 h in contact with PAMs compared to only 4 h (Figure 5E). At 4 h, a large quantity of cells was not combined to the PAMs, but cell/PAM aggregates could be observed from 6 h onwards. (Figure 5E). Since the cells completely surrounded the PAMs in Figure 5E, the PAMs could not directly be seen anymore. Using SEM, MIAMI

E/F SHH-siREST cells were observed to adhere onto the PAMs with their lamellipodia surrounding the PAMs to form three-dimensional complexes (Figure 5F). The percentage of cells adhered onto PAMs' surface at the end of the cell attachment protocol (72 h) was about 95.2 ± 2.7%, as measured using the CyQUANT DNA quantification method.

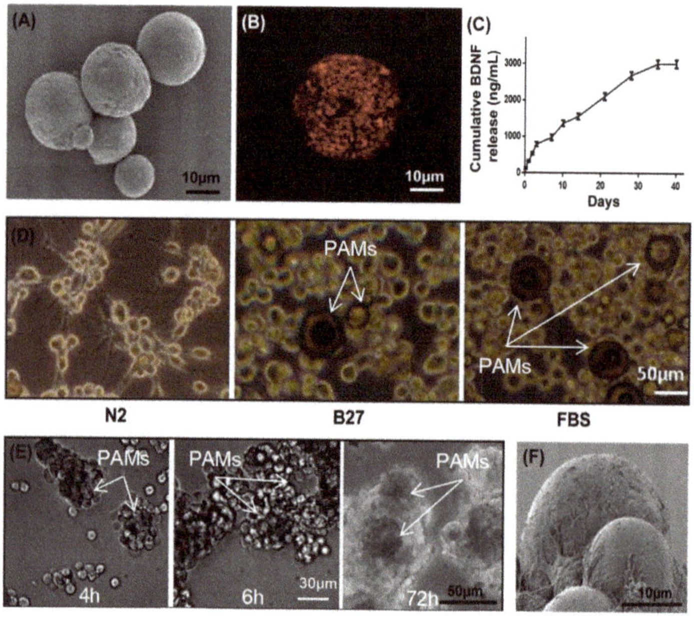

Figure 5. PAM characterization and adherence of GABA-like progenitors on PAMs. (**A**) PAMs observed using SEM and (**B**) laminin overlay onto PAMs observed by confocal microscopy after immunofluorescence. (**C**) Controlled release of BDNF from PAMs, measured using ELISA. (**D**) Different media were tested for the adher ence: 2% of B27, 15% of N2 and 3% of FBS. The last condition has been chosen for the rest of the experiments. (**E**) The cells were incubated with PAMs and pictures taken by brightfield microscopy at 4 h, 6 h and 72 h. (**F**) Observation of microspheres and cells-PAMs complexes by scanning electronic microscopy showing the cells adhering onto the PAMs. White arrows point at PAMs associated with cells.

3.5. *Characterization of MIAMI-SHH-siREST on PAMs*

MIAMI E/F cells secreted a few growth factors, but interestingly SHH/siREST exposure affected the secretory profile of MIAMI E/F cells, which secreted more hepatocyte growth factor (HGF) and had a tendency to secrete more leukemia inhibitory factor (LIF) and vascular endothelial growth factor-a (VEGFa). They also secreted SDF-1α (Figure 6A, Table A1). As observed using RT-qPCR, the GABA-like progenitors expressed less BDNF and nerve growth factor (NGF), but after 72 h of adherence of these cells on the PAMs, the expression of BDNF, NGF and VEGFa increased (Figure 6B). The expression of GAD67 also tended to increase, while DARPP32 remained very low (23.00 ± 12.82 mRNA arbitrary units) At the protein level, the secretome analysis confirmed the release by the cells adhered onto PAMs of the above mentioned growth factors in the media as well as FGF2, EGF and stem cell factor (SCF). There was more placental growth factor (PIGF-1) and of course BDNF when the cells adhered onto PAMs releasing BDNF compared to blank PAMs (Figure 6C, Table A1).

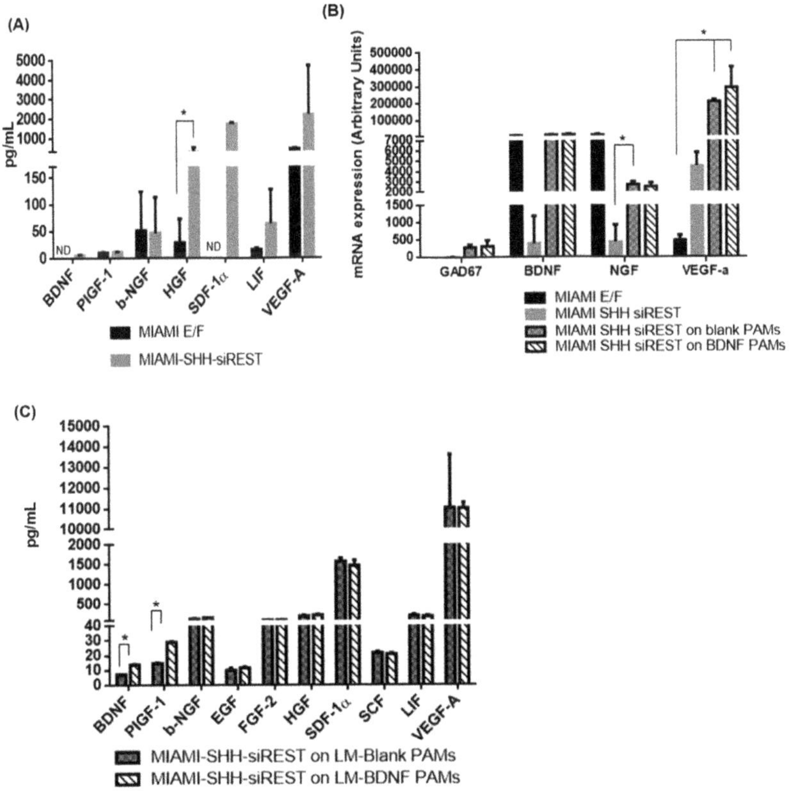

Figure 6. Characterization of the secretome profile of the GABA-like progenitors combined with PAMs. (**A**) Media collected from MIAMI E/F cells or MIAMI-SHH-SiREST cells were analyzed by the Luminex apparatus for the quantification of growth factors and cytokines ($N = 2$, $n = 2$). (**B**) The expression of GABAergic genes and GAD67 and growth factors (BDNF, NGF and VEGFa) was quantified by RT-qPCR 72H after adherence to the PAMs ($N = 3$). (**C**) Media collected 72 h after adherence between cells and PAMs were analyzed by the Luminex apparatus for the quantification of growth factors and cytokines ($N = 2$, $n = 2$) N represents one independent experiment with all the conditions, n the number of samples/condition. * Significantly different means at $P < 0.05$.

3.6. Behavior of PAMs and MIAMI-SHH-siREST in Huntington's Disease Model

MIAMI-SHH-siREST alone, pre-treated with BDNF, or complexed to PAMs or BDNF-PAMS were grafted in the recently reported ex vivo HD model [39] at Day 5 when 30% of GABAergic striatal cell degeneration has occurred. Immunofluorescence against human mitochondria was used to visualize MIAMI-SHH-siREST cells in the rat brain (Figure 7A) and antibody against human DARPP32 was used to localize GABAergic-like neurons 15 days after grafting in the HD organotypic slices (Figure 7B). Immunofluorescence staining was faint with cells alone, suggesting that some cells died. PAMs delivering or not BDNF improved cell survival and BDNF administered to the cells before transplantation also seemed to slightly improve survival. Moreover, the cells receiving BDNF showed an elongated morphology with neurite-like structures (Figure 7A). Some of the transplanted cells expressed DARPP32 in all the conditions studied except for the cells transplanted alone (Figure 7B). GAD67 and DARPP32 immunostaining of the GABAergic rat striatal cells was more intense after transplantation of MIAMI-SHH-siREST cells on PAMs particularly when delivering BDNF (Appendix A Figure A1).

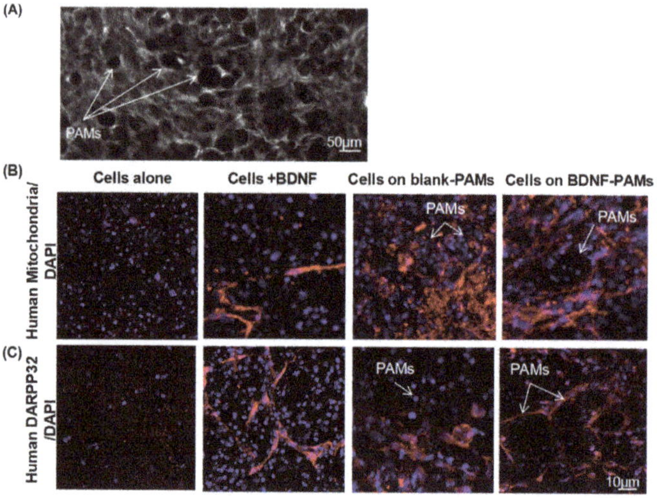

Figure 7. Behavior of MIAMI-SHH-siREST grafted into an ex vivo model of HD. (**A**) Black and white picture at lower magnification showing the PAMs with the transplanted cells within the tissue. All the cells are visualized by DAPI staining and the transplanted cells by immunohistofluorescence against human mitochondria. (**B**) Immunohistofluorescence against human mitochondria (in red) and (**C**) human DARPP32 (in red) for MIAMI cells 14 days after grafting, either alone, or exposed to BDNF for three days before grafting, or on LM-blank PAMs for three days before grafting or on LM-BDNF PAMs for three days before grafting. Red circles were suggested on the images to emphasize the localization of PAMs. DAPI staining (in blue) was used to visualize all the cells. Scale bar is 10 μm.

4. Discussion

Mesenchymal stromal cell-based neuronal therapies can provide a limitless, easily accessible source of cells but survival and differentiation remain a drawback. Combinatorial strategies with nano- and microvectors designed to improve stem cell differentiation and engraftment are necessary. In this study, we used a combinatorial cell and drug delivery approach. We showed that a subpopulation of MSCs, the MIAMI cells, can differentiate towards the neuronal GABAergic lineage by an epigenetic RNA interfering approach inhibiting REST expression combined to GABAergic inducers. Moreover, GABA-committed MIAMI cells adhered onto PAMs maintained the GABAergic phenotype and secreted neural tissue repair factors. They could furthermore be transplanted while attached to PAMs delivering BDNF in an ex vivo model of HD, and were able to survive and maintain this GABAergic neuronal phenotype, particularly in response to BDNF.

Our strategy to differentiate MIAMI cells into striatal-like neurons was developed based on the simplifying scheme that the normal course of neuronal differentiation may be separated into three successive steps: namely, (*i*) neural induction, (*ii*) regional commitment, and (*iii*) neuronal maturation. To obtain neural induction, we pre-treated MIAMI cells with EGF and FGF-2 as previously published by our group [23]. As expected, MIAMI E/F expressed Pax6 and slightly Dlx2, showing their progression towards a neural/neuronal phenotype. To induce a strong GABAergic commitment and further neuronal maturation, we chose to combine REST inhibition with adapted published protocols for pluripotent stem cells [41,42]. REST expression is progressively reduced during neuronal differentiation in neural stem cells [14] and its inhibition engaged MSCs into a neuronal pathway [15,38]. We chose to transiently inhibit REST with siREST carried by nanoparticles as previously performed [38]. Permanent down-regulation of REST with strategies such as ShREST would likely not be satisfactory since REST inhibition is only transient during normal brain development [43]. The siREST-LNCs inhibited REST expression for at least 5 days, probably due to their long retention time within the cells. After this transient REST inhibition together with SHH treatment followed by the combination of VPA and BDNF, we first obtained GABA-like progenitors and then immature neurons expressing striatal markers such as GAD67, DARPP32 and GAT1. The main difference observed after REST inhibition was a clear neuronal commitment of the GABAergic-like progenitors, which expressed the neuronal markers B3-tubulin and NFM. To our knowledge, our strategy is the first to describe a direct non-epigenetic transient modulation of REST with nanoparticles to direct MSCs toward a GABAergic lineage. Other approaches used to inhibit REST are burdened by intrinsic limitations such as the HIV origin of the product and host genome integrations of lentivirus or oligodeoxynucleotides, which may lead to unwanted side effects.

MIAMI cells committed or not towards a neuronal phenotype easily adhere onto PLGA PAMs with a biomimetic surface composed of extracellular matrix molecules in the presence of low serum concentrations [26,29,30]. Different serum free media were evaluated for the formation of PAM/cell complexes to avoid washing steps before implantation to dispose of the undesired serum. Moreover, in this way, the implantation procedure is simplified, and the cells implanted in their own conditioned media already containing some BDNF secreted by the cells and delivered from the PAMS. However, the GABA-like progenitor cells only adhered to PLGA-P188-PLGA PAMs with a LM biomimetic surface in the presence of 3% serum. In this study, we observed that MIAMI E/F derived GABA-like progenitors express more GAD67, a constitutive striatal marker, when adhered onto PAMs compared to the cells in a 2D culture. The laminin-covered PAMs stimulate this differentiation as it has been shown that laminin stimulates neuronal differentiation of MSCs [44] and of MIAMI cells [8]. Nevertheless, the 3D condition should also contribute as our previous results showed that PAMs offering a 3D biomimetic surface stimulate MIAMI cells neuronal differentiation [30]. In a similar manner neural stem/progenitor cells also more efficiently differentiated to neurons in a collagen/hyaluronan 3D matrix compared to a 2D culture condition [45].

MSCs have recently emerged as a promising cell population to protect degenerating neuronal cells and increase the function of the remaining cells in various neuronal disorders. In HD preclinical models, the functional recovery observed after MSC transplantation could be explained by the secretion of neurotrophic factors including BDNF, CNTF, NGF, insulin-like growth factor 1 and EGF [10,11]. Within this line, previous studies have shown that MIAMI cells, which are a homogeneous population of cells with an unique genetic profile, express several pluripotency markers (Oct4, Sox2, Nanog, SSEA4) and other proteins also expressed by embryonic stem cells, release more tissue repair factors than MSCs or embryonic stem cells and protect the neurovasculature [45]. Compared to naïve MIAMI cells, MIAMI cells pre-treated with E/F secreted more tissue repair factors such as VEGFA, NGF, LIF and HGF [30], and we observed that SHH/siREST treatment further increased the expression of HGF, but also of BDNF and SDF-1α. SDF-1α is an indispensable chemoattractant for neuron migration in different brain regions (CXCR4 regulates interneuron migration in the developing neocortex.). Moreover, it has been demonstrated that SDF-1 coexists in GABA-containing vesicles in the terminals of basket cells, regulates the strength of GABAergic input to nestin^{+}-type 2 neural progenitors in hippocampal dentate gyrus, and plays a crucial role in adult neurogenesis [46], while laminin also enhances the chemotactic activity of SDF-1 in the thymus [47].

We previously demonstrated that PAMs delivering neurotrophin-3 can improve MIAMI cells survival, thereby preserving neural function in a Parkinson's disease animal model [8,22]. PAMs encapsulating BDNF and composed of a triblock copolymer of PLGA-P188-PLGA were used in this study as we previously showed that it allowed for a complete release of functionally active growth factors from the microspheres [28,30]. Exposure of MIAMI E/F cells to SHH/siREST and adhesion to PAMs had a drastic effect on the expression of NGF and VEGF at the mRNA level, and we also observed an augmentation at the protein level of BDNF and PIGF-1 in the media when cells were attached to BDNF-loaded PAMs compared to blank PAMs. The increase in BDNF in the media was likely due to the release of BDNF from PAMs since no difference was observed at the mRNA level in MIAMI SHH/siREST attached to PAMs with or without encapsulated BDNF. VEGF and NGF are known to increase neurogenesis [48], while VEGF, NGF, and PIGF-1 also all share in common the possibility of being of benefit to enhance reparative angiogenesis [49,50]. In this regard, we demonstrated in another study that VEGF secreted by MIAMI cells can improve vascularization maintenance after grafting in an ex vivo model of Parkinson's disease [22]. These findings thus support the use of PAMs to enhance the benefit of MIAMI SHH/siREST in the context of HD. In this study, we tested this innovative strategy combining mesenchymal stromal cells and biomaterials in an ex vivo model of HD [39]. The benefit of this model is that it preserves cell graft-host tissue crosstalk, while maintaining the cytoarchitecture of the original tissue. We here demonstrated the screening capacity of this HD model for new therapeutic strategies. After grafting, GABA-like progenitors alone did not survive while if associated to PAMs the 3D support seemed to stimulate the survival of the cells after transplantation as already demonstrated by our group [22–24]. BDNF pre-treatment also stimulated cell survival while the combination of the BNDF delivery and the 3D polymeric support, provided by the PAMs, seemed to further enhance this survival. Moreover, these GABA-like progenitors submitted to BDNF, known to induce neuronal survival and differentiation [30–32] showed an elongated morphology with neurite-like structures suggesting the maintenance of their neuronal phenotype. In this regard, they also expressed DARPP32, a GABAergic marker. Finally, the combination of cells and PAMs delivering BDNF seemed to delay the degeneration of GABAergic neurons in the HD model, thereby demonstrating the potential benefits of this strategy for the treatment of HD. Further studies in a transgenic mouse model of HD are however needed to obtain confirmation that this combinatorial therapeutic strategy leads to neurorepair and provides a functional benefit.

5. Conclusions

We here demonstrated the therapeutic potential of this novel combinatorial strategy using LNC-delivered siREST and neuronally-committed MSCs transported by laminin-covered microcarriers delivering BDNF. We showed the capacity of MIAMI E/F cells to differentiate into GABAergic-like neurons by using REST inhibition and the appropriate media cues in vitro. GABA-like progenitors adhered to PAMs increased their mRNA expression of NGF/VEGF-a as well as their secretion of PIGF-1, three chemokines known for their potential to enhance reparative angiogenesis. Together, our results indicate that MIAMI cell/PAMs complexes survive after transplantation in an ex vivo HD model, maintain a GABAergic-like phenotype and may alleviate cell damage in the context of HD through chemokine secretion.

Author Contributions: Conceptualization, E.M.A. and C.N.M.-M.; Formal analysis, E.M.A. and G.J.D.; Funding acquisition, C.N.M.-M.; Investigation, E.M.A., G.J.D., S.K., L.S. and C.N.M.-M.; Methodology, E.M.A., G.J.D. and S.K.; Project administration, L.S. and C.N.M.-M.; Supervision, C.N.M.-M.; Validation, C.N.M.-M.; Writing—original draft, E.M.A.; Writing—review & editing, G.J.D. and C.N.M.-M.

Funding: This research was funded by La Région Pays de la Loire, grant Name: "MECASTEM".

Acknowledgments: The authors thank the SCAHU (Service commune d'animalerie hospital-universitaire), particularly Pierre Legras and Jérôme Roux for animal care, the SCIAM ("Service Commun d'Imageries et d'Analyses Microscopiques") of Angers for the confocal microscopy as well as the PACeM (Plateforme d'Analyse Cellulaire et Moléculaire) of Angers for the use of PCR facilities. We also wanted to thank Professor Paul C. Schiller for helpful critiques in experiments. This project is supported by the "Education Audiovisual" and the executive cultural agency of the European Union through the NanoFar Erasmus Mundus joint Doctoral program, by "Angers Loire Métropole" and by "la Région Pays de La Loire grant name MECASTEM" & "Inserm", France.

Conflicts of Interest: The authors declare no conflicts of interest.

Appendix A

Immunohistochemistry. A mouse anti-GAD67 (5 µg/mL, clone 1G10.2, Millipore SA, Guyancourt, France) and a mouse anti-DARPP32 (0.25 µg/mL, clone 15, BD Bioscience, Le Pont de Claix, France) antibodies were used to observe striatal-GP GABAergic neurons. Isotypic antibodies were used to control background staining. After 2 h of saturation in DPBS, 4% BSA, 1% triton and 10% NGS, slices were incubated 48 h with the primary antibody diluted in DPBS, 4% BSA and 1% triton at 4 °C.

After washes, slices were incubated with the biotinylated anti-mouse secondary antibody (2.5 μg/mL Vector Laboratories, Burlingame, CA, USA). Then slices were washed, and quenching of peroxidase was performed with 0.3% H_2O_2 (Sigma, Saint Quentin Fallavier, France) in DPBS, at room temperature for 20 min. After DPBS washes, slices were incubated with Vectastain ABC reagent (Vector Laboratories, Eurobio, Les Ulis, France) in DPBS at room temperature for 2 h. Sections were then washed and revealed with 0.03% H_2O_2, 0.4 mg/mL diaminobenzidine (DAB, Sigma, Saint Quentin Fallavier, France) in DPBS, 2.5% nickel chloride (Sigma, Saint Quentin Fallavier, France) and dehydrated before mounting.

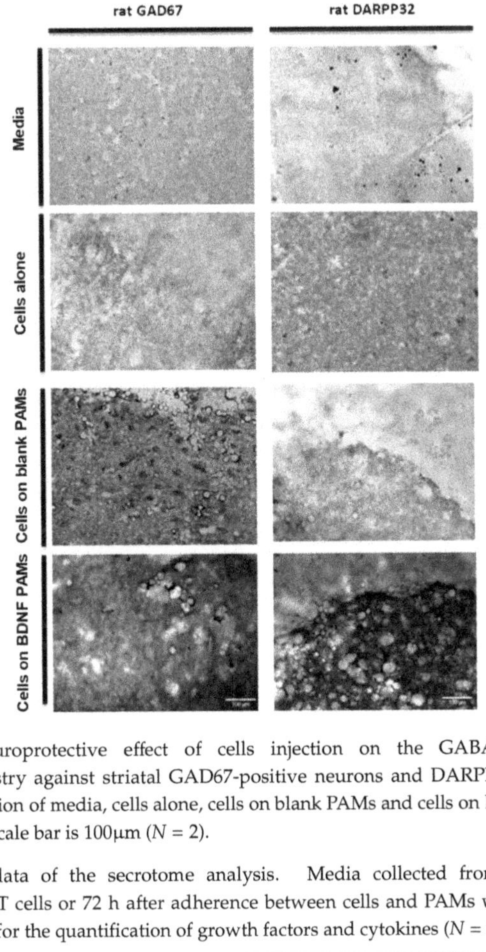

Figure A1. Neuroprotective effect of cells injection on the GABAergic degeneration. Immunohistochemistry against striatal GAD67-positive neurons and DARPP32-positive neurons at Day 25 after injection of media, cells alone, cells on blank PAMs and cells on BDNF PAMs in the ex vivo model of HD. Scale bar is 100μm (N = 2).

Table A1. Raw data of the secrotome analysis. Media collected from MIAMI E/F cells, MIAMI-SHH-SiREST cells or 72 h after adherence between cells and PAMs were analyzed by the Luminex apparatus for the quantification of growth factors and cytokines (N = 2, n = 2).

Heading	BDNF	PIGF-1	b-NGF	HGF	SDF-1	LIF	VEGF-A
MIAMI E/F media (Control)	4.55 ± 0.294	14.36 ± 0.20	0.00	124.93 ± 1.52	481.95 ± 3.27	49.35 ± 3.43	34.21 ± 0.676
Supernatant MIAMI E/F	1.85 ± 0.139	24.38 ± 1.25	51.07 ± 72.22	152.87 ± 42.87	385.73 ± 52.87	65.80 ± 2.54	482.53 ± 32.54
Step 1 Media (Control)	0.00	0.00	0.00	0.00	13.61 ±0.65	0.00	0.00
supernatant MIAMI E/F SHH siREST	5.95 ± 0.59	11.70 ± 0.26	46.68 ± 66.05	304.12 ± 216.21	1746.64 ± 63.87	63.78 ± 64.24	2205.48 ± 505.54
Step 2 Media without BDNF (Control)	0.00	0.00	0.00	0.00	14.76 ± 0.01	0.00	0.00
supernatant MIAMI E/F SHH siREST on LM-blanks PAMS	7.21 ± 0.19	14.73 ± 0.46	128.04 ± 12.21	192.08 ± 25.37	1567.67 ± 92.07	195 ± 39.81	11,028.92 ± 2576.19
supernatant MIAMI E/F SHH siREST on LM-BDNF PAMS	13.77 ± 0.35	29.02 ± 0.41	148.30 ± 6.18	219.43 ± 15.76	1467 ± 138.48	180.11 ± 20.07	11,003.58 ± 1189.29

References

1. MacDonald, M.E.; Ambrose, C.M.; Duyao, M.P.; Myers, R.H.; Lin, C.; Srinidhi, L.; Barnes, G.; Taylor, S.A.; James, M.; Groot, N.; et al. A novel gene containing a trinucleotide repeat that is expanded and unstable on Huntington's disease chromosomes. *Cell* **1993**, *72*, 971–983. [CrossRef]
2. Cattaneo, E.; Rigamonti, D.; Goffredo, D.; Zuccato, C.; Squitieri, F.; Sipione, S. Loss of normal huntingtin function: new developments in Huntington's disease research. *Trends Neurosci.* **2001**, *24*, 182–188. [CrossRef]
3. Klempíř, J.; Klempířová, O.; Štochl, J.; Špačková, N.; Roth, J. The relationship between impairment of voluntary movements and cognitive impairment in Huntington's disease. *J. Neurol.* **2009**, *256*, 1629–1633. [CrossRef] [PubMed]
4. Clabough, E.B. Huntington's Disease: The Past, Present, and future search for Disease Modifiers. *Yale J. Biol. Med.* **2013**, *86*, 217. [PubMed]
5. Fink, K.D.; Deng, P.; Torrest, A.; Stewart, H.; Pollock, K.; Gruenloh, W.; Annett, G.; Tempkin, T.; Wheelock, V.; Nolta, J.A. Developing stem cell therapies for juvenile and adult-onset Huntington's disease. *Regen. Med.* **2015**, *10*, 623–646. [CrossRef]
6. Ramaswamy, S.; Kordower, J.H. Gene therapy for Huntington's disease. *Neurobiol. Dis.* **2012**, *48*, 243–254. [CrossRef]
7. Snyder, B.R.; Chiu, A.M.; Prockop, D.J.; Chan, A.W.S. Human multipotent stromal cells (MSCs) increase neurogenesis and decrease atrophy of the striatum in a transgenic mouse model for Huntington's disease. *PloS One* **2010**, *5*, e9347. [CrossRef]
8. Delcroix, G.J.-R.; Garbayo, E.; Sindji, L.; Thomas, O.; Vanpouille-Box, C.; Schiller, P.C.; Montero-Menei, C.N. The therapeutic potential of human multipotent mesenchymal stromal cells combined with pharmacologically active microcarriers transplanted in hemi-parkinsonian rats. *Biomaterials* **2011**, *32*, 1560–1573. [CrossRef]
9. André, E.M.; Passirani, C.; Seijo, B.; Sanchez, A.; Montero-Menei, C.N. Nano and microcarriers to improve stem cell behaviour for neuroregenerative medicine strategies: Application to Huntington's disease. *Biomaterials* **2016**, *83*, 347–362. [CrossRef]
10. Emerich, D.F.; Cain, C.K.; Greco, C.; Saydoff, J.A.; HU, Z.Y.; Liu, H.; Lindner, M.D. Cellular delivery of human CNTF prevents motor and cognitive dysfunction in a rodent model of Huntington's disease. *Cell Transplant.* **1997**, 249–266. [CrossRef]
11. Jiang, Y.; Hailong, L.; Shanshan, H.; Tan, H.; Zhang, Y.; Li, H. Bone marrow mesenchymal stem cells can improve the motor function of a Huntington's disease rat model. *Neurol. Res.* **2011**, 331–337. [CrossRef] [PubMed]
12. Zheng, Y.; Huang, C.; Liu, F.; Lin, H.; Yang, X.; Zhang, Z. Comparison of the neuronal differentiation abilities of bone marrow-derived and adipose tissue-derived mesenchymal stem cells. *Mol. Med. Rep.* **2017**, *16*, 3877–3886. [CrossRef] [PubMed]
13. Ballas, N.; Mandel, G. The many faces of REST oversee epigenetic programming of neuronal genes. *Curr. Opin. Neurobiol.* **2005**, *15*, 500–506. [CrossRef] [PubMed]
14. Ballas, N.; Grunseich, C.; Lu, D.D.; Speh, J.C.; Mandel, G. REST and Its Corepressors Mediate Plasticity of Neuronal Gene Chromatin throughout Neurogenesis. *Cell* **2005**, *121*, 645–657. [CrossRef] [PubMed]
15. Yang, Y.; Li, Y.; Lv, Y.; Zhang, S.; Chen, L.; Bai, C.; Nan, X.; Yue, W.; Pei, X. NRSF silencing induces neuronal differentiation of human mesenchymal stem cells. *Exp. Cell Res.* **2008**, *314*, 2257–2265. [CrossRef] [PubMed]
16. Chen, Z.F.; Paquette, A.J.; Anderson, D.J. NRSF/REST is required in vivo for repression of multiple neuronal target genes during embryogenesis. *Nat. Genet.* **1998**, *20*, 136–142. [CrossRef] [PubMed]
17. Heurtault, B.; Saulnier, P.; Pech, B.; Proust, J.E.; Benoit, J.P. A novel phase inversion-based process for the preparation of lipid nanocarriers. *Pharm. Res.* **2002**, *19*, 875–880. [CrossRef] [PubMed]
18. Morille, M.; Montier, T.; Legras, P.; Carmoy, N.; Brodin, P.; Pitard, B.; Benoît, J.-P.; Passirani, C. Long-circulating DNA lipid nanocapsules as new vector for passive tumor targeting. *Biomaterials* **2010**, *31*, 321–329. [CrossRef]
19. Paillard, A.; Hindré, F.; Vignes-Colombeix, C.; Benoit, J.-P.; Garcion, E. The importance of endo-lysosomal escape with lipid nanocapsules for drug subcellular bioavailability. *Biomaterials* **2010**, *31*, 7542–7554. [CrossRef]
20. D'Ippolito, G. Marrow-isolated adult multilineage inducible (MIAMI) cells, a unique population of postnatal young and old human cells with extensive expansion and differentiation potential. *J. Cell Sci.* **2004**, *117*, 2971–2981. [CrossRef]

21. Roche, S.; D'Ippolito, G.; Gomez, L.A.; Bouckenooghe, T.; Lehmann, S.; Montero-Menei, C.N.; Schiller, P.C. Comparative analysis of protein expression of three stem cell populations: Models of cytokine delivery system in vivo. *Int. J. Pharm.* **2013**, *440*, 72–82. [CrossRef] [PubMed]
22. Daviaud, N.; Garbayo, E.; Sindji, L.; Martinez-Serrano, A.; Schiller, P.C.; Montero-Menei, C.N. Survival, Differentiation, and Neuroprotective Mechanisms of Human Stem Cells Complexed With Neurotrophin-3-Releasing Pharmacologically Active Microcarriers in an Ex Vivo Model of Parkinson's Disease. *Stem Cells Transl. Med.* **2015**, *4*, 670–684. [CrossRef] [PubMed]
23. Delcroix, G.J.-R.; Curtis, K.M.; Schiller, P.C.; Montero-Menei, C.N. EGF and bFGF pre-treatment enhances neural specification and the response to neuronal commitment of MIAMI cells. *Differentiation* **2010**, *80*, 213–227. [CrossRef] [PubMed]
24. Garbayo, E.; Raval, A.P.; Curtis, K.M.; Della-Morte, D.; Gomez, L.A.; D'Ippolito, G.; Reiner, T.; Perez-Stable, C.; Howard, G.A.; Perez-Pinzon, M.A.; et al. Neuroprotective properties of marrow-isolated adult multilineage-inducible cells in rat hippocampus following global cerebral ischemia are enhanced when complexed to biomimetic microcarriers: BMMs enhance MIAMI cell protection in cerebral ischemia. *J. Neurochem.* **2011**, *119*, 972–988. [CrossRef] [PubMed]
25. Tatard, V.M.; D'Ippolito, G.; Diabira, S.; Valeyev, A.; Hackman, J.; McCarthy, M.; Bouckenooghe, T.; Menei, P.; Montero-Menei, C.N.; Schiller, P.C. Neurotrophin-directed differentiation of human adult marrow stromal cells to dopaminergic-like neurons. *Bone* **2007**, *40*, 360–373. [CrossRef] [PubMed]
26. Tatard, V.M.; Sindji, L.; Branton, J.(G.); Aubert-Pouëssel, A.; Colleau, J.; Benoit, J.-P.; Montero-Menei, C.N. Pharmacologically active microcarriers releasing glial cell line—derived neurotrophic factor: Survival and differentiation of embryonic dopaminergic neurons after grafting in hemiparkinsonian rats. *Biomaterials* **2007**, *28*, 1978–1988. [CrossRef]
27. Tatard VM, V.-J.M.; Benoit JP, M.P. Montero-Menei CN In vivo evaluation of pharmacologically active microcarriers releasing nerve growth factor and conveying PC12 cells. *Cell Transplant.* **2004**, *13*, 573–583. [CrossRef] [PubMed]
28. Morille, M.; Van-Thanh, T.; Garric, X.; Cayon, J.; Coudane, J.; Noël, D.; Venier-Julienne, M.-C.; Montero-Menei, C.N. New PLGA-P188-PLGA matrix enhances TGF-β3 release from pharmacologically active microcarriers and promotes chondrogenesis of mesenchymal stem cells. *J. Control. Release Off. J. Control. Release Soc.* **2013**, *170*, 99–110. [CrossRef]
29. Karam, J.-P.; Bonafè, F.; Sindji, L.; Muscari, C.; Montero-Menei, C.N. Adipose-derived stem cell adhesion on laminin-coated microcarriers improves commitment toward the cardiomyogenic lineage. *J. Biomed. Mater. Res. A* **2015**, *103*, 1828–1839. [CrossRef]
30. Kandalam, S.; Sindji, L.; Delcroix, G.J.-R.; Violet, F.; Garric, X.; André, E.M.; Schiller, P.C.; Venier-Julienne, M.-C.; des Rieux, A.; Guicheux, J.; et al. Pharmacologically active microcarriers delivering BDNF within a hydrogel: Novel strategy for human bone marrow-derived stem cells neural/neuronal differentiation guidance and therapeutic secretome enhancement. *Acta Biomater.* **2017**, *49*, 167–180. [CrossRef]
31. Simmons, D.A.; Rex, C.S.; Palmer, L.; Pandyarajan, V.; Fedulov, V.; Gall, C.M.; Lynch, G. Up-regulating BDNF with an ampakine rescues synaptic plasticity and memory in Huntington's disease knockin mice. *Proc. Natl. Acad. Sci. USA* **2009**, *106*, 4906–4911. [CrossRef] [PubMed]
32. Baydyuk, M.; Xu, B. BDNF signaling and survival of striatal neurons. *Front. Cell. Neurosci.* **2014**, *8*, 254. [CrossRef] [PubMed]
33. Zuccato, C. Loss of Huntingtin-Mediated BDNF Gene Transcription in Huntington's Disease. *Science* **2001**, *293*, 493–498. [CrossRef] [PubMed]
34. Canals, J.M.; Pineda, J.R.; Torres-Peraza, J.F.; Bosch, M.; Martín-Ibañez, R.; Muñoz, M.T.; Mengod, G.; Ernfors, P.; Alberch, J. Brain-derived neurotrophic factor regulates the onset and severity of motor dysfunction associated with enkephalinergic neuronal degeneration in Huntington's disease. *J. Neurosci. Off. J. Soc. Neurosci.* **2004**, *24*, 7727–7739. [CrossRef] [PubMed]
35. Kells, A.P.; Henry, R.A.; Connor, B. AAV–BDNF mediated attenuation of quinolinic acid-induced neuropathology and motor function impairment. *Gene Ther.* **2008**, *15*, 966–977. [CrossRef]
36. Terzić, J.; Saraga-Babić, M. Expression pattern of PAX3 and PAX6 genes during human embryogenesis. *Int. J. Dev. Biol.* **1999**, *43*, 501–508. [PubMed]
37. Garel, S.; Yun, K.; Grosschedl, R.; Rubenstein, J.L.R. The early topography of thalamocortical projections is shifted in Ebf1 and Dlx1/2 mutant mice. *Development* **2002**, *129*, 5621–5634. [CrossRef]

38. Andre, E.M.; Pensado, A.; Resnier, P.; Braz, L.; da Costa, A.R.; Passirani, C.; Sanchez, A.; Montero-Menei, C.N. Characterization and comparison of two novel nanosystems associated with siRNA for cellular therapy. *Int. J. Pharm.* **2015**, *497*, 255–267. [CrossRef]
39. André, E.M.; Daviaud, N.; Sindji, L.; Cayon, J.; Perrot, R.; Montero-Menei, C.N. A novel ex vivo Huntington's disease model for studying GABAergic neurons and cell grafts by laser microdissection. *PLoS ONE* **2018**, *13*, e0193409. [CrossRef]
40. Resnier, P.; LeQuinio, P.; Lautram, N.; André, E.; Gaillard, C.; Bastiat, G.; Benoit, J.-P.; Passirani, C. Efficient in vitro gene therapy with PEG siRNA lipid nanocapsules for passive targeting strategy in melanoma. *Biotechnol. J.* **2014**, *9*, 1389–1401. [CrossRef]
41. Aubry, L.; Bugi, A.; Lefort, N.; Rousseau, F.; Peschanski, M.; Perrier, A.L. Striatal progenitors derived from human ES cells mature into DARPP32 neurons in vitro and in quinolinic acid-lesioned rats. *Proc. Natl. Acad. Sci. USA* **2008**, *105*, 16707–16712. [CrossRef] [PubMed]
42. Ma, L.; Hu, B.; Liu, Y.; Vermilyea, S.C.; Liu, H.; Gao, L.; Sun, Y.; Zhang, X.; Zhang, S.-C. Human Embryonic Stem Cell-Derived GABA Neurons Correct Locomotion Deficits in Quinolinic Acid-Lesioned Mice. *Cell Stem Cell* **2012**, *10*, 455–464. [CrossRef] [PubMed]
43. Palm, K.; Belluardo, N.; Metsis, M.; Timmusk, T. Neuronal expression of zinc finger transcription factor REST/NRSF/XBR gene. *J. Neurosci. Off. J. Soc. Neurosci.* **1998**, *18*, 1280–1296. [CrossRef]
44. Mruthyunjaya, S.; Manchanda, R.; Godbole, R.; Pujari, R.; Shiras, A.; Shastry, P. Laminin-1 induces neurite outgrowth in human mesenchymal stem cells in serum/differentiation factors-free conditions through activation of FAK-MEK/ERK signaling pathways. *Biochem. Biophys. Res. Commun.* **2010**, *391*, 43–48. [CrossRef] [PubMed]
45. Brännvall, K.; Bergman, K.; Wallenquist, U.; Svahn, S.; Bowden, T.; Hilborn, J.; Forsberg-Nilsson, K. Enhanced neuronal differentiation in a three-dimensional collagen-hyaluronan matrix. *J. Neurosci. Res.* **2007**, *85*, 2138–2146. [CrossRef]
46. Bhattacharyya, B.J.; Banisadr, G.; Jung, H.; Ren, D.; Cronshaw, D.G.; Zou, Y.; Miller, R.J. The chemokine stromal cell-derived factor-1 regulates GABAergic inputs to neural progenitors in the postnatal dentate gyrus. *J. Neurosci. Off. J. Soc. Neurosci.* **2008**, *28*, 6720–6730. [CrossRef]
47. Yanagawa, Y.; Iwabuchi, K.; Onoé, K. Enhancement of stromal cell-derived factor-1alpha-induced chemotaxis for CD4/8 double-positive thymocytes by fibronectin and laminin in mice. *Immunology* **2001**, *104*, 43–49. [CrossRef]
48. Jin, K.; Zhu, Y.; Sun, Y.; Mao, X.O.; Xie, L.; Greenberg, D.A. Vascular endothelial growth factor (VEGF) stimulates neurogenesis in vitro and in vivo. *Proc. Natl. Acad. Sci. USA* **2002**, *99*, 11946–11950. [CrossRef]
49. Luttun, A.; Tjwa, M.; Carmeliet, P. Placental growth factor (PlGF) and its receptor Flt-1 (VEGFR-1): novel therapeutic targets for angiogenic disorders. *Ann. N. Y. Acad. Sci.* **2002**, *979*, 80–93. [CrossRef]
50. Diao, Y.-P.; Cui, F.-K.; Yan, S.; Chen, Z.-G.; Lian, L.-S.; Guo, L.-L.; Li, Y.-J. Nerve Growth Factor Promotes Angiogenesis and Skeletal Muscle Fiber Remodeling in a Murine Model of Hindlimb Ischemia. *Chin. Med. J. (Engl.)* **2016**, *129*, 313–319. [CrossRef]

pharmaceutics

Article

MRI/Photoluminescence Dual-Modal Imaging Magnetic PLGA Nanocapsules for Theranostics

Yajie Zhang [1], Miguel García-Gabilondo [2], Anna Rosell [2,*] and Anna Roig [1,*]

1 Institut de Ciència de Materials de Barcelona (ICMAB-CSIC), Campus UAB, 08193 Bellaterra, Catalonia, Spain; yzhang@icmab.es
2 Neurovascular Research Laboratory, Vall d'Hebron Institut de Recerca, Universitat Autònoma de Barcelona, 08035 Barcelona, Catalonia, Spain; miguel.garcia@vhir.org
* Correspondence: anna.rosell@vhir.org (A.R.); roig@icmab.es (A.R.)

Received: 30 October 2019; Accepted: 17 December 2019; Published: 21 December 2019

Abstract: Developing multifunctional and biocompatible drug delivery nanoplatforms that integrate high drug loads and multiple imaging modalities avoiding cross-interferences is extremely challenging. Here we report on the successful chemical reaction of the high quantum yield biodegradable and photoluminescent polyester (BPLP) with the poly(lactic-co-glycolic acid) (PLGA) polymer to fabricate biocompatible photoluminescent nanocapsules (NCs). Furthermore, we transform the PLGA-BPLP NCs into a magnetic resonance (MR)/photoluminescence dual-modal imaging theranostic platform by incorporating superparamagnetic iron oxide nanoparticles (SPIONs) into the polymeric shell. In vitro phantoms confirmed the excellent MRI-r_2 relaxivity values of the NCs whilst the cellular uptake of these NCs was clearly observed by fluorescence optical imaging. Besides, the NCs (mean size ~270 nm) were loaded with ~1 wt% of a model protein (BSA) and their PEGylation provided a more hydrophilic surface. The NCs show biocompatibility in vitro, as hCMEC/D3 endothelial cells viability was not affected for particle concentration up to 500 μg/mL. Interestingly, NCs decorated with SPIONs can be exploited for magnetic guiding and retention.

Keywords: PLGA nanocapsules; magnetic resonance imaging; photoluminescence; drug delivery systems; magnetic targeting; multimodal imaging; theranostics

1. Introduction

Nanomedicine, which refers to the application of nanotechnology in medicine, offers valuable new tools for the diagnosis and treatment of many diseases. Nanoparticles are increasingly important by assisting to expedite the development of contrast agents, therapeutics, drug delivery vehicles, and theranostics in the context of nanomedicine [1]. Polymer-based nanoparticles are frequently proposed as drug carriers due to their biocompatibility and biodegradability, as well as the possibility of customizing their physicochemical properties for a specific drug or delivery route [2–4]. Among them, poly(lactic-co-glycolic acid) PLGA nanoparticles have gathered particular attention, since they encompass a number of interesting features: (i) FDA and European Medicine Agency approval in drug delivery systems for parenteral administration; (ii) well described formulations and methods of production adapted to various drugs i.e., hydrophilic or hydrophobic, small molecules, or macromolecules; (iii) protects the loaded drugs from degradation and possibility of sustained release; and, (iv) easy to modify to include targeted delivery or to provide tuned performance in a specific biological environment [5,6].

Theranostics is an emerging field that combines diagnostics and therapeutics into multifunction nanoparticle systems [7]. Drugs and nanocarriers in vitro/in vivo fate can be monitored while using noninvasive imaging techniques, such as magnetic resonance imaging (MRI) and fluorescent imaging, in order to optimize the route of delivery, biodistribution, and drug accumulation, among other factors [4,8,9]. PLGA nanoparticles can be used in diagnostic and therapeutic imaging by the addition of the imaging moieties during the particle synthesis. To date, large efforts have been devoted to conjugating PLGA with semiconducting quantum dots and organic dyes to create photoluminescent PLGA nanocarriers [9–11]. However, conventional approaches for physically blending imaging probes within the nanoparticles can lead to incomplete conclusions or misinterpretations on the nanoparticles' biodistribution and fate [12,13]. Moreover, the inevitable photobleaching and low dye-to-polymer labeling ratios of organic dyes and the innate toxicity of quantum dots prevent their practical use in vivo [14,15]. Recently, a series of biodegradable photoluminescent (PL) polyesters (BPLPs) could derived inherently photoluminescent PLGA-BPLP copolymers with excellent biocompatibility, tunable luminescence and degradation rates, and good thermal and mechanical properties, thus expanding the biomedical applications of PLGA to highly desired optical imaging [16,17].

Here, we present PLGA nanocapsules (NCs) as a dual-modal imaging theranostic platform for magnetic targeting protein delivery (Figure 1). We report on the protocol for the fabrication of magnetic PLGA-BPLP NCs with intrinsic photoluminescence and MRI capacity being endowed by the incorporation of superparamagnetic iron oxide nanoparticles (SPIONs). Moreover, the NCs are functionalized with poly(ethylene glycol) (PEG), providing a hydrophilic surface that could result in an enhanced stealth effect. Unlike solid PLGA nanoparticles [4,18,19], in our NCs, the functional moieties are incorporated into the polymeric shell matrix to minimize interferences with the cargo being placed inside the NC; this is especially important for delicate payloads, such as proteins, enzymes, or microRNAs. In this study, bovine serum albumin (BSA) has been used as a model protein for evaluating the protein loading capability and release kinetics of the proposed nanocarrier.

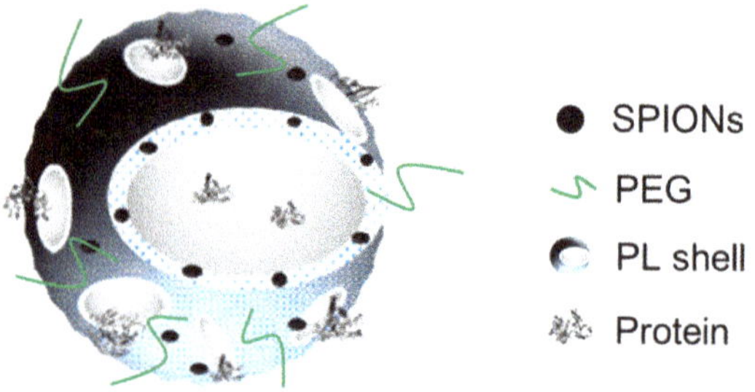

Figure 1. Schematic illustration of poly(lactic-co-glycolic acid)-biocompatible photoluminescent polymer nanocapsule (PLGA-BPLP NC) as a dual-modal imaging theranostic platform.

2. Experimental Section

All of the reagents were purchased from Sigma-Aldrich (Merck KGaA, Darmstadt, Germany) unless specified otherwise.

2.1. Synthesis of PLGA-BPLP Copolymer

The PLGA-BPLP copolymer was synthesized by a slight modification of Yang's method [16,17]. Step one involves the synthesis of a hydroxyl terminated BPLP pre-polymer. Briefly, citric acid, 1,8-octanediol, and L-cysteine with the molar ratio of 1:1:0.2 (5.76 g, 4.38 g, 0.72 g) were added into a flask of 50 mL with a stirring bar. Under a constant flow of argon, the reactants were melted by heating to 140 °C until a clear solution formed. The solution was allowed to react at this temperature for 80 min. and then stopped before the stirring bar stopped stirring completely in the increasingly viscous solution by adding 25 mL of 1,4-dioxane to dilute the produced prepolymer. The product was purified by dropwise precipitation from the 1,4-dioxane solution in water to remove the unreacted monomers. The final BPLP product was collected by centrifugation and then lyophilized.

In the second step, the PLGA-BPLP copolymer was synthesized while using the BPLP as a macroinitiator to react with L-lactide and glycolide via a ring-opening polymerization that was catalyzed by Tin (II) 2-ethylhexanoate ($Sn(OCt)_2$). Briefly, L-lactide, glycolide and BPLP with the molar ratios of (75:25):1 or (50:50):1 were added into a reaction tube. Subsequently, Sn $(OCt)_2$ (0.1 wt% of L-lactide and glycolide mixture) was added as a solution in dichloromethane which was then evaporated in vacuum. The tube was flushed with argon and capped and subsequently immersed in a 160 °C oil bath for 48 h. The obtained product was dissolved in chloroform and then purified by dropwise precipitation into pure ethanol to remove unreacted raw materials. Finally, the PLGA-BPLP copolymer was recovered by centrifugation and then dried in vacuum at room temperature. All of the PLGA-BPLP mentioned in the text were with molar ratios of (75:25):1, unless specified otherwise.

2.2. Synthesis of Oleic Acid Coated SPIONs

The SPIONs were synthesized by microwave assisted thermal decomposition in a microwave synthesizer (Discover SP, CEM Corporation, Matthews, NC, USA), and then coated by oleic acid (OA) [20]. The process to obtain homogeneous OA-SPIONs with an average diameter of 9 nm was as follows: 3.5 mmol of iron precursor Fe $(acac)_3$ was dissolved in 4.5 mL of benzyl alcohol in a microwave reaction glass tube. Microwave irradiation was initiated at 60 °C for 5 min. to fully dissolve the precursor and, subsequently, the irradiation was kept at 210 °C for 30 min., and the reaction was then stopped and cooled down to room temperature. 4 mL of oleic acid in toluene (0.8 mmol/mL) was added immediately into the as-synthesized SPIONs dispersion followed by incubation under ultrasound for 1 h. Subsequently, the obtained OA-SPIONs were separated by centrifugation in five-fold of acetone. The pellet was redispersed in 4 mL of toluene in a glass vial and a magnet was attached on the wall for 5 s, the un-adsorbed suspension containing SPIONs with smaller size was discarded and the adsorbed pellet was redispersed in 6 mL of dichloromethane (DCM), followed by centrifugation at 4000 rpm for 5 min. to sediment the unstable big particles. Finally, the stable OA-SPIONs (9 nm) dispersion in DCM was centrifugated in five-fold of acetone and the pellet product was dried under vacuum and redispersed in DCM at the concentration required for use.

2.3. Fabrication of Functional PLGA-BPLP NCs and Encapsulation of BSA

Functional PLGA NCs encapsulating BSA were prepared by a double emulsion solvent evaporation method. Briefly, 50 μL of inner aqueous phase (W1) containing BSA (30 mg/mL) was emulsified in 500 μL of DCM organic phase (O) that was composed of 50 mg of different proportions of PLGA (RG502H, Mn 12,000)/PLGA-BPLP/PLGA-PEG (PLGA Mn 7000, PEG Mn 5000) and a certain amount of OA-SPIONs by sonication (VC505, Sonics & Materials Inc., Newtown, CT, USA) at 200 W for 28 s to form the first emulsion (W1/O). Afterwards, 2 mL of external aqueous phase (W2) with polyvinyl alcohol (PVA) (20 mg/mL) was added and the second emulsion (W1/O/W2) was formed by sonication for another 28 s. The temperature during the whole emulsion process was kept at 4 °C by using an ice bath. The resulting double emulsion was poured into 50 mL of MilliQ water and mechanically stirred at RT for 2 h to allow for complete evaporation of the organic solvent DCM and the formation of NCs. Finally, the NCs were washed three times with MilliQ water and lyophilized in 6 mL of trehalose aqueous solution (2 mg/mL). The as-obtained powder was stored at 4 °C with desiccant silica gel.

The non-PEGylated and PEGylated NCs have been labelled, as follows: NC1.non PEGylated (90 wt% PLGA-BPLP + 10 wt% PLGA) and NC2.PEGylated (90 wt% PLGA-BPLP + 3 wt% PLGA + 7 wt% PLGA-PEG), both containing ~6 wt% of SPIONs, as gathered in Table 1.

Table 1. Summary of the two main formulations for the NCs. Oleic acid coated-SPIONs and PLGA-BPLP and PLGA-PEG were mixed in the organic phase during the double miniemulsion process.

NCs Type	Shell Polymers (wt%)			SPIONs Loading (wt%)	M_s (eum/g)	Size (DLS)		BSA Loading (wt%)	BSA EE%
	PLGA	PLGA-BPLP *	PLGA-PEG			d. nm	PdI		
NC1.non-PEGylated	10	90	/	5.7	4.0	272	0.11	0.96	38.5
NC2.PEGylated	3	90	7	6.0	4.2	265	0.05	1.01	41.3

* PLGA-BPLP from initial molar ratios (LA:GA):BPLP = (75:25):1.

2.4. Physicochemical Characterization of PLGA-BPLP, SPIONs, and NCs

2.4.1. Absorption Spectroscopies of the Polymers

Attenuated total reflectance-Fourier transform infrared (ATR-FTIR) characterization of the polymers was performed on a Bruker Vertex 70 FTIR spectrometer with a Pike Miracle Single-Bounce diamond crystal plate accessory at room temperature. The FTIR spectra were recorded over a wavelength range of 4000–500 cm^{-1} with a resolution of 4 cm^{-1}. UV-Vis absorbance of the fluorescent polymers was recorded on a Varian Cary-5000 UV-Vis spectrophotometer while using a quartz cuvette with an optical path of 1 cm.

2.4.2. Size Distribution and Disperse Stability of the NCs

Dynamic light scattering (DLS) (Zetasizer Nano ZS, Malvern Panalytical, Madrid, Spain) measurement of the hydrodynamic diameter and size distribution of NCs by intensity was performed by redispersing 0.5 mg of lyophilized NCs powder into 1 mL of MilliQ water.

Turbiscan (Turbiscan Lab, Formulaction, Toulouse, France) is used to detect the destabilization of the NCs suspension. 15 mL of 2 mg/mL NCs in the Turbiscan cell were scanned at all of the heights of the suspension with a time interval of 2 min. during 24 h. The back scattering signals at different heights of the cell were recorded and the delta of back scattering intensity was calculated by subtracting the reference time 0 s. The bottom part was defined as 1/5 of the liquid level.

Nanosight (NS300, Malvern Panalytical, Madrid, Spain) is used to measure the averaged size and concentration of the NCs water suspension. 0.2 mg/mL of NCs with 50 times dilution was pumped into the cell and the data were acquired and analyzed through the Nanoparticle Tracking Analysis (NTA) software along with the instrument.

2.4.3. Electron Microscopies

Field emitting scanning electron microscope (SEM, FEI Quanta 200 FEG, Thermo Fisher Scientific, OR, USA) and transmission electron microscope (TEM, JEM-1210, JEOL Ltd., Tokyo, Japan) were used to study the morphologies of SPIONs and NCs. 0.5 mg of lyophilized powder was redispersed into 1 mL of MilliQ water and centrifuged at 4000 rpm for 10 min. for the SEM sample preparation of NCs. Subsequently, the supernatant was discarded to remove the trehalose (used for cryopreserving during lyophilization) and 1 mL of fresh water was added, the pellet of NCs was redispersed in water with ultrasound. Finally, 6 L of the slightly turbid suspension was deposited onto a small slice of silicon wafer stuck on the top of a carbon layer and dried at room temperature overnight. The sample was sputtered with Au-Pd (20 mA 2 min, Emitech K550, Quorum Technologies Ltd., East Sussex, UK). The TEM samples were prepared by placing and drying one drop of the corresponding NCs or SPIONs dispersion on a copper grid at room temperature.

2.4.4. Magnetometry

Superconductive quantum interference device (SQUID, MPMS5XL, Quantum Design, San Diego, CA, USA) was used to measure the magnetization of NCs and SPIONs and calculate the SPIONs loading (wt%-SPIONs) of the magnetic NCs. Zero-field cooling and field cooling (ZFC-FC) measurement were used to determine the blocking temperature (T_B) of the SPIONs. A gelatin capsule filled with about

7 mg of samples, together with some cotton wool, was inserted into the SQUID magnetometer sample holder and the hysteresis loop was measured from −50 kOe to 50 kOe. The saturation magnetization of the NCs (M_S-NCs, emu/g) and of SPIONs (M_S-SPIONs, emu/g) was used to calculate wt%-SPIONs, as follows:

$$\text{wt\%-SPIONs} = M_S\text{-NCs}/M_S\text{-SPIONs} \times 100\%$$

2.4.5. Fluorescence Properties

Fluorescence spectra of the polymers and NCs were acquired on a spectrofluorometer (LS45, PerkinElmer Inc., Waltham, MA, USA). The excitation and emission slit widths were both set at 10 nm. The quantum yield of the polymers was measured by the Williams' method [16]. Briefly, a series of BPLP/PLGA-BPLP solutions in the corresponding solvents were prepared with gradient concentrations. Maximal excitation wavelength was determined, which generated the highest emission intensity. The fluorescence spectra were collected for the series of solutions in the 10 mm fluorescence cuvette (Figure S1). The integrated fluorescence intensity, which is the area of the fluorescence spectrum, was calculated and then noted. Afterwards, the UV-Vis absorbance spectra were collected with the same solutions and the absorbance at the maximal excitation wavelength within the range of 0.01–0.1 Abs units was noted (Figure S1). The graphs of integrated fluorescence intensity vs. absorbance were plotted. The quantum yield was calculated according to the equation:

$$\Phi s = \Phi_r \left(\frac{\text{Slope}_s}{\text{Slope}_r}\right)\left(\frac{n_s}{n_r}\right)^2$$

where, Φ = quantum yield; Slope = slope of the straight line obtained from the plot of intensity vs. absorbance; *n* = refractive index of the solvent; *s* = subscript denotes the sample; and, *r* = subscript denotes the reference used. Here, anthracene (Φ = 27% in ethanol when excited at 366 nm) was used as the reference.

The fluorescence intensities of different concentrations of NCs in water were quantified in a 96-well by a microplate reader (Spark, Tecan Group Ltd., Männedorf, Switzerland). The NCs were excited at the maximal excitation wavelength and the fluorescence signal was collected by area scan at the maximal emission wavelength.

Photostability was measured by continuously illuminating the polymer or NCs in a fluorescent confocal microscope (Leica SP5, Leica Microsistemas S.L.U., Barcelona, Spain) while using UV diode 405 nm excitation at different laser power (10%, 20%, 100%). The fluorescence images were acquired at a time interval of 1 s for 10 min., changes of mean fluorescence intensity of six region of interests (ROIs) in 10 min. were calculated while using software Las AF. The real laser power (W) at different percentage during 10 min. was monitored using a laser power meter.

2.5. MRI Phantoms of the NCs

In vitro agarose phantoms of NCs were prepared in Eppendorfs where a series of concentrations of NCs were vortexed and sonicated in agarose water solutions before the gel formed. The volume was kept at 1 mL with 0.63 wt% of agarose (Conda, Madrid, Spain). The corresponding iron doses (mmol/L) were calculated according to the wt% of SPIONs in each sample. T2 maps of the phantoms were acquired at 7 T in a 70/30 Bruker USR Biospec system (Bruker GmbH, Ettlingen, Germany), as follows: multi-slice multi-echo (MSME) sequence with echo time (TE) = 13 ms, repetition time (TR) = 4000 ms, field of view (FOV) = 5.5 × 11 mm, and three slices of 1 mm thickness. The quantitative T2 values were obtained from hand-drawn ROIs by using curve fitting in the Image Sequence Analysis (ISA) software along with the instrument.

2.6. In Vitro Toxicity Evaluation of the NCs

Two parallel methods assessed the toxicity of the NCs on human brain endothelial cells (hCMEC/D3): a viability assay based on WST-8 tetrazolium salt reduction (cell counting Kit-8, Dojindo) and direct cell counting. First, 10^4 viable cells per well were seeded on a 24-well plate pre-treated with collagen (rat tail type I, Corning) in 400 μL of endothelial growth medium (EGM2 from Lonza with 2% fetal bovine serum and half the amount of the growth factors that were included in the kit). After incubation for 72 h at 37 °C with 5% of CO_2, the cells were at 80–90% confluence and medium was changed to endothelial basal medium (EBM2, Lonza) containing NCs at 25, 50, 100, 500, and 1000 μg/mL. After 48 h, the cells were washed and incubated with 10% WST-8 solution for two hours. Culture supernatants were centrifuged at 20,000 rpm for 5 min. to remove NCs detritus that could interfere with the dye absorption before determining the absorbance at 450 nm. Cells were trypsinized and resuspended in growth medium and diluted 1:1 in Trypan Blue in order to perform cell counting in a Neubauer chamber. Cell viability and count are expressed as the percentage of absorbance or number of cells as compared with the control (vehicle without NCs). ANOVA test and Dunnett's multiple comparisons post-hoc test was performed vs. the control.

2.7. In Vitro Observation of NCs Cellular Uptake

The hCMEC/D3 endothelial cells were seeded in cover-slips that were pre-treated with collagen (2×10^4 cells/well in 24 well plates) and 24 h later were exposed to 50 μg/mL of empty fluorescent NCs. After an additional 24 h of culture, the wells were washed with PBS, fixated with 4% paraformaldehyde, and mounted with Vectashield antifade mounting medium (Vector Laboratories) with propidium iodide for nucleic acid counterstaining. Additionally, some of the cells were stained with PKH26 lipophilic fluorescent dye for cell membrane labeling according to the manufacturer's protocol. Images at 63× were obtained on a fluorescent confocal microscope (LSM 980 with Airyscan 2 detector, Zeiss, Oberkochen, Germany).

2.8. BSA Loading in NCs and Release Kinetics

The albumin content of the NC was directly determined while using the CBQCA protein assay kit (Invitrogen™ ref. C6667), which determines the protein concentration based on the production of fluorescent products measurable at $\lambda_{ex}/\lambda_{em}$ = 450 nm/550 nm via non-covalent interaction between CBQCA and primary aliphatic amines of proteins. This highly sensitive fluorescence-based method showed compatibility with DMSO, SPIONs, detergents, and other substances that interfere with other commonly used protein determination methods. Lyophilized NCs encapsulating albumin as well as empty NCs as control were fully dissolved in DMSO at 100 mg/mL. The protein contents in the NCs lysates were measured and calculated based on the difference in fluorescence with the control and a calibration curve drawn with standard albumin solutions, the protein contents in the BSA solutions used for encapsulation were also measured. As listed in Table 1 for the two types of NCs, at least two replicate NCs batches of each were measured for the BSA loading and encapsulation efficiency (EE%) calculation. All of the measurements were performed in duplicate for each NCs batch. The experimental BSA loading in the NCs is expressed as μg of BSA per mg of NCs (μg/mg) or the wt% of the NCs, and the BSA EE% is calculated, as follows:

$$\text{EE\% BSA} = \frac{\text{Experimental BSA loading}}{\text{Nominal BSA loading}} \times 100\%$$

For the release studies, lyophilized NCs were resuspended in phosphate buffered saline (PBS) (pH 7.4, Sigma ref. D1408) at 10 mg/mL in low protein binding microcentrifuge tubes (Thermo Scientific© ref. 90410). NCs solutions were incubated at 37 °C in a vertical rotator for different time measures to simulate the in vivo environment: right after the resuspension (time 0), and after three hours, six h, one day, and seven days of incubation. In all cases, an aliquot of 200 μL was frozen at −80 °C until protein determination. Before the CBQCA assay, the aliquots were centrifuged at 15,000× *g* rcf to separate supernatant and pellet. The amount of protein release was directly calculated as total released protein and indirectly from the remaining protein in the pellet as indirect measure. For this purpose, the pellets were fully dissolved in DMSO and then compared with the intact NCs fully dissolved in DMSO (100 mg/mL) used as the 100% release set up. Four BSA-loaded NCs batches and one of H_2O-NCs batch as control were used and measures done in duplicate. The release profiles were expressed in terms of cumulative release and plotted vs. time.

3. Results and Discussion

3.1. Photoluminescent PLGA-BPLP Copolymer

BPLPs are degradable oligomers synthesized from biocompatible monomers, including citric acid, aliphatic diols, and various amino acids via a convenient and cost-effective polycondensation reaction. BPLPs present some advantages over the traditional fluorescent organic dyes and quantum dots due to their cytocompatibility, minimal chronic inflammatory responses, controlled degradability, and excellent fluorescence properties [16]. Here, L-cysteine was selected and introduced into the polyester structure that was made of biocompatible monomers of citric acid and aliphatic 1,8-octanediol, since previously reported BPLP from this starting amino acid exhibited the highest quantum yield (62.3%) [16]. The fluorophore structure of BPLP was verified as a fused ring structure ((5-oxo-3,5-dihydro-thiazolopyridine-3,7-dicarboxylic acid, TPA) [21]. ATR-FTIR was also used to confirm the chemical structure of the as-synthesized BPLP (Figure 2A). Strong absorptions from the molecular backbone of the polyester were observed i.e., peaks at 1044 cm^{-1}, 1176 cm^{-1}, and 1716 cm^{-1} are attributed to the C=O stretch, C–O asymmetrical, and symmetrical stretches of the ester bond, respectively, the peaks at 2930 cm^{-1} and 2856 cm^{-1} are attributed to the C-H stretches of alkane from 1,8-octanediol and the band near 3467 cm^{-1} is from the –OH. NH bending of the secondary amide at 1527 cm^{-1} and –SH at 2575 cm^{-1} confirm that L-cysteine is chemically bound to the poly(diol citrate) chain. The shoulder band near 1635 cm^{-1} is attributed to the C=O stretching of the tertiary amide from the TPA ring. The average molecular weight (M_w) of BPLP measured by matrix-assisted laser desorption/ionization time of flight mass spectroscopy (MALDI-TOF-MS) was 1044 g/mol (Figure S2). The BPLP oligomer served as a macroinitiator to react with L-lactide and glycolide via a ring-opening polymerization to produce PLGA-BPLP [17]. The as-synthesized PLGA-BPLP (75:25):1 with molar ratios equal to 75:25 for L-lactide to glycolide and equal to 1:100 for BPLP to total L-lactide and glycolide was reported to have desirable glass transition temperature (Tg, 32.5 °C), mechanical properties, fluorescence properties, and degradation rate of the resulting product [17]. Moreover, we have proved here that the 75:25 formulations are more suitable for the fabrication of NCs by a double mini-emulsion method than the PLGA-BPLP (50:50):1 one, as shown in the following section.

The obtained PLGA-BPLP copolymer (Figure 2B inset) exhibits the inherent photoluminescence from the BPLP. The similarities of bands and shapes of the IR spectra of the as-synthesized PLGA-BPLP and commercial PLGA (Figure 2B), for instance, bands at 1084 cm^{-1}, 1165cm^{-1}, and 1747 cm^{-1} from the ester bonds in PLGA indicate their similar chemical structure, given the fact that BPLP is a very small portion of the PLGA-BPLP copolymer.

The fluorescence of the as-synthesized BPLP and PLGA-BPLP was evaluated and Figure 3A,B depict the excitation and emission spectra. The similar spectra further confirm the inherent photoluminescence of PLGA-BPLP from BPLP. Importantly, the fluorescence intensity of PLGA-BPLP only slightly decreased (10%) after 10 min. of continuous illumination under confocal microscope at 10% of laser power (0.40 ± 0.01 μW) (Figure 3C). When considering the laser power applied to observe stained cells is generally less than 10%, our photoluminescent polymer would exhibit good photostability at in vitro conditions. The calculated high quantum yields of BPLP (64%) and PLGA-BPLP (33%) from Figure 3D are consistent with the previously reported values [16,17]. The remarkable fluorescence properties that are shown here endow the PLGA-BPLP copolymer with high potential for the fabrication of functional photoluminescent NCs.

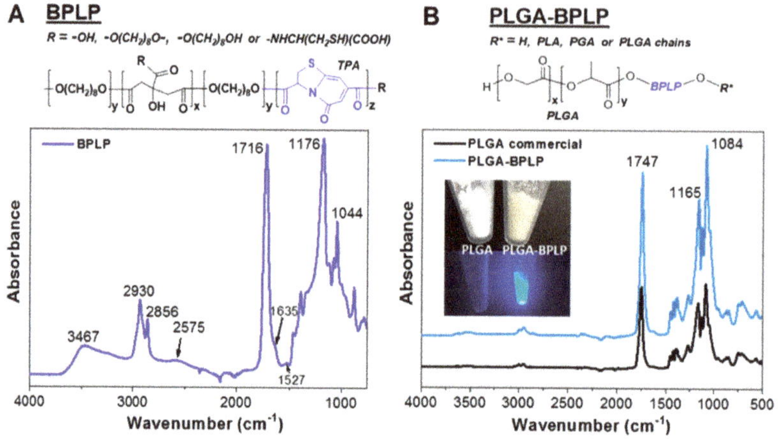

Figure 2. Attenuated total reflectance-Fourier transform infrared (ATR-FTIR) spectra of the as-synthesized biocompatible photoluminescent polymer (BPLP) and PLGA-BPLP copolymer confirming their chemical structures and the successful synthesis. (**A**) BPLP; (**B**) PLGA-BPLP and commercial PLGA as reference. Inset: fluorescence of the PLGA-BPLP under a UV lamp.

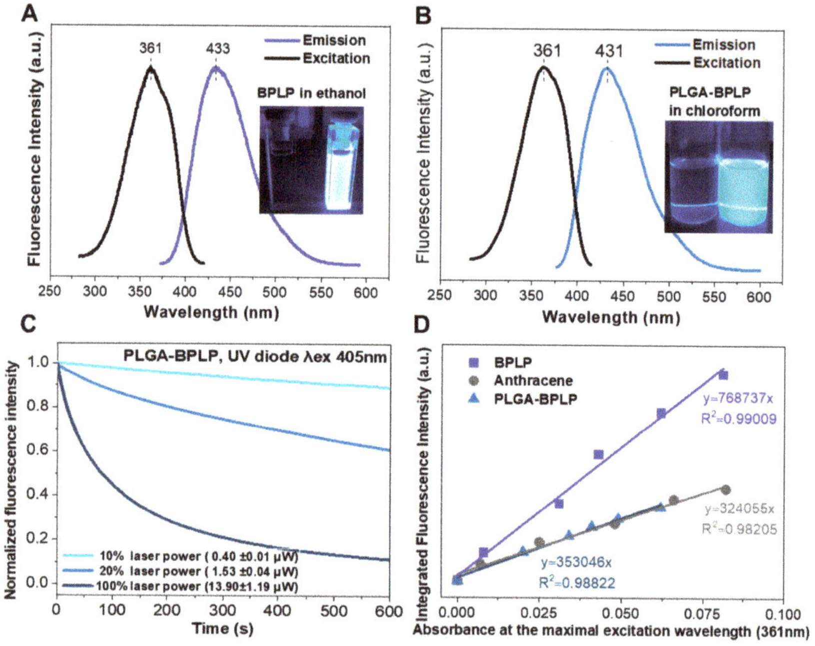

Figure 3. (**A**,**B**) excitation and emission spectra of the as-synthesized BPLP and PLGA-BPLP. Insets: fluorescence of BPLP and PLGA-BPLP dispersed in solutions under a UV lamp; (**C**) photostability evaluation of PLGA-BPLP powder under confocal microscope at different laser power, fluorescence intensity expressed as the percentage vs. the value at the initial time; and, (**D**) fluorescence intensity-absorbance curves of BPLP, PLGA-BPLP, and anthracene used as a reference used to calculate quantum yields.

3.2. Fabrication of PLGA-BPLP NCs Combining Other Functional Moieties

PLGA-based NCs with BSA encapsulated were prepared by a double emulsion solvent evaporation method with slight modifications from our previously reported method [22]. Functional moieties, such as PLGA-BPLP for the fluorescence imaging, PLGA-PEG to increase hydrophilicity and sealthness, and SPIONs for magnetic targeting and MRI (Figure 4A,B) can be incorporated to the organic phase during the NCs fabrication process (see Table 1). The first important remark is that NC morphology was not affected, even when using 100 wt% of PLGA-BPLP (75:25):1 (Figure S3A). A formulation with 90 wt% of PLGA-BPLP (75:25):1 was selected to allow for NC PEGylation by the addition of 7 wt% of PLGA-PEG and for the benefit of higher fluorescence intensity (Figure S4). Note that 7 wt% of PLGA-PEG (3 wt% PEG, Figure S3B) (NC2) was found to be the maximum amount that can be mixed in the organic phase during the NCs fabrication process, due to the amphiphilic property of PLGA-PEG. For a larger wt% of PLGA-PEG, the morphology and size of NCs were not maintained. As shown in the SEM image (Figure 4C), non-PEGylated and PEGylated NCs, as listed in Table 1, both contain ~6 wt% of SPIONs and depict homogeneous spherical morphologies and sizes (d.nm ~270) being similar to other reported PLGA systems that are suitable for intravenous administration [23]. The upper inset in Figure 4C shows a representative broken NC, exposing the hollow core where the protein drug is loaded. We have also found that the PLGA-BPLP (50:50):1 polymer from using initial molar ratio of LA:GA = 50:50 and BPLP:(LA + GA) = 1:100 was not as suitable for the fabrication of NCs. NCs with homogeneous morphology and narrow size distribution were attained only up to a maximum of 30 wt% of modified PLGA (Figure S3C) and without the possibility of further adding PLGA-PEG when using PLGA-BPLP (50:50):1 (Figure S3D). These results are in accordance with the higher glass transition temperature and better mechanical properties of PLGA-BPLP (75:25):1 over those of PLGA-BPLP (50:50):1 [17]. As expected, the NCs with a higher fraction of PLGA-BPLP (75:25):1 (90 wt%) show higher fluorescence intensity than the ones that were obtained with PLGA-BPLP (50:50):1 (30 wt%) at the same concentration (Figure S4).

Regarding magnetic loading, up to 6 wt% of SPIONs could be loaded without affecting the NCs morphology and yielding a saturation magnetization (M_S) value of around 4 emu/g NCs (Figure 4D). In addition, the lower blocking temperature (T_B, 33 K) of SPIONs in the NCs (Figure 4D inset) than the SPIONs of dry powder (55 K, Figure 4B inset) further demonstrates that the SPIONs are well dispersed in the polymer matrix. Note that a high magnetic loading is desirable for the magnetic retention of NCs. This is illustrated in the inset images of Figure 4D, where the darker-coloured water suspension of NCs with 6 wt% of SPIONs were adsorbed faster to the tube wall on the magnet side than the NCs with 1 wt% SPIONs loading at the same concentration, promisingly benefiting the magnetically targeted drug delivery as compared to the previously reported results [4,24]. Note that the superparamagnetic behaviour of NCs at room temperature (lack of coercivity, Figure S5) ensures no magnetic interactions among NCs in the absence of an external magnetic field, minimizing the risk of embolization during i.v. administration.

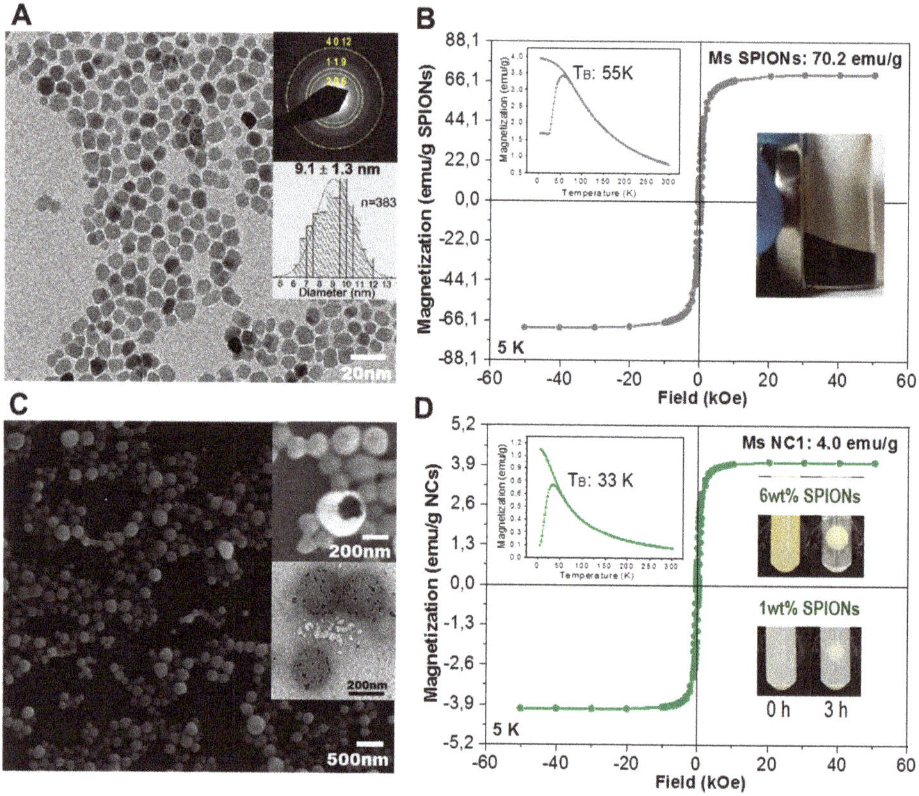

Figure 4. (**A**) Representative transmission electron microscope (TEM) image of the oleic acid (OA) coated SPIONs with the upper inset showing the selected area electron diffraction (SAED) pattern of maghemite and lower inset a size distribution histogram of the particles in the image; (**B**) hysteresis loop (5 K) and ZFC-FC (inset) for the OA-superparamagnetic iron oxide nanoparticles (OA-SPIONs), inserted picture shows the stable OA-SPIONs in dichloromethane attracted by an external magnet; (**C**) representative SEM image of lyophilized NCs with the upper inset showing the hollow core of a nanocapsule and lower inset a TEM image of three NCs with the SPIONs visible as black spots well distributed in the polymer matrix; (**D**) hysteresis loop (5 K) and ZFC-FC (inset) for the lyophilized nanocapsule batch NC1, inserted pictures show the water suspension of these NCs (2 mg/mL) where the 6 wt% loading of SPIONs were adsorbed faster to the tube wall on the magnet (diameter 8mm, surface field ~0.4 T) side than that of 1 wt% loading.

It is well reported that the surface PEGylation of engineered nanoparticles provides them with stealth character increasing blood circulation time since nanoparticles are less visible to the reticulo-endothelial system [6]. We evaluated the flocculation regime of the NCs with a Turbiscan to confirm successful surface modification with PEG, since the long hydrophilic PEG chains (Mn 5000) on the surface of the NCs are expected to increase the stability of the NCs in water suspension and decrease the sedimentation rate. Sedimentation of the NCs suspension was monitored for 24 h. Figure 5A shows that, as NCs sedimentation progresses, the back scattering signal of the bottom part of the suspension increases from an increasingly higher concentration of NCs, while the signal of the top part decreases. The sedimentation rates of non-PEGylated and PEGylated NCs were compared with or without a physiological concentration of BSA (0.5 mM) in Figure 5B, as expected the bottom back scattering signal of the PEGylated NCs media increases at a slower rate than the non-PEGylated ones both with and without BSA, which demonstrates a better dispersibility of the NCs due to the surface hydrophilic PEG chains. Note that the sedimentation rates of non-PEGylated and PEGylated NCs both slow down with the physiological concentration of BSA probably due to the interaction of NCs with the dense BSA solution. Additionally, in Figure 5A, the back scattering signal of the middle part did not vary with time, which means that the NCs were monodispersed at the physiological concentration of BSA and flocculation or coalescence did not occur during the 24 h period. This is in consistent with the DLS size distribution results that are shown in Figure 5C, both non-PEGylated and PEGylated NCs remained monodisperse at the physiological concentration of BSA, which is of great advantage for the i.v. administration and in vivo blood circulation. Nanosight was also used as an additional technique for the determination of size and the concentration of the NCs (Figure 5D). The results show a similar size distribution as obtained by DLS. From the number concentration of the NCs, we can determine a mean mass of 1.06×10^{-11} mg/NC.

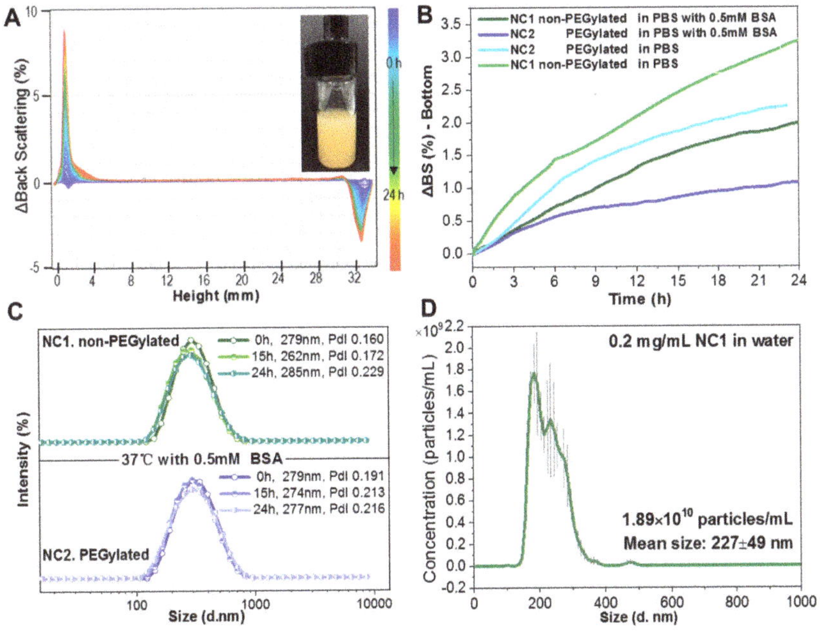

Figure 5. (**A**) Back scattering intensity change of the NC2 PBS suspension (shown inset) containing 0.5 mM of BSA at different height of the vial along 24 h measured by Turbiscan; (**B**) quantified back scattering intensity change of the bottom part of the vial along 24 h for NCs suspension in different media measured by Turbiscan; (**C**) dynamic light scattering (DLS) size distributions of NC1 and NC2 PBS suspensions with 0.5 mM of BSA along 24 h; (**D**) quantitative number concentration and size distribution of the nanocapsules measured by Nanosight (n = 3, mean ± SD with error bar).

3.3. Imaging Performance of the Magnetic Photoluminescent NCs

The fluorescence of the NCs was evaluated, and Figure 6A depicts the excitation and emission spectra. The spectra are similar to those of the PLGA-BPLP polymer shown in Figure 3B. Importantly, the incorporation of SPIONs in the polymer matrix does not quench the fluorescence of NCs. Note that a small displacement of the emission peak wavelength was observed for aqueous dispersed fluorescent NCs when compared to the emission peak of the polymer in a chloroform solution (Figure 3B), which we ascribe to the different interaction of the fluorescent probe with the two solvents. The fluorescence intensities of NCs show a linear dependency on the NCs concentration within a range of 0.1 to 1.0 mg/mL; at higher concentrations the fluorescence shows a trend towards saturation (Figure 6B).

NCs can be clearly imaged with a fluorescence confocal microscope and they show a very good photostability while using 10% of laser power, which is ideal for the observation of in vitro cellular uptake (Figure 6C). The strategy used here confers intrinsic photoluminescence to the PLGA NCs without introducing any cytotoxic quantum dots or photo-bleaching organic dyes when compared to other more conventional approaches that physically blend imaging probes within the carrier that can lead to misinterpretations on the tracing of the carrier [12,13], which may greatly expand the applications of this drug carrier.

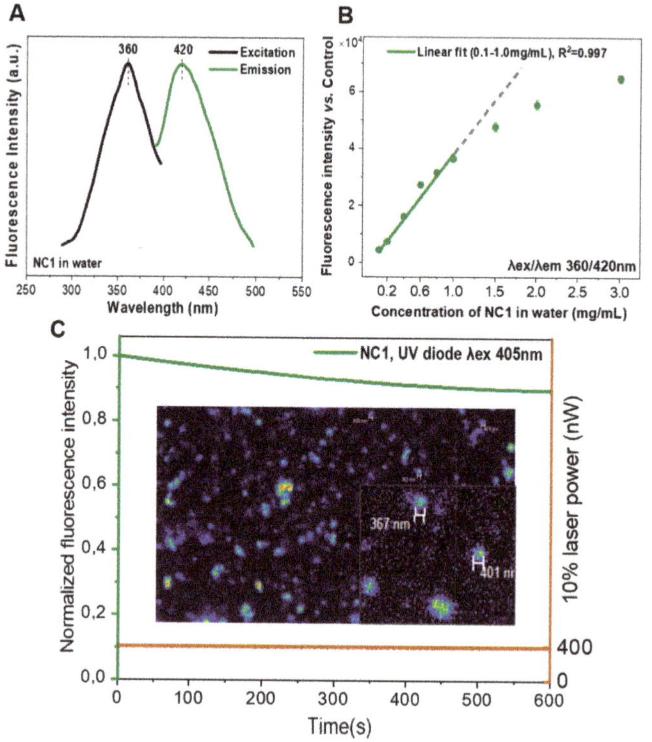

Figure 6. (**A**) Excitation and emission spectra of the nanocapsules (NCs) water suspension; (**B**) fluorescence intensity of different concentrations of NCs measured by microplate reader (n = 2, values represent mean ± sd and subtract values of control non-fluorescence NCs); and, (**C**) photostability evaluation of NCs under confocal microscope at 10% laser power, fluorescence intensity expressed as the percentage vs. the value at the initial time, inset: NC1 water suspension observed at 60× lens.

Phantom studies were conducted to confirm the MRI performance of the capsules. Phantoms of NCs that were dispersed in agarose gel at various concentrations were prepared (Figure 7A). Spin-spin relaxation time (T2) maps clearly exhibit signal decay in a concentration dependent manner. The calculated transverse relaxivity (r_2) values at 7 Tesla of both non-PEGylated NC1 (263 $mM^{-1}s^{-1}$) and PEGylated NC2 (237 $mM^{-1}s^{-1}$) are similar as those seen in Figure 7B, as expected, further demonstrating the similar loading and distribution of SPIONs in the polymer shell matrix for both systems. When compared with other clinically used SPIONs systems, such as Feridex (98 $mM^{-1}s^{-1}$) and Resovist (151 $mM^{-1}s^{-1}$) [25], the much higher r_2 value of our NCs formula is expected to be useful for in vivo MRI tracking of the NCs.

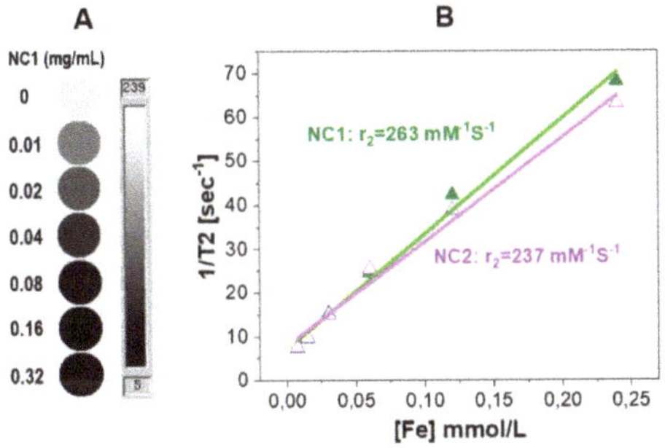

Figure 7. (**A**) T2 maps of a series of concentrations of NC1 in agarose phantoms; (**B**) r_2 relaxivity evaluation for the NCs.

3.4. Cell Viability after NCs Uptake

Photoluminescent NCs were incorporated by brain endothelial cells after several hours in culture, as seen in Figure 8A, with cytoplasmic localization of the NCs in perinuclear structures compatible with Golgi bodies and endosomes. This subcellular localization was confirmed by Z-stack images (Figure 8B). Importantly, this cellular uptake was biocompatible for endothelial cells, as the main exposed cells during NCs circulation in blood vessels, since viability tests did not show signs of cell toxicity at a wide range of NCs concentrations up to 500 µg/mL (Figure 8C,D) and 48 h exposure. Only extremely high doses (1000 µg/mL) with noticeable occupying space difficulties for cell culturing showed a significant reduction in cell viability and number.

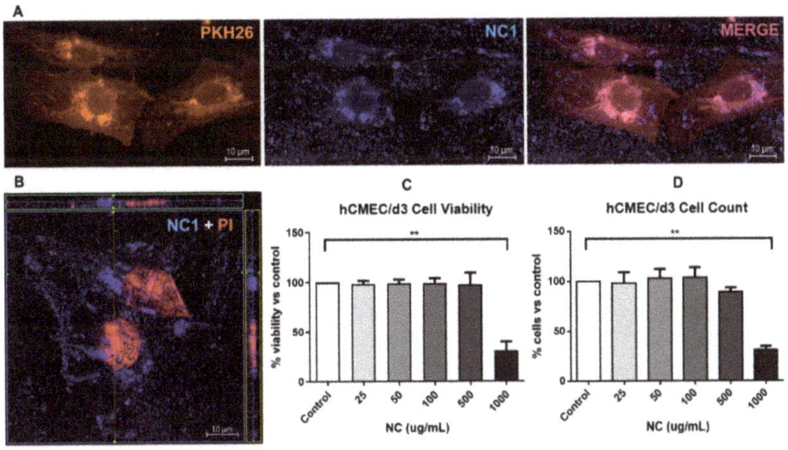

Figure 8. NCs uptake and cytotoxicity in human brain endothelial cells. (**A**) Representative images of hCEMC/d3 cells stained with membrane dye PHK26 and exposed to 50 µg/mL of NC1 for 24 h (63× magnification); (**B**) Orthogonal view of a Z-stack of Propidium Iodide (PI) stained cells (showing the cell nuclei) and the fluorescent NCs; (**C**,**D**) hCEMC/d3 cells were treated for 48 h with different concentrations of NC1 and cell viability was determined with WST-8 reduction assay or tripsinized and counted in a Neubauer chamber, (n = 3–4, values represent mean ± SEM, ** $p < 0.01$).

3.5. Protein Loading and In Vitro Release

Protein loading and encapsulation efficiency were determined by lysing NCs with DMSO and measuring total protein content. The BSA loading content was determined as ~10 µg BSA/mg NCs (1 wt%) with an EE% of around 40% for both NC2.PEGylated and NC1.non-PEGylated systems, as listed in Table 1, which indicated that the incorporation of PEG does not affect the protein encapsulation process. Note that a protein loading of 1 wt% is much higher than other reported values (0.03 wt% of vessel endothelial growth factor (VEGF) loaded PLGA NCs) [22] and the EE% of 40% is comparable to the PLGA NCs loaded with neurotrophin-3 or brain-derived neurotrophic factor (47%) [26].

BSA-loaded NCs were able to release protein cargo over time at physiological temperature in PBS media (32% protein release in one week). Figure 9A shows a fast BSA release within the first hours, but not after one day, which could be related to the protein degradation in ex vivo conditions of our assays. The amount of released protein in one week was similar when indirectly measured from the pellet retained protein (Figure 9B,C), although the release profile showed a more sustained pattern over time, which could be associated to the protein that was trapped within the PLGA polymer.

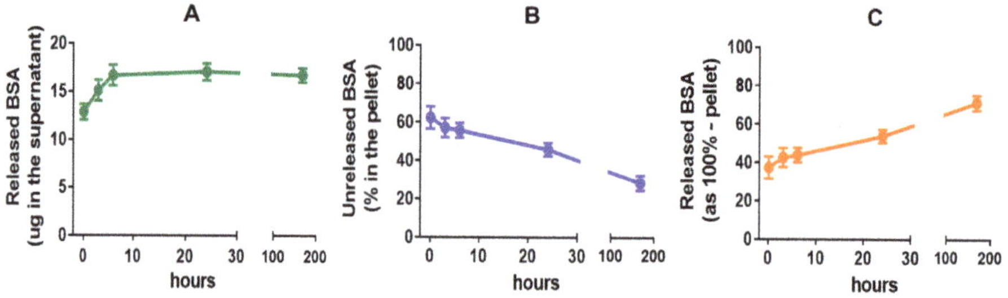

Figure 9. Protein release temporal profile of the NCs. (**A–C**) NC1 in PBS (10 mg/mL) were incubated at 37 °C in rotation and the released BSA quantified as total mass of protein released (**A**) together with BSA content in the remaining NCs pellet by DMSO lysis (**B**) and the released BSA calculated indirectly from the pellet values (**C**). The percentage was calculated versus intact unreleased NCs lysated also in DMSO (n = 8, values represent mean ± SEM).

By using the model protein BSA, here we were able to prove the loading capacity and release kinetics of the drug carrier, yet the preservation of protein functionality after the encapsulation process still needs to be investigated. Nevertheless, in a previous work, we have shown that VEGF could be encapsulated following a similar route and VEGF effect on cell proliferation could be determined [22], which implied the preserved protein functionality.

4. Conclusions

We here transform PLGA nanocapsules (NCs) into a highly sensitive, MRI/photoluminescence dual-modal imaging theranostic platform for drug delivery by integrating the biocompatible and photoluminescent polyester BPLP into the PLGA molecular structure, as well as by incorporating superparamagnetic iron oxide nanoparticles (SPIONs). Furthermore, we have shown that PEGylation provides a hydrophilic surface to the NCs slowing down their flocculation rated and without modiying the size, SPIONs content or protein loading capacity of the NCs. In all cases, the functional moieties are embedded in the PLGA shell with minimal interferences between them or with the therapeutic protein. The developed magnetic PLGA-BPLP NCs show biocompatibility in vitro. In this regard, the NCs did not affect the viability of endothelial cells in culture for concentration up to 500 µg/mL and 48 h incubation. Finally, we have shown that the NCs can contain 1 wt% of protein in their core achieved at fabrication level and that one third of the encapsulated protein is released in the first week. Interestingly, the NCs decorated with SPIONs can be exploited for magnetic retention and magnetic guiding.

Supplementary Materials: The following are available online at http://www.mdpi.com/1999-4923/12/1/16/s1. Figure S1: UV-Vis absorbance spectra and fluorescence spectra of the as-synthesized A/B) BPLP, C/D) PLGA-BPLP and E/F) reference anthracene at a series concentrations for fluorescence quantum yield calculation, Figure S2: Matrix-assisted laser desorption/ionization time of flight mass spectroscopy (MALDI-TOF-MS) was used to determine the average molecular weight (Mw) of BPLP. Average Mw calculated from the mass spectrum of BPLP is 1044 g/mol. For the analysis, the sample was spotted in a MALDI plate with 1,8,9-Antracenotriol (Ditranol) and analyzed using a mass spectrometer (Bruker Daltonics Ultraflex TOF/TOF) in reflectron mode, Figure S3: SEM images and DLS size distributions of NCs fabricated by different types of PLGA-BPLP, Figure S4: Fluorescence intensity of different concentrations of NCs measured by microplate reader at the maximal emission wavelengths (values have subtracted values of control non-fluorescence NCs), Figure S5: Hysteresis loop (300 K) and ZFC-FC (inset) measurements of the lyophilized nanocapsule batch NC2.

Author Contributions: Conceptualization, Y.Z. and A.R. (Anna Roig); Data curation, Y.Z. and M.G.-G.; Formal analysis, all authors; Investigation, Y.Z. and M.G.-G.; Methodology, Y.Z. and M.G.-G.; Resources, A.R. (Anna Rosell) and A.R. (Anna Roig); Supervision, A.R. (Anna Rosell) and A.R. (Anna Roig); Writing—original draft, Y.Z.; Writing—review & editing all authors. All authors have read and agreed to the published version of the manuscript.

Funding: This work was partially supported by the Spanish Ministry of Science, Innovation and Universities through the grants PCIN-2017-090, RTI2018-096273-B-I00, CTQ2017-87637-R, SAF2017-87670-R, SEV-2015-0496 and MDM-2017-0720 and by the Generalitat de Catalunya grants (2017SGR765 & 2017SGR1427). Yajie Zhang was supported by the China Scholarship Council (CSC). The Acciones Complementarias program from the Instituto de Salud Carlos III, Spain, co-financed by the European Regional Development Fund (AC17/00004 grant) as part of the MAGBBRIS project (Euronanomed III 8th joint call), and the RETICS INVICTUS RD16/0019/0021 were acknowledged. Miguel Servet program (CPII15/00003) from Fondo de Investigaciones Sanitarias-Instituto de Salud Carlos III and ERDF, supported A. Roig research contract.

Conflicts of Interest: The authors declare no conflict of interest.

References

1. Kim, D.; Shin, K.; Kwon, S.G.; Hyeon, T. Synthesis and biomedical applications of multifunctional nanoparticles. *Adv. Mater.* **2018**, *30*, 1802309. [CrossRef]
2. Cruz, L.J.; Stammes, M.A.; Que, I.; van Beek, E.R.; Knol-Blankevoort, V.T.; Snoeks, T.J.A.; Chan, A.; Kaijzel, E.L.; Lowik, C. Effect of PLGA NP size on efficiency to target traumatic brain injury. *J. Control. Release* **2016**, *223*, 31–41. [CrossRef]
3. Liu, J.; Xu, H.; Tang, X.; Xu, J.; Jin, Z.; Li, H.; Wang, S.; Gou, J.; Jin, X. Simple and tunable surface coatings via polydopamine for modulating pharmacokinetics, cell uptake and biodistribution of polymeric nanoparticles. *RSC Adv.* **2017**, *7*, 15864–15876. [CrossRef]
4. Cui, Y.; Zhang, M.; Zeng, F.; Jin, H.; Xu, Q.; Huang, Y. Dual-targeting magnetic PLGA nanoparticles for codelivery of paclitaxel and curcumin for brain tumor therapy. *ACS Appl. Mater. Interfaces* **2016**, *8*, 32159–32169. [CrossRef]
5. Danhier, F.; Ansorena, E.; Silva, J.M.; Coco, R.; Le Breton, A.; Préat, V. PLGA-based nanoparticles: An overview of biomedical applications. *J. Control. Release* **2012**, *161*, 505–522. [CrossRef] [PubMed]
6. Swider, E.; Koshkina, O.; Tel, J.; Cruz, L.J.; de Vries, I.J.M.; Srinivas, M. Customizing poly(lactic-co-glycolic acid) particles for biomedical applications. *Acta. Biomater.* **2018**, *73*, 38–51. [CrossRef] [PubMed]
7. Mir, M.; Ahmed, N.; ur Rehman, A. Recent applications of PLGA based nanostructures in drug delivery. *Coll. Surf. B Biointerfaces* **2017**, *159*, 217–231. [CrossRef] [PubMed]
8. Ouyang, Z.; Tan, T.; Liu, C.; Duan, J.; Wang, W.; Guo, X.; Zhang, Q.; Li, Z.; Huang, Q.; Dou, P. Targeted delivery of hesperetin to cartilage attenuates osteoarthritis by bimodal imaging with Gd2 (CO3) 3@ PDA nanoparticles via TLR-2/NF-κB/Akt signaling. *Biomaterials* **2019**, *205*, 50–63. [CrossRef] [PubMed]
9. Yang, R.; Xu, J.; Xu, L.; Sun, X.; Chen, Q.; Zhao, Y.; Peng, R.; Liu, Z. Cancer cell membrane-coated adjuvant nanoparticles with mannose modification for effective anticancer vaccination. *ACS Nano* **2018**, *12*, 5121–5129. [CrossRef]
10. Medina, D.X.; Householder, K.T.; Ceton, R.; Kovalik, T.; Heffernan, J.M.; Shankar, R.V.; Bowser, R.P.; Wechsler-Reya, R.J.; Sirianni, R.W. Optical barcoding of PLGA for multispectral analysis of nanoparticle fate in vivo. *J. Control. Release* **2017**, *253*, 172–182. [CrossRef]
11. Park, J.K.; Utsumi, T.; Seo, Y.E.; Deng, Y.; Satoh, A.; Saltzman, W.M.; Iwakiri, Y. Cellular distribution of injected PLGA-nanoparticles in the liver. *Nanomedicine* **2016**, *12*, 1365–1374. [CrossRef] [PubMed]
12. Abdel-Mottaleb, M.M.; Beduneau, A.; Pellequer, Y.; Lamprecht, A. Stability of fluorescent labels in PLGA polymeric nanoparticles: Quantum dots versus organic dyes. *Int. J. Pharm.* **2015**, *494*, 471–478. [CrossRef] [PubMed]
13. Cook, R.L.; Householder, K.T.; Chung, E.P.; Prakapenka, A.V.; DiPerna, D.M.; Sirianni, R.W. A critical evaluation of drug delivery from ligand modified nanoparticles: Confounding small molecule distribution and efficacy in the central nervous system. *J. Control. Release* **2015**, *220*, 89–97. [CrossRef] [PubMed]

14. Jaiswal, J.K.; Mattoussi, H.; Mauro, J.M.; Simon, S.M. Long-term multiple color imaging of live cells using quantum dot bioconjugates. *Nat. Biotechnol.* **2003**, *21*, 47. [CrossRef] [PubMed]
15. Jamieson, T.; Bakhshi, R.; Petrova, D.; Pocock, R.; Imani, M.; Seifalian, A.M. Biological applications of quantum dots. *Biomaterials* **2007**, *28*, 4717–4732. [CrossRef] [PubMed]
16. Yang, J.; Zhang, Y.; Gautam, S.; Liu, L.; Dey, J.; Chen, W.; Mason, R.P.; Serrano, C.A.; Schug, K.A.; Tang, L. Development of aliphatic biodegradable photoluminescent polymers. *Proc. Natl. Acad. Sci. USA* **2009**, *106*, 10086–10091. [CrossRef]
17. Hu, J.; Guo, J.; Xie, Z.; Shan, D.; Gerhard, E.; Qian, G.; Yang, J. Fluorescence imaging enabled poly(lactide-co-glycolide). *Acta Biomater.* **2016**, *29*, 307–319. [CrossRef]
18. Tao, W.; Zeng, X.; Wu, J.; Zhu, X.; Yu, X.; Zhang, X.; Zhang, J.; Liu, G.; Mei, L. Polydopamine-based surface modification of novel nanoparticle-aptamer bioconjugates for in vivo breast cancer targeting and enhanced therapeutic effects. *Theranostics* **2016**, *6*, 470. [CrossRef]
19. Kumar, P.; Van Treuren, T.; Ranjan, A.P.; Chaudhary, P.; Vishwanatha, J.K. In vivo imaging and biodistribution of near infrared dye loaded brain-metastatic-breast-cancer-cell-membrane coated polymeric nanoparticles. *Nanotechnology* **2019**, *30*, 265101. [CrossRef]
20. Gonzalez-Moragas, L.; Yu, S.-M.; Murillo-Cremaes, N.; Laromaine, A.; Roig, A. Scale-up synthesis of iron oxide nanoparticles by microwave-assisted thermal decomposition. *Chem. Eng. J.* **2015**, *281*, 87–95. [CrossRef]
21. Kasprzyk, W.; Bednarz, S.; Bogdał, D. Luminescence phenomena of biodegradable photoluminescent poly(diol citrates). *Chem. Commun.* **2013**, *49*, 6445–6447. [CrossRef]
22. Carenza, E.; Jordan, O.; Martinez-San Segundo, P.; Jiřík, R.; Starčuk Jr, Z.; Borchard, G.; Rosell, A.; Roig, A. Encapsulation of VEGF 165 into magnetic PLGA nanocapsules for potential local delivery and bioactivity in human brain endothelial cells. *J. Mater. Chem. B* **2015**, *3*, 2538–2544. [CrossRef]
23. Gref, R.; Lück, M.; Quellec, P.; Marchand, M.; Dellacherie, E.; Harnisch, S.; Blunk, T.; Müller, R. 'Stealth'corona-core nanoparticles surface modified by polyethylene glycol (PEG): Influences of the corona (PEG chain length and surface density) and of the core composition on phagocytic uptake and plasma protein adsorption. *Colloids Surf. B Biointerfaces* **2000**, *18*, 301–313. [CrossRef]
24. Butoescu, N.; Seemayer, C.A.; Palmer, G.; Guerne, P.-A.; Gabay, C.; Doelker, E.; Jordan, O. Magnetically retainable microparticles for drug delivery to the joint: Efficacy studies in an antigen-induced arthritis model in mice. *Arthritis Res. Ther.* **2009**, *11*, R72. [CrossRef] [PubMed]
25. Wang, Y.-X.J. Superparamagnetic iron oxide based MRI contrast agents: Current status of clinical application. *Quant. Imaging Med. Surg.* **2011**, *1*, 35. [PubMed]
26. Pakulska, M.M.; Donaghue, I.E.; Obermeyer, J.M.; Tuladhar, A.; McLaughlin, C.K.; Shendruk, T.N.; Shoichet, M.S. Encapsulation-free controlled release: Electrostatic adsorption eliminates the need for protein encapsulation in PLGA nanoparticles. *Sci. Adv.* **2016**, *2*, e1600519. [CrossRef]

pharmaceutics

MDPI

Article

Theranostic Sorafenib-Loaded Polymeric Nanocarriers Manufactured by Enhanced Gadolinium Conjugation Techniques

Tivadar Feczkó [1,2,3,*], Albrecht Piiper [3], Thomas Pleli [3], Christian Schmithals [3], Dominic Denk [3], Stephanie Hehlgans [4], Franz Rödel [4], Thomas J. Vogl [5] and Matthias G. Wacker [6]

1 Research Centre for Natural Sciences, Hungarian Academy of Sciences, Magyar tudosok krt. 2., H-1117 Budapest, Hungary
2 Research Institute of Biomolecular and Chemical Engineering, University of Pannonia, Egyetem u. 2., H-8200 Veszprém, Hungary
3 Department of Medicine 1, University Hospital Frankfurt, Theodor-Stern-Kai 7, D-60590 Frankfurt, Germany; Piiper@med.uni-frankfurt.de (A.P.); thomas_pleli@yahoo.com (T.P.); Christian.Schmithals@kgu.de (C.S.); domdenk@googlemail.com (D.D.)
4 Department of Radiotherapy and Oncology, University Hospital Frankfurt am Main, Theodor-Stern-Kai 7, D-60590 Frankfurt am Main, Germany; Stephanie.Hehlgans@kgu.de (S.H.); Franz.Roedel@kgu.de (F.R.)
5 Department of Diagnostic and Interventional Radiology, University Hospital Frankfurt, Theodor-Stern-Kai 7, D-60590 Frankfurt, Germany; Thomas.Vogl@kgu.de
6 Department of Pharmacy, National University of Singapore, 6 Science Drive 2, Singapore 117546, Singapore; phamgw@nus.edu.sg
* Correspondence: tivadar.feczko@gmail.com; Tel.: +36-88-624000/3508

Received: 15 August 2019; Accepted: 18 September 2019; Published: 23 September 2019

Abstract: Today, efficient delivery of sorafenib to hepatocellular carcinoma remains a challenge for current drug formulation strategies. Incorporating the lipophilic molecule into biocompatible and biodegradable theranostic nanocarriers has great potential for improving the efficacy and safety of cancer therapy. In the present study, three different technologies for the encapsulation of sorafenib into poly(D,L-lactide-*co*-glycolide) and polyethylene glycol-poly(D,L-lactide-*co*-glycolide) copolymers were compared. The particles ranged in size between 220 and 240 nm, with encapsulation efficiencies from 76.1 ± 1.7% to 69.1 ± 10.1%. A remarkable maximum drug load of approximately 9.0% was achieved. Finally, a gadolinium complex was covalently attached to the nanoparticle surface, transforming the nanospheres into theranostic devices, allowing their localization using magnetic resonance imaging. The manufacture of sorafenib-loaded nanoparticles alongside the functionalization of the particle surface with gadolinium complexes resulted in a highly efficacious nanodelivery system which exhibited a strong magnetic resonance imaging signal, optimal stability features, and a sustained release profile.

Keywords: gadolinium; drug release; polymeric nanocarrier; sorafenib; theranostic nanoparticles

1. Introduction

Liver cancer, of which the majority of cases are hepatocellular carcinoma (HCC), is a life-threatening disease and, according to global cancer statistics, the third leading cause of cancer-related mortality worldwide [1]. To date, sorafenib is the only drug able to prolong the lives of patients with HCC [2], although at the expense of severe side effects due to uptake of the drug into healthy tissues [3]. Sorafenib is a multi-kinase inhibitor targeting various receptor tyrosine kinases and rapidly accelerated fibrosarcoma (RAF) kinases. The pronounced lipophilicity of the molecule is responsible for its poor bioavailability and the distribution of the compound into healthy tissues [3]. As a consequence, patients are treated using high doses of the drug and suffer from a number of side effects.

Nanocarrier-based delivery of sorafenib has the potential to improve drug therapy significantly. Theranostic nanodelivery systems offer a versatile combination of therapeutic and diagnostic features, and have previously been applied to this task [4]. A variety of contrast agents, such as gadolinium diethylenetriamine pentaacetic acid (Gd-DTPA), shorten the longitudinal relaxation time, and have been widely used for both vascular and tumor magnetic resonance imaging (MRI) [4,5]. One major limitation of this technique lies in the short half-life of the contrast agent, as well as its poor specificity for the target site.

In this context, Gd-DTPA-conjugated human serum albumin (HSA) nanoparticles improved the contrast of MRI compared to free Gd-DTPA aqueous solution in vivo, due to negative contrasting of the tumors [5]. Even after conjugation of polyethylene glycol (PEG) to the particle surface, HSA nanoparticles exhibited only a short circulation time [6] and the scale-up potential of this technology is rather limited [7].

Block copolymers comprising polylactic-*co*-glycolic acid (PLGA) and PEG have been processed to nanoparticles in pilot scale using either microfluidic technologies [8] or emulsion techniques [9]. Recently, targeted nanotherapeutic formulations successfully passed phase 1 clinical trials [10]. A major shortcoming of nanocarrier delivery is limited drug loading, resulting in high excipient concentrations and administration volume [11].

Several preparation techniques of PLGA nanoparticle preparation have been described in the literature. Previous investigations reported the loading of sorafenib into PLGA nanoparticles, achieving a drug load of 1.4% oil-in-water single emulsion–solvent evaporation method [12]. Preparations made with a nanoprecipitation–dialysis technique using a block copolymer comprising dextran and PLGA resulted in a drug load of 5.3% [13]. To attain synergistic effects of cytostatic agents, co-delivery of drug molecules has been considered. By employing a sequential freeze–thaw method followed by ethanol coacervation, a core–shell construct was manufactured [14]. After preparation of a polyvinyl alcohol (PVA)-doxorubicin nanocore, a thin shell of HSA covering sorafenib was used as a second drug molecule. A drug load of 2.4% was reached. Lipid–polymer hybrid nanoparticles were synthesized for the co-delivery of doxorubicin and sorafenib to enhance efficacy in HCC therapy [15].

Other approaches have focused on theranostic drug delivery systems containing gadolinium (Gd) and sorafenib. Theranostic liposomal carriers with a drug content of 4.3% (*m/m*) have been produced [16]. Another system used a multiblock polymer comprising (poly(lactic acid)-poly(ethylene glycol)-poly(L-lysine)-diethylenetriamine pentaacetic acid and the pH-sensitive material poly(L-histidine)-poly(ethylene glycol)-biotin. A drug content of 2.4% (*m/m*) of sorafenib was reached, and the MRI signal intensity was more beneficial than that of Magnevist® in vivo [12].

In the present study, the impact of three different preparation techniques on the physicochemical features of poly(D,L-lactide-*co*-glycolide) (PLGA) and polyethylene glycol-copolymer (PEG-PLGA) nanocarriers were compared. Nanocarriers loaded with sorafenib were manufactured using the nanoprecipitation, single emulsion, and double emulsion–solvent evaporation methods. Nanoparticles manufactured using the single emulsion–solvent evaporation method were further optimized with regards to the intended target product profile. In vitro cytotoxicity and cellular uptake were investigated in HepG2 cells. Finally, the nanoparticles were modified on their surface using a gadolinium complex, resulting in a theranostic nanocarrier system. For this purpose, the encapsulation of HSA into nanoparticles and the covalent modification of the surface using HSA or polyethylene imine were considered.

2. Materials and Methods

2.1. Materials

Resomer® RG 502H (PLGA, lactide: glycolide: 50:50, inherent viscosity: 0.16–0.24 dL/g), Resomer® RG 752H (PLGA, lactide:glycolide: 75:25 inherent viscosity 0.14–0.22 dL/g), and block copolymer Resomer® RGP d5055 (PEG-PLGA, PEG content: 3–7% (*m/m*), inherent viscosity: 0.93 dL/g) were obtained from Evonik Industries AG (Essen, Germany). PVA (M_W = 30,000–70,000, 87–90% hydrolysed), polysorbate 80, Triton X-100, Pluronic F127, poly(methacrylic acid sodium salt) emulsifiers, dichloromethane (DCM), acetone, dimethyl sulfoxide (DMSO), sodium azide, D-trehalose dehydrate, mannitol, polyethyleneimine (PEI) (MW 25 kDa), 1-Ethyl-3-(3-dimethylaminopropyl)carbodiimide (EDC), *N*-hydroxysuccinimide (NHS), Gd-DTPA, and HSA were obtained from Sigma Aldrich (St. Louis, MO, USA). Sorafenib (free base) was purchased from LC Laboratories (Woburn, MA, USA). Magnevist® was purchased from Bayer AG (Leverkusen, Germany). The micro bicinchoninic acid (μBCA) protein assay kit was bought from Pierce Biotechnology, Inc. (Waltham, MA, USA).

2.2. Cell Culture Experiments in HepG2 Cells

The human hepatoma cell line HepG2 was grown in Dulbecco's modified Eagle's medium supplemented with 10% fetal calf serum (FCS), 100 U/mL penicillin, and 100 μg/mL streptomycin. The cells were cultured at 37 °C in a humidified atmosphere containing 5% carbon dioxide. The cells were trypsinised, resuspended, and precultured before use.

2.3. Preparation of Nanoparticles Using Nanoprecipitation

In brief, 5 to 10 mg of Resomer® RG 502 H, Resomer® RG 752H or Resomer® RGP d5055 and 0.5 to 4 mg of sorafenib were dissolved in between 0.5 and 1.0 mL of acetone under magnetic stirring. A water phase with a volume of 2.0 to 4.0 mL was composed of an aqueous solution (0.5 to 2.0% *w/v*) of emulsifying agent (polysorbate 80, poly(methacrylic acid sodium salt), Triton X-100, or Pluronic® F127), and added as a one-shot to the organic phase. Afterwards, the organic solvent was evaporated over a time period of 12 h at room temperature and 1 bar under constant stirring. The nanoparticles were centrifuged (Eppendorf 5424 R, Hamburg, Germany) at 37,565 *g* for 25 min, washed thrice and redispersed in an equal volume of purified water.

2.4. Preparation of Nanoparticles Using the Single Emulsion Technique

For the preparation of nanoparticles with the single emulsion–solvent evaporation method, the organic phase was prepared in two steps. An measure of 1 to 2 mg sorafenib was dissolved in 0.1 to 0.2 mL of acetone, and this solution was poured into a solution comprising 5 to 20 mg of Resomer® RG 502 H, Resomer® RG 752H, or Resomer® RGP d5055 in 1 to 2 mL DCM.

A volume of 4 to 8 mL of an aqueous solution of PVA (1 to 2% *w/v*) was added and sonicated using a Sonoplus HD2070, MS73 probe (Bandelin, Berlin, Germany) at an amplitude of 10% for 60 s. The organic solvent was evaporated over 2 h at atmospheric pressure and room temperature. The nanoparticles were purified as described above.

2.5. Preparation of Nanoparticles Using Double Emulsion–Solvent Evaporation Technique

The double emulsion–solvent evaporation technique was tested for the co-encapsulation of HSA into the sorafenib nanoparticles. The entrapped protein was further used for the covalent modification of the particle surface using the Gd-DTPA complex. In principle, the inner water phase was formed of 2.5 mg HSA in a volume of 0.1 mL of purified water. This solution was added to the organic phase, composed of 15 mg of Resomer® RG 752H or Resomer® RGP d5055 dissolved in 1.5 mL DCM, combined with 1.5 mg of sorafenib dissolved in 0.15 mL acetone. The first emulsification was performed by sonication using a Sonoplus HD2070, MS73 probe (Bandelin, Berlin, Germany) at an amplitude of 10% for 30 s. Afterwards, the water-in-oil emulsion was pipetted into 6.0 mL of an aqueous solution of PVA (1% (*w/v*)). A water-in-oil-in-water emulsion was formed by a second sonication step at an amplitude 15% for 45 s. The organic solvents were evaporated over a time period of 2 h under magnetic stirring at atmospheric pressure and room temperature. The nanoparticles were centrifuged at 37,565 *g* for 25 min (Eppendorf Centrifuge 5424 R), washed thrice and redispersed in 0.5 mL of phosphate buffer (pH 8), and, after gravimetric analysis, the suspensions were diluted to 20 mg·mL^{-1} nanoparticle concentration.

2.6. Particle Morphology and Particle Size Analysis

The morphology of the nanospheres was investigated after centrifugation and redispersion in the described medium. The samples were examined using a FEI Talos F200XG2 high-resolution analytical microscope operated at 200 keV (Thermo Fischer Scientific, Waltham, MA, USA).

Additionally, the particle size and size distribution were determined using a Zetasizer Nano ZS (Malvern Instruments, Malvern, UK) equipped with a backscatter detector at an angle of 173°. The particles were characterized for their intensity mean diameter and polydispersity index (PDI). Dynamic light scattering (also known as photon correlation spectroscopy) is based on the Brownian motion of particles dispersed in a liquid. The particle diameter is calculated from the intensity fluctuations of light scattered at the particle surface. The Zetasizer Nano ZS uses a HeNe gas laser to generate a signal of these intensity fluctuations, from which the size is calculated by applying the Stokes–Einstein equation. Consequently, the resulting size distribution is an intensity distribution.

2.7. Storage Stability of Nanoparticle Formulations

To improve the physical stability of the colloidal dispersion during storage, the nanoparticles were freeze-dried, and their storage stability was investigated after 6 months of storage. A concentration of 3% (*w/v*) of two lyoprotectors (trehalose dihydrate or mannitol) was evaluated. The solid concentration was adjusted to 7 mg·mL^{-1} of nanoparticles and the samples were put into a Christ Epsilon 2–7 freeze dryer (Martin Christ GmbH, Osterode am Harz, Germany). The lyophilisation was conducted in two drying steps.

Initially, the temperature was decreased to −60 °C for 1 h to freeze the samples. Afterwards, primary drying was initiated by evacuating the chamber to 0.94 mbar. In parallel, temperature was raised to −30 °C over a time period of 150 min. The pressure was then reduced to 0.006 mbar and the temperature was increased to −10 °C over a time period of 60 min. These conditions were maintained for 35 h. The second drying was accomplished at a temperature of 10 °C for 60 min and at 20 °C for 10 h. The freeze dried samples were stored at 4 °C for a total duration of 6 months, and reconstituted in the same volume of purified water. The nanocomposite size and size distribution after redispersion were characterized by triplicated measurement.

2.8. Nanoparticle Yield and Encapsulation Efficiency

The nanoparticle yield was determined by microgravimetry. The drug loading and encapsulation efficiency were investigated dissolving 10 mg nanoparticles in 1 mL of DMSO. The solution was diluted using DMSO to be within the detectable linear calibration range (1–20 µg/L). The absorbance of the solutions was measured spectrophotometrically (Hitachi U-3000, Tokyo, Japan) at 285 nm. A concentration range of HSA solution in DMSO was prepared in order to correct the absorbance of HSA-loaded nanoparticles, also measured at 285 nm.

2.9. Biorelevant In Vitro Drug Release Test Using the Centrifugation Method

The biorelevant in vitro drug release test was conducted using the centrifugation method. In brief, 1.5 mg of the sorafenib-loaded nanocomposites was re-suspended in 1 mL of human blood plasma containing 0.03% sodium azide as a preservative. The nanoparticle formulations in release medium were filled into 2 mL tubes, and incubated at a temperature of 37 °C in a Thermomixer (Eppendorf,

Hamburg, Germany) for 12 days at 700 rpm. At predetermined time points, 0.2 mL samples were collected. After each sampling time point the medium was replenished. The nanoparticles were separated from the plasma by centrifugation (Eppendorf Centrifuge 5424 R, 20 min at 37,565 *g*). The concentration of free sorafenib was determined using direct quantification of the drug remaining in the particle system. For this purpose, the nanoparticle pellets were dissolved in DMSO and the drug amount was detected spectrophotometrically.

2.10. Surface Modification Using Human Serum Albumin and Polyethyleneimine

The nanoparticle dispersions prepared by single emulsion technique were centrifuged and redispersed in phosphate buffer (pH 8), resulting in a nanoparticle concentration of 20 mg·mL^{-1}. To increase the number of free amino groups on the particle surface, HSA or PEI were covalently bound to the nanoparticle surface. A 50-fold molar excess of EDC and the same excess of NHS, both calculated to the molar polymer concentration, were dissolved in 0.5 mL phosphate buffer (pH 8) and incubated for 60 min, centrifuged and washed three times, and redispersed in 1.5 mL phosphate buffer (pH 8).

The obtained carbodiimide-activated nanoparticle dispersion was added into 0.5 mL phosphate buffer (pH 8) solution containing equimolar amounts of HSA or PEI, and shaken overnight at 20 °C in an Eppendorf Thermomixer (Hamburg, Germany) at 700 rpm. Afterwards, the nanoparticles were dialysed for 2 h against 400 mL of purified water, using a 100 kDa membrane to remove residues of the dissolved polymers HSA or PEI, respectively. The dialysis step was repeated with purified water.

2.11. Conjugation of Nanoparticles Using the Gd-DTPA Complex

The Gd-DTPA complex was covalently attached to either HSA or PEI after activation using the carbodiimide. In order to activate the carboxyl groups of the gadolinium complex, a 10-fold molar excess of Gd-DTPA (calculated on the amount of the Resomer®) was dissolved in 1 mL of buffer (pH 8) and combined with a 5-fold molar excess of EDC and a 5-fold molar excess of NHS (calculated on the amount of Gd-DTPA). The resulting solution was incubated for 50 min and added to the purified nanoparticle suspension overnight. The preparation was purified by dialysis twice, using a membrane with a MWCO of 3.5 kDa and 500 mL of purified water for 2 h. The dialysed nanoparticle suspension was centrifuged and redispersed in 1.5 mL phosphate buffer.

2.12. Quantification of Human Serum Albumin Using the Micro Bicinchoninic Acid Method

The crosslinked or co-encapsulated HSA content of the nanocomposites was determined by the μBCA method after centrifuging 0.1 mL nanoparticle dispersion and removing the supernatant, while the pellet was dissolved in 0.5 mL DMSO. This solution was diluted 10-fold with purified water, and incubated for 1 h at 60 °C with the same volume of freshly prepared μBCA reagents mixture. DMSO was added to the calibrating HSA solution with the same ratio. After cooling the colored mixtures to room temperature for 20 min, their absorbance was evaluated by spectrophotometry at 562 nm (Hitachi U-3000, Tokyo, Japan).

2.13. Quantification of Gadolinium Using Inductively Coupled Plasma Optical Emission Spectroscopy

The Gd concentration was measured using a Spectro Genesis ICP-OES (Kleve, Germany) simultaneous spectrometer with axial plasma observation. Multielemental standards (Merck, standard solutions for ICP, Darmstadt, Germany) were used for calibration. The limits of detection of the element were calculated according to Equation (1):

$$Limit\ of\ detection = Background\ signal + 3 \times SD_{Background} \times f_{dilution} \quad (1)$$

The purified nanoparticles were dissolved in 5 M hydrochloric acid and diluted to the desired calibration range.

2.14. In Vitro Investigation of Diagnostic Features by Magnetic Resonance Imaging

In vitro MRI was carried out using box analysis to compare the contrast achieved with the nanoparticle suspension with that of Magnevist® (Gd-DTPA complex with meglumine) solution. A calibration was achieved by diluting Magnevist® to 0.01–2.5 mg·mL^{-1} concentration. MR imaging was performed on a 3.0-T scanner (Siemens Magnetom Trio, Siemens Medical Solutions, Erlangen, Germany). The measurement conditions were T1-weighted 3D gradient echo sequences (fast low-angle shot) with the following parameters: TE (echo time) = 3.31 ms, TR (repetition time) = 8.67 ms, field of view = 100 × 78 mm, matrix acquisition = 640 × 480, slice thickness = 0.3 mm, flip angle = 16°, fat suppression = fat saturated, and bandwidth = 180 Hz/Px.

2.15. Labeling of Nanoparticles with a Fluorescent Dye

A volume of 1.0 mL of the nanoparticle suspension (12 mg·mL^{-1}) in phosphate buffer (pH 8) was added to 0.1 mL phosphate buffer (pH 8) containing a 25-fold molar excess of EDC and NHS (calculated on the amount of Resomer®) and incubated for 60 min, centrifuged, and washed, and redispersed in 1.0 mL phosphate buffer (pH 8). The obtained carbodiimide activated nanoparticle dispersion was given to 0.1 mL phosphate buffer (pH 8) solution containing 1 mg·mL^{-1} Cyanine5 amine fluorescent (Cy5) dye, and shaken for 1 h at 20 °C in an Eppendorf Thermomixer (Hamburg, Germany) at 700 rpm. The nanoparticle dispersion was then centrifuged in an Eppendorf Centrifuge 5424 R (Hamburg, Germany) at 37,565 *g* for 25 min, washed three times, and redispersed in phosphate-buffered saline to a nanoparticle concentration of 10 mg·mL^{-1}.

2.16. In Vitro Cellular Uptake and Cytotoxicity

Cellular uptake of the nanoparticles into the HepG2 cells was evaluated using flow cytometry. The cells were cultured in 24-well plates at a cell density of 2 × 10^5 cells per well at 37 °C and 5% CO_2 for 24 h. After cultivation, 100 μg nanoparticles/well (10 mg/mL nanoparticle suspension was diluted to 100 μg/mL) was pipetted to the cells and incubated for 24 h. The cells grown without nanoparticles were used as control. The cells were washed in phosphate-buffered saline (PBS), trypsinised, and redispersed in PBS containing 2% (*m/v*) of bovine serum albumin. Flow cytometry was performed on a Cytoflex S cytometer (Beckman Coulter, Brea, CA, USA).

To further analyze cellular uptake and intracellular localization, fluorescence microscopy of HepG2 cells plated on cover slides and incubated with either Cy5-labeled PLGA or PEG-PLGA NP for 4 h and 24 h was employed. Membrane staining was performed by using Alexa488 concanavalin A (2.5 μg/mL; Thermo Fisher Scientific, Schwerte, Germany) and nuclei were counterstained with 4′,6-diamidino-2-phenyl-indole (DAPI) solution (Merck, Darmstadt, Germany). Finally, slides were mounted with Vectashield mounting medium (Biozol, Eching, Germany) and images were obtained using an AxioImager Z1 microscope and Axiovision 4.6 software (Carl Zeiss, Jena, Germany).

The in vitro cytotoxicity in HepG2 cells was determined using 3-(4,5-dimethylthiazol-2-yl)-2,5-diphenyltetrazolium bromide (MTT) assay. Cells were seeded (50,000 cells/well) in 96-well plates. At 24 h pre-incubation, the media were replaced with 100 μL of fresh Dulbecco's Modified Eagle's Medium (DMEM) containing 10% FBS and sorafenib-loaded nanoparticles. Three different sorafenib concentrations (6.25 μg/mL, 12.5 μg/mL, and 25 μg/mL) were used, while the control samples contained the same amounts of free sorafenib in DMSO solution. After 24 h of incubation, a volume of 10 μL per well of MTT solution (5 mg MTT/mL) was added, followed by further incubation for 2 h. The supernatant was removed, and 0.2 mL MTT lysis solution was added into each well. The absorbance of cell suspension was determined at 595 nm using a spectrophotometer (EnVision 2104 Multilabel Reader, Perkin Elmer, Waltham, MA, USA). The percentage of viable cells was calculated by comparing the absorbance of treated cells against the untreated cells (negative control). The DMSO solution and the blank nanoparticle suspensions served as positive controls. The data are presented as mean and standard deviation with five replicates.

2.17. Statistics

All data are expressed as the mean value ± standard deviation (SD), which were calculated and plotted using Microsoft Excel (Microsoft, Redmond, WA, USA) and SigmaPlot 11.0 (Systat Software GmbH, Erkrath, Germany), respectively. All nanoparticle formulations were produced as three batches ($n = 3$).

3. Results and Discussion

In recent years, a variety of preparation methods have been evaluated for the synthesis of nanocarrier devices. The current study produced advanced theranostic drug carriers and compared the impact of manufacturing technology and surface modification on their physicochemical properties and in vitro features.

Initially, a particle size between 100 and 300 nm [17], a zeta potential of more than −15 mV [18,19], and a high drug load and particle yield were identified as key criteria for formulation development. Due to the enhanced permeability and retention (EPR) effect, these nanoparticles may be capable of targeting tumor tissues [17]. While smaller nanoparticles can be rapidly excreted by the kidneys, larger colloids with a size of more than 300 nm are quickly recognized by the macrophages of the reticuloendothelial system [11]. Among other aspects, the encapsulation efficiency and particle size play an important role in nanocarrier delivery. Selecting biodegradable polymers of the Resomer® type, three different techniques for the encapsulation of sorafenib were compared (Figure 1).

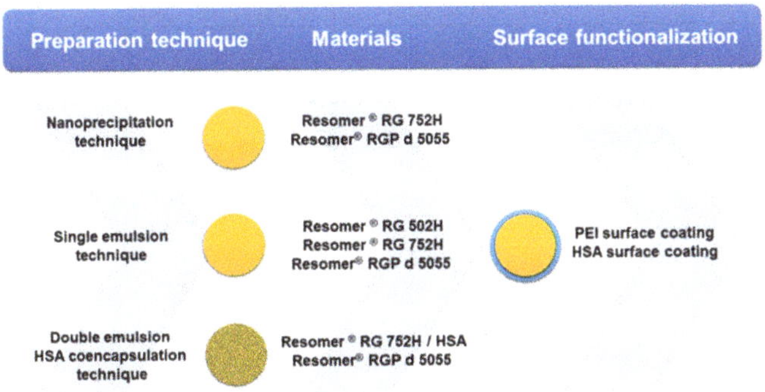

Figure 1. Scheme of the nanoparticle preparations with the preparation technique, materials used, and surface functionalization (from the left to the right). Each preparation technique was evaluated with the presented materials. Surface coating was undertaken for preparations manufactured by single emulsion technique only.

3.1. Manufacture of Sorafenib-Loaded Core Particles Using Nanoprecipitation Technique

The manufacture of sorafenib-loaded nanoparticles by nanoprecipitation using Triton X-100 or Pluronic® F127 stabilizers in aqueous solution resulted in a pronounced aggregation for all three polymers. Similar observations have been made at the medium scale, suggesting poor 'scalability' during the later stages of production [8]. Changing the stabilizer to polysorbate 80, similar aggregation occurred with Resomer® RG 502H and Resomer® RG 752H, respectively.

The use of Resomer® RGP d5055 (and polysorbate 80) resulted in nanoparticles within the desirable size range (153 ± 14 nm, Figure 2, upper micrographs). However, the PDI of more than 0.27, an encapsulation efficiency below 20%, and a particle yield between 20 and 40% were the major disadvantages of this formulation design. A lower density of the particle system due to the hydrophilic side chains of the polymer is the most likely explanation.

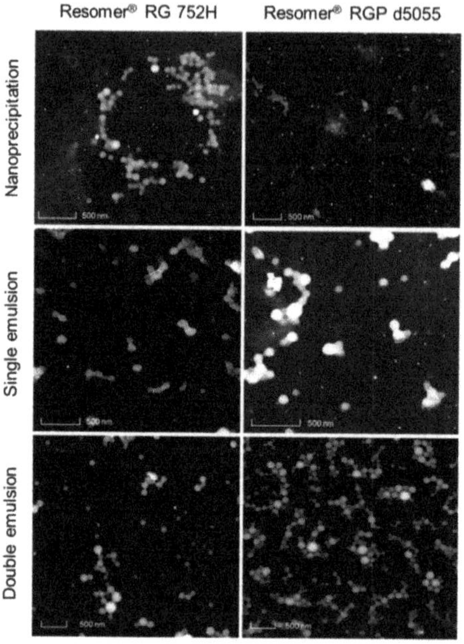

Figure 2. Scanning/transmission electron micrographs of nanoparticles comprising sorafenib in a Resomer® RG 752H (**upper left**) or Resomer® RGP d5055 (**upper right**) matrix prepared by nanoprecipitation, in a Resomer® RG 752H (**middle left**) and Resomer®, RGP d5055 (**middle right**) matrix prepared by single emulsion–solvent evaporation technique using polyvinyl alcohol (PVA) as a stabilizer, or in a Resomer® RG 752H-HAS (**lower left**) and Resomer®, RGP d5055-HSA (**lower right**) matrix by double emulsion–solvent evaporation technique using PVA as a stabilizer.

In comparison, the nanoprecipitation technique using PVA in combination with Resomer® RG 502H or Resomer® RGP d5055 led to particle systems broadly distributed in size, as indicated by elevated polydispersity indices ranging between 0.27 and 0.46. This was also confirmed by the intensity distributions, exhibiting a second and third fraction of larger particles in the micrometer range (Figure 3). The zeta potential of sorafenib-loaded Resomer® RG 502H or Resomer® RGP d5055 nanocomposites by nanoprecipitation and PVA emulsifier was found to be 13.2 ± 0.7 mV and 12.1 ± 1.2 mV, which also explains the poor stability resulting in a pronounced aggregation. Under similar conditions, crystallization of the drug accompanied by an increased PDI between 0.21 and 0.35 was reported by Lin et al. [18]. When applying the nanoprecipitation technique, a particle size in the desired range (196 ± 10 nm) as well as a reduced PDI of 0.21 ± 0.03, was achieved when using a polymer concentration of 10 mg·mL^{-1} of Resomer® RG 752H and 1 mg·mL^{-1} of sorafenib in the acetone phase.

Considering the difficulties in nanoparticle preparation when using the nanoprecipitation method, later efforts were focused on nanoparticle manufacture by single and double emulsion techniques.

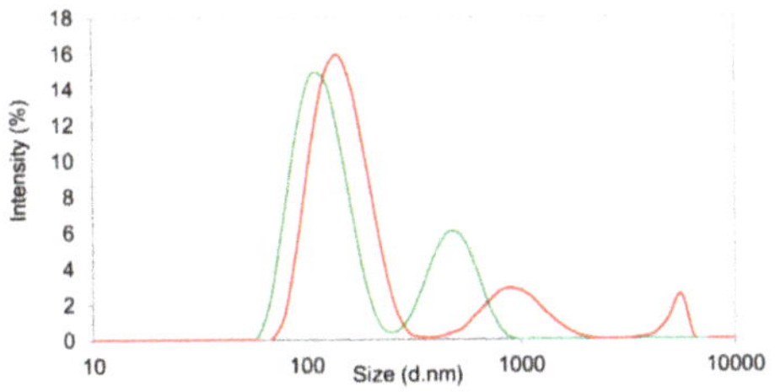

Figure 3. Size distribution by intensity of nanocomposites containing sorafenib in Resomer® RG 502H (green line) and Resomer® RGP d5055 (red line) matrix nanoparticles prepared by nanoprecipitation, using PVA as an emulsifier.

3.2. *Manufacture of Sorafenib-Loaded Core Particles by Single Emulsion Technique*

For the preparation of nanoparticles utilizing the single emulsion method, the hydrophilic phase was composed of a 1% or 2% (*w/v*) aqueous solution of PVA in purified water (Table 1). A polymer concentration of 10 mg·mL^{-1}, an initial drug amount of 10% (*m/m*), and a water-to-dichloromethane (DCM) ratio of 4:1 led to acceptable properties for each of the three polymers (Table 1). The morphology of the nanoparticles was tested by scanning/transmission electron microscopy. They were of spherical shape and within the expected size range (Figure 4).

Table 1. Properties of Resomer–sorafenib nanoparticles as a function of the encapsulating polymer and emulsifier (PVA) concentration.

Material	Resomer® RG 502H	Resomer® RG 752H	Resomer® RGP d5055	Resomer® RG 502H	Resomer® RG 752H	Resomer® RGP d5055
PVA (% *w/v*)	1	1	1	2	2	2
Mean size by intensity (nm)	235 ± 2.0	227.7 ± 3.3	228.3 ± 8.0	231.4 ± 15.6	231.3 ± 30.1	243.4 ± 40.4
PDI	0.14 ± 0.04	0.18 ± 0.01	0.12 ± 0.02	0.15 ± 0.06	0.19 ± 0.04	0.15 ± 0.14
Encapsulation efficiency (%)	70.4 ± 3.5	76.2 ± 2.1	76.7 ± 2.6	78.8 ± 4.4	76.6 ± 2.7	75.2 ± 6.7
Drug loading (%)	9.0 ± 0.5	10.2 ± 0.3	10.0 ± 0.4	12.0 ± 0.2	11.2 ± 0.1	8.9 ± 0.4
Zeta potential (mV)	−21.3 ± 2.4	−22.4 ± 2.4	−19.7 ± 1.1	−19.8 ± 3.3	−22.2 ± 1.8	−19.5 ± 1.6

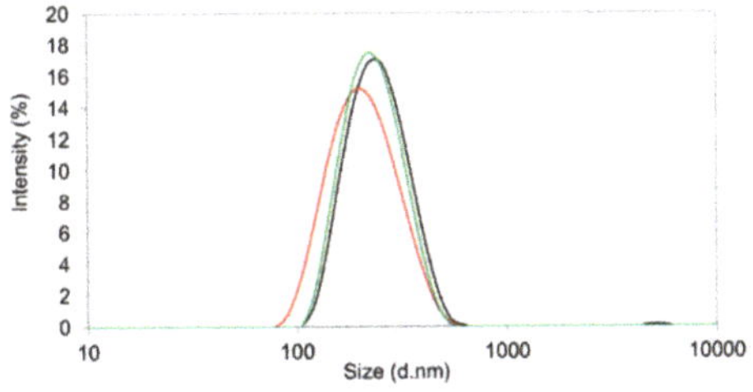

Figure 4. Size distribution by intensity of sorafenib-containing nanoparticles prepared from Resomer® RG 502H (black line) and Resomer® RG 752H (red line) polylactic-*co*-glycolic acid (PLGA) polymers as well as Resomer® RGP d5055 (green line) by single emulsion method.

With an encapsulation efficiency ranging between 70.4 and 78.8% (Table 1) and a monomodal size distribution (Figure 4), the emulsion method was superior compared to the nanoprecipitation technique. The drug loading (8.9–12.0% *m/m*) almost reached the initial drug ratio (10% *m/m*) (Table 1). Further increase of the drug amount from 10% (*m/m*) to 20% (*m/m*) resulted in elevated PDI values due to the formation of larger aggregates.

The best outcomes were achieved by using Resomer® RG 752H or Resomer® RGP d5055 as matrices (Figure 2, middle micrographs). Resomer® RG 752H resulted in a particle yield of 73.7%, an encapsulation efficiency of 76.6%, and a drug load of 11.2% (Figure 5). For Resomer® RGP d5055, similar promising results were achieved. A particle yield of 76.1%, an encapsulation efficiency of 75.2%, and a drug load of 8.9% were obtained using a 2% *m/v* emulsifier concentration. These nanocomposites were selected for further surface modification and cellular uptake and cytotoxicity studies (Figure 5). The zeta potential indicated high colloidal stability, which was also confirmed during further processing. It is likely that the surface charge resulted from the high number of carboxyl groups present in the PLGA polymers. However, there were no major differences between the zeta potential values of the employed polymers even when the nanocarriers were manufactured from the PEGylated derivative. This indicates that PEGylation did not significantly reduce the number of the carboxyl groups. These nanoparticle preparations were further processed and evaluated for their release behavior.

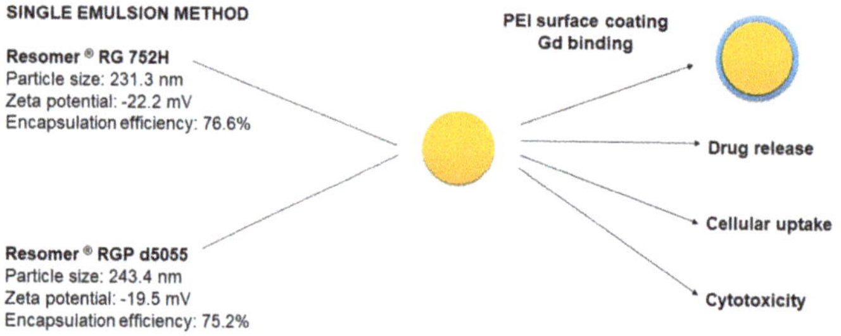

Figure 5. Selected formulations for surface modification and further tests.

3.3. Co-Encapsulation of HSA Using Double Emulsion Method or Surface Modification

To covalently bind the Gd-DTPA complex to the particle surface, amino groups were introduced into the polymeric matrices. In a first approach, HSA was co-encapsulated into the nanoparticles using the double emulsion–solvent evaporation method. Although there was no significant difference between the diameters of nanoparticles manufactured from Resomer RG 752H (mean diameter 210.6 nm, PDI 0.113) and Resomer®, RGP d5055 polymers (mean particle diameter 210.3 nm, PDI 0.099), incorporation of the protein resulted in smaller particle sizes than the single emulsion method. The detected size range was confirmed by electron microscopy (Figure 2, lower micrographs). All nanocomposites were of spherical shape (Figure 3). The particle yield was much lower compared to the methods described earlier. A particle yield of 50.4 ± 1.0% for Resomer® RG 752H and 49.0 ± 5.3% for Resomer® RGP d5055 was achieved. A similar amount of HSA was incorporated into both polymers (Resomer® RG 752H 10.6 ± 1.1%, Resomer® RGP d5055, 9.4 ± 1.2%). The formation of smaller particles with reduced density could be responsible for the decreased particle yield.

However, the incorporation method resulted in a high protein loading and an increased number of functional groups available for the EDC reaction. In comparison, an amount of 1.0 ± 0.3% HSA for Resomer® RG 752H and 1.9 ± 0.5% HSA for Resomer® RGP d5055 was bound to the particles using the surface coating technique.

3.4. Surface Modification with Gadolinium

The Gd-DTPA complex was conjugated to the amino groups of the HSA molecules after incorporation or covalent binding of the protein to Resomer® RG 752H and Resomer® RGP d5055 nanoparticles, respectively. As expected, the modification of nanocomposites with Gd-DTPA was limited by the availability of functional groups on the particle surface.

The nanoparticles modified on their surface with HSA were characterized by poor protein binding, with an amount of approximately 1.5 mg gadolinium per g Resomer® RG 752H, and 1.4 mg gadolinium per g Resomer® RGP d5055. Nanoparticles comprising HSA as part of their matrix structure exhibited a much higher binding of 2.3 mg gadolinium per g Resomer® RG 752H and 3.2 mg gadolinium per g Resomer® RGP d5055.

Nevertheless, further increase of the gadolinium content was achieved using a method described previously [20] with some modification. The binding of PEI to the surface resulted in a significant increase of the gadolinium content, with 15.7 mg gadolinium per g Resomer® RG 752H and 10.7 mg gadolinium per g Resomer® RGP d5055.

3.5. Evaluation of Contrast Signal Using In Vitro MRI

The MRI properties of theranostic nanocarriers were investigated in vitro. Magnevist® aqueous solutions were used as a reference. There was a linear correlation ($R^2 = 0.9724$) between the MRI signal and the gadolinium content in the range between 0.02 and 0.3 mg·mL^{-1} (with MRI intensities of 470–670 at 0.02 mg·mL^{-1}, and 1520–1750 at 0.3 mg·mL^{-1}). For nanoparticles exhibiting the highest content of gadolinium, the theoretical particle concentration (calculated from a maximum injectable particle concentration of 10 mg·mL^{-1} and the gadolinium load) fell into this calibration range. The signal intensities measured for the T1-weighed MRI were in good accordance with the inductively coupled plasma optical emission spectroscopy (ICP-OES). Earlier studies reported a loading of 4.7 to 8.5 mg gadolinium per g HSA nanoparticles after covalent binding of DTPA. This amount was found to be considerably high for in vivo MRI studies [5].

3.6. Storage Stability of Theranostic Nanocomposite Formulations

To freeze dry Resomer® RG 752H and Resomer® RGP d5055 nanoparticles manufactured by the single emulsion method, a concentration of 3% (*w/v*) of the lyoprotectors sucrose, trehalose, and mannitol has been previously applied [21]. The sorafenib-loaded nanoparticles with a drug load of approximately 10% were freeze dried in the presence of 3% (*w/v*) of trehalose or mannitol. The nanospheres remained stable over the time of storage. An increase in the PDI values was observed for formulations freeze-dried with mannitol. On this basis, trehalose at a concentration of 3% was identified to be the optimal lyoprotector for the developed particle system (Figure 6).

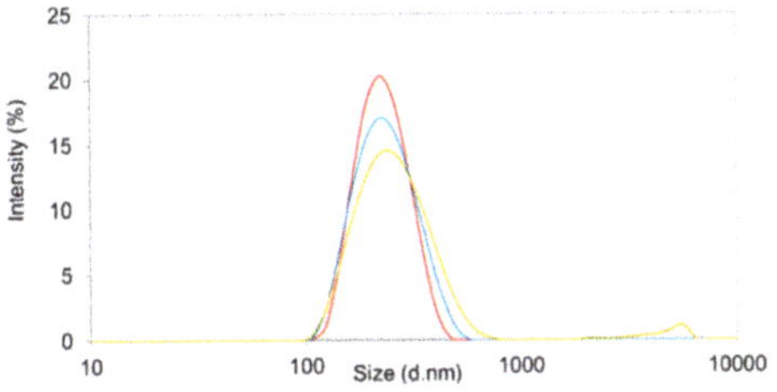

Figure 6. Size distribution by intensity of Resomer® RGP d5055 nanoparticles after preparation (red line), and following reconstitution after freeze-drying in the presence of 3% of trehalose (blue line) and 3% of mannitol (yellow line).

3.7. Evaluation of In Vitro Drug Release of Sorafenib from Theranostic Nanocomposites

For nanoparticles comprising polymers from the Resomer® family, a biphasic release pattern has been previously reported [22]. Preliminary investigations indicated a less pronounced burst release for Resomer® RG 752H compared to Resomer® RG 502H. After the initial release phase, a sustained release behavior was observed.

Consequently, the drug release of sorafenib-loaded Resomer®RGP d5055 and Resomer® RG 752H particles was investigated over a time period of 12 days in human blood plasma. Both of the nanocomposites were prepared under optimized conditions, that is, using the single emulsion method with an encapsulating polymer concentration of 10 mg·mL^{-1}, an initial drug amount of 10% (*m/m*), a water-to-DCM ratio of 4:1, and 2% (*w/v*) of PVA. The initial burst release was higher with nanoparticles manufactured from Resomer® RG 752H polymer (18.2 ± 2.9%) compared to Resomer®RGP d5055 copolymer, from which 8.8 ± 2.1% of the drug was released during the first hour.

Afterwards, a continuous release was observed for both composites (Figure 7). Surprisingly, the release from Resomer®RGP d5055 was substantially slower, reaching a plateau at 50.6 ± 9.2%. A specific interaction of sorafenib with the hydrophilic side chain of the block copolymer could be responsible for this behavior.

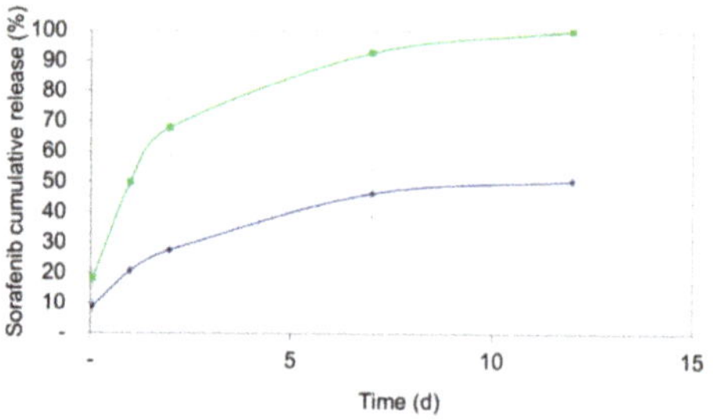

Figure 7. Sorafenib release from Resomer® RG 752H (green line) and Resomer® RGP d5055 -sorafenib (PEG-PLGA-SFB) nanocomposites (blue line) in human blood plasma. Data are presented as mean ± SD from three independent samples for each concentration.

3.8. Cellular Uptake and Cytotoxicity

For the cellular uptake and cytotoxicity studies, sorafenib-loaded Resomer®RGP d5055 and Resomer® RG 752H nanocomposites were prepared by the single emulsion method at a polymer concentration of 10 mg·mL^{-1}, an initial drug amount of 10% (*m/m*), a water-to-DCM ratio of 4:1, and 2% (*w/v*) of PVA. The internalization rates of sorafenib-containing nanoparticles in HepG2 cells were quantified by flow cytometry. Additionally, fluorescence microscopy was conducted to confirm the localization of the particles inside the cells (Figure 8).

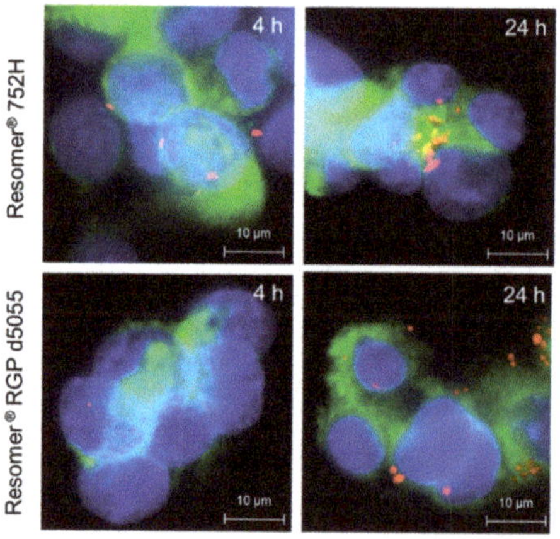

Figure 8. Cellular uptake of either Cy5-conjugated Resomer® 752H or Resomer® RGP d5055 nanoparticles (red). Membrane staining was performed by using Alexa488 concanavalin A (green). Nuclei were stained with 4′,6-diamidino-2-phenylindole (DAPIblue).

As expected, the uptake of drug-loaded nanoparticles prepared by using the PEGylated polymer was significantly lower (3.9 ± 3.3%, Figure 9B) than for Resomer® RG 752H nanoparticles (49.9 ± 4.9%, Figure 9C). The biocompatibility of nanomedicines is an important feature with regards to clinical success and commercialization.

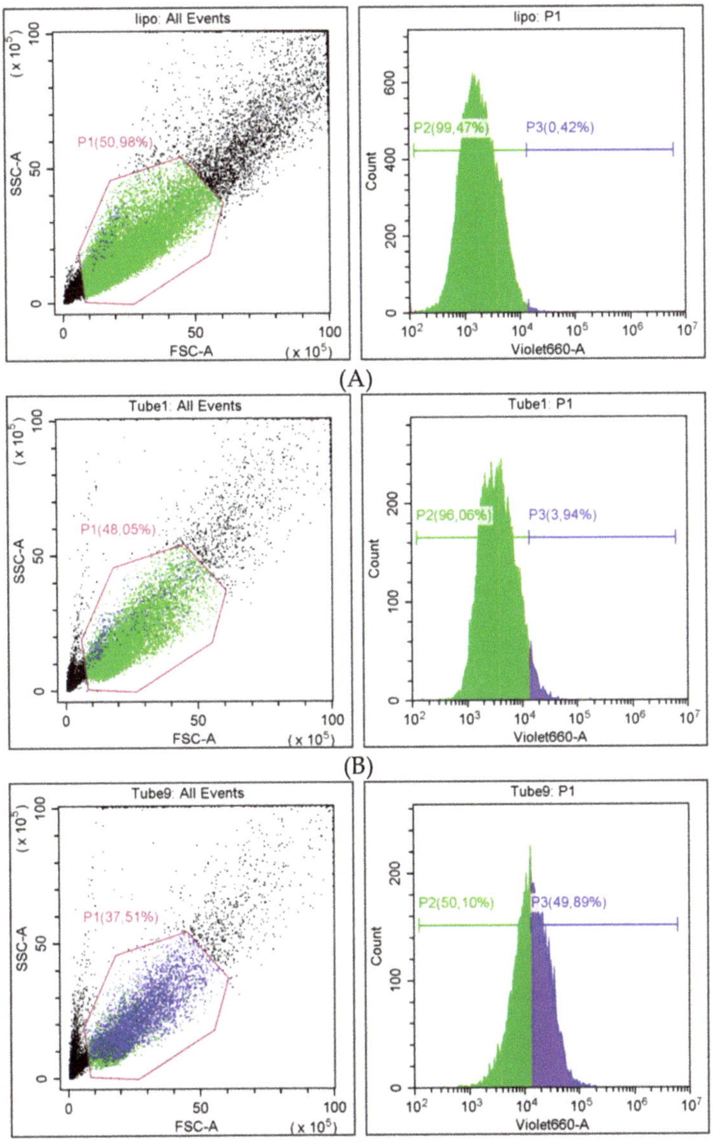

(A)

(B)

(C)

Figure 9. Flow cytometry diagrams of untreated cell control (**A**), sorafenib-loaded Resomer®RGP d5055 (**B**), and Resomer® RG 752H nanocomposites (**C**).

The cytotoxicity of the unloaded and the drug-loaded nanoparticles was investigated in HepG2 cells. Solutions of the drug in dimethyl sulfoxide (DMSO), the organic solvent alone, and an untreated negative control were used as references. As shown in Figure 10, the viability of the cells remained at approximately 90% when exposed to 100 µg/well of both types of unloaded nanoparticles (50,000 cells/well). The non-toxic features of Resomer® polymers have been verified previously [4].

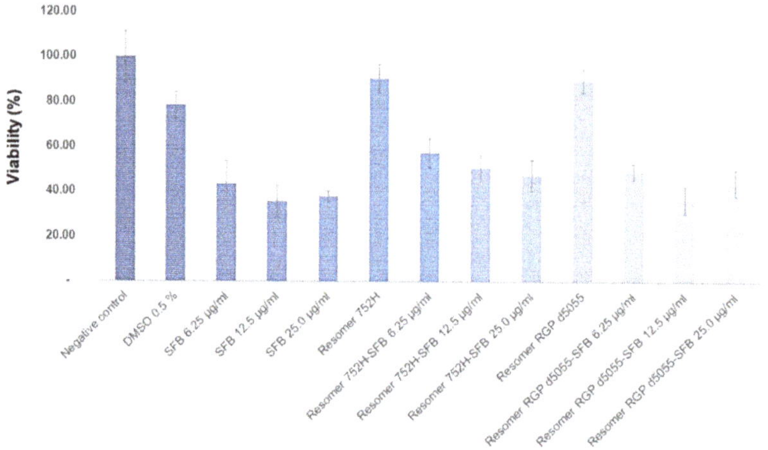

Figure 10. Viability of HepG2 cells treated with different concentrations of sorafenib (SFB) or Resomer® RG 752H or Resomer® RGP d5055 nanoparticles, and that of untreated cells (negative control).

As expected, sorafenib exhibited a concentration-dependent cytotoxicity in the HepG2 cells. Similar concentrations of sorafenib resulted in higher cytotoxicity for the free drug compared to the particle formulations (Figure 10), which could have been due to the slow release of the drug from the particle matrix.

4. Conclusions

Theranostic nanocomposites comprising PLGA or PEG-PLGA were loaded with the anti-tumor drug sorafenib and modified on their surface with the contrast agent Gd-DTPA. The single emulsion technique was found to be the most appropriate method for the effective preparation of monodisperse nanocomposites. These nanospheres exhibited superior properties compared to the particle systems described in the literature. Finally, the top-down manufacture combined with a modification of the particle surface using PEI and Gd-DTPA resulted in a strong MRI signal, optimal stability features, and a sustained release profile. On this basis, further investigations will focus on the in vivo performance of these nanocarriers.

Author Contributions: T.F. designed and achieved most of the experiments except cellular studies, and wrote the paper. A.P. designed the cellular studies and took part in the manuscript preparation. T.P., C.S., D.D. and T.J.V. performed the cytotoxicity and cellular uptake studies by MTT and flow cytometry, respectively. S.H. and F.R. designed and did the cellular uptake investigations by fluorescence microscopy. M.G.W. participated in designing the experiments and writing the manuscript.

Funding: This research was funded by Alexander von Humboldt Foundation (Ref. No.: 3.3-UNG/1161203 STP and 3.3-1161203-HUN-HFST-E) and the Ministry of National Economy (Hungary) via the Economic Development and Innovation Operation Programme (project No. BIONANO_GINOP-2.3.2-15-2016-00017).

Acknowledgments: The authors are grateful to the LOEWE initiative of the state of Hessen for financial support to the research center Translational Medicine and Pharmacology. Tivadar Feczkó acknowledges the funding of Alexander von Humboldt Foundation (Ref. No.: 3.3-UNG/1161203 STP and 3.3-1161203-HUN-HFST-E) and that of BIONANO_GINOP-2.3.2-15-2016-00017 project. S/TEM studies were performed at the electron microscopy laboratory of the University of Pannonia, established using grant no. GINOP-2.3.3-15-2016-0009 from the European Structural and Investments Funds and the Hungarian Government.

Conflicts of Interest: The authors declare no conflicts of interest.

References

1. Torre, L.A.; Bray, F.; Siegel, R.L.; Ferlay, J.; Lortet-Tieulent, J.; Jemal, A. Global cancer statistics, 2012. *CA Cancer J. Clin.* **2015**, *65*, 87–108. [CrossRef] [PubMed]
2. Llovet, J.M.; Ricci, S.; Mazzaferro, V.; Hilgard, P.; Gane, E.; Blanc, J.F.; de Oliveira, A.C.; Santoro, A.; Raoul, J.L.; Forner, A.; et al. Sorafenib in Advanced Hepatocellular Carcinoma. *New Engl. J. Med.* **2008**, *359*, 378–390.
3. Keating, G.M.; Santoro, A. Sorafenib: A review of its use in advanced hepatocellular carcinoma. *Drugs* **2009**, *69*, 223–240. [CrossRef] [PubMed]
4. Li, Y.J.; Dong, M.; Kong, F.M.; Zhou, J.P. Folate-decorated anticancer drug and magnetic nanoparticles encapsulated polymeric carrier for liver cancer therapeutics. *Int. J. Pharm.* **2015**, *489*, 83–90. [CrossRef] [PubMed]
5. Watcharin, W.; Schmithals, C.; Pleli, T.; Köberle, V.; Korkusuz, H.; Hübner, F.; Waidmann, O.; Zeuzem, S.; Korf, H.W.; Gelperina, S.; et al. Detection of hepatocellular carcinoma in transgenic mice by Gd-DTPA- and rhodamine 123-conjugated human serum albumin nanoparticles in T1 magnetic resonance imaging. *J. Control. Release* **2015**, *199*, 63–71. [CrossRef] [PubMed]
6. Fahrlander, E.; Schelhaas, S.; Jacobs, A.H.; Langer, K. PEGylated human serum albumin (HSA) nanoparticles: Preparation, characterization and quantification of the PEGylation extent. *Nanotechnology* **2015**, *26*, 145103. [CrossRef] [PubMed]
7. Wacker, M.; Zensi, A.; Kufleitner, J.; Ruff, A.; Schutz, J.; Stockburger, T.; Marstaller, T.; Vogel, V. A toolbox for the upscaling of ethanolic human serum albumin (HSA) desolvation. *Int. J. Pharm.* **2011**, *414*, 225–232. [CrossRef] [PubMed]
8. Villa Nova, M.; Janas, C.; Schmidt, M.; Ulshoefer, T.; Grafe, S.; Schiffmann, S.; de Bruin, N.; Wiehe, A.; Albrecht, V.; Parnham, M.J. Nanocarriers for photodynamic therapy-rational formulation design and medium-scale manufacture. *Int. J. Pharm.* **2015**, *491*, 250–260. [CrossRef] [PubMed]
9. Mohamed, F.; van der Walle, C.F. Engineering biodegradable polyester particles with specific drug targeting and drug release properties. *J. Pharm. Sci.* **2008**, *97*, 71–87. [CrossRef] [PubMed]
10. Von Hoff, D.D.; Mita, M.M.; Ramanathan, R.K.; Weiss, G.J.; Mita, A.C.; LoRusso, P.M.; Burris, H.A., 3rd; Hart, L.L.; Low, S.C.; Parsons, D.M.; et al. Phase I study of PSMA-targeted docetaxel-containing nanoparticle BIND-014 in patients with advanced solid tumors. *Clin. Cancer Res.* **2016**, *22*, 3157–3163. [CrossRef] [PubMed]
11. Wacker, M.G. Nanomedicines—A Scientific Toy or an Emerging Market. In *Polymer Nanoparticles for Nanomedicines: A Guide for their Design, Preparation and Development*; Vauthier, C., Ponchel, G., Eds.; Springer International Publishing: Cham, Schwitzerland, 2016; pp. 591–614.
12. Liu, J.; Boonkaew, B.; Arora, J.; Mandava, S.H.; Maddox, M.M.; Chava, S.; Callaghan, C.; He, J.; Dash, S.; John, V.T.; et al. Comparison of sorafenib-loaded poly(lactic/glycolic) acid and DPPC liposome anoparticles in the in vitro treatment of renal cell carcinoma. *J. Pharm. Sci.* **2015**, *104*, 1187–1196. [CrossRef] [PubMed]
13. Kim, D.H.; Kim, M.D.; Choi, C.W.; Chung, C.W.; Ha, S.H.; Kim, C.H.; Shim, Y.H.; Jeong, Y.I.; Kang, D.H. Antitumor activity of sorafenib-incorporated nanoparticles of dextran/poly(dl-lactide-coglycolide) block copolymer. *Nanoscale Res. Lett.* **2012**, *7*, 91. [CrossRef] [PubMed]
14. Malarvizhi, G.L.; Retnakumari, A.P.; Nair, S.; Koyakutty, M. Transferrin targeted core-shell nanomedicine for combinatorial delivery of doxorubicin and sorafenib against hepatocellular carcinoma. *Nanomedicine* **2014**, *10*, 1649–1659. [CrossRef] [PubMed]
15. Zhang, J.; Hu, J.; Chan, H.F.; Skibba, M.; Liang, G.; Chen, M. iRGD decorated lipid-polymer hybrid nanoparticles for targeted co-delivery of doxorubicin and sorafenib to enhance anti-hepatocellular carcinoma efficacy. *Nanomedicine* **2016**, *12*, 1303–1311. [CrossRef] [PubMed]
16. Xiao, Y.; Liu, Y.; Yang, S.; Zhang, B.; Wang, T.; Jiang, D.; Zhang, J.; Yu, D.; Zhang, N. Sorafenib and gadolinium co-loaded liposomes for drug delivery and MRI-guided HCC treatment. *Colloids Surf. B Biointerfaces* **2016**, *141*, 83–92. [CrossRef] [PubMed]
17. Maeda, H.; Wu, J.; Sawa, T.; Matsumura, Y.; Hori, K. Tumor vascular permeability and the EPR effect in macromolecular therapeutics: A review. *J. Control. Release* **2000**, *65*, 271–284. [CrossRef]
18. Lin Ts, T.; Gao, D.Y.; Liu, Y.C.; Sung, Y.C.; Wan, D.; Liu, J.Y.; Chiang, T.; Wang, L.; Chen, Y. Development and characterization of sorafenib-loaded PLGA nanoparticles for the systemic treatment of liver fibrosis. *J. Control. Release* **2016**, *221*, 62–70. [CrossRef] [PubMed]

19. Riddick, T.M. *Control of Colloid Stability through Zeta Potential*; Livingston: Wynnewood, PA, USA, 1968.
20. Ratzinger, G.; Agrawal, P.; Korner, W.; Lonkai, J.; Sanders, H.M.; Terreno, E.; Wirth, M.; Strijkers, G.J.; Nicolay, K.; Gabor, F. Surface modification of PLGA nanospheres with Gd-DTPA and Gd-DOTA for high-relaxivity MRI contrast agents. *Biomaterials* **2010**, *31*, 8716–8723. [CrossRef] [PubMed]
21. Holzer, M.; Vogel, V.; Mantele, W.; Schwartz, D.; Haase, W.; Langer, K. Physico-chemical characterisation of PLGA nanoparticles after freeze-drying and storage. *Eur J. Pharm Biopharm* **2009**, *72*, 428–437. [CrossRef] [PubMed]
22. Janas, C.; Mast, M.P.; Kirsamer, L.; Angioni, C.; Gao, F.; Mantele, W.; Dressman, J.; Wacker, M.G. The dispersion releaser technology is an effective method for testing drug release from nanosized drug carriers. *Eur J. Pharm Biopharm* **2017**, *115*, 73–83. [CrossRef] [PubMed]

pharmaceutics

MDPI

Article

Effect of Size and Concentration of PLGA-PEG Nanoparticles on Activation and Aggregation of Washed Human Platelets

Rana Bakhaidar [1], Joshua Green [2], Khaled Alfahad [2], Shazia Samanani [2], Nabeehah Moollan [2], Sarah O'Neill [1] and Zebunnissa Ramtoola [1,*]

[1] School of Pharmacy and Biological Sciences, Royal College of Surgeons in Ireland, 2 Dublin, Ireland; ranabakhaidar@rcsi.ie (R.B.); soneill@rcsi.ie (S.O.)

[2] School of Medicine, Royal College of Surgeons in Ireland, 2 Dublin, Ireland; joshuagreen@alumnircsi.com (J.G.); khaledalfahad@rcsi.ie (K.A.); shaziasamanani@rcsi.ie (S.S.); nabeehahmoollan@alumnircsi.com (N.M.)

* Correspondence: zramtoola@rcsi.ie; Tel.: +353-1-4028626

Received: 23 August 2019; Accepted: 26 September 2019; Published: 4 October 2019

Abstract: Nanotechnology is being increasingly utilised in medicine as diagnostics and for drug delivery and targeting. The small size and high surface area of nanoparticles (NPs), desirable properties that allow them to cross biological barriers, also offer potential for interaction with other cells and blood constituents, presenting possible safety risks. While NPs investigated are predominantly based on the biodegradable, biocompatible, and FDA approved poly-lactide-*co*-glycolide (PLGA) polymers, pro-aggregatory and antiplatelet effects have been reported for certain NPs. The potential for toxicity of PLGA based NPs remains to be examined. The aims of this study were to determine the impact of size-selected PLGA-PEG (PLGA-polyethylene glycol) NPs on platelet activation and aggregation. PLGA-PEG NPs of three average sizes of 112, 348, and 576 nm were formulated and their effect at concentrations of 0.0–2.2 mg/mL on the activation and aggregation of washed human platelets (WP) was examined. The results of this study show, for the first time, NPs of all sizes associated with the surface of platelets, with >50% binding, leading to possible internalisation. The NP-platelet interaction, however, did not lead to platelet aggregation nor inhibited aggregation of platelets induced by thrombin. The outcome of this study is promising, suggesting that these NPs could be potential carriers for targeted drug delivery to platelets.

Keywords: PLGA-PEG; nanoparticles; platelet; activation; aggregation; binding; uptake

1. Introduction

Biocompatible and biodegradable nanoparticles (NPs) continue to be extensively investigated for the diagnosis, prevention, and therapy of various diseases, for cellular and molecular imaging, and for tissue engineering applications [1–5]. NPs have been reported to target various therapeutic agents to the central nervous system (CNS), tumour sites, and the vascular compartment [6–10]. The small size and high surface area of NPs are desirable properties that allow them to cross biological barriers, but also offer potential for interaction and interference with other cells and blood constituents, presenting possible safety risks. While the NPs investigated are predominantly based on the biodegradable, biocompatible and FDA approved poly-lactide-*co*-glycolide (PLGA) polymers, irrespective of their intended destination, NPs introduced into the bloodstream will interact with endothelial cells as well as blood constituents. It is therefore important to understand the bio-distribution of such NPs in human blood and their interaction with various blood components such as platelets [6,11,12]. Previous studies report that grafting a polyethylene glycol (PEG) chain to a PLGA polymer backbone significantly increases the circulation half-life of the modified copolymer, as it remains camouflaged from the macrophage [13,14]. The longer duration of exposure to blood components can result in enhanced interaction potential with blood cells, and in particular with platelets, maximising the possibility of modulating platelet, activation, secretion, and aggregation [15,16].

NP physicochemical characteristics, such as size and net charge (zeta potential (ZP)) that are reported to be key factors enhancing their efficacy, can substantially affect their bio-distribution [17,18]. Variations in NP characteristics may result in a diverse range of platelet reactivity but unfortunately, only a few studies regarding the interaction of NPs with washed platelets (WPs) have been published [18–20]. We previously reported that biodegradable micro and NPs of PLGA, PLGA-PEG, and chitosan, of median diameter (D50%) of 2–9 μm and 100–500 nm, and with various surface morphology did not induce or inhibit platelet aggregation at particle concentrations of 0.1–500 μg/mL [20]. In contrast, silica NPs of 10 nm, 50 nm, and 150 nm, were shown to cause platelet hyper-aggregability, promoting the risk of thrombus deposition, while particles of 500 nm did not pose such a risk [21]. Mixed carbon NPs, carbon nanotubes, single-walled nanotubes, and multi-walled nanotubes were reported to stimulate platelet aggregation by the upregulation of GPIIb/IIIa receptors on platelets, inducing subsequent vascular thrombosis [19]. In a separate study, the contribution of silver NPs to the effective inhibition of integrin-mediated WPs responses, such as aggregation, secretion, and adhesion to immobilized fibrinogen or collagen, was established [22].

The pro-aggregatory effects of some NPs and the antiplatelet effects of others raise questions regarding the safety of biodegradable nanomaterials in terms of interference with the haemostatic equilibrium [20,23]. To date, the potential interactions of PLGA-PEG NPs with blood elements remains scant. In particular, the effect of their size on blood constituents, as well as any potential hazard they may pose, have not been studied. Platelet activation is a precisely regulated event, critical for maintaining normal blood flow. When designing drug-loaded NPs intended to be delivered to the systemic circulation, a major consideration is to maintain platelets in an inactive state [22].

The aims of this study were to determine the impact of size-selected PLGA-PEG NPs on platelet activation and aggregation. Coumarin-6-labelled PLGA-PEG NPs of three average sizes of 112, 348, and 576 nm were formulated using the solvent dispersion technique [20]. The effect of NP size and NP concentration of 0.01–2.2 mg/mL on the activation and aggregation profiles of washed platelets (WP) were examined using platelet aggregation assays and flow cytometry. Confocal imaging was carried out on NPs incubated with WP to characterise the interaction of NPs with platelets in the resting and activated states.

2. Materials and Methods

2.1. Materials

Albumin, from human serum, Coumarin-6 (98%), Polysorbate 80 (Tween® 80), Dextrose, Sodium Chloride (NaCl), Sodium bicarbonate ($NaHCO_3$), Potassium Chloride (KCl), Sodium Citrate (Tribasic, 10 dehydrate ($HOC(COONa)(CH_2COONa)_2$, KH_2PO_4 (monobasic potassium phosphate anhydrate or potassium dihydrogen phosphate), Calcium Chloride ($CaCl_2$), Magnesium Chloride hexahydrate $MgCl_2{\cdot}6H2O$, and HEPES (N-2-Hydroxyethylpiperazine-N′-2-Ethanesulfonic Acid) were purchased from Sigma-Aldrich Dublin, Ireland. PLGA-PEG containing PEG at 10% *w*/*w* (poly-D, L-lactic-*co*-glycolic acid-polyethylene glycol diblock copolymer 50:50 mPEG; 33 kDa with inherent viscosity: 0.05–0.15 dL/g, Lakeshore Biomaterials™, was purchased from Evonik Industries AG, Essen, Germany. Deionised water was used throughout the experiments.

2.2. Methods

2.2.1. Preparation and Characterisation of PLGA-PEG NPs

PLGA-PEG NPs containing a fluorescent marker, coumarin-6, were prepared using the solvent dispersion method [24]. Briefly, PLGA-PEG polymer was dissolved in acetone to form 10, 55, and 100 mg/mL solutions, and coumarin-6 at 0.05% *w*/*w* of the polymer was dissolved in the PLGA-PEG solutions. The polymer solutions were added dropwise to an external aqueous phase containing Tween 80® at 2% *w*/*v*, under constant stirring. The NPs formed were recovered by centrifugation at 11,000 rpm for 20 min (Rotina 35 R centrifuge, Hettich Zentrifugen, Tuttilingen, Germany). NPs were washed with deionised water, centrifuged, and the pellet resuspended in deionised water and stored at 4 °C. The average particle size (PS), polydispersity index (PDI), and zeta potential (ZP) of the NPs were determined by dynamic light scattering technique, using a Malvern Zetasizer Nano ZS 90 (Malvern Instruments, Worcestershire, UK). After centrifugation, NPs were suspended in deionised water, placed in disposable voltable zeta cells (DTS 1060) for analysis using DLS. Viscosity and refractive index of the dispersants were taken into account. For each sample, the average of five measurements was calculated. Results were expressed as an average of five measurements ± standard deviation.

The morphologies of NPs were examined under a MIRA3 variable pressure field emission scanning electron microscope (Tescan, Brno-Kohoutovice, Czech Republic) at an accelerating voltage of 5.0 kV and magnification of 30,000 to 50,000×. Samples of lyophilised NPs were applied onto aluminium stubs using a double-sided conductive tape and were sputter-coated with gold.

2.2.2. Preparation of Washed Platelets

Venous whole blood was obtained from healthy human volunteers, free from aspirin and other non-steroidal anti-inflammatory agents for the previous 7–10 days. Ethical approval was obtained from the Research Ethics Committee, Royal College of Surgeons in Ireland. The blood obtained was centrifuged at 180 g for 12 min, to separate the platelet-rich plasma (PRP) that was acidified to pH 6.5 and the plasma pelleted by centrifugation at 720 g for 12 min. Washed platelets were then adjusted to approximately 250×10^9 platelets per millilitre by resuspending the pellets in HEPES (N-2-Hydroxyethylpiperazine-N′-2-Ethanesulfonic Acid) platelet buffer and leaving to stand for 30 min before adding calcium chloride ($CaCl_2$) at a final concentration of 1.8 mM.

2.2.3. Effect of Size and Concentration of NPs on Platelet Aggregation

Platelet aggregation was determined by measuring the change in the optical density of stirred WPs in the absence or presence of 0–2.2 mg/mL PLGA-PEG NPs of different sizes, following the addition of the platelet agonist, thrombin. Platelets were incubated with NPs for 4 min prior to the addition of 0.1 U/mL thrombin. Platelet aggregation data was analysed using an eight channel platelet aggregometer,-PAP-8 (Bio/Data Corporation, Horsham, PA, USA). Aggregation results were expressed as final percent aggregation (% PA) at the end of the reaction time of 12 min. Data are presented as mean of n = 4–6 ± SEM.

2.2.4. Effect of Size and Concentration of NPs on Platelet Activation

WPs (250×10^3/μL) were incubated with 20 μL phycoerythrin (PE)-labelled CD62P antibody and mixed with PLGA-PEG NPs of different sizes, in the absence and presence of 0.1 U/mL thrombin and incubated for 4 min at 37 °C. At predetermined time points, samples were fixed using 1% *v*/*v* formaldehyde (FA) and platelet activation was measured by flow cytometry. Platelet activation was determined by measuring the level and extent of CD62P antibody binding to the platelet surface, expressed as percent positive cells (% PP) and mean fluorescence intensity (MFI). Platelets with and without CD62P were used as negative controls and platelets activated with thrombin at 0.1 U/mL for 4 min was the positive control. Data are average of $n \geq 3$ ± SEM.

2.2.5. Confocal Microscopy of the Effect of Incubation Time on The Interaction of PLGA-PEG NPs with Washed Platelets

Coumarin-labelled NPs of 112, 348, and 576 nm, at 0.1 and 2.2 mg/mL, were incubated with resting WPs in suspension for 1, 5, 15, and 30 min. At each time point, samples were fixed with 1% formaldehyde, stained red with phalloidin-tetramethylrhodamine isothiocyanate (phalloidin-TRIT-C) and examined using confocal laser scanning microscopy (CLSM), (Carl Zeiss, Jena, Germany). WPs were also incubated with coumarin-labelled NPs of the three different sizes at 2.2 mg/mL for 1, 5, and 30 min, and allowed to attach to fibrinogen-coated glass slides (20 µg/mL). At each time point, samples were fixed with 1% formaldehyde, platelets stained red with phalloidin-TRITC, and examined by CLSM.

2.2.6. Statistical Analysis

Statistical analysis was carried out using GraphPad Prism (version 7.00 for Windows; GraphPad Software, San Diego, CA, USA). One-way ANOVA and unpaired student T tests and One-way ANOVA followed by post hoc analysis using Dunnett's Tests were used to determine differences between samples and groups. A p value <0.05 was considered to be statistically significant.

3. Results

3.1. Characterisation of NPs Formulated

The size, PDI, and zeta potential of the NPs increased with increasing PLGA-PEG concentration (Table 1, Figure 1A–C). A PDI value of 0.10 indicating a nearly monosized distribution was observed for the smallest NPs of 112 nm, while larger PDI values of 0.54 and 0.70 were observed for the 348 and 576 nm NPs, respectively. A larger size distribution is expected for larger size NPs formulated using high polymer concentration. In Figure 1A, the SEM shows the ~100 nm NPs with a distinct spherical shapes and smooth surfaces, exhibiting a narrow range of sizes at different fields of view. In Figure 1B, the NPs of ~350 nm display an interparticular bridging/fusion. In Figure 1C, the NPs of ~600 nm average size maintain a spherical appearance similar to that for the NPs in Figure 1A. The SEMs did not show any internal or external porosity at different magnifications. Calculations by Image J software from SEM analysis showed an average size, which were lower than the sizes measured by DLS. SEM is a tool primarily utilised to analyse particle morphology. The obtained PS determined by microscopy techniques is usually considered the lower limits of PS [25].

Table 1. Physicochemical characteristics of nanoparticles (NPs) used in platelet aggregation studies. n = 5 batches ± SD. PLGA = poly-lactide-*co*-glycolide; PEG = poly ethylene glycol.

PLGA-PEG (mg/mL)	PS (nm)	PDI	ZP (mV)
10	111.55 ± 5.81	0.10 ± 0.02	−22.20 ± 5.71
55	348.00 ± 23.61	0.54 ± 0.04	−12.40 ± 6.30
100	576.19 ± 6.82	0.70 ± 0.06	−9.50 ± 14.56

PLGA: poly-lactide-*co*-glycolide; PEG: polyethylene glycol; PS: Particle size; PDI: Polydispersity Index, ZP: Zeta Potential.

(**A**)

(**B**)

(**C**)

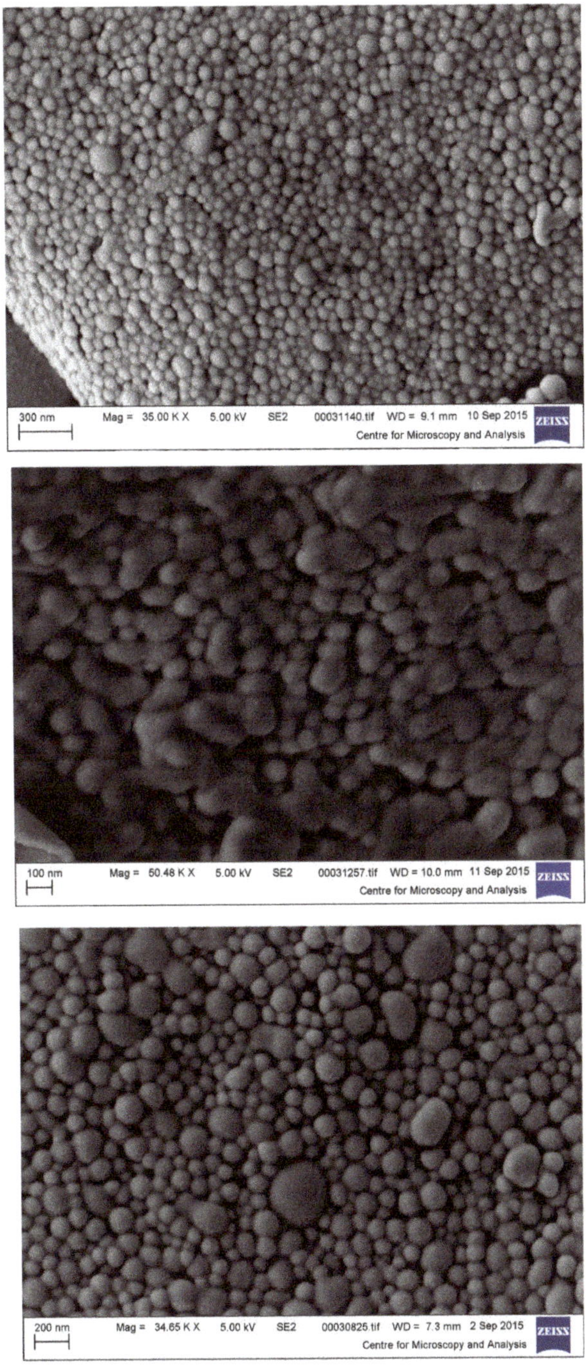

Figure 1. Scanning Electron Microscope Images of PLGA-PEG NPs of average size of (**A**) 112nm, (**B**) 348 nm, and (**C**) 576 nm. Each SEM image is representative of all images taken.

The 112 nm NPs had a net negative charge of −22.20 ± 5.71 mV indicating good colloidal stability, whereas the two larger NPs displayed moderately stable colloidal suspensions, with surface charges from −12.40 ± 5.47 to −9.50 ± 14.56 mV. The ZP values of the larger NPs decreased to within the range of −16 to −23 mV on dilution to the concentrations used in the study. The stability of the NPs generated was examined over a 28-day time period to ensure stability over the time period of use. It was noted that storage-induced instability was minimal, particularly in the batches formed of NPs of small and mid-level particle diameters (~100 nm and ~350 nm), with no significant changes in particle properties. Particle PS and PDI properties were retained over this period of 28 days and for the largest size NPs over at least 14 days, revealing a reasonable stability profile for the NPs. This approach in examining the stability of NPs over time is supported by recent studies by Fornaguera and Solans [26].

3.2. *Effect of Size and Concentration of NPs on Platelet Aggregation*

Incubation of NPs with platelets for 4 min in absence of the agonist, thrombin, showed that, irrespective of the size and concentration of NPs (0.05–2.2 mg/mL), spontaneous platelet aggregation was not detected (0 ± 0 % PA, $p > 0.05$, $n = 3$–6). The addition of 0.1 U/mL thrombin, at 4 min in absence of NPs, resulted in a % PA of 72.25–74.75 (Table 2). In presence of 112 nm NPs, no significant difference in % PA was observed after 12 min incubation at all NP concentrations tested (Table 2). Similarly, larger NPs of 348 and 576 nm showed no significant effect on the % PA over the 12 min incubation at the lower NP concentrations of 0.01–1.0 mg/mL and 0.01–1.5 mg/mL for the 348 and 576 nm NPs, respectively (Table 2). At the higher NP concentrations of 1.5 and 2.2 mg/mL, the % PA was significantly lower at 28.40 ± 3.47 and 27 ± 3.10, respectively, when incubated with 348 nm NPs (p value < 0.001). Similarly, at the highest concentration of 2.2 mg/mL, NPs of 576 nm resulted in a significantly reduced % PA of 49.00 ± 5.18 ($p < 0.05$) (Table 2).

Table 2. Effect of size and concentration of PLGA-PEG NPs on percent platelet aggregation (% PA). Data are the average of $n = 4$–6 experiments ± SEM. * $p < 0.05$, ** $p < 0.01$, *** $p < 0.001$ compared with 0.1 U/mL thrombin (control), one-way ANOVA, and unpaired student t-test, at each time point.

PLGA-PEG NPs (mg/mL)	% PA (112 nm)	% PA (348 nm)	% PA (576 nm)
0 (Control)	72.75 ± 1.89	74.75 ± 1.93	72.25 ± 0.62
0.05	72.00 ± 0.91	73.50 ± 1.88	75.00 ± 4.63
0.1	73.00 ± 1.23	76.20 ± 2.99	69.75 ± 8.22
0.25	69.00 ± 1.68	66.00 ± 5.11	74.00 ± 4.89
0.5	73.50 ± 0.96	70.25 ± 4.13	68.00 ± 4.97
1	72.75 ± 0.85	62.50 ± 3.59	71.25 ± 5.27
1.5	72.25 ± 1.55	28.40 ± 3.47***	68.75 ± 2.10
2.2	71.00 ± 1.80	27.00 ± 3.10***	49.00 ± 5.18*

Examination of the platelet aggregation time profile showed no delay in platelet aggregation for the smallest NPs of 112 nm at any of the concentrations tested (Figure 2A). However, the reduced % PA observed at the higher NP concentrations of 348 and 576 nm NPs was associated with a significant delay in platelet aggregation at the earlier time points following addition of thrombin to the sample. Platelet aggregation proceeded significantly more slowly for the 348 and 576 nm NPs at the higher NP concentrations of 1.5 and 2.2 mg/mL (Figure 2B,C), respectively.

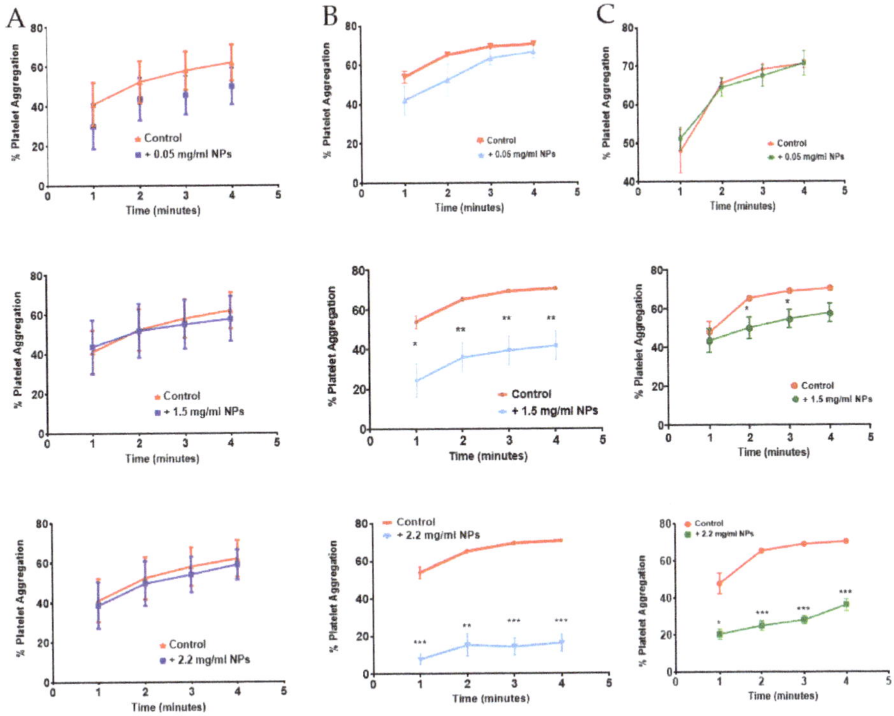

Figure 2. Effect of PLGA-PEG NPs of (**A**) 112 nm, (**B**) 348 nm, and (**C**) 576 nm at 0.05, 1.5, and 2.2 mg/mL on platelet aggregation profile for 4 min following the addition of thrombin. Data are the average of n = 4–6 experiments ± SEM. * $p < 0.05$, ** $p < 0.01$, *** $p < 0.001$ compared with 0.1 U/mL thrombin (control), one-way ANOVA, and unpaired student t-test, at each time point.

3.3. *Effect of Size of NPs on Platelet Activation Profile*

The percent positive platelets (% PP) and mean fluorescence intensity (MFI) associated with resting WPs in the absence of NPs were not significantly different when incubated with or without CD62P antibody (Figure 3A,B). On addition of the agonist, thrombin, the % PP and MFI significantly increased as expected, indicating activation of platelets with secretion of CD62P, which binds to the CD62P antibody ($p < 0.0001$ for % PP; $p < 0.05$ for MFI) (Figure 3A,B).

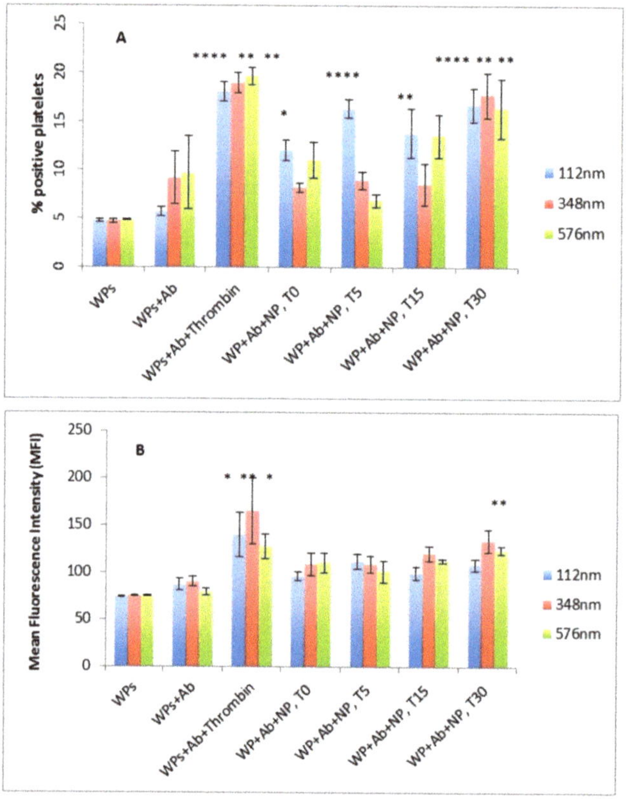

Figure 3. Effect of size of PLGA-PEG NPs at 2.2 mg/mL on platelet activation measured by flow cytometry (**A**) % positive platelets (% PP) (**B**) Mean fluorescence Intensity (MFI). Data are the average of n = 3–5 experiments ± SEM. * $p < 0.05$, ** $p < 0.01$, *** $p < 0.001$ compared with washed platelets (WPs) + Ab (control), one-way ANOVA followed by post hoc Dunnett's test.

In the presence of PLGA-PEG NPs, a significant increase in % PP for the smallest sized NPs of 112 nm was observed over the 30 min incubation period. However, for the larger NPs of 348 and 576 nm, a significant increase in % PP was only observed at the incubation time of 30 min, ($p < 0.05$, $n = 3$), (Figure 3A). No significant difference in MFI values for platelets incubated with the NPs was observed over the 30 min except for the 348 nm NPs at 30 min incubation (Figure 3B).

3.4. *Effect of size of NPs on Thrombin Activation of Washed Platelets*

In absence of NPs, an increase in the percentage of CD6P-positive platelets (% PP) and in CD62P binding level (MFI) were observed following the exposure of WPs to thrombin, as expected, confirming activation of WPs in response to the agonist. The presence of NPs at any of the sizes tested did not result in a significant change in the % PP or MFI following exposure of the WPs with thrombin. (Figure 4A,B).

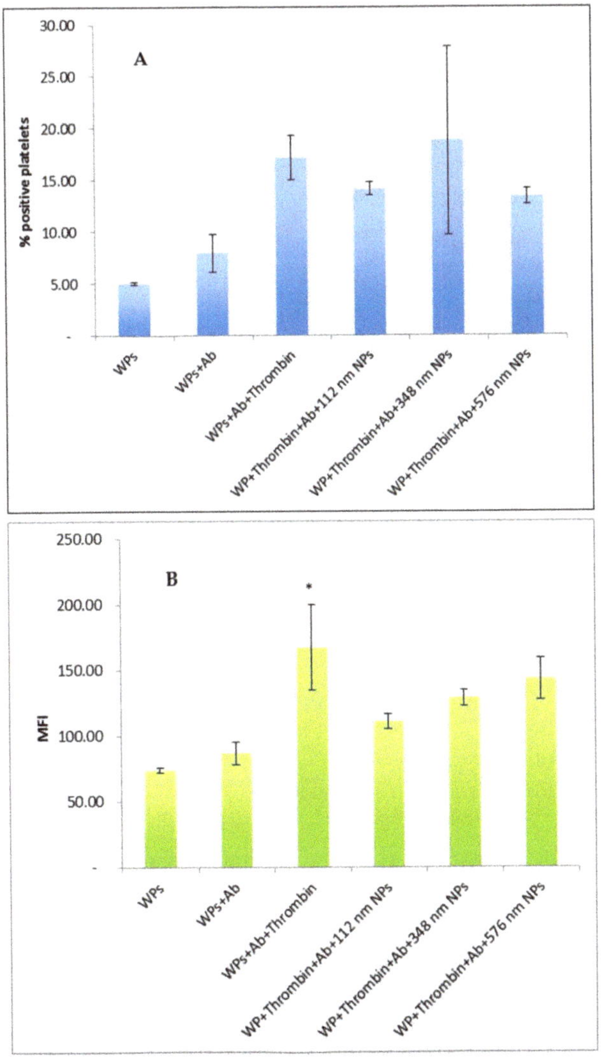

Figure 4. Effect of size of PLGA-PEG NPs at 2.2 mg/mL on thrombin activation of washed platelets measured by flow cytometry (**A**) % positive platelets (% PP) (**B**) Mean fluorescence Intensity (MFI). Data are the average of n = 3–5 experiments ± SEM. * $p < 0.05$, compared with WPs + Ab (control), one-way ANOVA followed by post hoc Dunnett's test.

3.5. *Confocal Microscopy of the Effect of Incubation Time on the Interaction of NPs and Washed Platelets*

A rapid interaction of NPs and WPs, within the first minute, was observed at all NP sizes tested and at both the low (0.01 mg/L) and high (2.2 mg/mL) concentrations of NPs (Figure 5). The percent of platelets bound to NPs of 112 nm was in the order of 44 ± 5.7% and 53.7 ± 7.1% at t = 1 min for low and high concentrations, respectively. Increasing the incubation time or concentration of NPs at any of the sizes tested, did not result in significant change in the percent of NPs-bound platelets. The observed interaction between resting platelets and PLGA-PEG NPs of the three sizes tested did not lead to any apparent morphological alteration of the shape of the platelets or development of pseudopods, indicating no effect on platelet activation. This observation was relevant to both low and high concentrations of NPs tested.

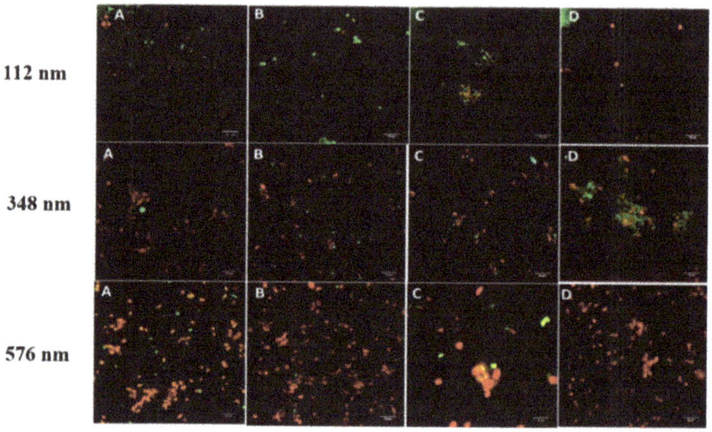

Figure 5. Confocal microscopy images of WPs (red) in suspension incubated with coumarin-PLGA-PEG NPs at 2.2 mg/mL) (**A**) 1 min, (**B**) 5 min, (**C**) 15 min, and (**D**) 30 min, at RT; Magnification 100×. Scale bar represents 10 µm.

Confocal microscopy of platelets layered onto fibrinogen at 20 µg/mL showed the typically formed filopodial projections and heterogeneity expected for adhered platelets. In the presence of NPs, of all of the three sizes and over the incubation times tested, similar attachment of platelets to fibrinogen with alteration in the shape of the platelets and development of extended filopodia projections were observed. The majority of NPs, at all sizes tested, were observed to be bound to platelets (Figure 6). This association of NPs to platelets was of the order of 53.7 ± 1.5% to 59 ± 14.1%, respectively, for 1 and 30 min incubation for 112 nm NPs, at 51 ± 6.1%, and 43 ± 11.3% for 1 and 30 min incubation times, respectively, for the 348 nm NPs. For the 576 nm NP, NP binding was 35.3 ± 5.7% at 1 min and increased to 52.7 ± 2.1% after 30 min incubation. The NP interaction with platelets was maintained upon activation through to adhesion (Figure 6).

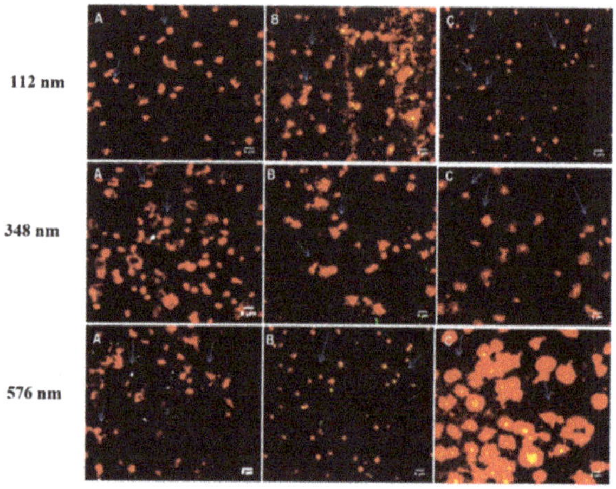

Figure 6. Confocal microscopy images of WPs (red) incubated with coumarin-PLGA-PEG NPs at 2.2 mg/mL and adhered to fibrinogen coated slides: (**A**) 1 min, (**B**) 5 min, and (**C**) 30 min. Magnification 100×. Scale bar represents 10 µm.

4. Discussion

The average NP sizes formulated were 112, 348, and 576 nm and were significantly different ($p < 0.05$). This ensured a reasonable comparison of the effects of size of NPs on how platelets may respond when in the presence of NPs with varying sizes. Addition of thrombin at 0.1 U/mL to washed human platelets produced a platelet aggregation of 72–74, which is similar to the percent platelet aggregation of 74 ± 7.8 reported [27]. Incubation of PLGA-PEG NPs, of any size or concentration tested, with WPs for 4 min did not cause any platelet aggregation. This is similar to the results reported for 50 µM silver NPs incubated with WPs for 2 min at 37 °C [22]. In the current study, the 112 nm NPs did not influence the thrombin-induced platelet aggregation profile. Similarly, the larger NPs showed no significant effect on the thrombin-induced platelet aggregation profiles at the lower NP concentrations of 0.01–1.0 mg/mL and 0.01–1.5 mg/mL for the 348 and 576 nm NPs, respectively. Interestingly, a tendency of inhibiting or slowing platelet aggregation was observed for the 348 nm and 576 nm PLGA-PEG NPs at their highest concentrations. This may be related to the presence of NPs in the medium, acting as a physical barrier preventing platelet-platelet contact, hence delaying and reducing platelet activation and aggregation. As the NPs have a negatively charged surface, they may interact with the positively charged thrombin, lowering the free thrombin concentration in the suspension medium, hence reducing the exposure of platelets to thrombin. The surface charge, rather than size, was reported to be the key factor governing this subcategory of NPs' interaction with platelets [28]. Physiological or non-physiological agents can initiate various signalling pathways in platelets. Platelet activation can also be induced by biomaterials interacting with the specific intrinsic coagulation pathway, which leads to thrombin generation and ultimately platelet activation. However, such mechanisms still remain unexplored [29].

The influence of size-selected NPs on the expression of CD62P on the surface of platelets was analysed by flow cytometry. CD62P is a 140 kD glycoprotein stored in the secretory α-granules of the platelets, mediating leukocyte rolling and platelet–leukocyte aggregation.

The percentage of positive platelets (% PP) for the antibody marker reflects the proportion of activated cells in the total platelet population; this was used to quantify platelet activation in the presence of the NPs. The incubation of the smallest NPs (~113 nm) with WPs resulted in a significant increase in platelet activation, almost similar to the activation observed for thrombin-induced platelet activation. Interestingly, no significant effect on platelet activation was observed in the presence of the larger NPs of 348 nm and the 576 nm. Smaller sized NPs have a higher surface area per unit mass, which result in enhanced interaction potential with platelets. Silica NPs at the smallest PS of 10 nm, at a concentration of 200 µg/mL, were reported to increase CD62P abundance on the platelet surface, while 50 nm-sized NPs did not show any influence [21]. Similarly, of the 10, 50, 113, and 150 nm-sized NPs of silica exposed to platelets, the smallest NPs (10 nm) induced the upregulation of CD62P when used at a low concentration of 10 µg/mL [15,21]. The tendency of 26 nm carboxylated-polystyrene NPs to activate platelets was reported, while particles of larger sizes showed no influence on platelet activation [30].

The presence of NPs of any of the three sizes tested did not influence the % PP or the CD62P abundance on the surface of thrombin-activated platelets. This suggests the investigated NPs did not limit the access of the PE-labelled marker to the WPs surface, irrespective of their sizes.

CLSM performed in this study showed a rapid physical interaction of NPs, of all three sizes examined, with platelets in resting state. This interaction appeared to progress to possible NP uptake by platelets over time. This NP-platelet interaction, however, did not lead to platelet aggregation nor inhibit aggregation of platelets induced by the agonist, thrombin. In contrast, a previous study regarding latex microspheres interaction with platelets suggested platelet activation to be a prerequisite in order to interact with and internalise those inert microparticles (MPs) [31].

The interaction and binding of the investigated NPs to the surface of platelets as well as possible internalisation of NPs into platelets is interesting. In view of their unique small sizes and their curvature,

NPs have been shown to enter various human cells and organs, a desirable characteristic for drug targeting [9,24,32].

In the present study, in the presence of PLGA-PEG NPs, variable morphologies of platelets from dendritic spread to the fully spread shape were displayed when platelets were layered onto fibrinogen. This indicates that the NPs did not affect platelet adhesion to fibrinogen. Similarly, silver NPs at different concentrations were reported to have no impact on platelet spreading on fibrinogen [22].

5. Conclusions

The results of this study show that while the smallest NPs demonstrated a tendency to result in platelet activation and the larger NPs, at the highest concentrations, to delay platelet aggregation, the three sizes of PLGA-PEG NPs examined had little or no effect on platelet aggregation. The NPs did not interfere with the activation status of platelets and did not inhibit platelet aggregation by the agonist thrombin. Importantly, the rapid binding of the studied PLGA-PEG NPs to platelets is promising, suggesting that these NPs could be potential carriers for targeted drug delivery to platelets in certain disease states. The data presented here supports the use of such NPs as targeting carriers of anti-platelet agents or thrombolytics. It is clear from this study that the question of whether these NPs would be consumed by resting and/or activated platelets remains unanswered. Future strategies may include a more sensitive analysis of the potential for these NPs, pre-loaded with therapeutic drug-candidates, to be internalised by platelets. A number of imaging modalities may be considered such as z stacking, the use of transmission electron microscopy (TEM), or using a combination of high resolution confocal microscopy with flow cytometric analysis that has the potential to provide a precise evaluation of whether NPs are internalised or adsorbed onto the platelets surface, as described by Vranic et al. [33]. With respect to the effects of the interactions of PLGA-PEG NPs at the studied sizes on normal platelet functions, the use of a more sensitive platelet aggregation assay, such as flow-induced platelet aggregation in the presence of PLGA-PEG NPs, should be considered in future studies. An example would include a flow-mimicking technique such as quartz crystal microbalance with dissipation (QCM-D), which operates under similar flow conditions encountered in microvasculature [34]. Further understanding of the interaction between platelets and NPs is required to aid in the development of a safe platelet-targeted drug delivery system in the future.

Author Contributions: Conceptualisation and Design: Z.R., S.O.; Methodology, Investigations, and Analysis: R.B., J.G., K.A., S.S., N.M.; Writing—original draft preparation: R.B., S.O., Z.R.; Review and editing: Z.R., S.O.; Supervision: S.O., Z.R.

Funding: This research was funded by King Abdulaziz University in Jeddah, Saudi Arabia and the Ministry of Higher Education in Saudi Arabia.

Acknowledgments: The authors gratefully acknowledge research funding by the King Abdulaziz University in Jeddah, Saudi Arabia and the Ministry of Higher Education in Saudi Arabia.

Conflicts of Interest: The authors declare no conflict of interest.

References

1. Nandagiri, V.K.; Gentile, P.; Chiono, V.; Tonda-Turo, C.; Matsiko, A.; Ramtoola, Z.; Montevecchi, F.M.; Ciardelli, G. Incorporation of PLGA nanoparticles into porous chitosan–gelatin scaffolds: Influence on the physical properties and cell behavior. *J. Mech. Behav. Biomed. Mater.* **2011**, *4*, 1318–1327. [CrossRef] [PubMed]
2. Couvreur, P.; Patrick, C. Nanoparticles in drug delivery: Past, present and future. *Adv. Drug Deliv. Rev.* **2013**, *65*, 21–23. [CrossRef] [PubMed]
3. Mohan, L.J.; Daly, J.S.; Ryan, B.M.; Ramtoola, Z. The future of nanomedicine in optimising the treatment of inflammatory bowel disease. *Scand. J. Gastroenterol.* **2019**, *54*, 18–26. [CrossRef] [PubMed]
4. Ma, C.; Jiang, L.; Wang, Y.; Gang, F.; Xu, N.; Li, T.; Liu, Z.; Chi, Y.; Wang, X.; Zhao, L.; et al. 3D Printing of Conductive Tissue Engineering Scaffolds Containing Polypyrrole Nanoparticles with Different Morphologies and Concentrations. *Mater.* **2019**, *12*, 2491. [CrossRef]

5. Du, J.; Sun, Y.; Shi, Q.-S.; Liu, P.-F.; Zhu, M.-J.; Wang, C.-H.; Du, L.-F.; Duan, Y.-R. Biodegradable Nanoparticles of mPEG-PLGA-PLL Triblock Copolymers as Novel Non-Viral Vectors for Improving siRNA Delivery and Gene Silencing. *Int. J. Mol. Sci.* **2012**, *13*, 516–533. [CrossRef] [PubMed]
6. Chan, J.M.; Zhang, L.; Yuet, K.P.; Liao, G.; Rhee, J.-W.; Langer, R.; Farokhzad, O.C. PLGA–lecithin–PEG core–shell nanoparticles for controlled drug delivery. *Biomater.* **2009**, *30*, 1627–1634. [CrossRef] [PubMed]
7. Mattu, C.; Pabari, R.; Boffito, M.; Sartori, S.; Ciardelli, G.; Ramtoola, Z.; Pabari, R. Comparative evaluation of novel biodegradable nanoparticles for the drug targeting to breast cancer cells. *Eur. J. Pharm. Biopharm.* **2013**, *85*, 463–472. [CrossRef]
8. Kirby, B.P.; Pabari, R.; Al Baharna, M.; Walsh, J.; Ramtoola, Z.; Chen, C.-N.; Chen, C. Comparative evaluation of the degree of pegylation of poly (lactic-co-glycolic acid) nanoparticles in enhancing central nervous system delivery of loperamide. *J. Pharm. Pharmacol.* **2013**, *65*, 1473–1481. [CrossRef]
9. O'Donnell, A.; Moollan, A.; Baneham, S.; Ozgul, M.; Pabari, R.M.; Cox, D.; Kirby, B.P.; Ramtoola, Z. Intranasal and intravenous administration of octa-arginine modified poly (lactic-co-glycolic acid) nanoparticles facilitates central nervous system delivery of loperamide. *J. Pharm. Pharmacol.* **2015**, *67*, 525–536. [CrossRef]
10. Liu, Y.; Zhang, B.; Yan, B. Enabling Anticancer Therapeutics by Nanoparticle Carriers: The Delivery of Paclitaxel. *Int. J. Mol. Sci.* **2011**, *12*, 4395–4413. [CrossRef]
11. Anselmo, A.C.; Mitragotri, S. Nanoparticles in the clinic. *Bioeng. Transl. Med.* **2016**, *1*, 10–29. [CrossRef] [PubMed]
12. Anselmo, A.C.; Modery-Pawlowski, C.L.; Menegatti, S.; Kumar, S.; Vogus, D.R.; Tian, L.L.; Chen, M.; Squires, T.M.; Gupta, A.S.; Mitragotri, S. Platelet-like nanoparticles: Mimicking shape, flexibility, and surface biology of platelets to target vascular injuries. *ACS Nano* **2014**, *8*, 11243–11253. [CrossRef] [PubMed]
13. Serda, R.E.; Godin, B.; Blanco, E.; Chiappini, C.; Ferrari, M. Multi-stage delivery nano-particle systems for therapeutic applications. *Biochim. Biophys. Acta, Gen. Subj.* **2011**, *1810*, 317–329. [CrossRef] [PubMed]
14. Özcan, I.; Segura-Sanchez, F.; Bouchemal, K.; Sezak, M.; Özer, Ö.; Güneri, T.; Ponchel, G. Pegylation of poly (γ-benzyl-L-glutamate) nanoparticles is efficient for avoiding mononuclear phagocyte system capture in rats. *Int. J. Nanomed.* **2010**, *5*, 1103. [CrossRef]
15. Santos-Martinez, M.J.; Inkielewicz-Stępniak, I.; Medina, C.; Rahme, K.; D'Arcy, D.M.; Fox, D.; Holmes, J.D.; Zhang, H.; Radomski, M.W. The use of quartz crystal microbalance with dissipation (QCM-D) for studying nanoparticle-induced platelet aggregation. *Int. J. Nanomed.* **2012**, *7*, 243–255. [CrossRef] [PubMed]
16. Dobrovolskaia, M.A.; Patri, A.K.; Zheng, J.; Clogston, J.D.; Ayub, N.; Aggarwal, P.; Neun, B.W.; Hall, J.B.; McNeil, S.E. Interaction of colloidal gold nanoparticles with human blood: Effects on particle size and analysis of plasma protein binding profiles. *Nanomed. Nanotechnol. Bio. Med.* **2009**, *5*, 106–117. [CrossRef]
17. Cheng, F.-Y.; Wang, S.P.-H.; Su, C.-H.; Tsai, T.-L.; Wu, P.-C.; Shieh, D.-B.; Chen, J.-H.; Hsieh, P.C.-H.; Yeh, C.-S. Stabilizer-free poly(lactide-co-glycolide) nanoparticles for multimodal biomedical probes. *Biomater.* **2008**, *29*, 2104–2112. [CrossRef] [PubMed]
18. Koziara, J.M.; Oh, J.J.; Akers, W.S.; Ferraris, S.P.; Mumper, R.J. Blood Compatibility of Cetyl Alcohol/Polysorbate-Based Nanoparticles. *Pharm. Res.* **2005**, *22*, 1821–1828. [CrossRef]
19. Radomski, A.; Jurasz, P.; Alonso-Escolano, D.; Drews, M.; Morandi, M.; Malinski, T.; Radomski, M.W. Nanoparticle-induced platelet aggregation and vascular thrombosis. *Br. J. Pharmacol.* **2005**, *146*, 882–893. [CrossRef]
20. Ramtoola, Z.; Lyons, P.; Keohane, K.; Kerrigan, S.W.; Kirby, B.P.; Kelly, J.G. Investigation of the interaction of biodegradable micro-and nanoparticulate drug delivery systems with platelets. *J. Pharm. Pharmacol.* **2011**, *63*, 26–32. [CrossRef]
21. Corbalan, J.J.; Medina, C.; Jacoby, A.; Malinski, T.; Radomski, M.W. Amorphous silica nanoparticles aggregate human platelets: Potential implications for vascular homeostasis. *Int. J. Nanomed.* **2012**, *7*, 631–639.
22. Shrivastava, S.; Bera, T.; Singh, S.K.; Singh, G.; Ramachandrarao, P.; Dash, D. Characterization of Antiplatelet Properties of Silver Nanoparticles. *ACS Nano* **2009**, *3*, 1357–1364. [CrossRef] [PubMed]
23. MMiller, V.M.; Hunter, L.W.; Chu, K.; Kaul, V.; Squillace, P.D.; Lieske, J.C.; Jayachandran, M. Biologic nanoparticles and platelet reactivity. *Nanomed.* **2009**, *4*, 725–733. [CrossRef] [PubMed]
24. Pabari, R.M.; Mattu, C.; Partheeban, S.; Almarhoon, A.; Boffito, M.; Ciardelli, G.; Ramtoola, Z. Novel polyurethane-based nanoparticles of infliximab to reduce inflammation in an in-vitro intestinal epithelial barrier model. *Int. J. Pharm.* **2019**, *565*, 533–542. [CrossRef] [PubMed]

25. Bootz, A.; Vogel, V.; Schubert, D.; Kreuter, J. Comparison of scanning electron microscopy, dynamic light scattering and analytical ultracentrifugation for the sizing of poly(butyl cyanoacrylate) nanoparticles. *Eur. J. Pharm. Biopharm.* **2004**, *57*, 369–375. [CrossRef]
26. Fornaguera, C.; Solans, C. Characterization of Polymeric Nanoparticle Dispersions for Biomedical Applications: Size, Surface Charge and Stability. *Pharm. Nanotechnol.* **2018**, *6*, 147–164. [CrossRef] [PubMed]
27. Zhou, L.; Schmaier, A.H. Platelet aggregation testing in platelet-rich plasma: Description of procedures with the aim to develop standards in the field. *Am. J. Clin. Pathol.* **2005**, *123*, 172–183. [CrossRef] [PubMed]
28. Ilinskaya, A.N.; A Dobrovolskaia, M. Nanoparticles and the blood coagulation system. Part I: Benefits of nanotechnology. *Nanomed.* **2013**, *8*, 773–784. [CrossRef]
29. Ragaseema, V.; Unnikrishnan, S.; Krishnan, V.K.; Krishnan, L.K. The antithrombotic and antimicrobial properties of PEG-protected silver nanoparticle coated surfaces. *Biomater.* **2012**, *33*, 3083–3092. [CrossRef]
30. Mayer, A.; Vadon, M.; Rinner, B.; Novak, A.; Wintersteiger, R.; Fröhlich, E. The role of nanoparticle size in hemocompatibility. *Toxicol.* **2009**, *258*, 139–147. [CrossRef]
31. Gupalo, E.; Kuk, C.; Qadura, M.; Buriachkovskaia, L.; Othman, M. Platelet–adenovirus vs. inert particles interaction: Effect on aggregation and the role of platelet membrane receptors. *Platelets* **2013**, *24*, 383–391. [CrossRef] [PubMed]
32. Shang, L.; Nienhaus, K.; Jiang, X.; Yang, L.; Landfester, K.; Mailänder, V.; Simmet, T.; Nienhaus, G.U. Nanoparticle interactions with live cells: Quantitative fluorescence microscopy of nanoparticle size effects. *Beilstein J. Nanotechnol.* **2014**, *5*, 2388–2397. [CrossRef] [PubMed]
33. Vranic, S.; Boggetto, N.; Contremoulins, V.; Mornet, S.; Reinhardt, N.; Marano, F.; Baeza-Squiban, A.; Boland, S. Deciphering the mechanisms of cellular uptake of engineered nanoparticles by accurate evaluation of internalization using imaging flow cytometry. *Part. Fibre Toxicol.* **2013**, *10*, 2. [CrossRef] [PubMed]
34. Samuel, S.P.; Santos-Martinez, M.J.; Medina, C.; Jain, N.; Radomski, M.W.; Prina-Mello, A.; Volkov, Y. CdTe quantum dots induce activation of human platelets: Implications for nanoparticle hemocompatibility. *Int. J. Nanomed.* **2015**, *10*, 2723–2734.

pharmaceutics

MDPI

Article

Microfluidics-Assisted Size Tuning and Biological Evaluation of PLGA Particles

Maria Camilla Operti [1], Yusuf Dölen [1,2], Jibbe Keulen [1], Eric A. W. van Dinther [1,2], Carl G. Figdor [1,2] and Oya Tagit [1,2,*]

1 Department of Tumor Immunology, Radboud Institute for Molecular Life Sciences, Radboud University Medical Center, 6500 HB Nijmegen, The Netherlands; MariaCamilla.Operti@radboudumc.nl (M.C.O.); Yusuf.Dolen@radboudumc.nl (Y.D.); jibbekeulen@gmail.com (J.K.); Eric.vanDinther@radboudumc.nl (E.A.W.v.D.); Carl.Figdor@radboudumc.nl (C.G.F.)

2 Oncode Institute, 3553 Utrecht, The Netherlands

* Correspondence: oya.tagit@radboudumc.nl

Received: 30 September 2019; Accepted: 6 November 2019; Published: 8 November 2019

Abstract: Polymeric particles made up of biodegradable and biocompatible polymers such as poly(lactic-co-glycolic acid) (PLGA) are promising tools for several biomedical applications including drug delivery. Particular emphasis is placed on the size and surface functionality of these systems as they are regarded as the main protagonists in dictating the particle behavior in vitro and in vivo. Current methods of manufacturing polymeric drug carriers offer a wide range of achievable particle sizes, however, they are unlikely to accurately control the size while maintaining the same production method and particle uniformity, as well as final production yield. Microfluidics technology has emerged as an efficient tool to manufacture particles in a highly controllable manner. Here, we report on tuning the size of PLGA particles at diameters ranging from sub-micron to microns using a single microfluidics device, and demonstrate how particle size influences the release characteristics, cellular uptake and in vivo clearance of these particles. Highly controlled production of PLGA particles with ~100 nm, ~200 nm, and >1000 nm diameter is achieved through modification of flow and formulation parameters. Efficiency of particle uptake by dendritic cells and myeloid-derived suppressor cells isolated from mice is strongly correlated with particle size and is most efficient for ~100 nm particles. Particles systemically administered to mice mainly accumulate in liver and ~100 nm particles are cleared slower. Our study shows the direct relation between particle size varied through microfluidics and the pharmacokinetics behavior of particles, which provides a further step towards the establishment of a customizable production process to generate tailor-made nanomedicines.

Keywords: PLGA; drug delivery systems; microfluidics; nanoparticles; microparticles

1. Introduction

As evidenced by the dramatic increase in the number of studies and clinical applications over time, polymeric particles play a crucial role in drug delivery, providing improved stability, targeted delivery, and sustained release of loaded therapeutic agents without causing off-target toxicities in vivo [1–5]. Drug delivery particles based on poly(lactic-co-glycolic acid) (PLGA) are among the most commonly studied vehicles due to excellent biocompatibility, tuneable degradation characteristics, and long clinical history of PLGA [6–8]. The remarkable physicochemical properties and high versatility of PLGA have proven to address several challenges in drug delivery. PLGA can be processed into almost any size and shape [7] and can encapsulate molecules of virtually any size such as small drugs [9–13], proteins [14–18], nucleic acids [19–23], and vaccines [24–29]. The colloidal features such as size and surface functionality have a direct influence on the cellular uptake, biodistribution, and thus therapeutic efficacy of the particles, dictating the in vivo fate of therapeutic cargo [30,31]. These colloidal features are particularly important in cancer immunotherapy as the efficacy of any given therapeutic agent is highly dependent on its ability to reach either the tumor microenvironment (TME) or the lymph nodes. The incomplete endothelial lining due to rapid angiogenesis at the tumor site results in the formation of large and irregular pores (0.1–3 μm) [32], through which nanoparticles can escape from the circulatory stream and accumulate at the TME via a so-called enhanced permeability and retention (EPR) effect [32]. Toy et al. demonstrated that, within the size range of 60–130 nm, smaller particles exhibited greater lateral drift towards the blood vessel walls, which is a prerequisite for interaction with the tumor vascular bed and essential for escape into the TME [33]. Furthermore, both blood clearance and uptake by the mononuclear phagocyte system (MPS) at the liver and spleen depend on the size and composition of PLGA carrier systems. The slit size in the inter-endothelial cells of the spleen is about 200 nm, which facilitates the leakage and circulation of smaller particles for longer time periods [34] or enables their entry to the lymphatic system through direct drainage to the lymph nodes [35]. On the other hand, larger particles, which are not eligible for intravenous administration due to their excessive size, are more likely to be taken up by immune cells at the injection site, such as in the case of subcutaneous or intramuscular pathways [36].

Surface functionality also influences drug pharmacokinetics in vivo. The use of polyethylene glycol (PEG) has been shown to significantly increase circulation time in several studies [36–38]. When attached to nanoparticles' surface (a process so-called PEGylation), the hydrophilicity of PEG chains recruits specific proteins from plasma, that cloak and limit the interactions of particles with MPS cells, hence prolonging blood circulation ('stealth effect') [39,40]. The decrease in the aggregation, opsonization, and phagocytosis of particles entails the extension of their circulation time. Morikawa et al. reported additional benefits of PEGylation such as reduced particle size and improved encapsulation efficiency for curcumin-loaded nanoparticles [41].

In addition to particle size and surface functionality that have a large influence on the therapeutic efficacy in vivo, particle uniformity in terms of size and encapsulated cargo is also an important aspect particularly for the clinical translation of polymeric drug delivery formulations [42,43]. The uniformity of particles depends largely on the utilized manufacturing approach. The conventional production techniques based on emulsion solvent diffusion, emulsion solvent evaporation, and nanoprecipitation are suitable for the production of sub-micron and micron-size PLGA particles [7,8]; however, these methods lack the precision and full control over the particle size and uniformity particularly for larger scale processes [8]. In recent years, microfluidics technology has emerged as an effective tool to produce particles in a highly controllable manner [44,45]. Major advantages of microfluidics-based particle manufacturing include the requirement of low sample volumes, high surface area, and reduced system footprint [46]. This technology allows for rapid fluid mixing at the nanoliter scale and production of PLGA particles with highly specific sizes as well as surface functionalities only by altering specific parameters such as concentration of the starting materials or fluid flow rates through micron-size channels [41,47–52]. Additionally, traditional small-scale laboratory synthesis techniques suffer from batch-to-batch variations while microfluidics technology provides a precise size control, a high degree of particle uniformity, and reproducibility. Furthermore, formulations can be scaled up by increasing the quantities of fluids pumped through the system or by parallelizing multiple microfluidic mixers [53].

In this study, we report on the use of microfluidics technology as a platform for the production of PEGylated PLGA particles. We explored the feasibility to generate different sizes of particles with low polydispersity index (PDI) by varying process and formulation parameters such as total flow rate and flow rate ratio of organic and aqueous phases as well as surfactant and polymer concentration. We tuned the particle size at sub-micron (~100 nm and ~200 nm) and micrometer (>1000 nm) length scales, which represent biologically-relevant cut-off values that influence the particle biodistribution and clearance in vivo. A fluorescent dye was used as a model drug to assess encapsulation efficiency, release profile, and uptake by mouse-derived immune cells. Clearance of systemically administered, fluorescently labeled particles was also studied on a mouse model through in vivo imaging.

2. Experimental

2.1. Materials

PLGA (Resomer RG 502H), with a 50:50 ratio of lactic acid:glycolic acid and Mw 7000–17,000 Da was obtained from Evonik Nutrition and Care GmbH (Darmstadt, Germany). PEG-PLGA copolymer (PEG M_n 5000, PLGA M_n 7000), polyvinyl alcohol (PVA, 9000–10,000 Mw, 80%, hydrolyzed) and cholesteryl BODIPY™ FL C_{12} were obtained from Thermo Fisher Scientific (Waltham, MA, USA). Near-infrared emitting fluorescent dye VivoTag-S 750 was purchased from Perkin Elmer Inc. (Waltham, MA, USA) and acetonitrile (ACN, 99.95%) was from VWR International (Radnor, PA, USA). Ultrapure Milli-Q® water (18.2 MΩ.cm) was used where necessary (Merck KGaA, Darmstadt, Germany). Roswell Park Memorial Institute (RPMI) 1640 medium, Anti-Anti (AA), and ß-mercaptoethanol were obtained from Gibco (Thermo Fisher Scientific, Waltham, MA, USA). Granulocyte-macrophage colony-stimulating factor (GM-CSF) was obtained from Peprotech Inc. (Rocky Hill, NJ, USA). X-Vivo medium and ultraglutamine were from Lonza Group (Basel, Switzerland). Fetal bovine serum (FBS) was purchased from Hyclone Laboratories Inc. (GE Healthcare, Chicago, IL, USA).

2.2. Equipment

The microfluidics system was set up by connecting syringe pumps (Harvard PHD-2000 infusion 70–200) to a Y-junction mixer with staggered herringbone ridges (NanoAssemblr™, Precision Nanosystems Inc., Vancouver, Canada) through 0.8 mm polytetrafluoroethylene (PFTE) tubing (ID 0.8 mm, OD 1.58 mm) obtained from Sigma-Aldrich (St. Louis, MO, USA). A detailed characterization of the mixing geometry was reported in [44]. Two 20 mL NORM-JECT Luer Lock syringes were connected to 0.8 mm diameter needles (Braun Sterican 0.8 × 120 mm), which were inserted in the fittings connected to the inlets of the mixing cartridge. A PFTE tubing connected to the chip outlet was used for sample collection.

2.3. Preparation of PLGA Particles

Prior to particle production, the pipes and the mixing chip were primed first with the solvents (ACN for the organic phase inlet and MilliQ for the aqueous phase inlet) and then with the appropriate phases for 1 min at organic: aqueous flow rates of 2:2. After the priming step, pumps were operated at the desired flow rates. The product obtained within the first 2 min was discarded. The collected particles were left stirring overnight (350 rpm) at room temperature for organic solvent evaporation. A schematic representation of the process is shown in Scheme 1.

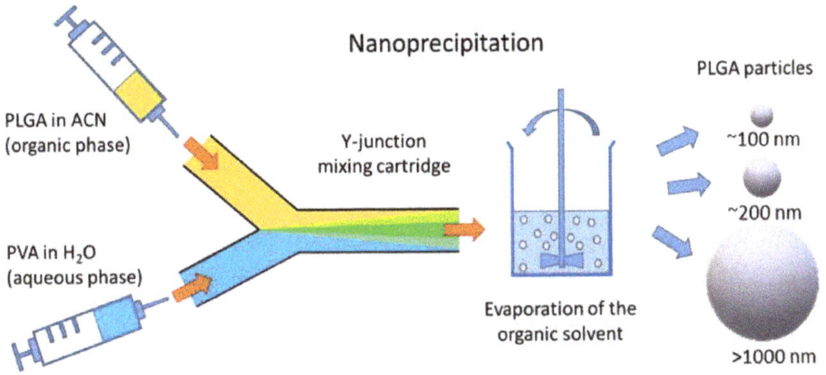

Scheme 1. Schematic illustration of the particle generation process via nanoprecipitation method using a Y-junction mixing cartridge. PLGA: poly(lactic-co-glycolic acid); ACN: acetonitrile; PVA: polyvinyl alcohol.

The particle size was tuned via varying the flow rates of organic and aqueous phases as well as PLGA and PVA concentrations. Upon determination of the optimized parameters for each target size, PEGylated PLGA particles encapsulating fluorescent dyes emitting either at visible or near-infrared regions were prepared by the following methods:

>1000 nm particles: the organic phase (5 mL ACN) contained 233.1 mg PLGA, 99.9 mg PEG-PLGA, and 50 µL of 1 mg/mL fluorescent dye in ethanol. A 3% PVA solution was used as the aqueous phase. Particles were produced at 6:2 organic:aqueous flow rates in triplicate.

~200 nm particles: the organic phase (5 mL ACN) contained 116.6 mg PLGA, 50 mg PEG-PLGA, and 50 µL of 1 mg/mL fluorescent dye in ethanol. A 1% PVA solution was used as the aqueous phase. Particles were produced at 2:6 organic:aqueous flow rates in triplicate.

~100 nm particles: the organic phase (8 mL ACN) contained 93.5 mg PLGA, 40 mg PEG-PLGA, and 80 µL of 1 mg/mL fluorescent dye in ethanol. A 1% PVA solution was used as the aqueous phase. Particles were produced at 4:6 organic:aqueous flow rates in triplicate.

After evaporation of the organic solvent, particles were washed three times with MilliQ water by centrifugation at 15,000 rpm for 35 min (for >1000 nm and ~200 nm particles) and by spin filtration using a 100,000 kDa MW cut-off centrifugal filter device (Millipore, Merck KGaA, Darmstadt, Germany) for ~100 nm particles. Particles were then lyophilized.

2.4. *Colloidal Characterization of PLGA Particles*

In total, 1.25 mg of lyophilized particles were dispersed in 1 mL of MilliQ water to determine the size distribution and polydispersity index (PDI) with dynamic light scattering using a Nanotrac Flex (Microtrac GmbH, Krefeld, Germany). Zeta potential measurements were obtained with a Zetasizer Nano ZS (Malvern Instruments, Malvern, United Kingdom). Prior to measurements, particles were suspended in a 50 mM NaCl solution. The average of three measurements were used to report the size, PDI, and Zeta potential for each sample.

Atomic force microscopy (AFM) images of particles were obtained with a Catalyst BioScope (Bruker, Billerica, MA, USA) coupled to a confocal microscope (TCS SP5II, Leica Mycrosystems, Wetzlar, Germany). Then, 50 µL of 10 mg/mL particle suspension was dried on clean glass substrates and particles were imaged in peak-force tapping mode using a silicon nitride cantilever with a nominal spring constant of 40 N/m (Bruker). AFM images were analyzed using NanoScope analysis software (Bruker, Billerica, MA, USA).

2.5. *Determination of Encapsulation Efficiency*

Dye loading was quantified by construction of a calibration curve using a BODIPY-C_{12} standard. A series of dye concentrations (in 0–0.25 mg/mL range) were prepared in a solvent composed of equal parts of ACN and MilliQ and fluorescence intensities were measured for each standard. The fit function applied to the linear portion of the curve was used for the calculation of the dye content of particles. The encapsulation efficiency was determined by comparing the total amount of dye in the lyophilized particles to the initial amount of dye supplied for the collected volume.

2.6. *In Situ Release Profile*

PLGA particles containing BODIPY-C_{12} were suspended in PBS in 5 mg/mL concentrations and were dialyzed at 37 °C for 14 days using dialysis tubes with a 1000 Da molecular weight cut-off membrane (GE Healthcare, Chicago, IL, USA). At different incubation times, the dialysis medium was collected for fluorescence measurements using an LS 55 Perkin Elmer fluorescence spectrometer (Waltham, MA, USA). Samples were excited at 488 nm and emission was recorded between 500 nm and 700 nm. After each measurement the dialysis medium was refreshed. Samples were studied in triplicate.

3. In Vitro Cellular Uptake Experiments

3.1. Generation of Bone Marrow-Derived Dendritic Cells (BMDCs)

Dendritic cells were generated from mouse bone marrow cells by culturing them in full RPMI medium containing 5 mL 200 mM L-glutamine (Gibco, Thermo Fisher Scientific, Waltham, MA, USA), 5 mL 100× antibiotics and antimycotics (Gibco, Thermo Fisher Scientific, Waltham, MA, USA), 10% fetal bovine serum (Gibco, Thermo Fisher Scientific, Waltham, MA, USA), and 500 µL β-mercapto ethanol (Sigma-Aldrich, St. Louis, MO, USA). Then, 5.0×10^6 cells in 13 mL of full medium were cultured in the presence of 20 ng/mL GM-CSF for 7 days. At day 3, 4 mL of complete medium containing 37.9 ng/mL GM-CSF was added. At day 6, 1 mL complete medium containing 158 ng/mL GM-CSF was added. Cells were used for uptake experiments at day 7.

3.2. Isolation of Myeloid-Derived Suppressor Cells (mMDSCs) and Polymorphonuclear Myeloid-Derived Suppressor Cells (pmnMDSCs)

Gr-1^{dim}Ly-$6G^-$ monocytic and Gr-1^{high}Ly-$6G^+$ polymorphonuclear myeloid-derived suppressor cells (mMDSCs and pmnMDSCs, respectively) were isolated from the spleen of tumor-bearing mice using an isolation kit (Miltenyi Biotec, Bergisch Gladbach, Germany) according to manufacturer's instructions. Briefly, the spleen was isolated under sterile conditions and meshed through a 100 µm cell strainer with a syringe plunger. The cell suspension was spun at 400× *g* for 5 min and resuspended in 3 mL of 1× ammonium chloride solution for the lysis of erythrocytes. After 5 min of incubation at room temperature, cells were washed with 10 mL of PBS. The cells were incubated with an Anti-Ly-6G-Biotin antibody and Anti-Biotin MicroBeads and were subsequently applied to a magnetic-activated cell sorting (MACS) column, which retained the pmnMDSCs. The flow-through containing mMDSCs were eluted as the positively selected cell fraction and were further purified by applying them to a second MACS column.

3.3. In Vitro Cellular Uptake

Firstly, 1.0×10^5 cells in 500 µL complete medium were transferred to 5 mL propylene round bottom tubes (Falcon). Then, 10 µg of particles containing BODIPY-C_{12} water were added to the round bottom tubes and were incubated for time periods of 1, 2, 4, 6, 24, and 48 h. After incubation, particle uptake was determined by flow cytometry analysis on a FACSVerse (BD Biosciences, Franklin Lakes, NJ, United Sates).

4. In Vivo Clearance Studies

All animal experiments were performed according to guidelines of Radboud University's Animal Experiment Committee and Central Authority for Scientific Procedures on Animals (project number 2015-019TIL, date September 2015) in accordance with the ethical standards described in the Declaration of Helsinki. Wild-type BALB/cAnNCrl mice, aged 8–12 weeks, were obtained from Charles River, Germany and maintained under specific pathogen-free conditions at the Central Animal Laboratory (Nijmegen, The Netherlands). Drinking water and food were provided ad libitum. Mice were warmed either in a heating chamber or under a heating lamp and 1 mg PLGA nanoparticles (~200 nm and ~100 nm) containing VivoTag-S 750 were injected in 200 µL of phosphate-buffered saline (PBS) solution through a lateral tail vein using a 1 mL syringe with a 29 G needle. Then, 0.5, 3, 24, and 48 h after injections mice were shaved and imaged in an IVIS Lumina II (Perkin Elmer) system. Mice were euthanized, and organs were dissected and imaged separately at 24 and 48 h. Imaging settings were: exposure time: 3 s; binning: medium; F/stop: 2; fluorescent excitation filter: 745 nm; fluorescent emission filter: 810–885 nm. A fluorescent background acquisition was performed for each time point. Living Image software (Caliper Life Sciences, Hopkinton, MA, USA) was used for data analysis. Background values were subtracted from measurement values. Same sized regions of interest (ROI) were applied on the liver and bladder for full body image analysis; also, same sized ROIs were applied on the isolated liver and spleen. Total flux (photon/s) per each ROI was calculated.

Statistical Analysis

An unpaired *t*-test was used to determine the significance of the difference in mean particle size values among the compared groups using GraphPad Prism.

5. Results and Discussion

5.1. Preparation of Particles

A commercially available NanoAssemblr™ cartridge [8] connected to syringe pumps was used for the production of PLGA particles through nanoprecipitation. In this method, water-miscible organic solvents are used to dissolve the polymer which, once at the interface between the aqueous medium and the organic solvent, will start to precipitate. Polymer deposition caused by fast diffusion of the solvent leads to instantaneous formation of a colloidal suspension and particle size is mainly determined by solvent diffusion coefficient (D) and mixing time (t_{mix}) [54]. The architecture of the mixing chip also plays a crucial for the generation of a hydrodynamic-focusing flow pattern and the mixing process can be controlled by varying the width of the focused stream (wf) as shown in Equation (1) [44]:

$$t_{mix} \approx \frac{wf^2}{4D} \quad (1)$$

By varying the ratio of flow rates, the lateral width of the focused stream can be adjusted very accurately, which leads to a highly controlled diffusion and mixing at the interface of organic and aqueous phases. In our study, acetonitrile (ACN) was used as the organic phase and particles were formed upon rapid mixing of organic and aqueous phases in micron-size channels. Due to rapid diffusion of ACN to the aqueous phase, PLGA precipitated and was subsequently stabilized by PVA present in the aqueous phase. Process parameters such as (total) flow rates of organic and aqueous phases as well as formulation parameters such as PLGA and PVA concentration were varied to tune the size of PLGA particles from sub-micron to micron-scales.

The first process parameter tested was the flow rate of the organic phase composed of a 33.3 mg/mL PLGA solution in ACN. The organic phase flow rate was varied between 2 mL/min and 6 mL/min while the aqueous flow rate was kept constant at 2 mL/min for a 1% PVA solution (Figure 1A). A gradual increase in the particle size from ~400 nm to ~900 nm was observed with an increase of the organic phase flow rate. This was most likely caused by precipitation of higher amounts of PLGA at a given mixing time when relatively higher organic phase flow rates were used. Similarly, increasing the flow rate of the aqueous phase for a given set of formulation parameters and organic phase flow rate resulted in the formation of smaller particles (Figure 1B). Doubling the aqueous flow rate to 4 mL/min resulted in particles almost half the size of those produced at 2 mL/min. However, increasing the aqueous flow rate up to 6 mL/min did not result in a notable change in particle size.

Another process parameter affecting particle size was the total flow rate of the organic and aqueous phases. With the equal flow rate of organic and aqueous phases, the increase of the total flow rates from 4 mL/min (2:2) to 8 mL/min (4:4) led to a decrease in the particle size (Figure 1C). Increasing the total flow rate further to 12 mL/min (6:6), however, did not cause any notable difference in the particle size other than a slight improvement in the PDI. Increasing the total flow rates from 8 mL/min to 12 mL/min apparently did not cause a relevant variation of solvent diffusion time to induce a significant change in particle size. The total flow rate determines both the mixing time of the two phases in the cartridge and the collection rate of the particles at the outlet channel. Faster mixing by increasing the total flow rates has also been reported in other studies to result in the formation of smaller particles [41,47].

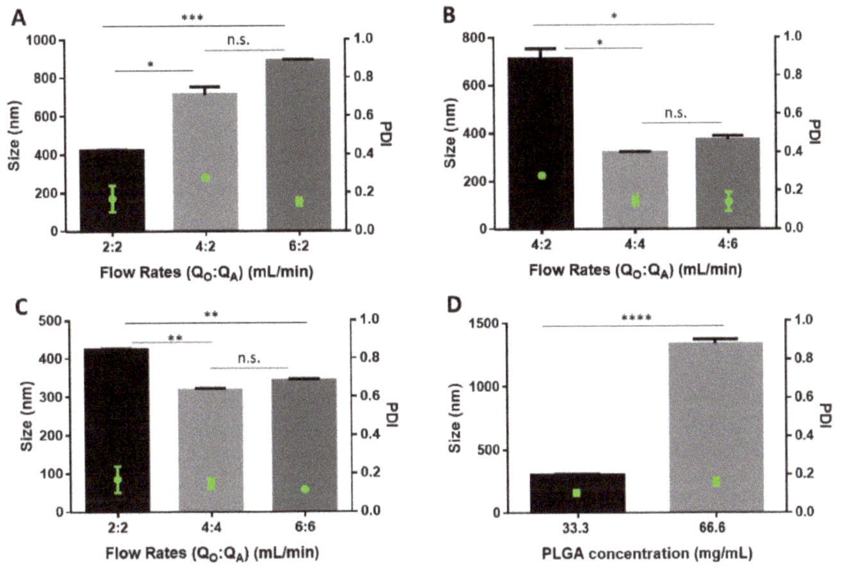

Figure 1. (**A–C**) Influence of process and formulation parameters on the particle size. The variation of particle size at different flow rates for 33.3 mg/mL PLGA as organic phase and 1% PVA as aqueous phase. (**D**) The influence of PLGA concentration on the particle size (organic:aqueous flow rate of 4:4 mL/min, 3% PVA as aqueous phase). Bars: intensity-averaged particle size; green dots: polydispersity index (PDI). Q_O: flow rate of organic phase, Q_A: flow rate of aqueous phase. Significant: *: $p < 0.05$; **: $p < 0.005$; ***: $p < 0.001$; ****: $p < 0.0001$. Non-significant (n.s.): $p > 0.05$.

Overall, particles in the sub-micron size range were obtained by varying only the flow parameters. Furthermore, we varied the PLGA and PVA concentrations in order to increase the particle size to micron scale. A representative data set in Figure 1D shows that particle size almost tripled when the PLGA concentration was doubled. In these batches, the PVA concentration was increased as well and organic:aqueous flow rates were kept equal (4:4 mL/min). The larger particle size obtained with a higher PLGA concentration can be explained by the higher viscosity of the organic phase, which resulted in a slower diffusion of ACN to the aqueous phase and an increased mixing time [55].

For all tested conditions, a good batch-to-batch reproducibility and a small PDI (≤0.2) was observed. After several trials, the optimal conditions and parameters were determined to obtain particles of sizes >1000 nm, ~200 nm, and ~100 nm. These parameters were then used to prepare particles with a surface functionalized with polyethylene glycol (PEG) and encapsulating a fluorescent dye (Table 1).

Table 1. Optimized formulation and process parameters to obtain PLGA particles with indicated size. PEG: polyethylene glycol.

Parameters	>1000 nm	~200 nm	~100 nm
PLGA (mg/mL)	66.6	33.3	16.7
PLGA:PEG-PLGA ratio	70:30	70:30	70:30
PVA (*w*/*v*%)	3%	1%	1%
Flow rates (mL/min) (QO:QA)	6:2	2:6	4:6
Fluorescent dye (*v*/*v*%)	1%	1%	1%

For fluorescent labeling, two different fluorescent dyes with visible (BODIPY™ FL C_{12}, green, 510 nm) and near-infrared (VivoTag-S 750, red, 750 nm) emission wavelengths were used. BODIPYs are low-polarity dyes with stable fluorescence emission properties [56] and are commonly used in fluorescence detection and photodynamic therapy applications [57]. VivoTag-S 750 is a near-infrared emitting (NIR) fluorochrome extensively used for in vivo imaging applications [58,59]. PLGA particles labeled with the green fluorescent dye were used for in situ release and in vitro uptake experiments, whereas those labeled with the red dye were used for in vivo imaging studies.

5.2. Colloidal Characterization

A detailed colloidal and functional characterization of PEGylated PLGA particles encapsulating BODIPY-C_{12} is shown in Figure 2. Both atomic force microscopy (AFM) images (Figure 2A–C) and dynamic light scattering (DLS) measurements (Figure 2D) revealed the formation of monodisperse particles (PDI ≤0.2) within the desired size range (i.e.; >1000 nm, ~200 nm, ~100 nm) also upon PEGylation and fluorescent dye encapsulation. It should be noted that although PDI values smaller than 0.3 are considered acceptable for drug delivery applications, more specific standards and guidelines have yet to be established by regulatory authorities.

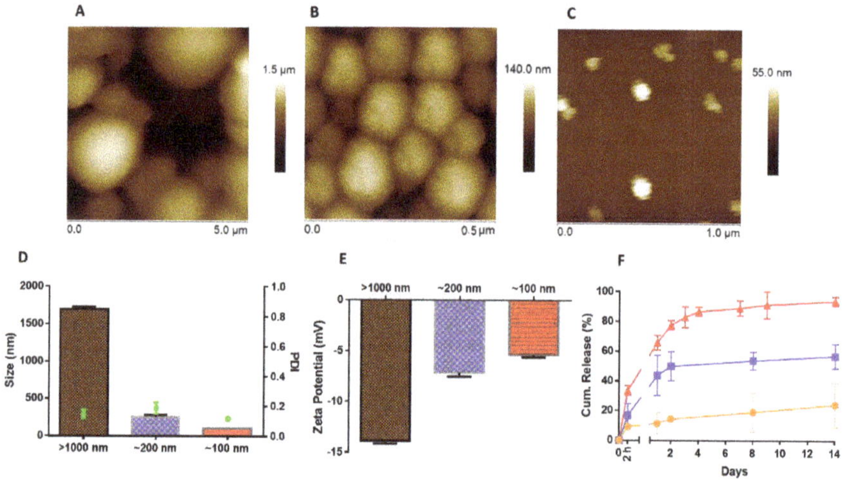

Figure 2. Atomic force microscopy height images of (**A**) >1000 nm, (**B**), ~200 nm, and (**C**) ~100 nm PLGA particles. The scale bars display the Z-range and the scan sizes are shown on the *x*-axis. (**D**) Size distribution and (**E**) ζ potential values of PLGA particles measured using dynamic light scattering. (**F**) Release profile of BODIPY-C_{12} from >1000 nm, ~200 nm, and ~100 nm particles. Data obtained for >1000 nm, ~200 nm, and ~100 nm particles are represented in orange, blue, and red, respectively.

The micron-sized particles showed an average size of 1690 nm (±60 nm), while the sub-micron particles were 250 nm (±48 nm) and 106 nm (±5 nm) in diameter. For ease of reporting, we will continue referring to them as >1000 nm, ~200 nm, and ~100 nm particles. Slightly negative ζ potential values that spanned a range between −5 mV and −15 mV were observed for all particle types (Figure 2E). While the variation of ζ potential was negligible among sub-micron particles (i.e., ~200 nm, ~100 nm), the micron-size particles displayed the most negative value. PLGA nanoparticles stabilized with PVA have also been reported to bear a slightly negative ζ potential (−5 mV) [60] due to the presence of residual PVA on the particle surface, which affects the number of carboxylic acid end groups [61]. The hydrophobic acetate moieties in partially hydrolyzed PVA can lead to its entrapment within the PLGA matrix on the particle surface, thereby masking the surface charges to almost neutral ζ potential values [62]. Therefore, the more negative ζ potential obtained for micron-size particles can be related to differences in the PVA content and the flow rate of the aqueous phase for micron-size particles compared to sub-micron particles, as well as higher PLGA content. Apparently, these variations in formulation and process parameters influence not only the particle size but also the particles' surface properties, which collectively determine the pharmacokinetic properties, cellular uptake, and particle biodistribution [63].

Encapsulation efficiency of BODIPY-C_{12} was determined using fluorescence spectroscopy by comparing the initial amount of dye supplied during particle production and the total amount of dye detected in the entire yield. The efficiency of dye encapsulation was significantly higher for smaller particles with the highest efficiency at 44% for ~100 nm particles, which decreased gradually to 21% and 13% for ~200 nm and >1000 nm particles, respectively. It should be noted that each particle type was produced using different formulation and process parameters, which equally play a role in determining the encapsulation efficiency [64]. An efficient encapsulation requires the rapid precipitation of polymer with the fluorescent dye to restrict the dye within the polymer matrix and prevent its diffusion to the aqueous phase. The micron-size particles with the lowest encapsulation efficiency were prepared using the highest polymer concentration and the highest organic:aqueous phase flow rate ratios. The high viscosity of the organic phase as well as the relatively low fraction of the aqueous phase probably resulted in a slower polymer precipitation, and therefore a lower encapsulation efficiency for micron-size particles.

The functional characterization of particles was achieved by monitoring the dye release. For this study, particles dispersed in PBS were dialyzed at 37 °C for a period of 14 days. At certain time points, fluorescence intensities of the dialysis media were measured using fluorescence spectroscopy. The dialysis medium was replaced with a fresh medium after each measurement. Each formulation type showed a distinct release profile that was strongly correlated with the particle size. After an initial burst, the release was steady for all particles (Figure 2F). The magnitude of burst release was higher for ~100 nm particles (approx. 30%), which was found as ~15% and ~10% for ~200 nm and >1000 nm particles, respectively.

In addition to the highest burst release, the overall release rate was also faster for ~100 nm particles such that they already released the majority of their content (~80%) by day 4. On the other hand, the released content of ~200 nm and >1000 nm particles barely reached 50% and 20%, respectively, by the end of the entire release period (two weeks). Indeed, several different release patterns such as mono-, bi-, and tri-phasic release are reported for PLGA particles, which are mainly regulated by the physicochemical properties of the cargo, PLGA type (molecular weight, lactide/glycolide ratio, etc.), and particle morphology (size, porosity, etc.) [65]. Since the same cargo and polymer type were used for the preparation of particles, the main reason for different release rates observed was the particle size. Larger surface area/volume ratio of smaller particles facilitated the faster diffusion of dye molecules located on or close to the surface. Such inverse relation between the release rate and particle size is in good agreement with the general trends reported in other studies [66–68].

5.3. *In Vitro Uptake Experiments*

Particle size and surface functionality are among the key parameters that influence their interactions with cells [69]. In this work, we used mouse-derived cells, namely bone marrow-derived dendritic cells (BMDCs), $CD103^+$ dendritic cells ($CD103^+$), and myeloid-derived suppressor cells (MDSCs) of monocytic (mMDSCs) and polymorphonuclear (pmnMDSCs) sub-types to study the uptake of fluorescently labeled, PEGylated PLGA particles of varying sizes (Figure 3).

Particles incubated with BMDCs for different time periods were analyzed using flow cytometry (Figure 3A). In our previous study we reported that, once taken up by the cells, the integrity of PLGA particles is compromised before 72 h of incubation [70]. Hence, in the present work, we monitored particle uptake up to 48 h in order to avoid possible variations in the mean fluorescence intensities (MFIs) due to particle degradation at later time points. For ease of comparison, MFI data were normalized to 1 for the values obtained at 48 h of incubation time. A clear correlation between the particle size and the trend in particle uptake was observed. For sub-micron particles (~100 nm and ~200 nm) a stable MFI was observed within the first 6 h, which increased at later time points (Figure 3A, red and blue curves). On the other hand, micron-size particles (>1000 nm) showed a time-dependent uptake behavior. For these particles, the intracellular MFI increased gradually within the first 6 h, reaching a plateau until the end of the incubation period (Figure 3A, orange curve).

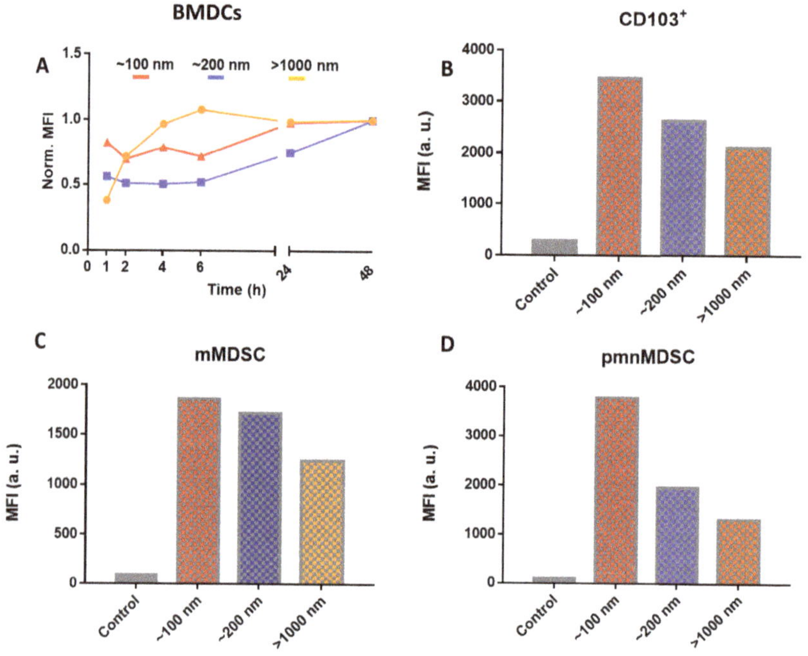

Figure 3. (**A**) Intracellular mean fluorescence intensity (MFI) values obtained for >1000 nm (orange), ~200 nm (blue), and ~100 nm (red) PLGA particles incubated with bone marrow-derived dendritic cells (BMDCs) for different time periods, and (**B**) correlation between the particle size and intracellular MFI after 2 h of incubation with CD103$^+$ dendritic cells, (**C**) monocytic myeloid-derived suppressor cells (MDSCs), and (**D**) polymorphonuclear MDSCs isolated from mouse. Data were obtained using flow cytometry.

Overall, the uptake of sub-micron particles was more efficient compared to micron-size particles. The MFI obtained at 1 h of incubation already corresponded to ~75% and ~50% of the total MFI for 100 nm particles (Figure 3A, red) and 200 nm particles (Figure 3A, blue), respectively. It should be noted that each type of PLGA particles had different dye loading ratios. In this uptake study, equal amounts of PLGA particles with different total fluorescence intensities were used. Although normalization of MFI enabled a fair comparison of the particle uptake trend for different sizes, we further investigated the particle uptake on other types of mouse-derived immune cells using particles with equal fluorescence intensities instead of equal particle mass in order to compensate for the variations in dye encapsulation (Figure 3B–D). Flow cytometry histograms of the data presented in Figure 3B–D are shown in Figure S1. A clear correlation between the particle size and intracellular MFI was observed for the cells that were either generated in vitro (Figure 3B) or were isolated from the spleen of tumor-bearing mice (Figure 3C,D). After 2 h incubation period, the uptake of ~100 nm particles was ~1.5 fold higher than the >1000 nm particles for CD103$^+$ dendritic cells (Figure 3B) and mMDSCs (Figure 3C), and the difference was even higher (almost three-fold) for pmnMDSCs (Figure 3D). Similar studies also reported the lower uptake efficiencies of micron-size particles compared to sub-micron particles by dendritic cells [71], which could be due to different uptake mechanisms associated with different particle sizes. The uptake of small particles (<100 nm) has been reported to be clathrin- and/or caveolin-mediated [72], whereas micropinocytosis and phagocytosis are the main mechanisms via which sub-micron particles (~200 nm) and micron-sized particles (>1000 nm) are taken up [73]. In addition to particle size, surface charge is an important parameter that influences the efficiency of particle uptake by cells. Prior to their internalization, particles need to attach on the cell surfaces that are decorated with negatively-charged proteoglycans [63]. Consequently, the uptake of positively-charged particles can be more efficient due to electrostatic interactions between the particles and cell surface. In our work, all particles had negative ζ potential values, which was highest for >1000 nm particles. Therefore, the poorer uptake efficiency observed for micron-size particles can be due to less-favorable surface charge in addition to larger particle size.

All the cell types used in our work are important regulators of immune response. BMDCs and $CD103^{+}$ are antigen presenting cells that can prime T cells to induce antigen-specific immune responses [74]. On the other hand, MDSCs play an important role in immune suppression in cancer as well as in tumor angiogenesis, drug resistance, and promotion of tumor metastases [75], representing an attractive potential therapeutic target, for things such as cancer immunotherapy. Among the studied PLGA formulations, ~100 nm particles with almost neutral surface charge would be the most efficient vehicle to deliver such things as antigens to dendritic cells (DCs) and immunomodulatory drugs to MDSCs for cancer immunotherapy.

5.4. *In Vivo Clearance of Particles*

Rapid clearance of particles from the bloodstream through the mononuclear phagocyte system (MPS) and reticuloendothelial system (RES) represents a major limitation to achieve preferential accumulation of particles in the target organs [76]. Tuning the size and surface properties of particles has proven useful in preventing their rapid removal from the bloodstream [77,78]. In order to evaluate particle clearance in vivo, we administered PLGA particles labeled with a near-infrared (NIR)-emitting dye (VivoTag-S 750) intravenously (i.v.) to mice. The acceptable particle size range for the i.v. injections have been reported as 10–1000 nm to prevent possible accumulation of larger particles in the lung capillaries [61,79,80]. Therefore, we used only sub-micron size particles for in vivo studies. The colloidal properties of ~100 nm and ~200 nm particles encapsulating the NIR-emitting dye were similar to those labeled with the green fluorescent dye (Figure S2). Whole-body and ex-vivo organ imaging were performed at different time points up to 48 h after i.v. administration of sub-micron PLGA particles (Figure 4).

Whole-body imaging revealed the presence of both particle types mainly in the liver and bladder already after 30 min following the injection (Figure 4A). The liver signal decreased gradually at later time points and was still above background levels at 48 h for both ~100 nm and ~200 nm particles. The presence of a high fluorescence signal in the bladder at 0.5 h was an interesting observation, which could be related to the excretion of free dye molecules that were burst-released. In fact, intact particles cannot pass through the renal filtration barrier since it has an effective size cut-off of ~10 nm [81]. The higher percentage of burst release displayed by ~100 nm particles in situ aligned well with the in vivo observations, in which the bladder signal resulting from the excretion of the free dye molecules was relatively lower for ~200 nm particles. Overall, the bladder signal did not provide a reliable measure of the systemic clearance of the particles due to the interference of the burst-released dye. Thus, we monitored the clearance of particles from the liver as representative of their systemic clearance. The influence of burst-released dye can be avoided via covalent attachment of the fluorescent dye.

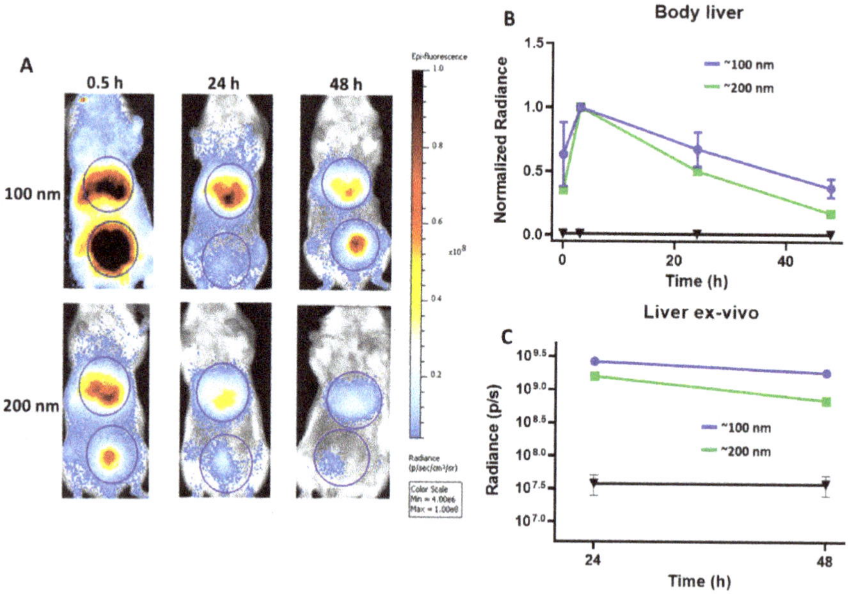

Figure 4. (**A**) Fluorescent whole-body images of mice obtained at different time points up to 48 h after intravenous (i.v.) administration of ~200 nm and ~100 nm PLGA particles. The liver and bladder are encircled. (**B**) Variations of liver fluorescence in whole-body images. The fluorescence intensities are normalized at maximum values observed at 3 h after administration of ~100 nm (blue) and ~200 nm (green) PLGA particles. (**C**) Variations of liver fluorescence intensities of ~100 nm (blue) and ~200 nm (green) PLGA particles obtained on an isolated liver at 24 h and 48 h after particle administration. Data obtained for untreated mice (negative control) are shown in black.

The variation of liver signal was monitored using the whole-body images obtained at different time points (Figure 4B). For a better comparison, the values were normalized to maximum intensities measured at 3 h. From this point on, the decay of liver signal was almost linear for ~100 nm particles with a slope of −0.014 ($R^2 = 0.9926$) (Figure 4B, blue curve), whereas the decay of ~200 nm particles better suited an exponential function (Figure 4B, green curve). Since the number of fit points was not sufficient for an accurate exponential fit, we used a linear fit for ~200 nm particles as well. These particles displayed a relatively faster decay with a slope of −0.018 ($R^2 = 0.9537$). Ex-vivo imaging of an isolated liver also showed similar variations in signal intensities, such that ~200 nm particles (Figure 4C, green curve) displayed a more pronounced decrease at 48 h compared to ~100 nm particles (Figure 4C, blue curve). Faster clearance of larger particles was also shown in other studies. When administered intravenously, particles larger than 200 nm activate human complement systems, are rapidly eliminated from the bloodstream, and gather mainly in the liver and spleen where the rate of accumulation is proportional to the size of the particles [30]. While particles with a diameter greater than 200 nm were most likely cleared by Kupffer cells, smaller particles displayed a decreased rate of clearance and an extended circulation time [82]. Of note, we did not observe a signal originating from the spleen in whole-body imaging. However, ex-vivo imaging of isolated spleen at 24 h and 48 h revealed signal intensities that were only slightly above the background noise, indicating that the spleen was not the preferential accumulation site for sub-micron PLGA particles (Figure S3). Indeed, it has been reported that the spleen receives barely 15% of the i.v. injected dose of nanomedicines [83]. Therefore, additional surface modification strategies may be needed for applications that require splenic accumulation of PLGA particles.

6. Conclusions

In this study, we demonstrated how PLGA particle size can be specifically tuned using a microfluidics system via modulating the formulation and process parameters. Through a series of optimization experiments, we obtained PEGylated PLGA particles in different sizes, which remarkably affected the characteristics of the particles in vitro and in vivo confirming the direct relation between the size and the pharmacokinetics behavior. This work can be considered as a further step towards the establishment of a production process that is able to generate tailor-made medicine for each individual clinical need.

Supplementary Materials: The following are available online at http://www.mdpi.com/1999-4923/11/11/590/s1, Figure S1: Flow cytometry histograms of the uptake experiments, Figure S2: Size distribution and ζ potential values of PLGA particles encapsulating a NIR-emitting fluorescent dye, Figure S3: Variations of spleen fluorescence intensities of ~100 nm and ~200 nm PLGA particles obtained on an isolated spleen at 24 h and 48 h after particle administration.

Author Contributions: M.C.O., Y.D., J.K., E.A.W.v.D.: Performed experiments, data curation, analysis, and writing. C.G.F.: Funding acquisition, supervision, review and editing. O.T.: Conceptualization, supervision, data interpretation, writing, review and editing.

Funding: This research was funded by EU grant PRECIOUS (686089).

Acknowledgments: C.F. received the NWO Spinoza grant, ERC Advanced Grant Pathfinder (269019) and Dutch cancer society award 2009-4402.

Conflicts of Interest: The authors declare no conflict of interest.

References

1. Khalid, M.; El-Sawy, H.S. Polymeric nanoparticles: Promising platform for drug delivery. *Int. J. Pharm.* **2017**, *528*, 675–691.
2. Kumari, A.; Yadav, S.K.; Yadav, S.C. Biodegradable polymeric nanoparticles based drug delivery systems. *Colloids Surf. B Biointerfaces* **2010**, *75*, 1–18. [CrossRef] [PubMed]
3. Ud Din, F.; Aman, W.; Ullah, I.; Qureshi, O.S.; Mustapha, O.; Shafique, S.; Zeb, A. Effective use of nanocarriers as drug delivery systems for the treatment of selected tumors. *Int. J. Nanomed.* **2017**, *12*, 7291. [CrossRef] [PubMed]
4. Chan, J.M.; Valencia, P.M.; Zhang, L.; Langer, R.; Farokhzad, O.C. Polymeric nanoparticles for drug delivery. In *Cancer Nanotechnology*; Springer: Berlin/Heidelberg, Germany, 2010; pp. 163–175.
5. Hines, D.J.; Kaplan, D.L. Poly (lactic-co-glycolic) acid—controlled-release systems: Experimental and modeling insights. *Crit. Rev. Ther. Drug Carr. Syst.* **2013**, *30*, 257–276. [CrossRef]
6. Danhier, F.; Ansorena, E.; Silva, J.M.; Coco, R.; Le Breton, A.; Préat, V. PLGA-based nanoparticles: An overview of biomedical applications. *J. Control. Release* **2012**, *161*, 505–522. [CrossRef]
7. Makadia, H.K.; Siegel, S.J. Poly lactic-co-glycolic acid (PLGA) as biodegradable controlled drug delivery carrier. *Polymers* **2011**, *3*, 1377–1397. [CrossRef]
8. Operti, M.C.; Fecher, D.; van Dinther, E.A.; Grimm, S.; Jaber, R.; Figdor, C.G.; Tagit, O. A comparative assessment of continuous production techniques to generate sub-micron size PLGA particles. *Int. J. Pharm.* **2018**, *550*, 140–148. [CrossRef]
9. Derakhshandeh, K.; Erfan, M.; Dadashzadeh, S. Encapsulation of 9-nitrocamptothecin, a novel anticancer drug, in biodegradable nanoparticles: Factorial design, characterization and release kinetics. *Eur. J. Pharm. Biopharm.* **2007**, *66*, 34–41. [CrossRef]
10. Fonseca, C.; Simoes, S.; Gaspar, R. Paclitaxel-loaded PLGA nanoparticles: Preparation, physicochemical characterization and in vitro anti-tumoral activity. *J. Control. Release* **2002**, *83*, 273–286. [CrossRef]
11. Jose, S.; Juna, B.; Cinu, T.; Jyoti, H.; Aleykutty, N. Carboplatin loaded Surface modified PLGA nanoparticles: Optimization, characterization, and in vivo brain targeting studies. *Colloids Surf. B Biointerfaces* **2016**, *142*, 307–314. [CrossRef]
12. Khan, I.; Gothwal, A.; Sharma, A.K.; Kesharwani, P.; Gupta, L.; Iyer, A.K.; Gupta, U. PLGA nanoparticles and their versatile role in anticancer drug delivery. *Crit. Rev. Ther. Drug Carr. Syst.* **2016**, *33*, 159–193. [CrossRef] [PubMed]

13. Sun, S.-B.; Liu, P.; Shao, F.-M.; Miao, Q.-L. Formulation and evaluation of PLGA nanoparticles loaded capecitabine for prostate cancer. *Int. J. Clin. Exp. Med.* **2015**, *8*, 19670. [PubMed]
14. Feczkó, T.; Tóth, J.; Dósa, G.; Gyenis, J. Optimization of protein encapsulation in PLGA nanoparticles. *Chem. Eng. Process. Process. Intensif.* **2011**, *50*, 757–765. [CrossRef]
15. Mohammadi-Samani, S.; Taghipour, B. PLGA micro and nanoparticles in delivery of peptides and proteins; problems and approaches. *Pharm. Dev. Technol.* **2015**, *20*, 385–393. [CrossRef] [PubMed]
16. Pirooznia, N.; Hasannia, S.; Lotfi, A.S.; Ghanei, M. Encapsulation of alpha-1 antitrypsin in PLGA nanoparticles: In vitro characterization as an effective aerosol formulation in pulmonary diseases. *J. Nanobiotechnol.* **2012**, *10*, 20. [CrossRef] [PubMed]
17. Rescignano, N.; Tarpani, L.; Tiribuzi, R.; Montesano, S.; Martino, S.; Latterini, L.; Kenny, J.M.; Armentano, I. Protein encapsulation in biodegradable polymeric nanoparticles: Morphology, fluorescence behaviour and stem cell uptake. *Macromol. Biosci.* **2013**, *13*, 1204–1212. [CrossRef]
18. Santander-Ortega, M.J.; Csaba, N.; González, L.; Bastos-González, D.; Ortega-Vinuesa, J.L.; Alonso, M.J. Protein-loaded PLGA–PEO blend nanoparticles: Encapsulation, release and degradation characteristics. *Colloid Polym. Sci.* **2010**, *288*, 141–150. [CrossRef]
19. Lü, J.-M.; Liang, Z.; Wang, X.; Gu, J.; Yao, Q.; Chen, C. New polymer of lactic-co-glycolic acid-modified polyethylenimine for nucleic acid delivery. *Nanomedicine* **2016**, *11*, 1971–1991. [CrossRef]
20. Patil, Y.B.; Swaminathan, S.K.; Sadhukha, T.; Ma, L.; Panyam, J. The use of nanoparticle-mediated targeted gene silencing and drug delivery to overcome tumor drug resistance. *Biomaterials* **2010**, *31*, 358–365. [CrossRef] [PubMed]
21. Harguindey, A.; Domaille, D.W.; Fairbanks, B.D.; Wagner, J.; Bowman, C.N.; Cha, J.N. Synthesis and Assembly of Click-Nucleic-Acid-Containing PEG–PLGA Nanoparticles for DNA Delivery. *Adv. Mater.* **2017**, *29*, 1700743. [CrossRef]
22. Cun, D.; Jensen, D.K.; Maltesen, M.J.; Bunker, M.; Whiteside, P.; Scurr, D.; Foged, C.; Nielsen, H.M. High loading efficiency and sustained release of siRNA encapsulated in PLGA nanoparticles: Quality by design optimization and characterization. *Eur. J. Pharm. Biopharm.* **2011**, *77*, 26–35. [CrossRef] [PubMed]
23. Colombo, S.; Cun, D.; Remaut, K.; Bunker, M.; Zhang, J.; Martin-Bertelsen, B.; Yaghmur, A.; Braeckmans, K.; Nielsen, H.M.; Foged, C. Mechanistic profiling of the siRNA delivery dynamics of lipid–polymer hybrid nanoparticles. *J. Control. Release* **2015**, *201*, 22–31. [CrossRef] [PubMed]
24. Park, Y.-M.; Lee, S.J.; Kim, Y.S.; Lee, M.H.; Cha, G.S.; Jung, I.D.; Kang, T.H.; Han, H.D. Nanoparticle-based vaccine delivery for cancer immunotherapy. *Immune Netw.* **2013**, *13*, 177–183. [CrossRef] [PubMed]
25. Prasad, S.; Cody, V.; Saucier-Sawyer, J.K.; Saltzman, W.M.; Sasaki, C.T.; Edelson, R.L.; Birchall, M.A.; Hanlon, D.J. Polymer nanoparticles containing tumor lysates as antigen delivery vehicles for dendritic cell–based antitumor immunotherapy. *Nanomed. Nanotechnol. Biol. Med.* **2011**, *7*, 1–10. [CrossRef]
26. Ma, T.; Wang, L.; Yang, T.; Ma, G.; Wang, S. Homogeneous PLGA-lipid nanoparticle as a promising oral vaccine delivery system for ovalbumin. *Asian J. Pharm. Sci.* **2014**, *9*, 129–136. [CrossRef]
27. Dölen, Y.; Kreutz, M.; Gileadi, U.; Tel, J.; Vasaturo, A.; van Dinther, E.A.; van Hout-Kuijer, M.A.; Cerundolo, V.; Figdor, C.G. Co-delivery of PLGA encapsulated invariant NKT cell agonist with antigenic protein induce strong T cell-mediated antitumor immune responses. *Oncoimmunology* **2016**, *5*, e1068493. [CrossRef]
28. Clawson, C.; Huang, C.-T.; Futalan, D.; Seible, D.M.; Saenz, R.; Larsson, M.; Ma, W.; Minev, B.; Zhang, F.; Ozkan, M. Delivery of a peptide via poly (d, l-lactic-co-glycolic) acid nanoparticles enhances its dendritic cell–stimulatory capacity. *Nanomed. Nanotechnol. Biol. Med.* **2010**, *6*, 651–661. [CrossRef]
29. Allahyari, M.; Mohit, E. Peptide/protein vaccine delivery system based on PLGA particles. *Hum. Vaccines Immunother.* **2016**, *12*, 806–828. [CrossRef]
30. Hoshyar, N.; Gray, S.; Han, H.; Bao, G. The effect of nanoparticle size on in vivo pharmacokinetics and cellular interaction. *Nanomedicine* **2016**, *11*, 673–692. [CrossRef]
31. Han, F.Y.; Thurecht, K.J.; Whittaker, A.K.; Smith, M.T. Bioerodable PLGA-based microparticles for producing sustained-release drug formulations and strategies for improving drug loading. *Front. Pharmacol.* **2016**, *7*, 185. [CrossRef]
32. Hashizume, H.; Baluk, P.; Morikawa, S.; McLean, J.W.; Thurston, G.; Roberge, S.; Jain, R.K.; McDonald, D.M. Openings between defective endothelial cells explain tumor vessel leakiness. *Am. J. Pathol.* **2000**, *156*, 1363–1380. [CrossRef]

33. Toy, R.; Hayden, E.; Shoup, C.; Baskaran, H.; Karathanasis, E. The effects of particle size, density and shape on margination of nanoparticles in microcirculation. *Nanotechnology* **2011**, *22*, 115101. [CrossRef] [PubMed]
34. Moghimi, S.M.; Parhamifar, L.; Ahmadvand, D.; Wibroe, P.P.; Andresen, T.; Farhangrazi, Z.; Hunter, A. Particulate systems for targeting of macrophages: Basic and therapeutic concepts. *J. Innate Immun.* **2012**, *4*, 509–528. [CrossRef] [PubMed]
35. Xie, Y.; Bagby, T.R.; Cohen, M.S.; Forrest, M.L. Drug delivery to the lymphatic system: Importance in future cancer diagnosis and therapies. *Expert Opin. Drug Deliv.* **2009**, *6*, 785–792. [CrossRef]
36. Bazile, D.; Ropert, C.; Huve, P.; Verrecchia, T.; Mariard, M.; Frydman, A.; Veillard, M.; Spenlehauer, G. Body distribution of fully biodegradable [14C]-poly (lactic acid) nanoparticles coated with albumin after parenteral administration to rats. *Biomaterials* **1992**, *13*, 1093–1102. [CrossRef]
37. Verhoef, J.J.; Anchordoquy, T.J. Questioning the use of PEGylation for drug delivery. *Drug Deliv. Transl. Res.* **2013**, *3*, 499–503. [CrossRef]
38. Suk, J.S.; Xu, Q.; Kim, N.; Hanes, J.; Ensign, L.M. PEGylation as a strategy for improving nanoparticle-based drug and gene delivery. *Adv. Drug Deliv. Rev.* **2016**, *99*, 28–51. [CrossRef]
39. Schöttler, S.; Becker, G.; Winzen, S.; Steinbach, T.; Mohr, K.; Landfester, K.; Mailänder, V.; Wurm, F.R. Protein adsorption is required for stealth effect of poly (ethylene glycol)-and poly (phosphoester)-coated nanocarriers. *Nat. Nanotechnol.* **2016**, *11*, 372. [CrossRef]
40. Simon, J.; Müller, L.K.; Kokkinopoulou, M.; Lieberwirth, I.; Morsbach, S.; Landfester, K.; Mailänder, V. Exploiting the biomolecular corona: Pre-coating of nanoparticles enables controlled cellular interactions. *Nanoscale* **2018**, *10*, 10731–10739. [CrossRef]
41. Morikawa, Y.; Tagami, T.; Hoshikawa, A.; Ozeki, T. The use of an efficient microfluidic mixing system for generating stabilized polymeric nanoparticles for controlled drug release. *Biol. Pharm. Bull.* **2018**, *41*, 899–907. [CrossRef]
42. Danaei, M.; Dehghankhold, M.; Ataei, S.; Hasanzadeh Davarani, F.; Javanmard, R.; Dokhani, A.; Khorasani, S.; Mozafari, M. Impact of particle size and polydispersity index on the clinical applications of lipidic nanocarrier systems. *Pharmaceutics* **2018**, *10*, 57. [CrossRef] [PubMed]
43. Bahari, L.A.S.; Hamishehkar, H. The impact of variables on particle size of solid lipid nanoparticles and nanostructured lipid carriers; a comparative literature review. *Adv. Pharm. Bull.* **2016**, *6*, 143. [CrossRef] [PubMed]
44. Chiesa, E.; Dorati, R.; Pisani, S.; Conti, B.; Bergamini, G.; Modena, T.; Genta, I. The microfluidic technique and the manufacturing of polysaccharide nanoparticles. *Pharmaceutics* **2018**, *10*, 267. [CrossRef] [PubMed]
45. Whitesides, G.M. The origins and the future of microfluidics. *Nature* **2006**, *442*, 368. [CrossRef]
46. Mark, D.; Haeberle, S.; Roth, G.; Von Stetten, F.; Zengerle, R. Microfluidic lab-on-a-chip platforms: Requirements, characteristics and applications. In *Microfluidics Based Microsystems*; Springer: Berlin/Heidelberg, Germany, 2010; pp. 305–376.
47. Amoyav, B.; Benny, O. Controlled and tunable polymer particles' production using a single microfluidic device. *Appl. Nanosci.* **2018**, *8*, 905–914. [CrossRef]
48. Dashtimoghadam, E.; Fahimipour, F.; Davaji, B.; Hasani-Sadrabadi, M.; Tayebi, L. Microfluidic-directed synthesis of polymeric nanoparticles for bone cancer therapy. *Dent. Mater.* **2016**, *1*, e59–e60. [CrossRef]
49. De Solorzano, I.O.; Uson, L.; Larrea, A.; Miana, M.; Sebastian, V.; Arruebo, M. Continuous synthesis of drug-loaded nanoparticles using microchannel emulsification and numerical modeling: Effect of passive mixing. *Int. J. Nanomed.* **2016**, *11*, 3397.
50. Khan, I.U.; Serra, C.A.; Anton, N.; Vandamme, T.F. Production of nanoparticle drug delivery systems with microfluidics tools. *Expert Opin. Drug Deliv.* **2015**, *12*, 547–562. [CrossRef]
51. Xie, H.; Smith, J.W. Fabrication of PLGA nanoparticles with a fluidic nanoprecipitation system. *J. Nanobiotechnol.* **2010**, *8*, 18. [CrossRef]
52. Xu, J.; Zhang, S.; Machado, A.; Lecommandoux, S.; Sandre, O.; Gu, F.; Colin, A. Controllable microfluidic production of drug-loaded PLGA nanoparticles using partially water-miscible mixed solvent microdroplets as a precursor. *Sci. Rep.* **2017**, *7*, 4794. [CrossRef]
53. Gdowski, A.; Johnson, K.; Shah, S.; Gryczynski, I.; Vishwanatha, J.; Ranjan, A. Optimization and scale up of microfluidic nanolipomer production method for preclinical and potential clinical trials. *J. Nanobiotechnol.* **2018**, *16*, 12. [CrossRef] [PubMed]

54. Huang, W.; Zhang, C. Tuning the Size of Poly (lactic-co-glycolic Acid)(PLGA) Nanoparticles Fabricated by Nanoprecipitation. *Biotechnol. J.* **2018**, *13*, 1700203. [CrossRef] [PubMed]
55. Halayqa, M.; Domańska, U. PLGA biodegradable nanoparticles containing perphenazine or chlorpromazine hydrochloride: Effect of formulation and release. *Int. J. Mol. Sci.* **2014**, *15*, 23909–23923. [CrossRef] [PubMed]
56. Tang, Q.; Si, W.; Huang, C.; Ding, K.; Huang, W.; Chen, P.; Zhang, Q.; Dong, X. An aza-BODIPY photosensitizer for photoacoustic and photothermal imaging guided dual modal cancer phototherapy. *J. Mater. Chem. B* **2017**, *5*, 1566–1573. [CrossRef]
57. Trofymchuk, K.; Valanciunaite, J.; Andreiuk, B.; Reisch, A.; Collot, M.; Klymchenko, A.S. BODIPY-loaded polymer nanoparticles: Chemical structure of cargo defines leakage from nanocarrier in living cells. *J. Mater. Chem. B* **2019**, *7*, 5199–5210. [CrossRef] [PubMed]
58. Devaraj, N.K.; Keliher, E.J.; Thurber, G.M.; Nahrendorf, M.; Weissleder, R. 18F labeled nanoparticles for in vivo PET-CT imaging. *Bioconjug. Chem.* **2009**, *20*, 397–401. [CrossRef]
59. Elsabahy, M.; Heo, G.S.; Lim, S.-M.; Sun, G.; Wooley, K.L. Polymeric nanostructures for imaging and therapy. *Chem. Rev.* **2015**, *115*, 10967–11011. [CrossRef]
60. Mura, S.; Hillaireau, H.; Nicolas, J.; Le Droumaguet, B.; Gueutin, C.; Zanna, S.; Tsapis, N.; Fattal, E. Influence of surface charge on the potential toxicity of PLGA nanoparticles towards Calu-3 cells. *Int. J. Nanomed.* **2011**, *6*, 2591.
61. De Jesus Gomes, A.; Lunardi, C.N.; Caetano, F.H.; Lunardi, L.O.; da Hora Machado, A.E. Phagocytosis of PLGA microparticles in rat peritoneal exudate cells: A time-dependent study. *Microsc. Microanal.* **2006**, *12*, 399–405. [CrossRef]
62. Pisani, E.; Fattal, E.; Paris, J.; Ringard, C.; Rosilio, V.; Tsapis, N. Surfactant dependent morphology of polymeric capsules of perfluorooctyl bromide: Influence of polymer adsorption at the dichloromethane–water interface. *J. Colloid Interface Sci.* **2008**, *326*, 66–71. [CrossRef]
63. Honary, S.; Zahir, F. Effect of zeta potential on the properties of nano-drug delivery systems—A review (Part 1). *Trop. J. Pharm. Res.* **2013**, *12*, 255–264.
64. Jyothi, N.V.N.; Prasanna, P.M.; Sakarkar, S.N.; Prabha, K.S.; Ramaiah, P.S.; Srawan, G. Microencapsulation techniques, factors influencing encapsulation efficiency. *J. Microencapsul.* **2010**, *27*, 187–197. [CrossRef] [PubMed]
65. Fredenberg, S.; Wahlgren, M.; Reslow, M.; Axelsson, A. The mechanisms of drug release in poly (lactic-co-glycolic acid)-based drug delivery systems—A review. *Int. J. Pharm.* **2011**, *415*, 34–52. [CrossRef] [PubMed]
66. Siepmann, J.; Elkharraz, K.; Siepmann, F.; Klose, D. How autocatalysis accelerates drug release from PLGA-based microparticles: A quantitative treatment. *Biomacromolecules* **2005**, *6*, 2312–2319. [CrossRef] [PubMed]
67. Chen, W.; Palazzo, A.; Hennink, W.E.; Kok, R.J. Effect of particle size on drug loading and release kinetics of gefitinib-loaded PLGA microspheres. *Mol. Pharm.* **2016**, *14*, 459–467. [CrossRef] [PubMed]
68. Dutta, D.; Salifu, M.; Sirianni, R.W.; Stabenfeldt, S.E. Tailoring sub-micron PLGA particle release profiles via centrifugal fractioning. *J. Biomed. Mater. Res. Part A* **2016**, *104*, 688–696. [CrossRef]
69. Moayedian, T.; Mosaffa, F.; Khameneh, B.; Tafaghodi, M. Combined effects of PEGylation and particle size on uptake of PLGA particles by macrophage cells. *Nanomed. J.* **2015**, *2*, 299–304.
70. Swider, E.; Maharjan, S.; Houkes, K.; van Riessen, N.K.; Figdor, C.; Srinivas, M.; Tagit, O. Förster Resonance Energy Transfer-Based Stability Assessment of PLGA Nanoparticles in Vitro and in Vivo. *ACS Appl. Bio Mater.* **2019**, *2*, 1131–1140. [CrossRef]
71. Foged, C.; Brodin, B.; Frokjaer, S.; Sundblad, A. Particle size and surface charge affect particle uptake by human dendritic cells in an in vitro model. *Int. J. Pharm.* **2005**, *298*, 315–322. [CrossRef]
72. Tonigold, M.; Mailänder, V. Endocytosis and intracellular processing of nanoparticles in dendritic cells: Routes to effective immunonanomedicines. *Future Med.* **2016**, *11*, 2625–2630. [CrossRef]
73. Cruz, L.J.; Tacken, P.J.; Fokkink, R.; Joosten, B.; Stuart, M.C.; Albericio, F.; Torensma, R.; Figdor, C.G. Targeted PLGA nano-but not microparticles specifically deliver antigen to human dendritic cells via DC-SIGN in vitro. *J. Control. Release* **2010**, *144*, 118–126. [CrossRef] [PubMed]
74. Helft, J.; Böttcher, J.; Chakravarty, P.; Zelenay, S.; Huotari, J.; Schraml, B.U.; Goubau, D.; e Sousa, C.R. GM-CSF mouse bone marrow cultures comprise a heterogeneous population of CD11c+ MHCII+ macrophages and dendritic cells. *Immunity* **2015**, *42*, 1197–1211. [CrossRef] [PubMed]

75. Gonzalez-Junca, A.; Driscoll, K.; Pellicciotta, I.; Du, S.; Lo, C.H.; Roy, R.; Parry, R.; Tenvooren, I.; Marquez, D.; Spitzer, M.H. Autocrine TGFβ is a Survival Factor for Monocytes and Drives Immunosuppressive Lineage Commitment. *Cancer Immunol. Res.* **2019**, *7*, 306–320. [CrossRef] [PubMed]
76. Albanese, A.; Tang, P.S.; Chan, W.C. The effect of nanoparticle size, shape, and surface chemistry on biological systems. *Annu. Rev. Biomed. Eng.* **2012**, *14*, 1–16. [CrossRef]
77. Liu, X.; Huang, N.; Li, H.; Jin, Q.; Ji, J. Surface and size effects on cell interaction of gold nanoparticles with both phagocytic and nonphagocytic cells. *Langmuir* **2013**, *29*, 9138–9148. [CrossRef]
78. Dreaden, E.C.; Austin, L.A.; Mackey, M.A.; El-Sayed, M.A. Size matters: Gold nanoparticles in targeted cancer drug delivery. *Ther. Deliv.* **2012**, *3*, 457–478. [CrossRef]
79. Jeon, H.-J.; Jeong, Y.-I.; Jang, M.-K.; Park, Y.-H.; Nah, J.-W. Effect of solvent on the preparation of surfactant-free poly (DL-lactide-co-glycolide) nanoparticles and norfloxacin release characteristics. *Int. J. Pharm.* **2000**, *207*, 99–108. [CrossRef]
80. Kreuter, J. Nanoparticle-based dmg delivery systems. *J. Control. Release* **1991**, *16*, 169–176. [CrossRef]
81. Zuckerman, J.E.; Choi, C.H.J.; Han, H.; Davis, M.E. Polycation-siRNA nanoparticles can disassemble at the kidney glomerular basement membrane. *Proc. Natl. Acad. Sci. USA* **2012**, *109*, 3137–3142. [CrossRef]
82. Kulkarni, S.A.; Feng, S.-S. Effects of particle size and surface modification on cellular uptake and biodistribution of polymeric nanoparticles for drug delivery. *Pharm. Res.* **2013**, *30*, 2512–2522. [CrossRef]
83. Jindal, A.B. Nanocarriers for spleen targeting: Anatomo-physiological considerations, formulation strategies and therapeutic potential. *Drug Deliv. Transl. Res.* **2016**, *6*, 473–485. [CrossRef] [PubMed]

pharmaceutics

Article

Terahertz Spectroscopy: An Investigation of the Structural Dynamics of Freeze-Dried Poly Lactic-co-glycolic Acid Microspheres

Talia A. Shmool [1], Philippa J. Hooper [1], Gabriele S. Kaminski Schierle [1], Christopher F. van der Walle [2] and J. Axel Zeitler [1,*]

[1] Department of Chemical Engineering and Biotechnology, University of Cambridge, Philippa Fawcett Drive, Cambridge CB3 0AS, UK; tas61@cam.ac.uk (T.A.S.); pjh200@cam.ac.uk (P.J.H.); gsk20@cam.ac.uk (G.S.K.S.)
[2] Biopharmaceutical Development, AstraZeneca, Granta Park, Cambridge CB21 6GH, UK; wallec@medimmune.com
* Correspondence: jaz22@cam.ac.uk; Tel.: +44-1223-334783

Received: 23 April 2019; Accepted: 11 June 2019; Published: 20 June 2019

Abstract: Biodegradable poly lactic-co-glycolic acid (PLGA) microspheres can be used to encapsulate peptide and offer a promising drug-delivery vehicle. In this work we investigate the dynamics of PLGA microspheres prepared by freeze-drying and the molecular mobility at lower temperatures leading to the glass transition temperature, using temperature-variable terahertz time-domain spectroscopy (THz-TDS) experiments. The microspheres were prepared using a water-in-oil-in-water (w/o/w) double-emulsion technique and subsequent freeze-drying of the samples. Physical characterization was performed by morphology measurements, scanning electron microscopy, and helium pycnometry. The THz-TDS data show two distinct transition processes, $T_{g,\beta}$ in the range of 167–219 K, associated with local motions, and $T_{g,\alpha}$ in the range of 313–330 K, associated with large-scale motions, for the microspheres examined. Using Fourier transform infrared spectroscopy measurements in the mid-infrared, we were able to characterize the interactions between a model polypeptide, exendin-4, and the PLGA copolymer. We observe a relationship between the experimentally determined $T_{g,\beta}$ and $T_{g,\alpha}$ and free volume and microsphere dynamics.

Keywords: terahertz spectroscopy; microspheres; drug delivery; formulation development; PLGA; molecular mobility

1. Introduction

Microencapsulation has been explored as a promising method for controlled drug release [1–3]. Polymeric microspheres, for example polylactic acid (PLA), polyglycolic acid (PGA), and poly(D,L-lactide-co-glycolide) (PLGA), can be used as effective drug-delivery systems protecting the encapsulated active agent and controlling the release rate over periods of hours to months [1,4]. By applying the correct methodology when preparing microspheres, challenges such as targeting the drug to a specific organ or tissue, and precisely controlling the rate of drug delivery to the target site can be addressed [3,5]. The technique of double water-in-oil-in-water (w/o/w) emulsion has been widely used for the encapsulation of hydrophilic drugs as this preparative method minimizes the loss of drug activity via contact with the organic solvent [2,6]. The basis of this methodology is to emulsify an aqueous solution of the active compound in an organic solution of the hydrophobic coating polymer. Then, this primary water-in-oil (w/o) emulsion is dispersed in a second aqueous phase, forming a double water-in-oil-in-water (w/o/w) emulsion. The solid microsphere is produced as the organic solvent evaporates. In this work, we prepared blank PLGA microspheres (containing no polypeptide) of grades 50:50 and 75:25 (lactide:glycolide ratio), using the double-emulsion process [2]. Using the same technique, we also loaded PLGA microspheres at low and high concentrations of exendin-4, a glucose-dependent insulinotropic polypeptide used to treat type 2 diabetes [7]. The microspheres were subsequently freeze-dried. The objective of this work was to prepare, characterize, and understand the structural dynamics of these peptide-loaded biodegradable polymeric microspheres.

To optimize the formulation of microspheres and achieve release delivery of the active drug, the microsphere product ought to be stable [1,8]. When determining the chemical stability of a material, molecular mobility is a key factor [9–11] and it has been shown that an increase in molecular mobility is directly linked to an increase in the chemical degradation of a material [9] and aggregation [12], and therefore its storage stability. Thus, it is critical to understand the molecular mobility behavior of a material and its dependence on temperature, and specifically the molecular dynamics of the material leading up to the glass transition temperature (T_g). It has been shown that a great number of polymers and amorphous pharmaceuticals, ranging from small molecules to high molecular weight peptides and proteins exhibit at least two dielectric relaxation processes: the primary, or α-relaxation process and a secondary, or β-relaxation process [13–15]. The former can be observed at temperatures above T_g and can be designated as $T_{g,\alpha}$, and the latter occurs at temperatures below T_g and is associated with a secondary glass transition process, which can be indicated as $T_{g,\beta}$ [16–20].

In previous work we investigated these processes in neat PLGA 50:50 and 75:25 of low, medium, and high molecular weights [21]. For all the copolymer PLGA film samples we observed three regions of relaxation behavior with distinct $T_{g,\beta}$ and $T_{g,\alpha}$ transition temperatures. We showed that with an increase in temperature, the copolymer chains can transition through different conformational environments, each constrained by its characteristic potential energy landscape, and that the movement of each PLGA chain is restricted by adjacent entangled chains. At $T_{g,\beta}$, the system has sufficient thermal energy and free volume to overcome the energy barrier arising from chain entanglement, allowing for local mobility to occur. Furthermore, we show that there is a relationship between free volume and the value of $T_{g,\beta}$ for the different film samples.

Fourier transform infrared spectroscopy (FTIR) is a vibrational spectroscopy used to gain structural information about a sample. The advantage of FTIR for structural analysis is it is a non-contact technique; only a small quantity of sample is needed ($\sim$1 mg mL^{-1}); the sample can be probed in different physical environments such as the solid state, liquid state, and when adsorbed to a surface; the sample preparation is minimal and a spectrum can be obtained in a few minutes [22]. For systems with multiple components such as polymer-peptide microspheres, FTIR can be used to identify the individual components, and interactions between the components which shift the peaks [23].

Terahertz time-domain spectroscopy (THz-TDS) is a valuable technique which can be used to detect $T_{g,\alpha}$ and $T_{g,\beta}$, associated with the α- and β-relaxation processes respectively [10,13]. It is a relatively recent technique which is used to investigate the molecular dynamics of relaxation processes at high frequencies [24]. The advantage of this technique is that it is a non-contact technique and can measure molecular mobility of a material over a broad temperature range over the spectral region of 0.1–3 THz. It can be used to investigate the microscopic mechanisms of amorphous polymer dynamics [25]. In the present study we examine samples of freeze-dried microspheres with different concentrations of encapsulated peptide. We investigate the molecular mobility of these materials and the behavior and trends these exhibit with respect to the temperature dependence using THz-TDS (Table 1), and physically characterise each system (Tables 2 and 3). Thus, the purpose of this work is to provide a comprehensive understanding of the relationship between the relaxation dynamics and the molecular structure of PLGA microspheres with varying exendin-4 peptide concentrations, and to rationalize the behavior of these materials in relation to $T_{g\alpha}$ and $T_{g\beta}$.

2. Materials and Methods

2.1. Materials

Throughout this work the polymers are referred to by their monomer ratio used. For instance, PLGA 75:25 refers to a copolymer consisting of 75% lactic acid and 25% glycolic acid. Medium-MW (10–25 kDa) PLGA 50:50 and medium-MW (20–30 kDa) PLGA 75:25 were purchased from Evonik Corporation (Birmingham, AL, USA). The exendin-4 peptide (4.2 kDa) was provided by MedImmune Limited (Cambridge, UK).

2.2. Microencapsulation Preparation

Blank PLGA microspheres, were prepared for the study using the water-in-oil-in-water (w/o/w) double-emulsion technique: First, 500 mL of a 0.5% aqueous solution of polyvinyl alcohol (PVA) (87–90% hydrolyzed, 13–23 kDa) were emulsified in 12.5 mL of a 5% (w/v) PLGA dissolved in dichloromethane (DCM) by stirring at a rotation speed of 22,000 pm for 15 seconds using an ultra-turrax (IKA T-25, Cole-Parmer, UK). This primary emulsion was emulsified into 0.5% aqueous PVA under stirring at 200 rpm, on a stirring plate, for four hours. For the preparation of the peptide-loaded microspheres, exendin-4 was dissolved into a buffer consisting of citrate, citric acid, (pH = 4.5) and 0.5% PVA. This peptide solution was emulsified into the 5% (w/v) dispersion of PLGA in DCM, which was stirred at rotation speed 22,000 rpm for 15 s using the ultra-turrax. Different peptide loadings in the microspheres were achieved using exendin-4 concentrations of 1% and 10% (w/v) in the primary emulsion. Conceptually, the same method was followed to produce the peptide-loaded microspheres, as that for the blank microspheres. The solid microspheres were collected by centrifugation at 4000 *g* for 5 min, then washed with distilled water three times, and centrifuged once more at 4000 *g* for 5 min. Finally, after removal of the aqueous supernatant and dispersion, the microspheres were freeze-dried using a lyophilizer. First, prior to lyophilization, an annealing step was performed by cooling the shelf to 233 K for 240 min, raising the temperatures to 257 K for 200 min and cooling the shelf again to 233 K for 170 min at a pressure of 160 mbar. Then, lyophilization was performed using the following steps: primary drying was completed at 233 K for 30 min, and then the temperature was raised to 253 K for 2440 min at a pressure of 133 mbar; secondary drying was subsequently performed at 313 K for 960 min, also at 133 mbar. The vials were closed under a pressure of 266 mbar at 298 K using a rubber stopper, removed from the lyophilizer, and crimped with aluminum seals. Vials were stored at 278 K until all further measurements and analysis. Exendin-4 concentration was determined using the bicinchoninic acid protein assay kit (MilliporeSigma, St. Louis, MO, USA) following the manufacturer's instructions, with bovine serum albumin (BSA) used as the standard. The exendin-4 concentration was analyzed using a UV-VIS spectrophotometer (Agilent Cary 60 UV-Vis spectrophotometer, Agilent Technologies, Santa Clara, CA, USA) at 562 nm. The encapsulation efficiency-based BCA assay values were determined on the liquid samples. Based on these measurements, we calculated the encapsulation efficiency to be: 15.4% and 37.12% for the low and high peptide loading of PLGA 75:25 microspheres, respectively, and 42.88% and 36.72% for the low and high peptide loading of PLGA 50:50 microspheres, respectively. This experiment was completed one time. The water content for each lyophilized microsphere was determined using Karl Fischer (Mettler Toledo, Leicester, UK) coulometric titration.

2.3. Helium Pycnometry Measurement

The lyophilized microsphere samples were analyzed as powders using a helium pycnometer (Micromeritics Accupyc II 1340, Norcross, GA, USA) to determine the density and specific volume of the microparticles. Approximately 20 mg of each sample were placed in the instrument compartment and measurements were performed with helium gas at 298 K at a pressure of 10 mbar. No pretreatment conditions were required. The measurements were repeated for each sample five times. The average volume of the five repeat measurements was used to determine the density of each material. For comparison, polymer films of PLGA 50:50 and PLGA 75:25 were prepared using the vacuum compression molding (VCM) tool (MeltPrep, Graz, Austria), and these were also analyzed using helium pycnometry.

2.4. Morphology Measurement

The particle size and shape of the microsphere particles were characterized, for unlyophilized liquid samples and for lyophilized powder samples, using a Morphologi G3 instrument (Malvern Panalytical Ltd., Malvern, UK). Each unlyophilized liquid sample was dispersed in 1 mL of 5% aqueous PVA solution to allow for spatial separation of the particles and reduction of agglomerates. The lyophilized powder samples were prepared for measurement using a dry powder disperser. The instrument captured images of the individual particles and the morphological properties for each particle were determined from the images by image analysis using the Morphologi G3 software (Malvern Panalytical Ltd., Malvern, UK).

2.5. Scanning Electron Microscopy

The range of morphologies of the microspheres were characterized using scanning electron microscopy (SEM), with a Zeiss CrossBeam 540 instrument, equipped with a Gemini 2 column (Carl Zeiss Microscopy GmbH, Jena, Germany). To qualitatively analyze the shape and surface of the samples, the lyophilized microspheres were sputter coated with gold using an Emitech 550 (Emitech Ltd., Ashford, UK). The samples were then examined under vacuum at an acceleration voltage of 1 kV, and imaged using secondary electrons via an Everhart-Thornley detector (Carl Zeiss SMT GmbH, Oberkochen, Germany).

2.6. Differential Scanning Calorimetry (DSC)

A Q2000 Differential Scanning Calorimeter (TA Instruments, New Castle, DE, USA) was used to determine the calorimetric glass transition temperature ($T_{g,DSC}$, defined by the onset temperature) for each material. 2–3 mg of sample material were placed in hermetically sealed aluminum pans under a constant flow of nitrogen atmosphere (flow rate of 50 mL min^{-1}) and heated at a rate of 10 K min^{-1} to 358 K, and subsequently cooled down to 293 K at 40 K min^{-1}. Finally, the samples were heated from 293 K through T_g to 358 K again at a rate of 10 K min^{-1}. The temperature and heat flow of the instrument were calibrated using indium (T_m = 430 K, ΔH_{fus} = 29 J g^{-1}).

2.7. Fourier Transform Infrared Spectroscopy

FTIR was used to examine the change in the secondary structure of the PLGA 75:25 microspheres at 278 K. For FTIR analysis, each microsphere material (300 µg) was mixed with 100 mg potassium bromide (KBr) using an agate mortar and pressed into 7 mm self-supporting disks using a load of 10 Tons. FTIR spectra were acquired using a Cary 680 FTIR spectrometer (Agilent Technologies, Santa Clara, CA, USA) with 60 scans and a resolution of 1 cm^{-1}. At least four spectra were measured for each material. The recorded spectra were normalized based on the total area under the curve [26].

2.8. *Terahertz Time-Domain Spectroscopy (THz-TDS)*

2.8.1. Sample Preparation and Experimental Methodology

For each sample 70 mg were weighed in under atmospheric protection in a glove bag (AtmosBag, Merck UK, Gillingham, Dorset, UK) which was purged with dry nitrogen gas (relative humidity $< 1\%$) to avoid moisture sorption from atmospheric water vapor during preparation. The lyophilized powder samples were pressed into 13 mm diameter disks using a load of 1.5 tons. The tablets were between 300–650 µm in thickness each, and were placed between two z-cut quartz windows of 2.05 mm thickness. This sandwich structure was sealed in the sample holder, and used immediately following preparation for THz-TDS measurements.

The THz-TDS spectra were acquired using a commercial TeraPulse 4000 instrument across the spectral range of 0.2–2.2 THz (TeraView, Cambridge, UK). The sample temperature (90–360 K) was controlled using a continuous flow cryostat with liquid nitrogen as the cryogen (Janis ST-100, Wilmington, MA, USA) as outlined previously [27]. The cryostat cold finger accommodated both the reference (two z-cut quartz windows) as well as the sample (quartz/sample/quartz sandwich structure as described above). The two z-cut quartz windows that were used for the reference (same thickness and diameter dimensions as sample) were directly pressed to one another without any spacer in between the two windows to avoid internal reflections in the time-domain signal. The cryostat cold finger was moved vertically using a motorized linear stage to switch between sample and reference at each measurement temperature. For each temperature, the sample and the reference were measured at the center position, with 1000 waveforms co-averaged for each acquisition, resulting in a measurement time of approximately 1 min for each sample.

The temperature of the sample was measured using a silicon diode mounted to the copper cold finger of the cryostat. The temperature controller used was a Lake Shore model 331 (Westerville, OH, USA). For each series of measurements, a sample and a reference were loaded into the cryostat, the cryostat chamber was evacuated to 10 mbar and the cold finger was cooled to a temperature of 90 K. The cryostat was allowed to equilibrate for 10–15 min at 90 K and the first set of sample and reference measurements was acquired. Subsequently the cold finger was heated using temperature intervals of 10 K (at a rate of $2\,\mathrm{K\,min^{-1}}$). At each desired temperature point the system was allowed to equilibrate for 3 min before a set of sample and reference measurements were acquired.

2.8.2. Data Analysis

To calculate the absorption coefficient and the refractive index of the sample a modified method for extracting the optical constants from terahertz measurements based on the concept introduced by Duvillaret et al. was used [27,28]. The changes in dynamics of the polymer sample were analyzed by investigating the change in the absorption coefficient at a frequency of 1 THz as a function of temperature using the methodology introduced in [21].

3. Results

3.1. *Differential Scanning Calorimetry Data*

The calorimetric $T_{g,DSC}$ was determined for each sample and the resulting values are listed in Table 1. We observed no significant difference in T_g between samples. Additionally, we observed one T_g for each material, indicating that no phase separation occurred for these samples.

Table 1. Gradient, m, of the linear fit ($y = mx + c$) for the respective temperature regions as outlined in Section 2.8.2 as well as the respective glass transition temperatures determined based on the terahertz analysis and by DSC.

Material	Peptide Loading in Aqueous Phase (% m/v)	Region 1 (cm^{-1} K^{-1})	Region 2 (cm^{-1} K^{-1})	Region 3 (cm^{-1} K^{-1})	$T_{g\beta}$ (K)	$T_{g\alpha}$ (K)	$T_{g,DSC}$ (K)
PLGA 50:50	0 (blank)	0.0026 ± 0.0040	0.0237 ± 0.0010	0.070 ± 0.018	167	318	318
PLGA 50:50	1 (low)	0.0010 ± 0.0012	0.021 ± 0.00062	0.074 ± 0.0044	168	320	316
PLGA 50:50	10 (high)	0.014 ± 0.0013	0.023 ± 0.0012	0.095 ± 0.0052	219	330	317
PLGA 75:25	0 (blank)	0.0087 ± 0.0022	0.026 ± 0.00095	0.043 ± 0.0043	179	313	317
PLGA 75:25	1 (low)	0.0060 ± 0.0016	0.022 ± 0.0014	0.057 ± 0.0024	192	320	320
PLGA 75:25	10 (high)	0.0078 ± 0.0015	0.022 ± 0.0013	0.067 ± 0.0021	215	327	322

3.2. Morphology Measurement Analysis

For each material we report the number of particles counted and the circular equivalent (CE) diameter $D[n, 0.1]$, $D[n, 0.5]$ and $D[n, 0.9]$ percentiles. As shown from Table 2, preparing the microspheres using an oil-in-water emulsification yields products with a relatively broad size distribution for both the blank microspheres and the exendin-4 loaded microspheres before lyophilization. Following lyophilization, the particle size distribution is significantly narrower than before. This could be due to the removal water and moisture [5,9]. The Karl Fischer measurements we conducted revealed that the residual moisture for each vial was less than 1%. Notably, upon lyophilization, as water molecules are removed peptides and copolymers, for example, exendin-4, and PLGA microspheres, can form an extensive hydrogen bonding network [11,29,30] leaving few sites for bonding with water molecules. Additionally, the system is exposed to various stresses of temperature and pressure, and these could cause two or more adjacent pores to merge, as common pore walls rupture, and adjacent particles can combine. Thus, the removal of water in the drying step of lyophilization would affect the size distribution of the dry particles produced, and we can obtain a more narrow particle size distribution compared to a sample in solution [31]. It is worth noting that exendin-4 is inherently flexible and thus, chemical stability is of primary concern; however, given that exendin-4 is a peptide, it lacks a defined protein domain with a characteristic architecture that can undergo unfolding upon exposure to acute freezing and dehydration stresses of lyophilization [32,33]. Finally, when comparing the CE values of the lyophilized samples, the blank PLGA 50:50 samples have a higher CE compared to the 50:50 loaded samples, while the blank 75:25 samples have a lower CE compared to the 75:25 loaded samples. This can be explained by considering emulsion stability which dictates microsphere size. For example, exendin-4 would change the emulsion stability through surfactant-like activity, and the viscosity of the emulsion can increase with higher polymer concentration, polymer MW, and hydrophobicity [34]. Specifically, as the polymer is more hydrophobic, as is the case for PLGA 75:25 due to its a higher lactide fraction, more energy would be required to generate smaller droplets; while for PLGA 50:50 it is expected that smaller droplets could be generated with less energy input; however these parameters were not optimized for and beyond the scope of this work [33,34].

Table 2. Morphology data for unlyophilized samples and lyophilized samples. CE is the circular equivalent diameter: the diameter of a circle with the same area as the 2D projection image of the particle.

Ratio	Lyophilized	Peptide Loading in Aqueous Phase (% m/v)	Particles Counted (Number)	CE $D[n, 0.1]$ (μm)	CE $D[n, 0.5]$ (μm)	CE $D[n, 0.9]$ (μm)
50:50	No	0	3218	77	119	176
50:50	No	1	1108	45	76	110
50:50	No	10	1108	45	76	110
50:50	Yes	0	763	73	100	136
50:50	Yes	1	2321	75	116	159
50:50	Yes	10	3839	68	110	156
75:25	No	0	1909	55	106	167
75:25	No	1	2479	68	114	170
75:25	No	10	361	111	178	242
75:25	Yes	0	8465	63	112	163
75:25	Yes	1	903	94	154	197
75:25	Yes	10	1477	82	147	186

3.3. Helium Pycnometry Analysis

The pycnometry data shows that increasing exendin-4 loading increases the density of the material (Table 3). Our helium pycnometry measurements agree with values reported in the literature [4].

Table 3. Helium pycnometry data for lyophilized microsphere samples and PLGA 50:50 and 75:25 copolymer films.

Material	Peptide Loading in Aqueous Phase (% m/v)	Density (g/cm^3)
PLGA 50:50 Film	0	1.39
Microsphere PLGA 50:50 blank	0	0.97
Microsphere PLGA 50:50	1	1.66
Microsphere PLGA 50:50	10	2.09
PLGA 75:25 Film	0	1.36
Microsphere PLGA 75:25 blank	0	1.41
PLGA 75:25	1	1.56
PLGA 75:25	10	1.62

3.4. Scanning Electron Microscopy (SEM) Characterization

Using SEM, qualitative analysis of a representative set of images indicated that the microspheres examined are predominantly spherical in shape and exhibit a smooth and porous surface (Figure 1, see Supplementary Materials). The internal structure of fractured spheres revealed a porous interior. The qualitative analysis of the data showed that for the blank PLGA microspheres, there is no significant difference visually in the porosity of the blank 50:50 and 75:25 systems, respectively. In contrast, for both the 50:50 and 75:25 systems, the microspheres with high loading of peptide appeared more porous compared to the microspheres which contained a low loading of peptide. See Supplementary Materials for SEM data for blank PLGA 50:50 and 75:25 microspheres, and low polypeptide loaded PLGA 50:50 and 75:25 microspheres.

Figure 1. Representative SEM micrographs for low exendin-4 loaded PLGA 75:25 microspheres shown in (**a**–**c**), and high exendin-4 loaded PLGA 75:25 microspheres shown in (**d**–**f**).

3.5. Fourier Transform Infrared Spectroscopy

FTIR was performed for the PLGA 75:25 microsphere samples (Figure 2). The absorbance of the peaks at around 3300 cm^{-1} and 2950 cm^{-1} increased in intensity with an increase in polypeptide loading. The peak at 3300 cm^{-1} originates from the -OH stretch in exendin-4, and the peak at 2950 cm^{-1} is assigned to the C-H stretching modes in exendin-4 [26,35]. The amide I band at 1600–1700 cm^{-1} can be attributed to the –C=O stretch, with contributions from the out-of-phase –CN stretch, –CCN deformation, and –NH in-plane bend and is sensitive to the structure of the protein backbone. Additionally, there is an increase in absorption in the amide I region around 1600–1700 cm^{-1}, and a shift in the frequency of the mode for the sample with a high polypeptide loading, which is indicative of changes in the hydrogen bonding network [22]. Thus, the FTIR spectra confirm the increase in exendin-4 loading between the blank, low and high polypeptide formulations, and, with this, a change in the hydrogen bonding network. Notably, it was not possible to investigate an only exendin-4 control, as a stable product could not be freeze-dried.

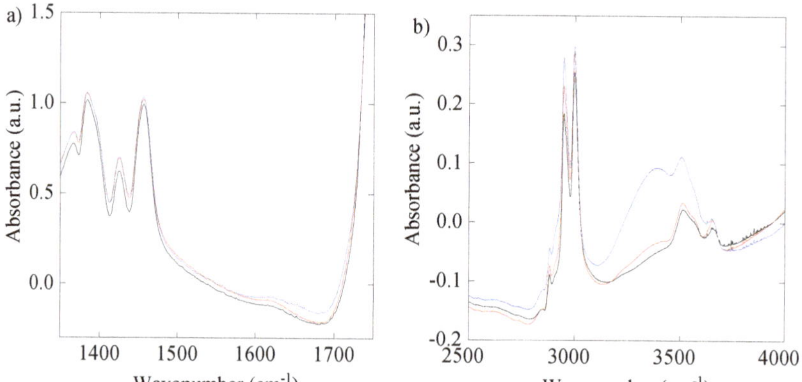

Figure 2. FTIR spectra of blank microspheres (solid black line), low polypeptide loaded (solid red line) and high polypeptide loaded (solid blue line) PLGA 75:25 microspheres. (**a**) shows the wavenumber range of 1300 − 1700 cm^{-1} and (**b**) shows the wavenumber range of 2500–4000 cm^{-1}.

3.6. Analysis of THz-TDS Data

The terahertz spectra of all the microspheres showed an increase in absorption with frequency and temperature over the entire investigated range in line with previous measurements of amorphous molecular solids (Figure 3). As expected for non-crystalline materials, no discrete spectral features were present and the spectra were dominated by the monotonous increase with frequency that is characteristic for the rising flank of the peak due to the vibrational density of states (VDOS) [24]. In contrast, the refractive index subtly decreases with increasing frequency. To further investigate the relationship between the increase of absorption coefficient and temperature we examined the temperature-dependent changes in absorption losses at a frequency 1 THz in more detail. Given the lack of distinct spectral features we chose the frequency of 1 THz. The rationale for this choice is based on the fact that the signal-to-noise ratio of the measurement at this frequency is high and that we know from our previous work that at a frequency of 1 THz the minimum in losses in the spectral response is exceeded for all samples studied and hence the absorption is clearly dominated by the VDOS [27].

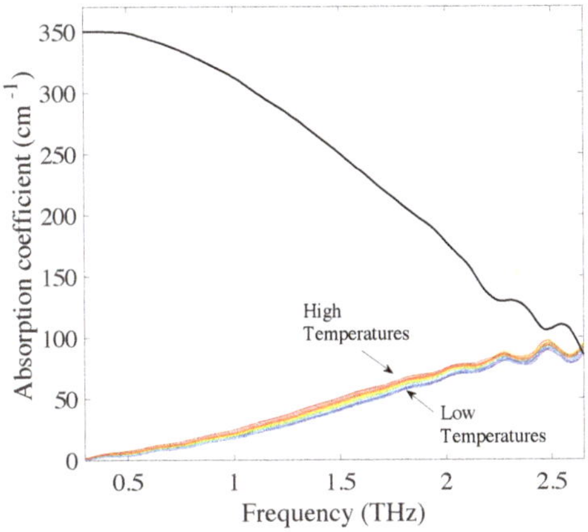

Figure 3. Absorption coefficient spectra of high MW PLGA 75:25 in the temperature range of 100–350 K, with 10 K temperature increments between spectra. Solid black line indicates the maximum absorption coefficient.

The changes in absorption at a frequency of 1 THz with temperature for the microsphere samples of PLGA 50:50 and PLGA 75:25, are plotted in Figures 4 and 5 respectively. For all the materials the absorption coefficient was found to increase in a linear fashion with increasing temperature and several distinct temperature regions can be identified for each material. $T_{g,\beta}$ was defined as the intersection point of the two best-fit linear fits at low temperatures for all cases, and $T_{g,\alpha}$ was defined as the intersection point of the two best-fit linear lines at high temperatures, as outlined in [21].

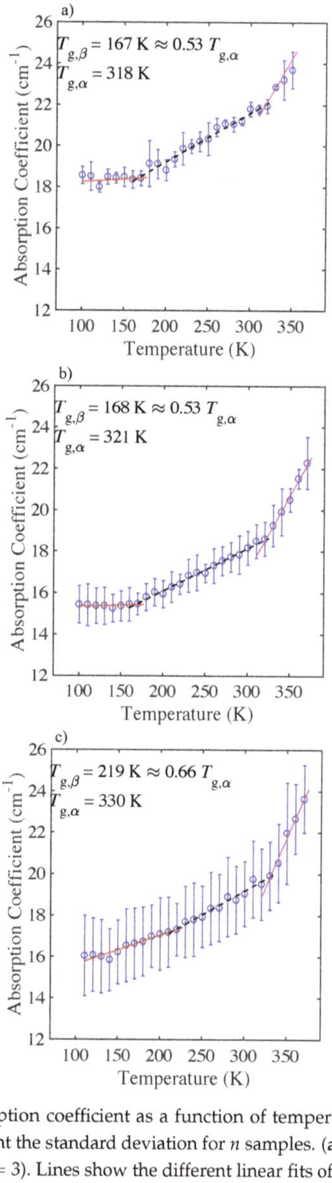

Figure 4. Mean terahertz absorption coefficient as a function of temperature at 1 THz for PLGA 50:50 microspheres. Error bars represent the standard deviation for n samples. (**a**) blank (n = 3), (**b**) low (n = 4), and (**c**) high peptide loading (n = 3). Lines show the different linear fits of the respective regions.

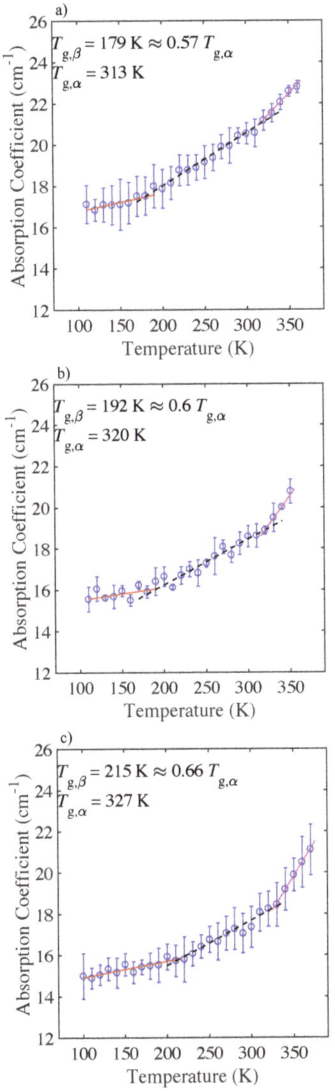

Figure 5. Mean terahertz absorption coefficient as a function of temperature at 1 THz for PLGA 75:25 microspheres. Error bars represent the standard deviation for *n* samples (**a**) blank ($n = 3$), (**b**) low ($n = 3$), and (**c**) high peptide loading ($n = 3$). Lines show the different linear fits for the different regions.

For all of the microsphere samples the change in absorption with temperature can be observed to take place over three distinct regions and two transition temperatures, $T_{g\beta}$ and $T_{g\alpha}$, as determined using the methodology outlined above (see Figure 4 and 5 and Table 1). In this work we attempt to explain the origin of the transition temperatures by proposing a physical picture of the change in the microsphere dynamics with temperature and relate this to the free volume approach, and to peptide and polymer interactions. It is worth noting that the values of $T_{g,\alpha}$, as determined from the THz-TDS experiments, are in good agreement with our own calorimetric measurements, $T_{g,DSC}$, as well as the values reported in the literature for these materials [36].

4. Discussion

4.1. Understanding Peptide and Copolymer Interactions

It has been shown that the α-relaxation process is associated with large-scale mobility, whereas the secondary or β-relaxation process is thought to be associated with local mobility, or small-scale mobility [24, 37]. To date, there remains a discussion in the literature as to the origin and molecular mechanisms associated with the α- and β-relaxation processes. One secondary relaxation, the Johari-Goldstein (JG) β-relaxation, also referred to as the slow β-relaxation, is considered a universal feature of all amorphous materials [38,39]. This process is observed at higher frequencies than the α-relaxation, and has been associated predominantly with the intermolecular degrees of freedom of a material [16,38,40]. Notably, recent experimental and theoretical work clearly highlights that the potential energy surface (PES) model proposed by Goldstein almost half a century ago is the most intuitive and comprehensive model to understand the molecular dynamics in amorphous systems and that intra- and intermolecular processes are always fundamentally coupled by means of the PES [41,42]. Goldstein explained that as a liquid flows it can move on the PES from one minimum to another minimum, and each minimum is associated with an energy barrier, yet as the liquid moves the volume nor energy of the liquid changes. When the liquid is cooled down to a glass, the liquid structure is trapped in a deep minimum, with some level of mobility remaining as a distribution of relaxation times [42]. Thus, the work of Goldstein provides an intuitive illustration that a liquid can exist in numerous transient structures and it is the potential energy barriers which determine the molecular motions of the viscous liquid which forms a glass: a picture which can be applied to bulk amorphous systems.

The different molecular motions of a copolymer chain can be tracked with changes in temperature. At low temperatures, the copolymer chains are completely disordered and are densely packed, and the motions of the copolymer are restricted. Upon heating to $T_{g,\beta}$, the activation energy and free volume is sufficient for the copolymer to undergo local motion, giving rise to the β-relaxation processes [40]. Previously, we have shown that the $T_{g,\beta}$ is fundamentally linked to the onset of motions in an organic molecule that results in changes of the dihedral angle of one or multiple bonds in the system. Specifically, for PLGA, this could involve local motions of small segments of the copolymer backbone and side chain groups [37,43]. With a further increase in temperature, the copolymer chains are more loosely packed and the activation energy and free volume of the system increases further, allowing for intermolecular large-scale copolymer motions to occur at the temperature indicated by $T_{g,\alpha}$ associated with the α-relaxation process.

Exendin-4 is a polypeptide composed of 39 amino acids, and its structure is thought to resemble a random coil chain [7,44]. The short-range local interactions of a polypeptide influence the conformational preferences of its amino acid chain [45–47]. In general terms, it is well established that the partial double bond character of a peptide bond gives rise to its planar structure, and free rotation is restricted about this bond [47]. We hypothesize that with sufficient thermal energy and free volume present in the system, two local motions could take place for exendin-4, which would give rise to the β-relaxation process: (1) local side chain rotations, and (2) local rotation about two single bonds [45,48]. Given that the amino acids of the exendin-4 polypeptide chain are linked by peptide bonds, rotational freedom arises from the single bonds between an amino group and the α-carbon atom and the carbonyl group of the peptide backbone [47].

However, the rotation of these bonds is limited by steric hindrance [47,48]. Based on the chemical structure of the exendin-4 this polypeptide can act as a hydrogen bond acceptor via its carbonyl groups (–C=O) or as a hydrogen bond donor via its amine group (–NH), and its hydroxyl groups (–OH) can act as both the hydrogen bond donor or acceptor. The linear PLGA—$[C_3H_4O_2]_x[C_2H_2O_2]_y$—chains include the methyl side groups of poly(lactic acid) (PLA) and oxygen atoms at every third position of the copolymer backbone, and C=O bonds which introduce significant structural rigidity to the copolymer backbone. Specifically, hydrogen bonds can form: (1) between the carbonyl groups of PLGA and the amine groups of exendin-4, (2) between the amine groups of exendin-4 and the hydroxyl group of PLGA, and (3) between the hydroxyl groups and the carbonyl groups of exendin-4 and PLGA. Additionally, van der Waals as well as dipole-dipole interactions between the peptide groups and the hydroxyl groups can serve to stabilize the system [49]. These strong interactions, the dynamics of which are infrared active, can reduce molecular mobility and improve the physical stability of these systems [3,29,30]. Notably, it is conceivable that with additional free volume and thermal energy input large-scale rotational motions of the backbone dihedral angles in the polypeptide chain could occur, which could lead to large-scale changes in chain conformation, contributing to the α-relaxation.

4.2. Tracking the Dynamics of Microspheres Using THz-TDS

We observe three regions with two distinct transition points, $T_{g,\beta}$ and $T_{g,\alpha}$ for both the PLGA 50:50 and PLGA 75:25 microspheres (Figure 4). For the PLGA 50:50 formulation, the high exendin-4 loaded microspheres have a significantly higher $T_{g,\beta} = 219\,\text{K}$, than the blank and low polypeptide loaded microspheres ($T_{g,\beta} = 167\,\text{K}$, and $T_{g,\beta} = 168\,\text{K}$, respectively). In the polymer dispersion, exendin-4 can form hydrogen bonds with PLGA. Intuitively, an increase in exendin-4 loading would correspondingly increase the hydrogen bonding interactions between the polypeptide and PLGA [48], reduce the molecular mobility of the system, which would raise the value of $T_{g,\beta}$ [9,50]. Indeed, we observe a significant increase in $T_{g,\beta}$ for the high polypeptide loaded microspheres, compared to the $T_{g,\beta}$ value observed for the blank and low polypeptide loaded microspheres. Notably, for the blank and low exendin-4 loaded 50:50 microspheres, the onset for local mobility occurs at approximately the same value of $T_{g,\beta}$. This could suggest that the low polypeptide loaded microsphere behave similarly to the blank microspheres, due to the limited interaction between peptide and polymer [46,51]. For the blank microspheres and low exendin-4 loaded microspheres, the onset for local mobility occurs at approximately the same value of $T_{g,\beta}$, ($T_{g,\beta} = 167\,\text{K}$). For the high exendin-4 loaded microspheres, $T_{g,\beta}$ increases significantly ($T_{g,\beta} = 219\,\text{K}$). As more polypeptide is added to the PLGA matrix it forms an extensive hydrogen bonding network, which in turn reduces the configurational entropy of the system [44,45,51]. The resultant interactions between exendin-4 and PLGA appear to be stronger compared with the interactions between adjacent PLGA chains.

For the samples of the PLGA 75:25 microspheres, we observe a similar trend to the PLGA 50:50 microspheres. The value of $T_{g,\beta}$ increases from 179 K for the blank microspheres, increases further to 192 K for the low exendin-4 loaded microspheres, and rises to $T_{g,\beta} = 215\,\text{K}$, for the high exendin-4 loaded microspheres (Figure 5). Notably, we have previously shown that the methyl side group of PLGA 75:25 inhibits polymer mobility and introduces steric hinderance [43]. Thus, for PLGA 75:25, the steric hinderance caused by the lactide monomer can restrict the intermolecular interactions between exendin-4 and PLGA, limiting the sites that are able to participate in hydrogen bonding, most likely making the carbonyl group of PLGA more favorable for hydrogen bonding. With limited sites for hydrogen bonding, less hydrogen bonds can form between exendin-4 and PLGA, resulting in PLGA chain entanglement, lower mobility, and reduced free volume. Our measurement that the PLGA 75:25 microspheres exhibit higher $T_{g,\beta}$ values compared to the 50:50 microspheres could therefore be explained by steric effects.

Finally, we observe that for both PLGA 50:50 and PLGA 75:25 the value of $T_{g,\alpha}$ increases in the order of blank < low polypeptide loaded < high polypeptide loaded microspheres (Figure 5). This suggests that at high temperature, sufficient activation energy and free volume must be available to facilitate mobility of the polymer and polypeptide. Thus, with increase in exendin-4 loading, due to steric hinderance the threshold for mobility increases, as reflected in the raised value of $T_{g,\alpha}$ (Figure 6). Ideally if it were possible to produce a stable freeze-dried exendin-4 product, it may be feasible to determine whether the exendin alone could aggregate and lead to the changes observed.

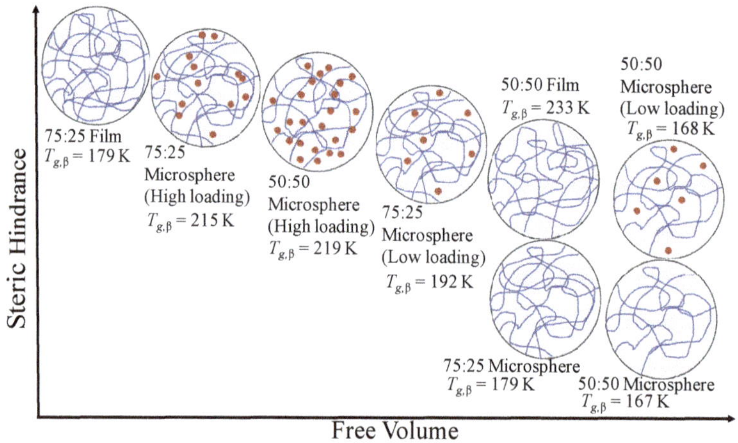

Figure 6. Behavior of the different materials with increasing steric hinderance and free volume. Blue lines represent the PLGA copolymer (20–30 kDa MW), and solid red circles represent the exendin-4 molecules (4.2 kDa MW). Low loading and high loading refers to 1 mg/mL and 10 mg/mL of polypeptide loaded in PLGA microspheres, respectively. With an increase in free volume and decrease in steric hinderance, the values of $T_{g,\beta}$ is decreased.

5. Conclusions

We have produced and characterized lyophilized blank PLGA microspheres and lyophilized PLGA microspheres containing two different loadings of the polypeptide exendin-4. We studied the dynamics and relaxations and the glass transition behavior of these PLGA microspheres by performing variable temperature THz-TDS measurements. A monotonous increase of absorption coefficient with temperature was observed for all the materials examined, and all of the microspheres exhibit three temperature regimes, with a distinct $T_{g,\beta}$ and $T_{g,\alpha}$. We explain our experimental results using the concepts of free volume and discuss the interactions of the polypeptide and copolymer matrix and steric effects. We define $T_{g,\beta}$ as the point at which the material has sufficient amount of activation energy and free volume to allow for local motions to occur, and $T_{g,\alpha}$ as the point at which large-scale movement can take place, and relate the onset of the transition temperatures to the interaction strength between the polymer and the peptide. Our work provides a physical explanation for the behavior of these microspheres leading to T_g, and agrees with the PES concept outlined by Goldstein [42]. This work provides a framework for understanding the dynamics of complex systems, such as lyophilized microspheres, and considers the parameter of $T_{g,\beta}$ as a valuable criterion for preparing stable formulations. Finally, this work demonstrates that THz-TDS is an effective method to measure the molecular dynamics and temperature-dependent behavior of a polymer-polypeptide microsphere system.

Supplementary Materials: The following are available online at http://www.mdpi.com/1999-4923/11/6/291/s1: Figure S1: Representative SEM micrographs for blank PLGA 50:50 microspheres shown in (a–c) and blank PLGA 75:25 microspheres shown in (d–f), Figure S2: Representative SEM micrographs for low exendin-4 loaded PLGA 50:50 microspheres shown in (a–c) and high exendin-4 loaded PLGA 50:50 microspheres shown in (d–f), Figure S3: FTIR spectra of blank microspheres (solid black line), low polypeptide loaded (solid red line) and high polypeptide loaded (solid blue line) PLGA 75:25 microspheres, Figure S4: MDSC thermogram of PLGA 50:50 blank microsphere, Figure S5: MDSC thermogram of PLGA 50:50 low peptide loading microsphere, Figure S6: MDSC thermogram of PLGA 50:50 high peptide loading microsphere, Figure S7: MDSC thermogram of PLGA 75:25 blank microsphere, Figure S8: MDSC thermogram of PLGA 75:25 low peptide loading microsphere, Figure S9: MDSC thermogram of PLGA 75:25 high peptide loading microsphere, Figure S10: Terahertz absorption spectra of a blank PLGA 75:25 microsphere sample over 0.3–2.8 THz in the temperature range of 100–360 K, Figure S11: Refractive index spectra of a blank PLGA 75:25 microsphere sample over 0.1–2.8 THz in the temperature range of 100–360 K, Figure S12: Terahertz absorption spectra of a blank PLGA 50:50 microsphere sample over 0.3–2.8 THz in the temperature range of 100–350 K, Figure S13: Refractive index spectra of a blank PLGA 50:50 microsphere sample over 0.1–2.8 THz in the temperature range of 100–350 K, Figure S14: Terahertz absorption spectra of a low polypeptide loaded PLGA 75:25 microsphere sample over 0.3–2.8 THz in the temperature range of 100–350 K, Figure S15: Refractive index spectra of a low polypeptide loaded PLGA 75:25 microsphere sample over 0.1–2.8 THz in the temperature range of 100–350 K, Figure S16: Terahertz absorption spectra of a low polypeptide loaded PLGA 50:50 microsphere sample over 0.3–2.8 THz in the temperature range of 100–370 K, Figure S17: Refractive index spectra of a low polypeptide loaded PLGA 50:50 microsphere sample over 0.1–2.8 THz in the temperature range of 100–350 K, Figure S18: Terahertz absorption spectra of a low polypeptide loaded PLGA 75:25 microsphere sample over 0.3–370 THz in the temperature range of 100–350 K, Figure S19: Refractive index spectra of a low polypeptide loaded PLGA 75:25 microsphere sample over 0.1–3 THz in the temperature range of 100–370 K, Figure S20: Terahertz absorption spectra of a high polypeptide loaded PLGA 50:50 microsphere sample over 0.3–2.8 THz in the temperature range of 100–370 K, Figure S21: Refractive index spectra of a high polypeptide loaded PLGA 50:50 microsphere sample over 0.1–2.8 THz in the temperature range of 100–370 K, Figure S22: Terahertz absorption spectra of a medium MW PLGA 50:50 over 0.3–2.2 THz in the temperature range of 90–360 K, Figure S23: Refractive index spectra of a medium MW PLGA 50:50 over 0.3–2.2 THz in the temperature range of 90–360 K, Figure S24: Terahertz absorption spectra of a medium MW PLGA 75:25 sample over 0.3–1.9 THz in the temperature range of 90–350 K, Figure S25: Refractive index spectra of a medium MW PLGA 75:25 sample over 0.3–1.9 THz in the temperature range of 90–350 K. All raw data are available for download at https://doi.org/10.17863/CAM.40737.

Author Contributions: Conceptualization and methodology, T.A.S., C.F.v.d.W. and J.A.Z.; infrared spectroscopy, P.J.H.; formal analysis, T.A.S. and P.J.H.; data curation, T.A.S.; writing—original draft preparation, T.A.S.; writing—review and editing, all authors; supervision, J.A.Z., C.F.v.d.W. and G.S.K.S.; project administration, J.A.Z.; funding acquisition, J.A.Z.

Funding: The authors acknowledge funding from AstraZeneca UK Limited (MedImmune Limited) and the UK Engineering and Physical Sciences Research Council (EP/N022769/1). T.A.S. would like to thank the AJA-Karten Trust and the AIA-Kenneth Lindsay Trust for their financial support.

Conflicts of Interest: The authors declare no conflict of interest. The funders had no role in the design of the study; in the collection, analyses, or interpretation of data; in the writing of the manuscript, or in the decision to publish the results.

Abbreviations

The following abbreviations are used in this manuscript:

CE	Circular equivalent
DCM	Dichloromethane
DSC	Differential scanning calorimetry
FTIR	Fourier transform infrared spectroscopy
JG β-relaxation	Johari-Goldstein secondary relaxation
PES	potential energy surface
PGA	Polyglycolic acid
PLA	Polylactic acid
PLGA	Poly(D,L-lactic-co-glycolic acid), Poly(lactide-co-glycolide)
SEM	Scanning electron microscopy
THz-TDS	Terahertz time-domain spectroscopy
VDOS	Vibrational density of states

References

1. Pakulska, M.M.; Donaghue, I.E.; Obermeyer, J.M.; Tuladhar, A.; McLaughlin, C.K.; Shendruk, T.N.; Shoichet, M.S. Encapsulation-Free Controlled Release: Electrostatic Adsorption Eliminates the Need for Protein Encapsulation in PLGA Nanoparticles. *Sci. Adv.* **2016**, *2*, e1600519. [CrossRef] [PubMed]
2. Nihant, N.; Schugens, C.; Grandfils, C.; Jérome, R.; Teyssié, P. Polylactide Microparticles Prepared by Double Emulsion/Evaporation Technique. I. Effect of Primary Emulsion Stability. *Pharm. Res.* **1994**, *11*, 1479–1484. [CrossRef] [PubMed]
3. Huang, J.; Wigent, R.J.; Schwartz, J.B. Drug-Polymer Interaction and its Significance on the Physical Stability of Nifedipine Amorphous Dispersion in Microparticles of an Ammonio Methacrylate Copolymer and Ethylcellulose Binary Blend. *J. Pharm. Sci.* **2008**, *97*, 251–262. [CrossRef] [PubMed]
4. Vay, K.; Scheler, S.; Friess, W. New Insights into the Pore Structure of Poly(D,L-lactide-co-glycolide) Microspheres. *Int. J. Pharm.* **2010**, *402*, 20–26. [CrossRef] [PubMed]
5. D'Souza, S.; Faraj, J.A.; Giovagnoli, S.; DeLuca, P.P. Development of Risperidone PLGA Microspheres. *J. Drug Deliv.* **2014**, 620464. [CrossRef] [PubMed]
6. Freitas, S.; Merkle, H.P.; Gander, B. Microencapsulation by Solvent Extraction/Evaporation: Reviewing the State of the Art of Microsphere Preparation Process Technology. *J. Control. Release* **2005**, *102*, 313–332. [CrossRef] [PubMed]
7. Liu, B.; Dong, Q.; Wang, M.; Shi, L.; Wu, Y.; Yu, X.; Shi, Y.; Shan, Y.; Jiang, C.; Zhang, X.; et al. Preparation, Characterization, and Pharmacodynamics of Exenatide-Loaded Poly(DL-lactic-co-glycolic acid) Microspheres. *Chem. Pharm. Bull.* **2010**, *58*, 1474–1479. [CrossRef]
8. Alqurshi, A.; Chan, K.L.A.; Royall, P.G. In-Situ Freeze-Drying—Forming Amorphous Solids Directly Within Capsules: An Investigation of Dissolution Enhancement for a Poorly Soluble Drug. *Sci. Rep.* **2017**, *7*, 2910. [CrossRef]
9. Moorthy, B.S.; Iyer, L.K.; Topp, E.M. Characterizing Protein Structure, Dynamics and Conformation in Lyophilized Solids. *Curr. Pharm. Des.* **2015**, *21*, 5845–5853. [CrossRef]
10. Sibik, J.; Elliott, S.R.; Zeitler, J.A. Thermal Decoupling of Molecular-Relaxation Processes from the Vibrational Density of States at Terahertz Frequencies in Supercooled Hydrogen-Bonded Liquids. *J. Phys. Chem. Lett.* **2014**, *5*, 1968–1972. [CrossRef]
11. Hancock, B.C.; Shamblin, S.L.; Zografi, G. Molecular Mobility of Amorphous Pharmaceutical Solids Below Their Glass Transition Temperatures. *Pharm. Res.* **1995**, *12*, 799–806. [CrossRef] [PubMed]
12. Stephens, A.D.; Nespovitaya, N.; Zacharopoulou, M.; Kaminski, C.F.; Phillips, J.J.; Kaminski Schierle, G.S. Different Structural Conformers of Monomeric α-Synuclein Identified after Lyophilizing and Freezing. *Anal. Chem.* **2018**, *90*, 6975–6983. [CrossRef] [PubMed]
13. Capaccioli, S.; Ngai, K.L.; Thayyil, M.S.; Prevosto, D. Coupling of Caged Molecule Dynamics to JG β-relaxation: I. *J. Phys. Chem. B* **2015**, *119*, 8800–8808. [CrossRef] [PubMed]
14. Lodge, T.P.; Muthukumar, M. Physical Chemistry of Polymers: Entropy, Interactions, and Dynamics. *J. Phys. Chem.* **1996**, *100*, 13275–13292. [CrossRef]
15. Diddens, D.; Heuer, A. Chain End Mobilities in Polymer Melts–A Computational Study. *J. Chem. Phys.* **2015**, *142*, 014906. [CrossRef]
16. Yu, H.B.; Wang, W.H.; Samwer, K. The β Relaxation in Metallic Glasses: An Overview. *Mater. Today* **2013**, *16*, 183–191. [CrossRef]
17. Alegria, A.; Colmenero, J. Dielectric Relaxation of Polymers: Segmental Dynamics Under Structural Constraints. *Soft Matter* **2016**, *12*, 7709–7725. [CrossRef]
18. Cerveny, S.; Bergman, R.; Schwartz, G.A.; Jacobsson, P. Dielectric α- and β-Relaxations in Uncured Styrene Butadiene Rubber. *Macromolecules* **2002**, *35*, 4337–4342 [CrossRef]
19. Roy, A.K.; Inglefield, P.T. Solid State NMR Studies of Local Motions in Polymers. *Prog. Nucl. Magn. Reson. Spectrosc.* **1990**, *22*, 569–603. [CrossRef]

20. Barnes, M.D.; Fukui, K.; Kaji, K.; Kanaya, T.; Noid, D.W.; Otaigbe, J.U.; Pokrovskii, V.N.; Sumpter, B.G. *Advances in Polymer Science Polymer Physics and Engineering*, 1st ed.; Springer: Berlin/Heidelberg, Germany, 2001; pp. 317–319.
21. Shmool, T.A.; Zeitler, J.A. Insights Into the Structural Dynamics of Poly Lactic-co-glycolic Acid at Terahertz Frequencies. *Polym. Chem.* **2019**, *10*, 351–361. [CrossRef]
22. Barth, A.; Zscherp, C. What Vibrations Tell About Proteins. *Q. Rev. Biophys.* **2002**, *35*, 369–430. [CrossRef] [PubMed]
23. van de Weert, M.; van't Hof, R.; van der Weerd, J.; Heeren, R.M.; Posthuma, G.; Hennink, W.E.; Crommelin, D.J. Lysozyme Distribution and Conformation in a Biodegradable Polymer Matrix as Determined by FTIR Techniques. *J. Control. Release* **2000**, *68*, 31–40. [CrossRef]
24. Sibik, J.; Zeitler, J.A. Direct Measurement of Molecular Mobility and Crystallisation of Amorphous Pharmaceuticals Using Terahertz Spectroscopy. *Adv. Drug Deliv. Rev.* **2016**, *100*, 147–157. [CrossRef] [PubMed]
25. Ngai, K.L.; Capaccioli, S.; Prevosto, D.; Wang, L.M. Coupling of Caged Molecule Dynamics to JG β-Relaxation II: Polymers. *J. Phys. Chem. B* **2015**, *119*, 12502–12518. [CrossRef] [PubMed]
26. Wang, M.; Lu, X.; Yin, X.; Tong, Y.; Peng, W.; Wu, L.; Li, H.; Yang, Y.; Gu, J.; Xiao, T.; et al. Synchrotron Radiation-Based Fourier-Transform Infrared Spectromicroscopy for Characterization of the Protein/Peptide Distribution in Single Microspheres. *Acta Pharm. Sin. B* **2015**, *5*, 270–276. [CrossRef]
27. Sibik, J.; Zeitler, J.A. Terahertz Response of Organic Amorphous Systems: Experimental Concerns and Perspectives. *Philos. Mag.* **2015**, *96*, 842–853. [CrossRef]
28. Duvillaret, L.; Garet, F.; Coutaz, J.L. A Reliable Method for Extraction of Material Parameters in Terahertz Time-Domain Spectroscopy. *IEEE J. Sel. Top. Quantum Electron.* **1996**, *2*, 739–746. [CrossRef]
29. Qiu, Y.; Chen, Y.; Zhang, G.G.Z.; Yu, L.; Mantri, R.V. *Developing Solid Oral Dosage Forms: Pharmaceutical Theory and Practice*, 2nd ed.; Academic Press: London, UK, 2017; pp. 49–52.
30. Nair, R.; Nyamweya, N.; Gonen, S.; Martinez-Miranda, L.J.; Hoag, S.W. Influence of Various Drugs on the Glass Transition Temperature of Poly(vinylpyrrolidone): A Thermodynamic and Spectroscopic Investigation. *Int. J. Pharm.* **2001**, *225*, 83–96. [CrossRef]
31. Cohen, S.; Bernstein, H. (Eds.) *Microparticulate Systems for the Delivery of Proteins and Vaccines*; CRC Press: Boca Raton, FL, USA, 1996.
32. Rey, L.; May, J.C. (Eds.) *Freeze-Drying/Lyophilization of Pharmaceutical and Biological Products*, 3rd ed.; Informa Healthcare: London, UK, 2010; pp. 172, 174, 277, 364–368.
33. Mondal, S.; Varenik, M.; Bloch, D.; Atsmon-Raz, Y.; Jacoby, G.; Adler-Abramovich, L.; Shimon, L.; Beck, R.; Miller, Y.; Regev, O.; et al. A Minimal Length Rigid Helical Peptide Motif Allows Rational Design of Modula Surfactants. *Nat. Commun.* **2017**, *8*, 14018. [CrossRef]
34. Ravivarapu, H.; Lee, H.; DeLuca, P. Enhancing Initial Release of Peptide from Poly(D,L-lactide-co-glycolide) (PLGA) Microspheres by Addition of a Porosigen and Increasing Drug Load. *Pharm. Dev. Technol.* **2000**, *5*, 287–296. [CrossRef]
35. Coates, J. Interpretation of Infrared Spectra, A Practical Approach. *Encycl. Anal. Chem.* **2006**. [CrossRef]
36. Keles, H.; Naylor, A.; Clegg, F.; Sammon, C. Investigation of Factors Influencing the Hydrolytic Degradation of Single PLGA Microparticles. *Polym. Degrad. Stab.* **2015**, *119*, 228–241. [CrossRef]
37. Carraher, C.E.J. *Polymer Chemistry*; Marcel Dekker: New York, NY, USA, 2003; pp. 65–72.
38. Johari, G.P.; Goldstein, M. Viscous Liquids and the Glass Transition. II. Secondary Relaxations in Glasses of Rigid Molecules. *J. Chem. Phys.* **1970**, *53*, 2372–2388. [CrossRef]
39. Williams, G.; Watts, D.C. Molecular Motion in the Glassy State. The Effect of Temperature and Pressure on the Dielectric β Relaxation of Polyvinyl Chloride. *Trans. Faraday Soc.* **1971**, *67*, 1971–1979. [CrossRef]
40. Williams, M.L.; Landel, R.F.; Ferry, J.D. The Temperature Dependence of Relaxation Mechanisms in Amorphous Polymers and Other Glass-forming Liquids. *J. Am. Chem. Soc.* **1955**, *77*, 3701–3707. [CrossRef]
41. Ruggiero, M.T.; Krynski, M.; Kissi, E.O.; Sibik, J.; Markl, D.; Tan, N.Y.; Arslanov, D.; van der Zande, W.; Redlich, B.; Korter, T.M.; et al. The Significance of the Amorphous Potential Energy Landscape for Dictating Glassy Dynamics and Driving Solid-State Crystallisation. *Phys. Chem. Chem. Phys.* **2017**, *19*, 30039–30047. [CrossRef]

42. Goldstein, M. Viscous Liquids and the Glass Transition: A Potential Energy Barrier Picture. *J. Chem. Phys.* **1969**, *51*, 3728–3739. [CrossRef]
43. Hamilton, W.C.; Edmonds, J.W.; Trippe, A. Methyl Group Rotation and the Low Temperature Transition in Hexamethylbenzene. A Neutron Diffraction Study. *Discuss. Faraday Soc.* **1969**, *48*, 192–204. [CrossRef]
44. Kimmich, R. *Principles of Soft-Matter Dynamics: Basic Theories, Non-invasive Methods, Mesoscopic Aspects*; Springer: Heidelberg, Germany, 2012; pp. 28–30.
45. Schell, D.; Tsai, J.; Scholtz, J.M.; Pace, C.N. Hydrogen Bonding Increases Packing Density in the Protein Interior. *Proteins* **2006**, *63*, 278–282. [CrossRef]
46. Kim, H.L.; McAuley, A.; McGuire, J. Protein Effects on Surfactant Adsorption Suggest the Dominant Mode of Surfactant-Mediated Stabilization of Protein. *J. Pharm. Sci.* **2014**, *103*, 1337–1345. [CrossRef]
47. Berg, J.M.; Tymoczko, J.L.; Stryer, L. *Biochemistry*, 5th ed.; W. H. Freeman: New York, NY, USA, 2002; pp. 64–70.
48. McCammon, A.; Harvey, S.C. *Dynamics of Proteins and Nucleic Acids*; Cambridge University Press: Cambridge, UK, 1987; pp. 116–147.
49. Improta, R.; Berisio, R.; Vitagliano, L. Contribution of Dipole–Dipole Interactions to the Stability of the Collagen Triple Helix. *Protein Sci.* **2008**, *17*, 955–961. [CrossRef] [PubMed]
50. Newman, A. *Pharmaceutical Amorphous Solid Dispersions*; Wiley: Hoboken, NJ, USA, 2015; pp. 29–32.
51. Hildebrand, P.W.; Günther, S.; Goede, A.; Forrest, L.; Frmmel, C.; Preissner, R. Hydrogen-Bonding and Packing Features of Membrane Proteins: Functional Implications. *Biophys. J.* **2008**, *94*, 1945–1953. [CrossRef] [PubMed]

MDPI
St. Alban-Anlage 66
4052 Basel
Switzerland
Tel. +41 61 683 77 34
Fax +41 61 302 89 18
www.mdpi.com

Pharmaceutics Editorial Office
E-mail: pharmaceutics@mdpi.com
www.mdpi.com/journal/pharmaceutics

www.ingramcontent.com/pod-product-compliance
Lightning Source LLC
LaVergne TN
LVHW070214100526
838202LV00015B/2045